Recent Trends in Biofilm Science and Technology

Recent Trends in Biofilm Science and Technology

Edited by

Manuel Simões, PhD
Assistant Professor
Laboratory for Process Engineering, Environment, Biotechnology and Energy (LEPABE)
Department of Chemical Engineering
Faculty of Engineering
University of Porto (FEUP)
Porto, Portugal

Anabela Borges, PhD
Junior Researcher
Laboratory for Process Engineering, Environment, Biotechnology and Energy (LEPABE)
Department of Chemical Engineering
Faculty of Engineering
University of Porto (FEUP)
Porto, Portugal

Lúcia Chaves Simões, PhD
Junior Researcher
Centre of Biological Engineering
University of Minho
Braga, Portugal

ELSEVIER

ACADEMIC PRESS
An imprint of Elsevier

Academic Press is an imprint of Elsevier
125 London Wall, London EC2Y 5AS, United Kingdom
525 B Street, Suite 1650, San Diego, CA 92101, United States
50 Hampshire Street, 5th Floor, Cambridge, MA 02139, United States
The Boulevard, Langford Lane, Kidlington, Oxford OX5 1GB, United Kingdom

Library of Congress Cataloging-in-Publication Data
A catalog record for this book is available from the Library of Congress

British Library Cataloguing-in-Publication Data
A catalogue record for this book is available from the British Library

ISBN: 978-0-12-819497-3

For information on all Academic Press publications visit our website at
https://www.elsevier.com/books-and-journals

Publisher: Andre Gerhard Wolff
Acquisitions Editor: Linda Versteeg-buschman
Editorial Project Manager: Tracy I. Tufaga
Production Project Manager: Niranjan Bhaskaran
Cover Designer: Miles Hitchen

Typeset by TNQ Technologies

Contents

Contributors xiii
Preface xvii
Acknowledgments xix

1. Biofilm formation and resistance

Astrid Gędas and Magdalena A. Olszewska

1.1 Biofilm mode of growth 1
1.2 Biofilm formation is a multistep process 3
1.3 Biofilm-specific resistance to antimicrobials 9
1.4 Conclusions 14
Acknowledgments 15
References 15

2. Nuclear magnetic resonance to study bacterial biofilms structure, formation, and resilience

Ana Cristina Abreu and Ignacio Fernández

2.1 Introduction 23
2.2 Biofilm formation and structure 25
2.3 The composition of extracellular polymeric substances and how it affects biofilm architecture 28
2.4 Applications of nuclear magnetic resonance spectroscopy to study biofilms 31
2.4.1 Several analytical techniques to study and characterize soluble parts of biofilms 33
2.4.2 Solid-state nuclear magnetic resonance to determine the insoluble constituents of biofilms 36
2.4.3 Imaging techniques to define biofilm structures and dynamics 37
2.4.4 Explore diffusion and mass transport within a biofilm 40
2.4.5 Diffusion-ordered nuclear magnetic resonance spectroscopy applications to determine molecular size 45

2.5 Nuclear magnetic resonance–based metabolomics approach to study biofilms 47
- 2.5.1 Designing a metabolomics experiment 49
- 2.5.2 Multivariate data analysis in metabolomics 53
- 2.5.3 Recent advances on nuclear magnetic resonance–based metabolomics applied to biofilms 54

2.6 Conclusion 57
References 58

3. Design and fabrication of biofilm reactors

D.M. Goeres, S. Pedersen, B. Warwood, D.K. Walker, A.E. Parker, M. Mettler and P. Sturman

3.1 Definition of a biofilm reactor 71
3.2 Design process 73
3.3 Implementing the design process: industrial surfaces biofilm reactor 79
- 3.3.1 Step 1. Identify the list of key attributes for the biofilm reactor 79
- 3.3.2 Step 2. Using the list of attributes identified in step 1, design a reactor and build a prototype 81
- 3.3.3 Step 3. Laboratory validation and design optimization 83
- 3.3.4 Step 4. Repeatability testing of the final design 84
- 3.3.5 Step 5. Incorporate a manufacturing approach to ready the reactor for commercialization 84

3.4 Conclusions 85
Acknowledgments 86
References 87

4. Oral biofilms

Marta Ribeiro and Manuel Simões

4.1 Defining the problematic: an introduction 89
4.2 The oral cavity and its microbiota 90
4.3 Dental plaque 90
4.4 Disease-associated oral biofilms 92
- 4.4.1 Dental caries 92
- 4.4.2 Periodontitis 93
- 4.4.3 Periimplantitis 94

4.5 Non-oral infections associated with oral bacteria 94
4.6 Conclusions 96
Acknowledgments 96
References 96

5. The role of filamentous fungi in drinking water biofilm formation

Ana F.A. Chaves, Lúcia Chaves Simões, Russell Paterson, Manuel Simões and Nelson Lima

5.1	**Drinking water concerns**	101
5.2	**Microbiology of drinking water distribution systems**	102
	5.2.1 Filamentous fungi in drinking water	104
	5.2.2 Biofilms	109
5.3	**Drinking water distribution systems maintenance**	114
	5.3.1 Disinfection	115
5.4	**Bacterial and fungal interactions**	118
	Acknowledgments	119
	References	120

6. Microalgal and cyanobacterial biofilms

Ana Cláudia Barros, A.L. Gonçalves and Manuel Simões

6.1	**Microalgae and cyanobacteria**	127
6.2	**Applications of microalgae and cyanobacteria**	127
6.3	**Microalgal/cyanobacterial cultivation**	130
6.4	**Microalgal harvesting techniques**	131
	6.4.1 Chemical flocculation	134
	6.4.2 Autoflocculation	134
	6.4.3 Bioflocculation	134
	6.4.4 Electrocoagulation–flocculation	135
	6.4.5 Gravitational sedimentation	135
	6.4.6 Flotation	135
	6.4.7 Centrifugation	136
	6.4.8 Filtration	136
6.5	**Factors affecting microalgal/cyanobacterial biofilms**	136
	6.5.1 Light	137
	6.5.2 CO_2 concentration	137
	6.5.3 pH	137
	6.5.4 Nutrients	138
	6.5.5 Temperature	138
	6.5.6 Surface properties	138
	6.5.7 Hydrodynamic conditions	139
	6.5.8 Extracellular polymeric substances	139
	6.5.9 Microalgal species	139
	6.5.10 Presence of other microorganisms	140
6.6	**The role of microalgal/cyanobacterial biofilms in wastewater treatment processes**	140
6.7	**Conclusions**	145
	Acknowledgments	145
	References	145

7. Biofilms in membrane systems for drinking water production

Nadia M. Farhat, Szilard S. Bucs and Johannes S. Vrouwenvelder

7.1	**Introduction**		157
7.2	**Methods to evaluate biofilm growth potential of feedwater**		160
7.3	**Conventional biofouling control strategies**		162
	7.3.1	Preventive biofouling control	162
	7.3.2	Risk of chemical dosage: antiscalants, acids, and biocides	164
	7.3.3	Conventional curative biofouling control: cleanings	166
7.4	**New control strategies**		167
	7.4.1	Membrane modification	167
	7.4.2	Feed spacer modification	168
	7.4.3	Advanced cleaning strategies	169
7.5	**Future perspectives**		170
	Acknowledgment		171
	References		171

8. Biofilm fuel cells

Marcela N. Gatti, Facundo Quiñones and Ruben H. Milocco

8.1	**Processes involved in the biofilm of a microbial fuel cell**		179
	8.1.1	Bacteria-catalyzed reactions	179
	8.1.2	Mechanisms involved in releasing electrons on the anode surface	180
	8.1.3	The faradaic charge transfer processes	181
	8.1.4	The double-layer capacitance	182
	8.1.5	Mass transfer	183
8.2	**Microbial fuel cell structures**		185
8.3	**Integration of main processes in a microbial fuel cell model**		188
	8.3.1	Parameter identification	189
	8.3.2	Electrical analogy for small-signal operation	190
	8.3.3	Maximum steady-state power delivered by microbial fuel cell	193
	8.3.4	Sensitivity analysis of the maximum power	194
	8.3.5	Achievable maximum power	196
8.4	**Dimensional electrodes**		198
	8.4.1	Problem formulation	198
	8.4.2	Substrate diffusion	198
	8.4.3	Charge transfer	200
	8.4.4	Simulation results	201
8.5	**Conclusions**		202
	References		202

9. Application of lactic acid bacteria and their metabolites against foodborne pathogenic bacterial biofilms
E. Giaouris

9.1 Introduction 205
9.2 Antibiofilm activities of lactic acid bacteria and their metabolites against foodborne bacterial pathogens 207
9.2.1 Antibiofilm activity of the in situ lactic acid bacteria presence 207
9.2.2 Antibiofilm activity of lactic acid bacteria cell-free culture supernatants 212
9.2.3 Antibiofilm activity of purified lactic acid bacteriocins 219
9.2.4 Antibiofilm activity of lactic acid bacteria exopolysaccharides 222
9.2.5 Antibiofilm activity of lactic acid bacteria biosurfactants 224
9.2.6 Biofilm inhibition trough quorum sensing interference by lactic acid bacteria 224
9.3 Conclusions 225
References 227

10. Role of equipment design in biofilm prevention
Thierry Bénézech and Christine Faille

10.1 Introduction 233
10.2 Simple equipment geometries 233
10.3 Complex equipment design 236
10.3.1 Immersed surfaces 236
10.3.2 Air–liquid–material interfaces 238
10.3.3 Splash areas 239
10.4 Material properties 242
10.4.1 Physicochemistry 242
10.4.2 Topography 243
10.5 Conclusion 244
Acknowledgments 245
References 245

11. Biofilm control with enzymes
Anabela Borges, Ana Meireles, Filipe Mergulhão, Luís Melo and Manuel Simões

11.1 Biofilms and problems associated with their control 249
11.2 Biofilm structure and mechanisms of bacterial resistance 251
11.3 Emergent strategies of biofilm control and eradication 253
11.4 Antibiofilm enzymes 256

11.4.1 Biofilm disruption by enzymes 257
11.4.2 Example applications 259
11.5 Conclusions 262
Acknowledgments 263
References 263

12. The potential of phytochemical products in biofilm control

Mafalda Andrade, Joana Malheiro, Fernanda Borges, Maria José Saavedra and Manuel Simões

12.1 Antimicrobial properties of phytochemicals 273
12.1.1 Phytochemical classes 274
12.1.2 Modes of antimicrobial action 276
12.2 Phytochemicals as biofilm-controlling agents 279
12.3 Conclusions 281
Acknowledgments 287
References 287

13. Photoinactivation of biofilms

Alex Fiori Silva, João Vitor de Oliveira Silva, Paula Aline Zanetti Campanerut-Sá and Jane Martha Graton Mikcha

13.1 Photodynamic therapy 295
13.1.1 Historical remarks 295
13.1.2 Basic principles and mechanism of photosensitization 295
13.1.3 Photosensitizers and light sources 296
13.2 Photoinactivation of biofilms 298
13.3 Concluding remarks 303
References 303

14. The potential of drug repurposing to face bacterial and fungal biofilm infections

Isabel M. Oliveira, Anabela Borges and Manuel Simões

14.1 Introduction 307
14.2 Antimicrobial activity among drugs used for noninfectious human diseases 308
14.3 Drug repurposing—an alternative strategy against biofilm infections 314
14.4 Conclusions 320
Acknowledgments 320
References 320

15. *In silico* development of quorum sensing inhibitors
Rita P. Magalhães, Tatiana F. Vieira, André Melo and Sérgio F. Sousa

15.1 Biofilms in health 329
15.2 Mechanisms of biofilm formation 330
15.3 Quorum sensing 332
15.4 *In silico* methods 335
15.4.1 Computer-aided drug design: history and methods 335
15.4.2 Molecular docking 337
15.4.3 Virtual screening 341
15.4.4 Quantitative structure–activity relationships 345
15.5 Conclusions 349
Acknowledgment 349
References 349

16. Challenges and perspectives in reactor scale modeling of biofilm processes
Hermann J. Eberl and Matthew J. Wade

16.1 Introduction 359
16.2 Mathematical modeling of biofilm reactors 363
16.2.1 Chemostat modeling 364
16.2.2 One-dimensional biofilm models 368
16.2.3 A biofilm reactor model 371
16.3 Modeling challenges and perspectives 373
16.3.1 Mathematical and computational challenges 373
16.3.2 Implicit model assumptions and level of detail in process descriptions 375
16.3.3 Some parallels and connections between modeling and experimental studies 377
16.4 Conclusion 379
Acknowledgments 380
References 380

Index 385

Contributors

Ana Cristina Abreu, Department of Chemistry and Physics, Research Centre CIAIMBITAL, University of Almería, Almería, Spain

Mafalda Andrade, LEPABE — Laboratory for Process Engineering Environment, Biotechnology and Energy, Department of Chemical Engineering, Faculty of Engineering, University of Porto, Porto, Portugal

Ana Cláudia Barros, LEPABE — Laboratory for Process Engineering Environment, Biotechnology and Energy, Department of Chemical Engineering, Faculty of Engineering, University of Porto, Porto, Portugal

Thierry Bénézech, Univ. Lille, CNRS, INRAE, ENSCL, UMET, Villeneuve d'Ascq, France

Anabela Borges, LEPABE – Laboratory for Process Engineering Environment, Biotechnology and Energy, Department of Chemical Engineering, Faculty of Engineering, University of Porto, Porto, Portugal

Fernanda Borges, Cardiff School of Pharmacy and Pharmaceutical Sciences, Cardiff University, Cardiff, Wales, United Kingdom

Szilard S. Bucs, Water Desalination and Reuse Center (WDRC), Division of Biological and Environmental Science and Engineering (BESE), King Abdullah University of Science and Technology (KAUST), Thuwal, Saudi Arabia

Paula Aline Zanetti Campanerut-Sá, Department of Clinical Analysis and Biomedicine, State University of Maringá, Maringá, Paraná, Brazil

Ana F.A. Chaves, Faculty of Engineering, University of Porto, Porto, Portugal

João Vitor de Oliveira Silva, Department of Clinical Analysis and Biomedicine, State University of Maringá, Maringá, Paraná, Brazil

Hermann J. Eberl, Department of Mathematics and Statistics, University of Guelph, Guelph, ON, Canada

Christine Faille, Univ. Lille, CNRS, INRAE, ENSCL, UMET, Villeneuve d'Ascq, France

Nadia M. Farhat, Water Desalination and Reuse Center (WDRC), Division of Biological and Environmental Science and Engineering (BESE), King Abdullah University of Science and Technology (KAUST), Thuwal, Saudi Arabia

Ignacio Fernández, Department of Chemistry and Physics, Research Centre CIAIMBITAL, University of Almería, Almería, Spain

Marcela N. Gatti, Grupo de Control Automático y Sistemas, Facultad de Ingeniería. Universidad Nacional del Comahue, Neuquén, Argentina

Astrid Gędas, Department of Industrial and Food Microbiology, Faculty of Food Science, University of Warmia and Mazury in Olsztyn, Olsztyn, Poland

E. Giaouris, Department of Food Science and Nutrition, School of the Environment, University of the Aegean, Ierou Lochou 10 and Makrygianni, Lemnos, Greece

D.M. Goeres, Center for Biofilm Engineering, Montana State University, Bozeman, MT, United States

A.L. Gonçalves, LEPABE — Laboratory for Process Engineering Environment, Biotechnology and Energy, Department of Chemical Engineering, Faculty of Engineering, University of Porto, Porto, Portugal

Nelson Lima, Centre of Biological Engineering, University of Minho, Braga, Portugal

Rita P. Magalhães, UCIBIO/REQUIMTE, BioSIM – Departamento de Biomedicina, Faculdade de Medicina da Universidade do Porto, Porto, Portugal

Joana Malheiro, LEPABE — Laboratory for Process Engineering Environment, Biotechnology and Energy, Department of Chemical Engineering, Faculty of Engineering, University of Porto, Porto, Portugal; CIQUP/Department of Chemistry and Biochemistry, Faculty of Sciences, University of Porto, Porto, Portugal; Cardiff School of Pharmacy and Pharmaceutical Sciences, Cardiff University, Cardiff, Wales, United Kingdom

Ana Meireles, LEPABE — Laboratory for Process Engineering Environment, Biotechnology and Energy, Department of Chemical Engineering, Faculty of Engineering, University of Porto, Porto, Portugal

André Melo, LAQV/REQUIMTE, Departamento de Química e Bioquímica, Faculdade de Ciências da Universidade do Porto Porto, Portugal

Luís Melo, LEPABE — Laboratory for Process Engineering Environment, Biotechnology and Energy, Department of Chemical Engineering, Faculty of Engineering, University of Porto, Porto, Portugal

Filipe Mergulhão, LEPABE — Laboratory for Process Engineering Environment, Biotechnology and Energy, Department of Chemical Engineering, Faculty of Engineering, University of Porto, Porto, Portugal

M. Mettler, Center for Biofilm Engineering, Montana State University, Bozeman, MT, United States

Jane Martha Graton Mikcha, Department of Clinical Analysis and Biomedicine, State University of Maringá, Maringá, Paraná, Brazil

Ruben H. Milocco, Grupo de Control Automático y Sistemas, Facultad de Ingeniería. Universidad Nacional del Comahue, Neuquén, Argentina

Isabel M. Oliveira, LEPABE – Laboratory for Process Engineering Environment, Biotechnology and Energy, Department of Chemical Engineering, Faculty of Engineering, University of Porto, Porto, Portugal

Magdalena A. Olszewska, Department of Industrial and Food Microbiology, Faculty of Food Science, University of Warmia and Mazury in Olsztyn, Olsztyn, Poland

A.E. Parker, Center for Biofilm Engineering, Montana State University, Bozeman, MT, United States; Department of Mathematical Sciences, Montana State University, Bozeman, MT, United States

Russell Paterson, Centre of Biological Engineering, University of Minho, Braga, Portugal

S. Pedersen, BioSurface Technologies Corporation, Bozeman, MT, United States

Facundo Quiñones, Grupo de Control Automático y Sistemas, Facultad de Ingeniería. Universidad Nacional del Comahue, Neuquén, Argentina

Marta Ribeiro, LEPABE — Laboratory for Process Engineering Environment, Biotechnology and Energy, Department of Chemical Engineering, Faculty of Engineering, University of Porto, Porto, Portugal

Maria José Saavedra, CITAB-Centre for Research and Technology of Agro-Environmental and Biological Sciences, Veterinary Science Department, University of Trás-os-Montes e Alto Douro, Vila Real, Portugal

Alex Fiori Silva, Federal Institute of Paraná, Paranavaí, Paraná, Brazil

Lúcia Chaves Simões, Centre of Biological Engineering, University of Minho, Braga, Portugal; LEPABE — Laboratory for Process Engineering Environment, Biotechnology and Energy, Department of Chemical Engineering, Faculty of Engineering, University of Porto, Porto, Portugal

Manuel Simões, LEPABE – Laboratory for Process Engineering Environment, Biotechnology and Energy, Department of Chemical Engineering, Faculty of Engineering, University of Porto, Porto, Portugal

Sérgio F. Sousa, UCIBIO/REQUIMTE, BioSIM – Departamento de Biomedicina, Faculdade de Medicina da Universidade do Porto, Porto, Portugal

P. Sturman, Center for Biofilm Engineering, Montana State University, Bozeman, MT, United States

Tatiana F. Vieira, UCIBIO/REQUIMTE, BioSIM – Departamento de Biomedicina, Faculdade de Medicina da Universidade do Porto, Porto, Portugal

Johannes S. Vrouwenvelder, Water Desalination and Reuse Center (WDRC), Division of Biological and Environmental Science and Engineering (BESE), King Abdullah University of Science and Technology (KAUST), Thuwal, Saudi Arabia

Matthew J. Wade, School of Engineering, Newcastle University, Newcastle-upon-Tyne, United Kingdom

D.K. Walker, Center for Biofilm Engineering, Montana State University, Bozeman, MT, United States

B. Warwood, BioSurface Technologies Corporation, Bozeman, MT, United States

Preface

The ability of microorganisms to adhere on surfaces and form biofilms is a privilege. These aggregates of cells enclosed in a matrix of extracellular polymeric substances show a physiology that is distinctly different from that of the same cells in planktonic state. Biofilm growth is arguably the most relevant growth state for the majority of microorganisms, particularly bacteria. Its complexity relative to planktonic growth means that we still have a poor understanding of how microorganisms behave in such a complex structure. Biofilms are as versatile as they are ubiquitous. Intentional and unintentional biofilms concern a broad range of areas, comprising special attention in the industrial/environmental and biomedical areas. As consequence, research on biofilm science and technology is an evolving research area.

This book contributes with new insights regarding the biofilm mode of life, giving the readers a significant content focusing the recent advances on multidisciplinary biofilm research. The book is strategically outlined with data on biofilm formation by diverse microorganisms—bacteria, microalgae, and filamentous fungi. Top-notch methods for biofilm analysis and characterization are described in terms of analytical chemistry and mathematical modeling. Advanced strategies for biofilm control are detailed in several chapters as well as the in silico analysis for the development of biofilm-targeting molecules. Biofilms are further conveniently described for their biotechnological potential, particularly for wastewater treatment and for bioenergy production as biofilm fuel cells.

Acknowledgments

This work was financially supported by Base Funding—UIDB/00511/2020 of the Laboratory for Process Engineering, Environment, Biotechnology and Energy (LEPABE)—funded by national funds through the FCT/MCTES (PIDDAC); projects PTDC/BII-BTI/30219/2017 - POCI-01-0145-FEDER-030219, POCI-01−0145-FEDER-028397, POCI-01−0145-FEDER-033298, and POCI-01−0145-FEDER-035234 funded by FEDER funds through COMPETE2020—Programa Operacional Competitividade e Internacionalização (POCI) and by national funds (PIDDAC) through the FCT/MCTES. FCT under the scope of the strategic funding of UIDB/04469/2020 unit and BioTecNorte operation (NORTE-01-0145-FEDER-000004) funded by the European Regional Development Fund under the scope of Norte2020 - Programa Operacional Regional do Norte. Sabbatical grant SFRH/BSAB/150379/2019.

FCT Fundação para a Ciência e a Tecnologia | REPÚBLICA PORTUGUESA | COMPETE 2020 | PORTUGAL 2020 | UNIÃO EUROPEIA Fundo Europeu de Desenvolvimento Regional

Chapter 1

Biofilm formation and resistance

Astrid Gędas, Magdalena A. Olszewska
Department of Industrial and Food Microbiology, Faculty of Food Science, University of Warmia and Mazury in Olsztyn, Olsztyn, Poland

1.1 Biofilm mode of growth

Overall, microorganisms exist in two modes of growth: a unicellular, in which single, free-living cells (planktonic) prevail in the microbial population, and a multicellular one, in which the cells bind to each other (cohesion) and to the substratum (adhesion) (Bjarnsholt, 2013). In nature, microbes barely occur as planktonic cells but instead exist as communities of sessile cells that grow as biofilms (Berlanga and Guerrero, 2016; Rabin et al., 2015; Donlan, 2002). Biofilms can be defined as aggregated microbial communities surrounded by a matrix of self-produced extracellular polymeric substances (EPS), which develop on a wide variety of inert or organic surfaces (Armbruster and Parsek, 2018; Kim and Lee, 2016).

Biofilm development depends upon different factors, including those associated with environmental conditions, specific strain attributes, and material surface properties (Chmielewski and Frank, 2003). In fact, biofilm formation is usually enhanced in harsh environmental conditions, such as nutrient-deficient or toxic media (Rendueles and Ghigo, 2015). Besides, microbes within the biofilm can coordinate their behavior for promoting growth and producing EPS (Moradali et al., 2017). The ability to form biofilms seems to be universal among microorganisms. Microbial communities exhibiting this ability may be composed not only of single species but of multiple species as well (O'Toole et al., 2000; López et al., 2010). In most biofilms, however, microorganisms account for less than 10% of the biofilm dry mass, whereas the EPS may account for over 90% (Flemming and Wingender, 2010; Tsagkari and Sloan, 2018). Indeed, this self-produced matrix is responsible for the cohesion and adhesion of cells, but more importantly, for the development of a microenvironment that allows the microbes for cell–cell interaction and communication and serves as a

Recent Trends in Biofilm Science and Technology. https://doi.org/10.1016/B978-0-12-819497-3.00001-5

reservoir of metabolic substances, nutrients, and energy for biofilm inhabitants (Flemming and Wingender, 2010).

This highly hydrated EPS matrix is mainly composed of polysaccharides (PSs), but it can also consist of proteins, lipids, extracellular DNA (eDNA), and other biopolymers (Das et al., 2016; Kim and Lee, 2016; Kostakioti et al., 2013). The PSs synthesized by microbial cells differ significantly in their composition and thus in their chemical and physical properties (Limoli et al., 2015). For instance, in gram-negative bacteria, some PSs are neutral or polyanionic and hence rendered more anionic, whereas in gram-negative bacteria, the EPS show mostly cationic nature due to teichoic acid and certain quantities of proteins (Donlan, 2002; Vu et al., 2009). Enzymes may also play an important role in a biofilm life cycle, i.e., they can break down EPS polymers and provide carbon and energy during starvation or cause biofilm degradation during detachment and dispersal (Rabin et al., 2015). Several studies have shown the significance of enzymes particularly in releasing cells from biofilms to start a new biofilm life cycle (Petrova and Sauer, 2016). In biofilms, e.g., these formed by *Vibrio cholerae* and *Pseudomonas aeruginosa*, nonenzymatic proteins, such as lectins, which take part in formation and stabilization of the matrix, can also be found (Fong and Yildiz, 2015). Moreover, a crucial role in biofilm formation has recently been demonstrated for eDNAs which, among others, take part in cell adhesion. In the case of *Staphylococcus aureus*, eDNA is responsible for matrix structure and enables cell–cell as well as cell–surface interactions (Boles and Horswill, 2011). eDNA is also essential in cell-to-cell connection and in *Pseudomonas* biofilm stabilization especially at the initial stages of biofilm development, when the amount of EPS components is small (Kostakioti et al., 2013).

Not only composition but also the quantity of EPS changes depending on the type of microorganisms, the age of the biofilm, and current environmental conditions (Donlan, 2002). Importantly, EPS form the structure and architecture of the biofilm and protect cells from adverse and disruptive environmental conditions (Vu et al., 2009). Several studies have shown that biofilms can increase microbial resistance to dehydration (Roy et al., 2018), UV radiation (Kirmusaoğlu, 2016), extreme temperature and pH (Achinas et al., 2019; Kirmusaoğlu, 2016), high salinity (Kirmusaoğlu, 2016), nutrients deficiency (Roy et al., 2018; Chmielewski and Frank, 2003), various sanitizers (Lu et al., 2019), antibiotics (Lu et al., 2019; Kirmusaoğlu, 2016), etc. Therefore, biofilm formation is of serious and ongoing concern in a wide range of fields from industrial processes such as food processing, to health-related fields such as medicine and dentistry (Van Houdt and Michiels, 2010). Aside from economic and health problems, biofilms play a positive role and can arouse commercial interest in the immobilization technology (Nie et al., 2016; Żur et al., 2016).

1.2 Biofilm formation is a multistep process

Biofilm formation is a dynamic process that can be divided into several steps, i.e., initial attachment, irreversible attachment, early development of biofilm architecture, maturation, and dispersion and which involves a highly coordinated cascade of gene expression (Reen, 2019) (Table 1.1).

The first step in the process can be either active or passive, depending on cell adherence mechanism (Dufour et al., 2012; O'Toole et al., 2000; Srey et al., 2013). Adhesive surface structures, such as fimbria and pili, are promotive to the active adhesion. Flagella facilitate bacteria transfer to a specific attachment site and together with pili help overcoming the repulsive barriers encountered by the cell when it approaches the surface (Berne et al., 2015; Chmielewski and Frank, 2003). However, they can also play an important role in surface sensing by which microbes sense and respond to contact with the surface (O'Toole and Wong, 2016). In the case of a biofilm-forming bacterium *P. aeruginosa,* flagellar mechanosensing is regulated by a secondary messenger, bis-(3–5)-cyclic dimeric guanosine monophosphate (c-di-GMP), which affects the activity of a regulator of flagellar gene expression, FleQ. Normally, FleQ suppresses the expression of *pel* gene, which is responsible for the synthesis of biofilm exopolysaccharides. When c-di-GMP binds FleQ, the inhibition of *pel* gene expression decreases, thus promoting biofilm formation (Belas, 2014; O'Toole and Wong, 2016). This small cytoplasmic signaling molecule—c-di-GMP—is a key regulator of the motility-to-biofilm transition for many other bacteria, e.g., *Escherichia coli*, *Vibrio cholerae*, and *Salmonella enterica* serovar Typhimurium (Guttenplan and Kearns, 2013; Valentini and Filloux, 2016). Importantly, several genes encoding adhesins associated with biofilm formation were identified within the gram-positive and nonmotile bacteria like *Enterococcus*, such as SagA, Acm (*Enterococcus faecium*), and Ace (*Enterococcus faecalis*) that facilitate adherence to eukaryotic extracellular matrix components (Kostakioti et al., 2013). Moreover, a new study has identified bacterial membrane vesicles (MVs) produced by *E. faecium* and suggested that MV-associated proteins are involved in virulence and antimicrobial resistance including biofilm-promoting proteins and extracellular matrix–binding proteins (Wagner et al., 2018). Indeed, at the beginning of biofilm development, adhesins play a major role in anchoring to the surface (Esteban et al., 2014). Biofilm formation by *Staphylococcus epidermidis* and *S. aureus* requires also other surface proteins, such as biofilm-associated protein (Bap) and the accumulation-associated protein (Aap) (Conrady et al., 2008). Fibronectin is also associated with enhancing cell adhesion. There are two types of fibronectin synthesized by *S. aureus*—fibronectin-binding protein A (FnBPA) and fibronectin-binding protein B (FnBPB), which differ by ligands. A group of adhesins that bind collagen (CAN) may be distinguished as well (Esteban et al., 2014). In turn, Cairns et al. (2013) showed that the gram-positive model organism *Bacillus subtilis* flagellum also

TABLE 1.1 Genes related with biofilm formation.

Step	Gene	Function	Species	References
Initial attachment	*icaADCB*	Regulation of synthesis polysaccharide intercellular adhesins	*Staphylococcus epidermidis*, *Staphylococcus aureus*	Büttner et al. (2015)
	cupA-C	Involved in the assembly of cell surface fimbriae and in biofilm formation	*Pseudomonas aeruginosa*	Mikkelsen et al., (2009)
	abpA	Adhesin synthesis; amylase binding	*Streptococcus gordonii*	Loo C.Y. (2003)
	sspA/B	Adhesin synthesis; collagen binding	*Str. gordonii*	Loo C.Y. (2003)
	embp	Intercellular adhesion	*Sta. epidermidis*	Büttner et al. (2015)
	Aap	Role in primary attachment; accumulation-associated protein	*Sta. epidermidis*	Schaeffer et al. (2015)
	flhDC	Flagella production, motility regulator	*Salmonella*	Raman et al. (2019)
	degU	Transcription of protein in flagellum DegS-DegU system of mechanosensing	*Bacillus subtilis*	Cairns et al. (2013)
Irreversible attachment	*lsrR/K*	Quorum sensing regulation; affect biofilm architecture	*Escherichia coli*	Li et al. (2007)
	afeI	AHL synthase, role in quorum sensing system	*Acidithiobacillus ferrooxidans*	Farah et al. (2005)
	σ^{28}	Flagellum-specific sigma factor; control expression of proteins forming the filament and motor	*Salmonella typhimurium*	Das et al. (2018)
	Flu	Outer membrane surface protein Ag43; controls colony form variation and bacterial aggregation	*E. coli*	Schembri and Klemm (2001)
	fnbA	Fibronectin binding protein A	*Sta. aureus*	Stutz et al. (2011)
	fnbB	Fibronectin binding protein B	*Sta. aureus*	Stutz et al. (2011)

Early development of biofilm architecture	*luxS*	Role in the quorum sensing autoinducer-2 synthesis	*Salmonella serovar Dublin*	Ju et al. (2018)
	comABCDE	Quorum sensing signaling system essential for genetic competence	*Streptococcus mutans*	Li et al. (2002)
	rpoS	Stimulates adhesion; flagellar biosynthesis; regulates a number of stress-related genes	*E. coli*	Dudin et al. (2014)
	fsrBDC	Essential for the detection of gelatinase activity; quorum sensing dependent regulatory system	*Enterococcus faecalis*	Teixeira et al. (2012)
	lasR/lasI	Synthesis and regulation of AHL systems in QS	*P. aeruginosa*	Ahmed et al. (2019)
	rhlR/rhlI	Synthesis and regulation of AHL systems in QS	*P. aeruginosa*	Ahmed et al. (2019)
	algD-A	Alginate synthesis, essential for adherence and biofilm formation	*P. aeruginosa*	Limoli et al. (2015)
	Agr	Encode the staphylococcal quorum sensing two-component system	*Sta. aureus*	Liu et al., 2016
	argR	Arginine regulator; role in biofilm formation and development	*Str. gordonii*	Robinson et al. (2017)
Maturation and dispersion	*dspB*	PIA-degrading enzyme	*Sta. epidermidis*	Tan et al. (2015)
	lytA	Autolysin gene	*Streptococcus pneumoniae*	Sanz et al., 2017
	atlE	Autolysin gene	*Sta. epidermidis*	Büttner et al. (2015)
	hapA	Hemagglutinin/protease; promotes detachment	*Vibrio cholera*	Karatan and Watnick (2009)

acts as a mechanosensor upon contact with the surface and that the inhibition of flagellar rotation involving the flagellar stator gene, *motB*, results in the activation of the DegS-DegU two-component signal transduction system with sensory kinase that controls biofilm formation. Moreover, a recent study has shown that c-di-GMP levels are higher in matrix producers than motile cells and differ between sporulating and competent cell types and thus might also play an important role during *B. subtilis* biofilm formation (Weiss et al., 2019).

In turn, the passive adhesion is facilitated by gravity, diffusion, and fluid dynamics (Chmielewski and Frank, 2003). Initial attachment is a reversible stage due to the loose binding with the abiotic or biotic surface, where van der Waals and electrostatic forces are involved, and at this stage, cells might still detach and return to the planktonic mode (Armbruster and Parsek, 2018). However, after 5–30 s of weak interaction, when cells lie flat on the surface forming a monolayer and resist to physically dislodge, the attachment becomes irreversible (Chmielewski and Frank, 2003). Importantly, material properties affect bacterial adhesion. Surface charge plays an important role in cell–surface interaction. Most bacterial cells are negatively charged; therefore the adhesion may be disrupted by a negative charge of the surface, whereas an opposite effect can be observed with the positively charged surface (Song et al., 2015). Moreover, surface charge may influence bacterial motility and biofilm cells morphology. A study by Rzhepishevska et al. (2013) showed that *P. aeruginosa* developed a mushroom-shaped structure on negatively charged poly(3-sulfopropylmethacrylate) (SPM) and zwitterionic poly(2-(methacryloyloxy)ethyl)dimethyl-3-sulfoproyl) ammonium hydroxide) (MEDSAH), whereas a flat biofilm on positively charged surfaces. Additionally, high levels of c-di-GMP were found in the mushroom structure, which may suggest that—through physiological adaptation—bacteria can attach even to the negatively charged surfaces. A more recent study has indicated, however, that *Pseudomonas putida* NBRC 100650 easily adhered and formed biofilm on the hydrophobic polyvinylidenefluoride (PVDF) surface. In contrast, biofilm was rarely observed on the hydrophilic polyvinyl alcohol (PVA) surface (Saeki et al., 2016). Surface energy, roughness, and topography are also significant factors in cell attachment (Achinas et al., 2019; Song et al., 2015). Nonetheless, irreversible attachment is characterized by permanent bonding such as dipole–dipole interaction, hydrogen bonds, or ionic covalent bonding, and thus is hard to break (Chmielewski and Frank, 2003; Srey et al., 2013).

Following this stage, cells start to multiply and form microcolonies and simultaneously produce a polymer matrix (Taraszkiewicz et al., 2013). For many gram-negative bacteria, this stage is characterized by high levels of c-di-GMP because it is associated with intensified production of a biofilm matrix (Armbruster and Parsek, 2018). From now on, bacteria display a coordinated behavior through cell–cell communication commonly known as quorum sensing (QS), which is important for controlling, e.g., sporulation, competence, antibiotic production, virulence factor secretion, and biofilm formation

(Rutherford and Bassler, 2012). For this purpose, they secrete autoinducers (AIs), which accumulate in the environment, as the bacterial population density increases and alters gene expression (Reen, 2019). For biofilm development, the genes involved in biofilm formation and maturation are activated, as population density reaches a critical value. There are three groups of signaling molecules in bacteria: acyl homoserine lactones (AHLs) that are specific for gram-negative bacteria, peptides called autoinducing peptides (AIPs) that are specific for gram-positive bacteria, and a universal signaling molecule called autoinducer-2 (AI-2) (Taraszkiewicz et al., 2013). When concentration of AHLs is high, the bonding with cytoplasmic receptors occurs that further governs the gene expression. Gram-negative bacteria use the LuxI/LuxR system, in which AIs are synthesized by LuxI-type enzymes and detected by LuxR-type transcriptional regulators (Bassler, 2002). There are many gram-negative bacteria identified to use LuxI/LuxR to control a wide range of biological behaviors, including those encoding virulence factors and antibiotics biosynthesis, plasmid transfer, bioluminescence, and biofilm formation (Ziemichód and Skotarczak, 2017). Each species produces a specific AHL or a combination of AHL, and the members of the same species respond only to the signal molecule (Li and Tian, 2012). When AIP concentration is high, it binds to a two-component sensor kinase receptor transmitting signal and consequently activating the expression of genes, e.g., an accessory gene regulator (Agr) and RNAIII (Lu et al., 2019; Taraszkiewicz et al., 2013). For example, the agr global regulator, besides the transition between planktonic and sessile modes, controls cell detachment, contributing to survival via cell dispersal and colonization of new niches by *S. aureus* (Grande et al., 2014). A recent understanding of the molecular mechanisms underlying *agr* QS was reviewed by Tan et al. (2018). Agr operon controls Agr system and consists of AgrA, AgrB, AgrC, and AgrD. When cell density is high, AIPs bind to the membrane-located AgrC, which further activates AgrA and consequently promotes the target gene expression. There are two promoters. One of them, P3, may also activate RNAIII expression, which prompts the upregulation of exo-toxins and exo-enzymes (Lu et al., 2019; Otto, 2013). Ciulla et al. (2018) described a potential method for the treatment of *S. aureus* biofilm infections with using RNA III-inhibiting peptide (RIP). This peptide is capable of blocking the QS mechanisms by competing with RNA III-activating peptide, and it leads to the inhibition of *S. aureus* pathogenesis. In turn, AI-2 is produced by a range of gram-positive and gram-negative bacteria and detected by a variety of them, thus allowing for intra- and interspecies communication, including development of dental plaque (Rickard et al., 2006), and affecting gut microbiota composition (Thompson et al., 2015). It has recently been shown that AI-2 produced by *E. faecalis* biofilms attracts *E. coli* cells, resulting in enhanced aggregation and microcolony formation by *E. coli* and in increased stress resistance of both species (Laganenka and Sourjik, 2017).

Once the first layer of the biofilm is developed, biofilm often grows to a mushroom- or tower-shaped structure, as observed for *P. aeruginosa*, *S. aureus* (Rabin et al., 2015), or honeycomb-like organizational structures of *Listeria monocytogenes* biofilms (Guilbaud et al., 2015) (Fig. 1.1). Biofilm structure development depends on many factors, including changes in the amount of nutrients or atmospheric pressure (Reen, 2019). At this stage, a monolayer biofilm, where there are cell–surface interactions only, transforms into a multilayer biofilm, where bacteria are attached to both the surface and other biofilm cells (Karatan and Watnick, 2009). Along with maturation, an increasingly complex biofilm structure is being formed together with water channels, which distribute nutrients and wastes (Lu et al., 2019). The arrangement of bacteria in the structure is not accidental and determined by their metabolism and tolerance to oxygen. For instance, anerobic bacteria are located in deeper layers, thus avoiding exposure to oxygen (Rabin et al., 2015).

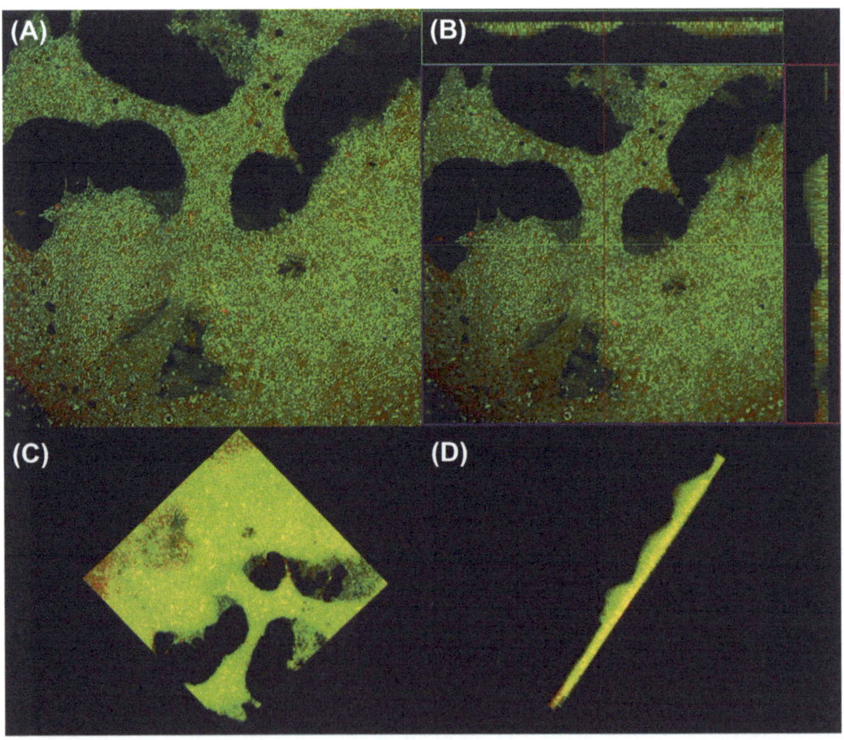

FIGURE 1.1 3D projections (A–D) from confocal laser scanning microscope (CLSM) images of the biofilm formed by *Listeria monocytogenes* serotype 1/2a showing the honeycomb-like morphotype (obtained with the Zeiss Zen software—authors' own contribution). The biofilm was grown in an eight-well chamber slide system in TSB at 37 °C and stained with the Live/Dead BacLight.

This is also associated with different cell growth rates—fast growers are usually located on the periphery and slow growers in deeper layers, which indicates that the periphery of biofilm microcolonies is compositionally and mechanically dynamic (Chmielewski and Frank, 2003).

Environmental stress or nutrient deficit leads to the last and critical stage in the biofilm life cycle, wherein cells leave the biofilm and return to the planktonic state or settle in a new niche (Dufour et al., 2012). In general, there are three main ways of bacterial cells removal, i.e., desorption, detachment, and dispersion. Desorption means a direct transfer of microorganisms to the bulk liquid and is observed throughout the biofilm development process (Petrova and Sauer, 2016). Detachment may be categorized into four mechanisms: abrasion, grazing, erosion, and sloughing. Abrasion means detachment caused by collisions with particles from the bulk liquid. Feeding activity of eukaryotic organisms leads to the release of a biofilm, called grazing. Peripheral cells in the structure are constantly exposed to fluid shear, which results in the continuous removal of biofilm portions. This kind of detachment is called erosion. In contrast, sloughing is a rapid removal of intact pieces or entire biofilm mass. Besides shear forces, the chemical or enzymatic degradation is also involved in this process (Kim and Lee, 2016; Petrova and Sauer, 2016). Enzymes are produced by community inhabitants, which break down polysaccharides holding the biofilm together, thereby actively releasing bacteria for the colonization of new niches in a process called dispersion. For instance, *P. aeruginosa* produces alginate lyase, *E. coli* uses *N*-acetyl-heparosan lyase, and *Streptococcus equi* uses hyaluronidase to break down a biofilm matrix (Kregiel and Antolak, 2016). Another example of a detachment agent can be a hydrolase called dispersin B (DspB), which is produced by *Aggregatibacter actinomycetemcomitans*. It catalyzes the hydrolysis of poly-*N*-acetyl-D-glucosamines (PGAs) and, as a result, causes the biofilm to lose its structural integrity (Ragunath et al., 2015). DspB effectively inhibits biofilm formation by both gram-positive and gram-negative bacteria, for instance, by *S. epidermidis* and *Actinobacillus pleuropneumoniae* (Kaplan et al., 2004). For this reason, Ghalsasi and Sourjik (2016) proposed using engineered *E. coli* to produce dispersin B as a targeting system to biofilm removal. The last way of cell release from biofilm is dispersion, wherein the sensing of certain signals is involved. The regulatory network is crucial in physiological changes, which facilitate bacteria release. Dispersion can be divided into two classes by the source of signals. In native dispersion, signaling molecules are self-synthesized, whereas changes in the external environment are the cues for environmentally induced dispersion (Petrova and Sauer, 2016).

1.3 Biofilm-specific resistance to antimicrobials

To survive, microbes have developed cell protection or resistance mechanisms against the harsh environmental conditions (Možina et al., 2013). Along with the transformation of planktonic cells into a sessile form, biofilm resistance becomes a more complex phenomenon with a range of genes that showed increased expression in biofilms (Table 1.2). Biofilm displays therefore unique properties, making the community inhabitants even 1000 times more resistant than the planktonic cells (Chadha, 2014). Due to this fact, human infections related to biofilms are often difficult or impossible to eradicate and turn into serious chronic conditions (Aswathanarayan and Vittal, 2013). A group of bacteria comprising *P. aeruginosa*, *E. faecium*, *S. aureus*, *Klebsiella pneumoniae*, *Acinetobacter baumannii*, and *Enterobacter* species is characterized by tolerance and resistance to antimicrobial agents and referred to as "ESKAPE pathogens" (Santajit and Indrawattana, 2016). *P. aeruginosa* is especially important owing to its capability to form biofilms, where the bacteria embed themselves in the EPS and escape from antibiotic treatment (Ciofu and Tolker-Nielsen, 2019). Aside from the healthcare, antimicrobial resistance via attachment of cells to the surface has serious consequences also in the petroleum industry, drinking water distribution system, paper industry, metalworking industry, and food-processing industry (Singh et al., 2017). However, because biofilm-related health issues generate significant global health and economic costs, the link between biofilm and antibiotic resistance deserves a profound interest. Several studies have revealed certain antibiotics can induce biofilm formation, as recently reviewed by Ciofu and Tolker-Nielsen (2019) or Song et al. (2016). Other researchers have focused on the association between biofilm production and multidrug resistance (MDR) and demonstrated antibiotic resistance to be higher among biofilm producers than nonproducers, e.g., *P. aeruginosa* or *A. baumanii* (Gurung et al., 2013). Zeighami et al. (2019) have recently revealed the high frequency of biofilm producers among extensively drug-resistant (XDR) *A. baumannii* clinical isolates, with a high prevalence of biofilm-related genes of *csuE* and *pgaB*. Moreover, although Qi et al. (2016) found that MDR *A. baumannii* isolates were weaker biofilm producers than the non-MDR isolates, the enhancement in resistance occurred irrespective of the quantity of biofilm produced.

Different species in the community are able to cooperate with each other rather than compete, which is also unique to biofilms. These biofilm aggregates could easily give rise to a horizontal gene transfer, thus spreading resistance genes (Bordi and de Bentzmann, 2011). Haaber et al. (2017) suggests that mobile genetic elements (MGEs), which carry antibiotic resistance genes (ARGs) in *S. aureus* that forms biofilm on medical implants, are transferred by conjugation or bacteriophage transduction, with the latter being potentially higher in biofilms compared with the planktonic cells. In turn, gene operons are involved in biofilm-associated tolerance and resistance. *E. coli* uses a four-gene operon, *pgaABCD*, which encodes the exopolysaccharide, poly-N-acetylglucosamine, and plays an important role in biofilm formation and resistance. Besides *E. coli*, poly-*N*-acetyl-glucosamine was also described in *S. aureus*, *S. epidermidis*, and *Actinobacillus* species. Cerca and Jefferson

TABLE 1.2 Genes related with biofilm resistance.

Gene	Function	Species	References
csuE, pgaB	Biofilm-related genes	*Acinetobacter baumannii*	Zeighami et al. (2019)
ARGs	Antibiotic resistance genes	*Staphylococcus aureus*	Haaber et al. (2017)
pgaABCD operon	Involved in biofilm-associated tolerance and resistance; encodes the exopolysaccharides	*Escherichia coli*	Haaber et al. (2017)
ndvB	Encodes glucose polymers	*Pseudomonas aeruginosa*	Olsen (2015)
rpoS	Controls a specific physiological response to a specific stress and enhanced resistance of biofilm cells to antimicrobial agents	*E. coli*	Možina et al. (2013)
PA1874-1877	Encodes the efflux pump	*P. aeruginosa*	Zhang and Mah (2008)
mexXY	Encodes MexXY efflux pump	*P. aeruginosa*	Morita et al. (2012)
acrD	Encodes AcrD pump; efflux pump membrane transporter	*E. coli*	Morita et al. (2012)
amrAB-oprA	Encodes AmrAB-OprA efflux pump	*Burkholderia Pseudomallei*	Morita et al. (2012)
adeABC	Encodes AdeABC efflux pump	*A. baumannii*	Beheshti et al. (2014)
arcAB, oqxAB	Encodes AcrAB and OqxAB efflux pumps; important in nitrofurantoin resistance	*Klebsiella pneumoniae*	Xu et al. (2019)
tetV	Encodes TetV; causing resistance to tetracycline	*Mycobacterium tuberculosis*	Kanji et al. (2019)
rv1258c (tap)	Encodes Tap efflux pump; conferring low resistance to aminoglycosides and tetracycline; mutations in gene cause resistance to pyrazinamide, isoniazid, streptomycin	*M. tuberculosis*	Kanji et al., (2019); Liu et al. (2019)

Continued

TABLE 1.2 Genes related with biofilm resistance.—cont'd

Gene	Function	Species	References
relA	Overexpression causes increased persister frequency	*E. coli*	Renbarger et al. (2017)
Hip	Deletion results in suppressed persister formation during the stationary phase	*E. coli*	Renbarger et al. (2017)
relA, spoT	Production of ppGpp in stress conditions	*P. aeruginosa, E.coli*	Wood et al. (2013)
tisB	Encodes toxin in type I TisB/IstR-1 system	*E. coli*	Fisher et al. (2017)
mqsRA	Encodes toxin (RNase) in MqsR/MqsA system; mediate biofilm formation	*E. coli*	Kwan et al. (2013); Van Acker and Coenye (2016)

(2008) revealed that the expression of *pga* and polysaccharide synthesis was induced in *E. coli* by several chemicals, i.e., ethanol, NaCl, and glucose, and biofilm formation was also induced by one of them (glucose). Besides, many of the studies on biofilm-induced resistance have been done with *P. aeruginosa,* as reviewed by, e.g., Olsen (2015), where biofilm-specific resistance genes and operons were identified. It was, for instance, suggested that glucose polymers encoded by the *ndvB* present in *P. aeruginosa* prevent antibiotics from reaching their ribosomal target sites by isolating them in the periplasm and by the activation of ethanol oxidation genes (Bordi and de Bentzmann, 2011; Olsen, 2015).

In biofilm environment, cells are confronted with different physical, chemical, or biological stress conditions, and they often cope with them via activation of a general stress response, in which a master regulator, RpoS, is involved that controls a specific physiological response to a specific stress and enhanced resistance of biofilm cells to antimicrobial agents (Možina et al., 2013). This largely falls into innate biofilm resistance mechanisms, where limited diffusion of an antibiotic agent through biofilm matrix, decreased oxygen and nutrient availability accompanied by altered metabolic activity, formation of persisters, and other specific molecules are considered (Ciofu and Tolker-Nielsen, 2019). Besides the innate resistance, induced resistance has also been recognized, where direct exposure to antimicrobial agents triggers a specific response. In the case of *P. aeruginosa* biofilm, overproduction of β-lactamase occurs as a result of β-lactam antibiotic exposure, causing the hydrolysis of this molecule (Aswathanarayan and Vittal, 2013). However, it is worth keeping in mind that the resistance to antibiotics in biofilms is due to both the innate and induced mechanisms. It has been shown that QS affects the tolerance of *P. aeruginosa* biofilms to tobramycin, kanamycin, and hydrogen peroxide (Olsen, 2015). It might be related with QS participation in eDNA production, which inhibits penetration of some antibiotics into the biofilm's structure. Under extreme conditions, the bacterial survival is strongly sustained by efflux pumps. There are single or multicomponent systems to get rid of toxins and waste products, and these pumps are widely involved in antibiotic resistance, even that related with MDR (Možina et al., 2013). The complete deletion of PA1874-1877 genes encoding the efflux pump in *P. aeruginosa* PA14 results in its increased sensitivity to tobramycin, gentamicin, and ciprofloxacin, in particular when this mutant is grown in a biofilm (Zhang and Mah, 2008). Thus, the efflux pump is more highly expressed in biofilm cells than in planktonic counterparts, suggesting that these genes play an important role for biofilm resistance to antibiotics. In addition, Zhang and Mah (2008) discovered that a mutant strain combining the *ndvB* mutation with PA1874-1877 gene deletion was more resistant than a single mutant, which shows the interaction of two different mechanisms of antibiotic resistance. *P. aeruginosa* has also a MexXY pump, which is a significant determinant of the resistance of aminoglycosides (such as amikacin, tobramycin). Besides MexXY pump, AcrD of *E. coli*, AmrAB-OprA of *Burkholderia pseudomallei*, and AdeABC of *A. baumannii* are aminoglycoside efflux pumps as well (Morita at al., 2012). In the case of *K. pneumoniae*, AcrAB and OqxAb efflux pumps play an important role in nitrofurantoin resistance (Xu et al., 2019). *Mycobacterium tuberculosis* resistance is also caused by the efflux pumps system. Among others, there is TetV causing resistance to tetracycline and Tap (Rv1258c) conferring low resistance to aminoglycosides and tetracycline (Kanji et al., 2019).

Certainly, the so-called biofilm cell heterogeneity has an important impact on increased resistance to antimicrobial agents (Singh et al., 2017). A subpopulation called persister cells is frequently found in the biofilm; although it represents a relatively small part of the population, its frequency in biofilm is still higher than in the planktonic communities (Stewart, 2002). The metabolism of persisters is slowed down, perhaps even in the sporelike state. By this inhibition of metabolism, they are not antibiotic targets because they are inactive (Mah, 2012). A recent study has suggested the presence of a long-retention effect or "memory effect" in the persister cell state of not only gram-negative bacteria (*E. coli*, *Acinetobacter*, *Salmonella*) but of gram-positive bacteria as well (*Staphylococcus*, *Bacillus*) (Miyaue et al., 2018). In particular, it was revealed that *E. coli* forms more persister cells in biofilm culture than in the liquid culture and that these persisters can be sustained in higher numbers than those from liquid culture for up to 4 weeks at 37°C in a fresh, antibiotic-containing medium, even after withdrawal from the biofilm culture. Hence, this memory effect may be caused by molecular signal inscribed into the persister cells during biofilm culture. Importantly, the number of persisters formed depends on the growth phase and is higher in the stationary phase (Renbarger et al., 2017). To obtain a high level of persisters in a bacterial culture of *P. aeruginosa* and *S. aureus*, an improved method based on carbonyl cyanide *m*-chlorophenylhydrazone (CCCP) treatment can be used (Grassi et al., 2017). This highly efficient chemical method uses the uncoupling agent CCCP to induce persister cells in a culture by decreasing ATP production (Kwan et al., 2013). Persisters can also promote the spread of resistance plasmids of *Salmonella* Typhimurium and *E. coli* in the gut (Bakkeren et al., 2019). The persister gene *hip* (high persistence) was first discovered in 1983 in *E. coli* (Moyed and Bertrand, 1983). Deletion of the *hip* operon results in suppressed persister formation during the stationary phase. On the other hand, persister frequency can increase via overexpression of *relA*. This causes a higher level of (p)ppGpp (the alarmone guanosine tetraphosphate) (Renbarger et al., 2017). In stress conditions such as nutrient limitation, ppGpp is produced by RelA and SpoT. It reduces DNA replication, protein synthesis, and indirectly stress response sigma factor and activates type II and type I toxin–antitoxin (TA) system (Wood et al., 2013). TA system is common in bacterial genome and accounts for the formation of persisters from regular cells. It consists of two genes. The first encodes a stable toxin to disturb cellular function. The second encodes an unstable antitoxin to mediate toxins effect (Kwan et al., 2013). There are known six different types of the TA system, categorized by various mechanisms of neutralization. The SOS response by *E. coli* induces the expression of the pore-forming toxin gene *tisB* in type I TisB/IstR-1 system (Fisher et al., 2017). Another example of TA system is MqsR/MqsA, where MqsR is a toxin, i.e., RNase (Kwan et al., 2013). Additionally, it was found that MqsRA mediates biofilm formation of *E. coli* (Van Acker and Coenye, 2016). Persister cells are tolerant not only to antibiotics but also to metals. Persisters of copper-tolerant *P. aeruginosa* ST308 pose a potential risk of human contamination because bacteria may survive in drinking water installations (Jeanvoine et al., 2019). This strongly highlights the need for monitoring of the hospital environment and food processing environment for persister cells especially among bacterial biofilms because they provide an increased overall level of resistance to the community and spread of antibiotic resistance (Fisher et al., 2017). Thus, a better understanding of the internal and environmental basis for the formation of persister cells will allow to create and improve effective methods to combat biofilm.

1.4 Conclusions

In conclusion, biofilms are regulated by a variety of physiological, environmental, and genetic factors, which are still not completely understood. It is a very complex phenomenon, ranging from biofilm structure characteristics to various biofilm-associated resistance mechanisms. Consequently, due to the complexity of biofilm formation and biofilm-related antimicrobial resistance, medical and industrial implications are still difficult to control. Therefore, more research into biofilms is still necessary to fully understand this phenomenon and to develop more effective methods for biofilm prevention and eradication.

Acknowledgments

This work was supported by the stipend provided by the Rector of University of Warmia and Mazury in Olsztyn for Magdalena Olszewska for the academic year 2018/19 at University of Porto, Portugal, and by National Science Centre, Poland, through project SONATA 7 (grant no. 2014/13/D/NZ9/04803).

References

Achinas, S., Charalampogiannis, N., Euverink, G.J.W., 2019. A brief recap of microbial adhesion and biofilms. Applied Sciences 9 (14), 2801.

Ahmed, S.A.K.S., Rudden, M., Smyth, T.J., Dooley, J.S.G., Marchant, R., 2019. Natural quorum sensing inhibitors effectively downregulate gene expression of *Pseudomonas aeruginosa* virulence factors. Applied Microbiology and Biotechnology 103 (8), 3521–3535.

Armbruster, C.R., Parsek, M.R., 2018. New insight into the early stages of biofilm formation. Proceedings of the National Academy of Sciences 115 (17), 4317–4319.

Aswathanarayan, J.B., Vittal, R.R., 2013. Microbial biofilms and their control by various antimicrobial strategies. In: Méndez-Vilas, A. (Ed.), Microbial Pathogens and Strategies for Combating Them: Science, Technology and Education. Formatex, pp. 124–133.

Bakkeren, E., Huisman, J.S., Fattinger, S.A., Hausmann, A., Furter, M., Egli, A., Slack, E., Sellin, M.E., Bonhoeffer, S., Regoes, R.R., Diard, M., Hardt, W.-D., 2019. *Salmonella* persisters promote the spread of antibiotic resistance plasmids in the gut. Nature 573 (7773), 276–280.

Bassler, B.L., 2002. Small talk: cell-to-cell communication in bacteria. Cell 109 (4), 421–424.

Beheshti, M., Telebi, M., Ardebili, A., Bahador, A., Lari, A.R., 2014. Detection of AdeABC efflux pump genes in tetracycline-resistant *Acinetobacter baumannii* isolates from burn and ventilator-associated pneumonia patients. Journal of Pharmacy and BioAllied Sciences 6 (4), 229–232.

Belas, R., 2014. Biofilms, flagella, and mechanosensing of surfaces by bacteria. Trends in Microbiology 22 (9), 517–527.

Berlanga, M., Guerrero, R., 2016. Living together in biofilms: the microbial cell factory and its biotechnological implications. Microbial Cell Factories 15 (165), 1–11.

Berne, C., Ducret, A., Hardy, G.G., Brun, Y.V., 2015. Adhesins involved in attachment to abiotic surfaces by Gram-negative bacteria. Microbiology Spectrum 3 (4), 1–45.

Bjarnsholt, T., 2013. The role of bacterial biofilms in chronic infections. Acta Pathologica, Microbiologica et Immunologica Scandinavica 136, 1–51.

Boles, B.R., Horswill, A.R., 2011. Staphylococcal biofilm disassembly. Trends in Microbiology 19 (9), 449–455.

Bordi, C., de Bentzmann, S., 2011. Hacking into bacterial biofilms: a new therapeutic challenge. Annals of Intensive Care 1 (19), 1–8.

Büttner, H., Mack, D., Rohde, H., 2015. Structural basis of *Staphylococcus epidermidis* biofilm formation: mechanisms and molecular interactions. Frontiers in Cellular and Infection Microbiology 5 (14), 1–15.

Cairns, L.S., Marlow, V.L., Bissett, E., Ostrowski, A., Stanley-Wall, N.R., 2013. A mechanical signal transmitted by the flagellum controls signaling in *Bacillus subtilis*. Molecular Microbiology 90 (1), 6–21.

Cerca, N., Jefferson, K.K., 2008. Effect of growth conditions on poly-N –acetylglucosamine expression and biofilm formation in *Escherichia coli*. FEMS Microbiology Letters 283 (1), 36–41.

Chadha, T., 2014. Bacterial biofilms: survival mechanisms and antibiotic resistance. Journal of Bacteriology and Parasitology 5 (3), 1–4.

Chmielewski, R.A.N., Frank, J.F., 2003. Biofilm formation and control in food processing facilities. Comprehensive Reviews in Food Science and Food Safety 2, 22–32.

Ciofu, O., Tolker-Nielsen, T., 2019. Tolerance and resistance of *Pseudomonas aeruginosa* biofilms to antimicrobial agents—how *P. aeruginosa* can escape antibiotics. Frontiers in Microbiology 3 (10), 913.

Ciulla, M., Di Stefano, A., Marinelli, L., Cacciatore, I., Di Biase, G., 2018. RNAIII inhibiting peptide (RIP) and derivatives as potential tools for the treatment of *S. aureus* biofilm infections. Current Topics in Medical Chemistry 18, 1–12.

Conrady, D.G., Brescia, C.C., Horii, K., Weiss, A.A., Hassett, D.J., Herr, A.B., 2008. A zinc-dependent adhesion module is responsible for intercellular adhesion in staphylococcal biofilms. Proceedings of the National Academy of Sciences 105 (49), 19456–19461.

Das, C., Mokashi, C., Mande, S.S., Saini, S., 2018. Dynamics and control of flagella assembly in *Salmonella typhimurium*. Frontiers in Cellular and Infection Microbiology 8 (36), 1–13.

Das, T., Ibugo, A.I., Klare, W., Manefield, M., 2016. Role of pyocyanin and extracellular DNA in facilitating *Pseudomonas aeruginosa* biofilm formation. In: Microbial Biofilms – Importance and Applications. Intech, pp. 23–42 (Chapter 2).

Donlan, R.M., 2002. Biofilms: microbial life on surfaces. Emerging Infectious Diseases 8 (9), 881–890.

Dudin, O., Geidelmann, J., Ogasawara, H., Ishihama, A., Lacour, S., 2014. Repression of flagellar genes in exponential phase by CsgD and CpxR, two crucial modulators of *Escherichia coli* biofilm formation. Journal of Bacteriology 196 (3), 707–715.

Dufour, D., Leung, V., Lévesque, C.M., 2012. Bacterial biofilm: structure, function, and antimicrobial resistance. Endodontic Topics 22, 2–16.

Esteban, J., Pérez-Tanoira, R., Pérez-Jorge-Peremarch, C., Gómez-Barrena, E., 2014. Bacterial adherence to biomaterials used in surgical procedures. Microbiology for Surgical Infections 41–57 (Chapter 3).

Farah, C., Vera, M., Morin, D., Haras, D., Jerez, C.A., Guiliani, N., 2005. Evidence for a functional quorum-sensing type AI-1 system in the extremophilic bacterium *Acidithiobacillus ferrooxidans*. Applied and Environmental Microbiology 71 (11), 7033–7040.

Fisher, R.A., Gollan, B., Helaine, S., 2017. Persistent bacterial infections and persister cells. Nature Reviews Microbiology 15 (8), 453–464.

Flemming, H.C., Wingender, J., 2010. The biofilm matrix. Nature Reviews Microbiology 8, 623–633.

Fong, J.N.C., Yildiz, F.H., 2015. Biofilm matrix proteins. Microbiology Spectrum 3 (2), 1–27.

Ghalsasi, V.V., Sourjik, V., 2016. Engineering *Escherichia coli* to disrupt poly-N-acetylglucosamine containing bacterial biofilms. Current Synthetic and Systems Biology 4 (130), 1–5.

Grande, R., Nistico, L., Sambanthamoorthy, K., Longwell, M., Iannitelli, A., Cellini, L., Di Stefano, A., Stoodley, L.H., Stoodley, P., 2014. Temporal expression of *agrB, cidA*, and *alsS* in the early development of *Staphylococcus aureus* UAMS-1 biofilm formation and the structural role of extracellular DNA and carbohydrates. Pathogens and Disease 70, 414–422.

Grassi, L., Di Luca, M., Maisetta, G., Rinaldi, A.C., Esin, S., Trampuz, A., Batoni, G., 2017. Generation of persister cells of *Pseudomonas aeruginosa* and *Staphylococcus aureus* by chemical treatment and evaluation of their susceptibility to membrane-targeting agents. Frontiers in Microbiology 8, 1917.

Guilbaud, M., Piveteau, P., Desvaux, M., Brisse, S., Briandet, R., 2015. Exploring the diversity of *Listeria monocytogenes* biofilm architecture by high-throughput confocal laser scanning microscopy and the predominance of the honeycomb-like morphotype. Applied and Environmental Microbiology 81 (5), 1813–1819.

Gurung, J., Khyriem, A.B., Banik, A., Lyngdoh, W.V., Choudhury, B., Bhattacharyya, P., 2013. Association of biofilm production with multidrug resistance among clinical isolates of *Acinetobacter baumannii* and *Pseudomonas aeruginosa* from intensive care unit. Indian Journal of Critical Care Medicine: Peer-reviewed, Official Publication of Indian Society of Critical Care Medicine 17 (4), 214–218.

Guttenplan, S.B., Kearns, D.B., 2013. Regulation of flagellar motility during biofilm formation. FEMS Microbiology Reviews 37 (6), 849–871.

Haaber, J., Penadés, J.R., Ingmer, H., 2017. Transfer of antibiotic resistance in *Staphylococcus aureus*. Trends in Microbiology 25 (11), 893–905.

Jeanvoine, A., Meunier, A., Puja, H., Bertrand, X., Valot, B., Hocquet, D., 2019. Contamination of a hospital plumbing system by persister cells of a copper-tolerant high-risk clone of *Pseudomonas aeruginosa*. Water Research 157, 579–586.

Ju, X., Li, J., Zhu, M., Lu, Z., Lv, F., Zhu, X., Bie, X., 2018. Effect of the *luxS* gene on biofilm formation and antibiotic resistance by *Salmonella serovar* Dublin. Food Research International 107, 385–393.

Kanji, A., Hasan, R., Hasan, Z., 2019. Efflux pump as alternate mechanism for drug resistance in *Mycobacterium tuberculosis*. Indian Journal of Tuberculosis 66 (1), 20–25.

Kaplan, J.B., Ragunath, C., Velliyagounder, K., Fine, D.H., Ramasubbu, N., 2004. Enzymatic detachment of *Staphylococcus epidermidis* biofilms. Antimicrobial Agents and Chemotherapy 48 (7), 2633–2636.

Karatan, E., Watnick, P., 2009. Signals, regulatory networks, and materials that build and break bacterial biofilms. Microbiology and Molecular Biology Reviews 73 (2), 310–347.

Kim, S.-K., Lee, J.-H., 2016. Biofilm dispersion in *Pseudomonas aeruginosa*. Journal of Microbiology 54 (2), 71–85.

Kırmusaoğlu, S., 2016. Staphylococcal biofilms: pathogenicity, mechanism and regulation of biofilm formation by quorum-sensing system and antibiotic resistance mechanisms of biofilm-embedded microorganisms. In: Microbial Biofilms – Importance and Applications. Intech, pp. 189–209 (Chapter 10).

Koastakioti, M., Hadjifrangiskou, M., Hultgren, S.J., 2013. Bacterial biofilms: development, dispersal, and therapeutic strategies in the dawn of the postantibiotic era. Cold Spring Harbor Perspectives Medicine 3 (4), a010306.

Kregiel, D., Antolak, H., 2016. Biofilms in beverage industry. In: Microbial Biofilms – Importance and Applications. Intech, pp. 171–186 (Chapter 9).

Kwan, B.W., Valenta, J.A., Benedik, M.J., Wood, T.K., 2013. Arrested protein synthesis increases persister-like cell formation. Antimicrobial Agents and Chemotherapy 57 (3), 1468–1473.

Laganenka, L., Sourjik, V., 2017. Autoinducer 2-dependent *Escherichia coli* biofilm formation is enhanced in a dual-species coculture. Applied and Environmental Microbiology 84 (5) e02638-17.

Li, J., Attila, C., Wang, L., Wood, T.W., Valdes, J.J., Bentley, W.E., 2007. Quorum sensing in *Escherichia coli* is signaled by AI-2/LsrR: effects on small RNA and biofilm architecture. Journal of Bacteriology 189 (16), 6011–6020.

Li, Y.-H., Tang, N., Aspiras, M.B., Lau, P.C., Lee, J.H., Ellen, R.P., Cvitkovitch, D.G., 2002. A quorum-sensing signaling system essential for genetic competence in *Streptococcus mutans* is involved in biofilm formation. Journal of Bacteriology 184 (10), 2699–2708.

Li, Y.-H., Tian, X., 2012. Quorum sensing and bacterial social interactions in biofilms. Sensors 12 (3), 2538.

Limoli, D.H., Jones, C.J., Wozniak, D.J., 2015. Bacterial extracellular polysaccharides in biofilm formation and function. Microbiology Spectrum 3 (3), 1–30.

Liu, J., Shi, W., Zhang, S., Hao, X., Maslov, D.A., Shur, K.V., Bekker, O.B., Danilenko, V.N., Zhang, Y., 2019. Mutations in efflux pump Rv1258c (Tap) cause resistance to pyrazinamide, isoniazid, and streptomycin in *M. tuberculosis*. Frontiers in Microbiology 10 (216), 1–7.

Liu, Q., Yeo, W.-Y., Bae, T., 2016. The SaeRS two-component system of Staphylococcus aureus. Genes (Basel) 7 (10), 81.

Loo, C.Y., 2003. Oral Streptococcal genes that encode biofilm formation. In: Wilson, M., Devine, D. (Eds.), Medical Implications of Biofilms, pp. 189–211.

López, D., Vlamakis, H., Kolter, R., 2010. Biofilms. Cold Spring Harbor Perspectives in Biology 2 (7), a000398.

Lu, L., Hu, W., Tian, Z., Yuan, D., Yi, G., Zhou, Y., Cheng, Q., Hu, J., Li, M., 2019. Developing natural products as potential anti-biofilm agents. Chinese Medicine 14 (11), 1–17.

Mah, T.-F., 2012. Biofilm-specific antibiotic resistance. Future Microbiology 7 (9), 1061–1072.

Mikkelsen, H., Ball, G., Giraud, C., Filloux, A., 2009. Expression of *Pseudomonas aeruginosa CupD* fimbrial genes is antagonistically controlled by RcsB and the EAL-containing PvrR response regulators. PloS One 4 (6), e6018.

Miyaue, S., Suzuki, E., Komiyama, Y., Kondo, Y., Morikawa, M., Maeda, S., 2018. Bacterial memory of persisters: bacterial persister cells can retain their phenotype for days or weeks after withdrawal from colony–biofilm culture. Frontiers in Microbiology 9 (1396), 1–6.

Moradali, M.F., Ghods, S., Rehm, B.H.A., 2017. *Pseudomonas aeruginosa* lifestyle: a paradigm for adaptation, survival, and persistence. Frontiers in Cellular and Infection Microbiology 7, 39.

Morita, Y., Tomida, J., Kawamura, Y., 2012. MexXY multidrug efflux system of *Pseudomonas aeruginosa*. Frontiers in Microbiology 3 (408), 1–13.

Moyed, H.S., Bertrand, K.P., 1983. HipA, a newly recognized gene of *Escherichia coli* K-12 that affects frequency of persistence after inhibition of murein synthesis. Journal of Bacteriology 155 (2), 768–775.

Možina, S.S., Klančnik, A., Raspor, P., 2013. Mechanisms of microbial resistance in biofilms. In: Simões, M., Mergulhão, F. (Eds.), Biofilms in Bioengineering, First ed., Mechanisms of Microbial Resistance in Biofilms. Nova Science Publishers, pp. 311–332 (Chapter 12).

Nie, M., Nie, H., He, M., Lin, Y., Wang, L., Jin, P., 2016. Immobilization of biofilms of *Pseudomonas aeruginosa* NY3 and their application in the removal of hydrocarbons from highly concentrated oil-containing wastewater on the laboratory scale. Journal of Environmental Management 173, 34–40.

O'Toole, G., Kaplan, H.B., Kolter, R., 2000. Biofilm formation as microbial development. Annual Review of Microbiology 54 (1), 49–79.

O'Toole, G., Wong, G.C.L., 2016. Sensational biofilms: surface sensing in bacteria. Current Opinion in Microbiology 30, 139–146.

Olsen, O., 2015. Biofilm-specific antibiotic tolerance and resistance. European Journal of Clinical Microbiology and Infectious Diseases 34 (5), 877–886.

Otto, M., 2013. Staphylococcal infections: mechanisms of biofilm maturation and detachment as critical determinants of pathogenicity. Annual Review of Medicine 64 (1), 1–14.

Petrova, O.E., Sauer, K., 2016. Escaping the biofilm in more than one way: desorption, detachment or dispersion. Current Opinion in Microbiology 30, 67–78.

Qi, L., Li, H., Zhang, C., Liang, B., Li, J., Wang, L., Du, X., Liu, X., Qiu, S., Song, H., 2016. Relationship between antibiotic resistance, biofilm formation, and biofilm-specific resistance in *Acinetobacter baumannii*. Frontiers in Microbiology 7 (483), 1–10.

Rabin, N., Zheng, Y., Opoku-Temeng, C., Du, Y., Bonsu, E., O Sintim, H., 2015. Biofilm formation mechanisms and targets for developing antibiofilm agents. Future Medicinal Chemistry 7 (4), 493–512.

Ragunath, C., DiFranco, K., Shanmugam, M., Gopal, P., Vyas, V., Fine, D.H., Cugini, C., Ramasubbu, N., 2015. Surface display of *Aggregatibacter actinomycetemcomitans* autotransporter Aae and dispersin B hybrid act as antibiofilm agents. Molecular Oral Microbiology 31 (4), 329–339.

Raman, V., Van Dessel, N., O'Connor, O., Forbes, N., 2019. The motility regulator *flhDC* drives intracellular accumulation and tumor colonization of *Salmonella*. Journal of ImmunoTherapy of Cancer 7 (44), 1–16.

Reen, F.J., 2019. Microbial biofilms. In: Reference Module in Life Sciences, pp. 1–10.

Renbarger, T.L., Baker, J.M., Sattley, W.M., 2017. Slow and steady wins the race: an examination of bacterial persistence. AIMS Microbiology 3 (2), 171–185.

Rendueles, O., Ghigo, J.M., 2015. Mechanisms of competition in biofilm communities. Microbiology Spectrum 3 (3), 1–18.

Rickard, A.H., Palmer Jr., R., Blehert, D.S., Campagna, S.R., Semmelhack, M.F., Egland, P.G., Bassler, B.L., Kolenbrander, P.E., 2006. Autoinducer 2: a concentration-dependent signal for mutualistic bacterial biofilm growth. Molecular Microbiology 60 (6), 1446–1456.

Robinson, J.C., Rostami, N., Casement, J., Vollmer, W., Rickard, A.H., Jakubovics, N.S., 2017. ArcR modulates biofilm formation in the dental plaque colonizer *Streptococcus gordonii*. Molecular Oral Microbiology 33 (2), 143–154.

Roy, R., Tiwari, M., Donelli, G., Tiwari, V., 2018. Strategies for combating bacterial biofilms: a focus on anti-biofilm agents and their mechanisms of action. Virulence 9 (1), 522–554.

Rutherford, S.T., Bassler, B.L., 2012. Bacterial quorum sensing: its role in virulence and possibilities for its control. Cold Spring Harbor Perspectives in Medicine 2 (11), a012427.

Rzhepishevska, O., Hakobyan, S., Ruhal, R., Gautrot, J., Barbero, D., Ramstedt, M., 2013. The surface charge of anti-bacterial coatings alters motility and biofilm architecture. Biomaterials Science 1 (6), 589–602.

Saeki, D., Nagashima, Y., Sawada, I., Matsuyama, H., 2016. Effect of hydrophobicity of polymer materials used for water purification membranes on biofilm formation dynamics. Colloids and Surfaces A: Physicochemical and Engineering Aspects 506, 622–628.

Santajit, S., Indrawattana, N., 2016. Mechanisms of antimicrobial resistance in ESKAPE Pathogens. BioMed Research International 2016, 1–8.

Sanz, J.C., Rios, E., Rodriguez-Avial, I., Ramos, B., Marin, M., Cercenado, E., 2017. Identification of *Streptococcus pneumoniae lytA, plyA* and *psaA* genes in pleural fluid by multiplex real-time PCR. Enfermedades infecciosas y microbiologia clinica 36 (7), 428–430.

Schaeffer, C.R., Woods, K.M., Longo, G.M., Kiedrowski, M.R., Paharik, A.E., Büttner, H., Christer, M., Boissy, R.J., Horswill, A.R., Rohde, H., Fey, P.D., 2015. Accumulation-associated protein enhances *Staphylococcus epidermidis* biofilm formation under dynamic conditions and is required for infection in a rat catheter model. Infection and Immunity 83 (1), 214–226.

Schembri, M.A., Klemm, P., 2001. Coordinate gene regulation by fimbriae-induced signal transduction. The EMBO Journal 20 (12), 3074–3308.

Singh, S., Singh, S.K., Chowdhury, I., Singh, R., 2017. Understanding the mechanism of bacterial biofilms resistance to antimicrobial agents. The Open Microbiology Journal 11, 53–62.

Song, F., Koo, H., Ren, D., 2015. Effects of material properties on bacterial adhesion and biofilm formation. Journal of Dental Research 94 (8), 1027–1034.

Song, T., Duperthuy, M., Wai, S.N., 2016. Sub-optimal treatment of bacterial biofilms. Antibiotics 5 (23), 1–18.

Srey, S., Jahid, I.K., Ha, S.-D., 2013. Biofilm formation in food industries: a food safety concern. Food Control 31 (2), 572–585.

Stewart, P.S., 2002. Mechanisms of antibiotic resistance in bacterial biofilms. International Journal of Medical Microbiology 292, 107–113.

Stutz, K., Stephan, R., Tasara, T., 2011. SpA, ClfA, and FnbA genetic variations lead to staphaurex test-negative phenotypes in bovine mastitis *Staphylococcus aureus* isolates. Journal of Clinical Microbiology 49 (2), 638–646.

Tan, L., Li, S.R., Jiang, B., Hu, X.M., Li, S., 2018. Therapeutic targeting of the *Staphylococcus aureus* accessory gene regulator (agr) system. Frontiers in Microbiology 9 (55), 1–11.

Tan, Y., Ma, S., Liu, C., Yu, W., Han, F., 2015. Enhancing the stability and antibiofilm activity of DspB by immobilization on carboxymethyl chitosan nanoparticles. Microbiological Research 178, 35–41.

Taraszkiewicz, A., Fila, G., Grinholc, M., Nakonieczna, J., 2013. Innovative strategies to overcome biofilm resistance. Biomed Research International 2013, 1–13.

Teixeira, N., Santos, S., Marujo, P., Yokohata, R., Iyer, V.S., Nakayama, J., Hancock, L.E., Serror, P., de Fátima Silva Lopes, M., 2012. The incongruent gelatinase genotype and phenotype in *Enterococcus faecalis* are due to shutting off the ability to respond to the gelatinase biosynthesis-activating pheromone (GBAP) quorum-sensing signal. Microbiology 158 (2), 519–528.

Thompson, J.A., Oliveira, R.A., Djukovic, A., Ubeda, C., Xavier, K.B., 2015. Manipulation of the quorum sensing signal AI-2 affects the antibiotic-treated gut microbiota. Cell Reports 10 (11), 1861–1871.

Tsagkari, E., Sloan, W.T., 2018. Biofilm growth in drinking water systems under stagnant conditions. In: E-Proceedings. Protection and Restoration of the Environment XIV, Greece, 3–6 July 2018, pp. 707–717.

Valentini, M., Filloux, A., 2016. Biofilms and cyclic di-GMP (c-di-GMP) signaling: lessons from *Pseudomonas aeruginosa* and other bacteria. The Journal of Biological Chemistry 291 (24), 12547–12555.

Van Acker, H., Coenye, T., 2016. The role of efflux and physiological adaptation in biofilm tolerance and resistance. The Journal of Biological Chemistry 291 (24), 12565–12572.

Van Houdt, R., Michiels, C.W., 2010. Biofilm formation and the food industry, a focus on the bacterial outer surface. Journal of Applied Microbiology 109 (4), 1117–1131.

Vu, B., Chen, M., Crawford, R.J., Ivanova, E.P., 2009. Bacterial extracellular polysaccharides involved in biofilm formation. Molecules 14 (7), 2535–2554.

Wagner, T., Joshi, B., Janice, J., Askarian, F., Škalko-Basnet, N., Hagestad, O.C., Mekhlif, A., Wai, S.N., Hegstad, K., Johannessen, M., 2018. *Enterococcus faecium* produces membrane vesicles containing virulence factors and antimicrobial resistance related proteins. Journal of Proteomics 187, 28–38.

Weiss, C.A., Hoberg, J.A., Liu, K., Tu, B.P., Winkler, W.C., 2019. Single-cell microscopy reveals that levels of cyclic di-GMP vary among *Bacillus subtilis* subpopulations. Journal of Bacteriology 201 (16) e00247-19.

Wood, T.K., Knabel, S.J., Kwan, B.W., 2013. Bacterial persister cell formation and dormancy. Applied and Environmental Microbiology 79 (23), 7116–7121.

Xu, Q., Jiang, J., Zhu, Z., Xu, T., Sheng, Z.K., Ye, M., Wang, M., 2019. Efflux pumps AcrAB and OqxAB contribute to nitrofurantoin resistance in an uropathogenic *Klebsiella pneumoniae* isolate. International Journal of Antimicrobial Agents 53 (2), 223–227.

Zeighami, H., Valadkhani, F., Shapouri, R., Samadi, E., Haghi, F., 2019. Virulence characteristics of multidrug resistant biofilm forming *Acinetobacter baumannii* isolated from intensive care unit patients. BMC Infectious Diseases 19 (629), 1–9.

Zhang, L., Mah, T.-F., 2008. Involvement of a novel efflux system in biofilm-specific resistance to antibiotics. Journal of Bacteriology 190 (13), 4447–4452.

Ziemichód, A., Skotarczak, B., 2017. QS – systems communication of Gram-negative bacterial cells. Acta Biologica 24, 39–49.

Żur, J., Wojcieszyńska, D., Guzik, U., 2016. Metabolic responses of bacterial cells to immobilization. Molecules 21 (7), 958.

Chapter 2

Nuclear magnetic resonance to study bacterial biofilms structure, formation, and resilience

Ana Cristina Abreu, Ignacio Fernández
Department of Chemistry and Physics, Research Centre CIAIMBITAL, University of Almería, Almería, Spain

2.1 Introduction

Despite the accentuated quest for new therapeutic solutions, the incidence of nosocomial and community-acquired infections caused by multidrug resistant (MDR) bacteria continues to increase worldwide (Li and Webster, 2018). MDR bacteria are of additional concern if they appear as complex surface-associated communities, referred to as biofilms (Frieri et al., 2017; Hall and Mah, 2017). It has been estimated that 60%−80% of human microbial infections are caused by bacteria growing as biofilms (Zhang and Powers, 2012). These cells structures are commonly associated with indwelling medical devices (Brooun et al., 2000; Simões et al., 2008; Busetti et al., 2010), such as venous and urinary catheters, arthroprostheses, fracture fixation devices, and heart valves (Pinto et al., 2011; Griffith et al., 2000; Weber et al., 2010; Cozad and Jones, 2003; Otter et al., 2011; Costerton et al., 2005), but can also be non−device-related, causing chronic inflammatory diseases such as cystic fibrosis, chronic obstructive pulmonary disease, otitis media, and prostatitis (Singh et al., 2000; Lynch et al., 2007a,b; Shin et al., 2009).

In comparison with planktonic cells, biofilm cells display unique phenotypic traits, being the most outstanding of them their characteristic ability to withstand to both antimicrobial agents and host immune factors (Zhu et al., 2013). This is due to the heterogeneous and complex structures of biofilms, which allow for sophisticated singular and collective behaviors, which increase antibiotic resistance (Fish et al., 2016). These chronic tissue-related and device-related infections are thus difficult to treat and expose the patient to the

Recent Trends in Biofilm Science and Technology. https://doi.org/10.1016/B978-0-12-819497-3.00002-7

risk of recurrence (Lebeaux et al., 2014). After formation, a biofilm cannot be easily eliminated by standard clinical procedures, and the strategy for treating these infections is often limited to the removal of the contaminated device (Donlan, 2001; Di Luca et al., 2014). To fight against antibiotic resistance, prevention only is far from being an acceptable strategy (Baquero et al., 2014). As resistant pathogens capable of biofilm formation continue to emerge and propagate, understanding and circumventing biofilm resistance to antibiotics is a paramount requirement.

Decades of extensive research in aspects such as morphology, physiology, and genomics of biofilm formation and resilience were not enough to provide effective biofilm control strategies (Kumar et al., 2017; Chong et al., 2018). Despite the prevalent roles that biofilms play in many fields, such as human infection, few reliable quantitative information is available regarding biofilm matrix composition. Unfortunately, a complete biochemical profile of biofilms is difficult to obtain. As most biofilms are multispecies cultures and, therefore, are highly heterogeneous with respect to structure and to the biological and chemical composition (Stewart and Franklin, 2008), there hardly exist standardized qualitative or quantitative analytical methods for detailed and comprehensive characterization of these biofilms.

The heterogeneous matrix of extracellular polymeric substances (EPS) produced by biofilm cells is known to play a key role on its resistance to degradation and removal (Costa et al., 2018). However, even though polysaccharides and proteins are recognized as the main components of EPS (Metzger et al., 2009), their biochemical characteristics remain unclear. In fact, several top-notch biofilm researchers recently alerted for the lack of knowledge on biofilm EPS composition (Seviour et al., 2019). Without this knowledge, we are unable to effectively manage biofilms, neither their formation nor their eradication. The use of improved analytical methods to explain the roles of EPS and unravel biochemical production pathways is a clear research need. Unfortunately, no single analytical technique meets all the requirements for comprehensive metabolic profiling of complex biological systems. Knowledge about biofilms under realistic or natural conditions at different length scales (meso- and microscale) is missing, for example, regarding their impact on modeling and numerical description of physical properties of biofilms (detachment, deformation, superficial area, pore volumes, etc.). Interactions between fluids and the diverse structures in a biofilm are essential. Comprehensive knowledge of mass transport into and out of the biofilm matrix (molecular diffusion, surface, and structure interactions) is the basis for realistic and meaningful modeling, which is complicated by the typically spatially heterogeneous composition of a biofilm matrix. As biochemical reactions and reactivity depend on the biofilm's structure, metabolism is expected to be also spatially heterogeneous. These facts lead to the necessity of time and spatially resolved studies of structure, transport, and chemical composition.

A holistic analytical approach is thus essential to provide information on the biofilm metabolome and to gain a deeper insight on biofilm formation and structure. Also, obtaining a mechanistic and structural understanding of metabolic changes imposed by multicellular and drug resistance behavior is essential to formulate new strategies for clinical application and for the design of treatments. This would have a great impact considering the widespread of MDR bacteria, which threatens all achievements of modern medicine.

This chapter addresses several imaging and analytical methods, especially those based on nuclear magnetic resonance (NMR), to strike out in new directions to study biofilms and to summarize current knowledge. Because of its noninvasive nature, NMR spectroscopy is a unique tool for studying molecular dynamics in chemical and biological systems. Solution- and solid-state NMR methods have been used to study the chemical composition and molecular mobility of biofilm EPS and extracts. Multinuclear NMR has been employed to study bulk metabolism in cells artificially immobilized in gel or biopolymer matrices and natural, symbiotic bacteria in plant nodules (Hesse et al., 2000). NMR/MRI has been used to measure flow and diffusion in biofilm systems as well as planktonic cell metabolism. Also, recent advancements from case studies reviewed herein have shown the potential of NMR-based metabolomics to shed light on numerous biological problems related to biofilms (Zhang and Powers, 2012). Metabolomics offer unique advantages by allowing a fingerprinting of the state of the metabolome and bringing to light key intra- and extracellular metabolites involved in cellular pathways (amino acid uptake, lipid catabolism, etc.) and processes of development, communication, pathogenesis, persistence, and resistance on biofilms (Pinu and Villas-Boas, 2017; Duncan et al., 2019). Characterizing such metabolites is essential to identify new drug targets and chemical leads vital for the drug discovery process. Some authors already achieved interesting results in this field (Booth et al., 2011; Ammons et al., 2014; Sun et al., 2012).

Briefly, the following aspects will be discussed in this chapter: (1) biofilm development and structure, (2) current knowledge on EPS composition and how it affects biofilm recalcitrance, (3) applications of NMR to study biofilms, with respect to (i) the characterization of both soluble and insoluble part of EPS matrix, (ii) assess biofilm structures and dynamics, (iii) understand diffusion and mass transport within biofilm systems, and finally, (4) an overview is made on NMR-based metabolomics applications in this field.

2.2 Biofilm formation and structure

Bacterial biofilms are complex architectures, three-dimensional (3D) communities found nearly everywhere in nature, and, more importantly, associated with many human diseases. The formation and structure of bacterial biofilms have been extensively reviewed and will only be briefly summarized herein. As described by Zhang and Powers (2012), a biofilm is composed of three parts: a living or nonliving substance that provides a moist surface for attachment of the highly organized microbial structure; a slimlike matrix made of extracellular DNA, proteins, and polysaccharides (β(1−6)-linked N-acetylglucosamine polymer) that embeds the microorganism; and an aggregate of microorganisms in a community that exchange fluids, nutrients, and chemical signals, such as metabolites. A central tenet of biofilm formation is its dynamic nature. Most current models depict biofilm formation as a sequential and dynamic process, which involves (1) reversible, (2) irreversible attachment of bacterial cells to a surface, (3) maturation, and (4) dispersion. Fig. 2.1 describes this process in more detail. The irreversible attachment of bacterial cells to a surface is achieved by the production of EPS by bacteria.

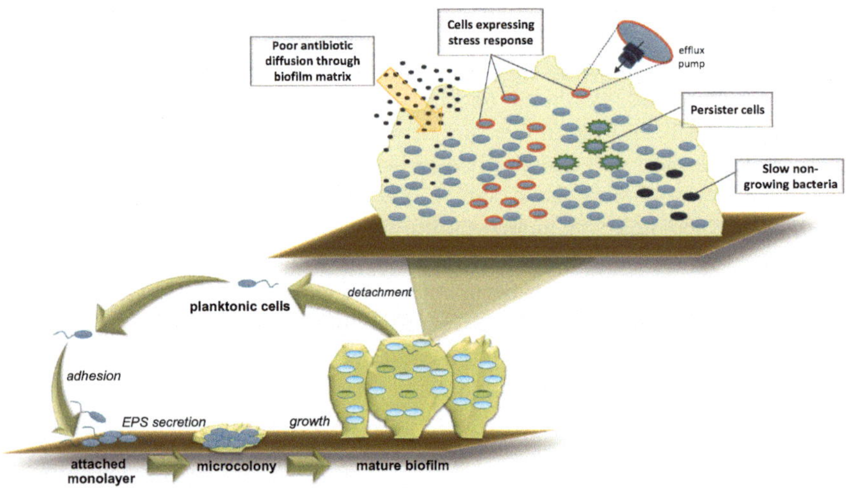

FIGURE 2.1 Biofilm formation and recalcitrance against antimicrobial therapy. Biofilm formation begins when free-floating bacterial cells attach to a surface (Shen et al., 2011). Then, it is believed to occur in a sequential process that includes initial reversible and then irreversible adhesion to a surface and/or other microbes previously attached to the surface, cell−cell communication (quorum sensing), formation of microcolonies, extracellular polymeric substances (EPS) production, and, finally, differentiation of microcolonies into exopolymeric-encased and mature biofilms (Costerton et al., 1999; Simões et al., 2010). Within a biofilm, cells are more resistant to antimicrobial therapy. This is often attributed to the failure of the antimicrobial agents to penetrate the biofilm matrix (Fux et al., 2005; Al-Fattani and Douglas, 2006). By being in different layers of the biofilm, cells will be affected differently by antimicrobials, depending on their diffusion through the biofilm matrix and on their mechanism of action (e.g., penicillins kill only growing bacteria) (Fernández et al., 2011; Stewart and William Costerton, 2001). The heterogenous structure of biofilms also allows for different gradients of nutrients and oxygen that get to the cells, thus inducing distinct growth states. Other theories include a reduced susceptibility of biofilm microorganisms compared with their freely suspended counterparts (Stewart, 1996). Efflux pumps, induced specifically under biofilm conditions, may also be involved in biofilm recalcitrance by removing antibiotics from the bacterial intracytoplasmic space (Lebeaux et al., 2014). Lastly, the existence of persister cells, a small population of cells with a highly protected phenotype, is well described (Brooun et al., 2000; Lai et al., 2009; Stewart, 2002).

EPS participate in the formation of microbial aggregates and are responsible for binding cells and other particulate materials together (cohesion) and to the surface (adhesion) (Simões, 2005).

Cells within biofilms face gradients of nutrients, oxygen, pH, and metabolic by-products varying with depth and are known to express differing phenotype and metabolism in response to these environmental gradients and to numerous changes in gene regulation (Shen et al., 2011). Thus, biofilm cells become phenotypically and metabolically different from their planktonic counterparts (Beitelshees et al., 2018). These changes have drastic effects upon biofilm properties and may have a direct relation to their resistance to antibiotics in a medical scenario (Majors et al., 2005).

The structure of biofilms provides an ideal environment for gene transfer and cell-to-cell interactions (Wu et al., 2015). Cell-to-cell signaling, termed quorum sensing, controls a variety of physiological functions including motility, conjugation, competence, sporulation, virulence, cell attachment and detachment, and biofilm differentiation and formation (Kaper and Sperandio, 2005). Also, biofilms offer enhanced opportunities for cooperative interactions such as horizontal gene transfer and cometabolism (Burmølle et al., 2006). In vivo studies support that in multispecies biofilms, individual species are added sequentially during their progressive formation, and variation in their accretion leads to differences in composition and spatial distribution (Peyyala et al., 2011; Teixeira et al., 2007). However, bacteria do not always cooperate with each other. Biofilms are also sites of intense competition for nutrients and space (Franklin et al., 2015).

Research on biofilms has been intensively focused on its relationship to bacterial infections and drug resistance (Stewart, 1996; Xu et al., 2000; Mah and O'Toole, 2001; Smith, 2005). Within a biofilm, organisms can tolerate hostile environmental conditions, including desiccation, grazing, predation, antimicrobials presence, and host immune responses (Olson et al., 2002; Verstraeten et al., 2008), thus being until up to 1000 times more resistant to antimicrobial therapy relative to their planktonic counterparts (Yeom et al., 2013). This phenomenon, known as "recalcitrance of biofilm toward antibiotics," is complex and is due to several phenomena contributing to a high tolerance and resistance, as illustrated in Fig. 2.1. The obvious slow or incomplete penetration of the antibiotic in the biofilm is an important contributor to biofilm resistance. However, several studies have demonstrated that reductions in the diffusion coefficients of antibiotics within biofilms are insufficient to account solely for the observed changes in susceptibility (Gilbert et al., 2002). Other factors include the appearance of cells expressing an adaptive stress response or slow-/nongrowing bacteria and of the development of specific populations, the so-called "persister cells," that differentiate into a highly protected, dormant, and nondividing phase and are now considered to explain most of the biofilm recalcitrance toward antibiotics (Lewis, 2008).

2.3 The composition of extracellular polymeric substances and how it affects biofilm architecture

As already discussed, most microbes in nature are not found as homogeneous suspensions of free cells but are attached to solid surfaces and to one another within a protective film of secreted polymers (Majors et al., 2005). The architecture and composition of a biofilm and its EPS matrix are strongly influenced by many factors, including contact surface and environmental properties, such as hydrodynamic conditions, shear forces, temperature, and the availability of nutrients, as well as by the presence of other bacterial species embedded in the biofilm matrix, bacterial motility, and intercellular communication. Biofilms tend to be polymicrobial (Jahn and Nielsen, 1995), and different members of the microbial community contribute to their own EPS that then merge into a complex mixture (Flemming and Wingender, 2010). Also, a particular strain may even have the ability to produce different EPS depending on the environmental conditions (Bogino et al., 2013).

Biofilm architecture is an important factor in the biology and virulence of biofilm-forming bacteria (Lynch et al., 2007a,b). Thus, characterizing the architecture of biofilms is the first step toward its understanding. The backbone of a biofilm is its EPS matrix, which forms a hydrogel that surrounds and attaches cells with each other or the interface creating an "immobilized but dynamic microbial environment" (Garny et al., 2010). EPS components are kept together by weak physicochemical interactions such as Van der Waals forces (including hydrogen bonding) and electrostatic interactions (Flemming and Wingender, 2010). The EPS network confers mechanical stability to the biofilm and plays a crucial role in most matrix functions, including water retention, protection from environmental stresses, adsorption of compounds, and nutrient availability.

As mentioned before, despite the prevalent roles that biofilms play in human infection, there has been little reliable quantitative information available regarding biofilm matrix composition and architecture (Reichhardt et al., 2015a). EPS have been called "the dark matter of biofilms" because of the large range of matrix biopolymers and the difficulty in analyzing them (Flemming and Wingender, 2010). Our knowledge on the identification and functions of extracellular proteins, eDNA, and lipids in the biofilm matrix is still in development. This is mostly because most studies assessing the efficacy of antimicrobial agents on biofilms only use cell count approaches, which provides no information regarding the in situ physical characteristics of the biofilm, such as the thickness, structure, and density of the EPS component, as this information is destroyed during sampling and plating. The complexity and tiny quantities of individual polymers makes EPS isolation and characterization an extremely difficult task. Also, current sample extraction methods cannot capture representative depth-dependent metabolite concentrations, because metabolism changes occur rapidly (within milliseconds) during the environmental perturbation that accompanies extraction. Finally, most current techniques used to measure temporal and spatial metabolite profiles in these delicate structures are invasive or destructive. Invasive extraction methods can damage the sample, rendering it useless for subsequent measurements. Thus, in situ spatially resolved analytical techniques are required for the metabolic characterization of biofilms (Majors et al., 2005).

In the past decades, several achievements have been made in the characterization of EPS composition that will be briefly discussed in this section. The heterogeneous structure of biofilm matrices mainly comprises water (up to 97% of matrix), cells (2%–5% of matrix), polysaccharides (1%–2% of matrix), proteins and glycopeptides (1%–2% of matrix), DNA and RNA (1%–2% of matrix), and lipids, with minor contents of nucleic acids and other biopolymers such as humic substances (Yu et al., 2011; Fysun et al., 2019). More general information on EPS composition in terms of major constituents and their function on the biofilm's architecture can be found in Table 2.1. Recent advances on the characterization of such EPS constituents for biofilms found in medical, industrial, or environmental/agriculture settings are also given. The analytical techniques used for the characterization of these components on EPS will be further discussed in this chapter. A detailed description of the roles of each component of EPS can be easily found elsewhere. For example, Bogino et al. (2013) described the production, composition, and functional roles of exopolysaccharides (EPS) in several beneficial and pathogenic plant-associated bacteria.

Besides the gaps on biofilm composition, there are many other aspects about biofilms that must be considered. For instance, adhesins, amyloid-forming proteins, and exopolysaccharides are known to be the main responsible to generate the morphological structures of the biofilm matrix and to shape their aspect. However, questions regarding how their actual 3D patterns are controlled and why they formed have remained elusive. Serra et al. (2013), for example, described that biofilms grown for several days on agar surfaces, i.e., in microcolonies, can adopt different elaborate 3D structures, which have been termed "wrinkled," "rugose," or "rdar" (for red, dry, and rough). Also, less is known about the regulation of matrix components to understand how the production of an individual component is coordinated with that of the others (Branda et al., 2006). Further studies along this line will greatly enhance our understanding of the process of biofilm formation. Moreover, understanding the interactions among the various matrix components could lead to the development of inhibitors able to disrupt the stability of the 3D matrix (Cugini et al., 2019). As an example of what is possible, recent in silico docking analysis targeting the *Streptococcus mutans* water-insoluble glucan-producing GtfC generated a selective inhibitor of biofilm production (Nijampatnam et al., 2018).

TABLE 2.1 Examples of major constituents of extracellular polymeric substances and their function on the structure of the biofilm.

Constituent	Functions	Examples
Extracellular polysaccharides	• Allow the initial steps of cells adhesion and long-term attachment to surfaces • Enable bridging between cells and their immobilization on the biofilm • Form complex networks (the biofilm matrix) • Confer resistance to host defenses and various antimicrobial agents • Are source of sugars for the synthesis of matrix extracellular polymeric substances (EPS).	• Alginate, Pel and Psl in *Pseudomonas aeruginosa* (Colvin et al., 2012) • Pea, Peb, alginate, and a cellulose-like polymer in *Pseudomonas putida* (Chang et al., 2007) • Poly-*N*-acetylglucosamine in many bacteria (Branda et al., 2005) • Cellulose (glucose polymer with β-1,4 glycosidic linkage), e.g., in *Salmonella* Typhimurium and *Escherichia coli* (Serra et al., 2013) • Dextran [α-D-Glc(1,4)], levan-type [β-D-Fru(2,6)] and inulin-type [β-D-Fru(2,1)] fructans in lactic acid bacteria (Torino et al., 2015) • A polygalactan with a backbone of α-D-(1 → 6)-linked galactosyl, β-d-(1 → 4)-linked galactosyl, β-D-(1 → 2,3)-linked galactosyl residues, and a tail end of β-D (1 →)-linked galactosyl residue from *Lactobacillus plantarum* 70810 (Wang et al., 2014) • Several types of glucan by *Streptococcus mutans* (Lynch et al., 2007a,b) and *Haemophilus influenzae* (Domenech et al., 2016) • Galactomannon, galactosaminogalactan, α-1,3 glucans in *Aspergillus fumigatus* (Reichhardt et al., 2015a) • *Kingella kingae* produces a linear polymer of galactofuranose residues in alternating β(1 → 3) −β(1 → 6) linkages (Bendaoud et al., 2011)
Structural proteins	• Involved in adhesion to surfaces and host cells, and in the formation and stabilization of the polysaccharide matrix • Constitute a link between the bacterial surface and extracellular EPS • Function as cytotoxins for both plant cells and bacteria	• LecA and LecB in *P. aeruginosa* (Diggle et al., 2006) • TasA in *Bacillus subtilis* (Branda et al., 2006) • Glucan-binding proteins in biofilms of *S. mutans* (Lynch et al., 2007a,b) • Several lectins in *Azospirillum brasilense* (Mora et al., 2008) • Amyloid adhesins (Larsen et al., 2007) • Curli in *E. coli* and other Enterobacteriaceae (Lim et al., 2012) • The fimbriae-associated adhesin Fap1 (Wu et al., 1998) and the protein FimA (a potential virulence factor) in *Streptococcus parasanguinis* FW213 (Burnette-Curley et al., 1995) • LapA, a cell surface protein in *Pseudomonas fluorescens* (Branda et al., 2005)

TABLE 2.1 Examples of major constituents of extracellular polymeric substances and their function on the structure of the biofilm.—cont'd

Constituent	Functions	Examples
Extracellular enzymes	• Enable the digestion of exogenous macromolecules for nutrient acquisition • Contribute to the degradation of structural EPS, allowing the detachment and dispersal of biofilm cells • Certain enzymes released by pathogenic bacteria may act as virulence factors	• Polysaccharide-degrading enzymes, e.g., endocellulase, chitinase, α- and β-glucosidase, β-xylosidase (Flemming and Wingender, 2010) • Alginate lyase (AlgL) (Farrell and Tipton, 2012) and LasB elastase (Park et al., 2012) in *P. aeruginosa* • DNase, cis-2-decenoic acid in *Staphylococcus aureus* (Park et al., 2012) • Serine Esp protease in *Staphylococcus epidermidis* (Iwase et al., 2010) • AHL lactonase (AiiA) in *Bacillus thuringiensis* (Liu et al., 2008) • α-Amylase from *B. subtilis* (Kalpana et al., 2012) • Diguanylate synthase (Dgc) and a phosphodiesterase (PdeA) in *Gluconacetobacter xylinus* (Branda et al., 2005) • Dispersin B in *Actinobacillus actinomycetemcomitans* (Kaplan et al., 2003)
Lipids	• Act as biosurfactants due to surface active properties, contributing to dispersal and bioavailability of hydrophobic substances • Contribute to bacterial attachment and detachment • Can have antibacterial or antifungal properties	• Glycolipids and phospholipids • Lipopolysaccharides involved in the adherence of *Thiobacillus ferrooxidans* to pyrite surfaces (Flemming and Wingender, 2010) • *Serratia marcescens* produces extracellular lipids with surface-active properties (the "serrawettins") (Matsuyama et al., 2011), • Rhamnolipids in the EPS matrix of *P. aeruginosa* (Abdel-Mawgoud et al., 2010).

2.4 Applications of nuclear magnetic resonance spectroscopy to study biofilms

Detailed metabolic information is critical whether to understand and exploit beneficial biofilms as to combat pathogenic, antibiotic-resistant, and disease-associated forms. This section describes a range of possible applications and recent developments in several analytical and microscopic techniques, especially those based on NMR spectroscopy, and how they can contribute to the study of biofilm structures.

Briefly, NMR relies on the quantum physical property of angular momentum intrinsic to a single nucleus and the response of that nucleus to a magnetic field to study the behavior of macroscopic systems, which consist of large ensembles of nuclei over varying length and time scales (Vogt, 2013). NMR has always been one of the main characterization techniques for solid polymeric materials, since its first applications in the 1970s. Nowadays, NMR is used in a variety of ways and contexts to study biofilms. The advantages of NMR to study biofilms are many; since it is a nondestructive, noninvasive, and nonsample consuming technique, it can be applied to opaque and heterogeneous samples both in static and dynamic forms, and thus it can measure physical features or characteristic behaviors that are challenging, which are difficult to be directly observed with other methods. The disadvantage of NMR is its inherent low sensitivity, although partially palliated by the use of modern cryoprobes, requiring careful optimization to reduce measurement times and lower concentration detection thresholds.

NMR offers multifaceted and noninvasive approaches to study biofilms. Particularly, solid-state NMR (SS-NMR) allows to study nonsoluble polymers at a molecular level in almost all their states and with minimal sample preparation procedure, in a nondestructive manner. Apart from the classic ^{1}H NMR, heteronuclear spectroscopy in its one- and multidimensional versions reveals not only functional and chemical groups but also molecular structure and conformation. Biofilms have also been investigated by NMR flow and diffusion in porous media and flow cells to study water dynamics and biofilm growth at different time and length scales. The application of one-dimensional (1D) and two-dimensional (2D) NMR relaxation measurement methods, diffusion NMR and magnetic resonance imaging (MRI) to better understand the structure and transport changes that occur during biofilm growth and to assess the extent and distribution of this growth will be discussed in this section. MRI is a popular imaging technique in biological, medical, and clinical applications (Kirtil and Oztop, 2016), that provides images of the internal structure without any disruption to the sample and has been the technique of choice to study flow velocity in biofilms (Van As and Lens, 2001). It has been used to spatially resolve biofouled porous systems and monitor flow changes in biofouled bead packs.

Finally, metabolomics applications have the possibility of providing mechanistic insights into the function and ecology of microbial communities and biofilms. NMR-based metabolomics approaches allow direct, time-resolved monitoring of metabolite concentrations, metabolic pathways, and flux rates for in situ studies of live cell suspensions. Chemometric data processing approaches combined after NMR acquisition routines will be explored in more detail in the next section.

2.4.1 Several analytical techniques to study and characterize soluble parts of biofilms

In most reports, to characterize EPS constituents, they must be first separated from cellular compounds, which are generally subsequently analyzed with destructive methods. There is no universal EPS isolation method. Centrifugation, filtration, heating, blending, sonication, and treatment with complexing agents and with ion exchange resins, e.g., by using a Dowex-resin (with Na^+-form, strongly acidic), have been described (Flemming and Wingender, 2010). Following extraction, a common concentration step is to precipitate solubilized EPS by adding ethanol or acetone; however, this method primarily precipitates polysaccharides and thus leads to an underestimation of important components of EPS. Common EPS isolation techniques inherently select for water-soluble EPS and lose insoluble EPS, including cellulose, which is an important constituent of the matrices of many bacteria. This is because isolation of cellulose requires harsh conditions, such as treatment with acetic acid and nitric acid at 95°C (Flemming and Wingender, 2010), and thus, it can destroy other cellular compounds. So, the extraction yield of EPS constituents is rather low: for example, Jahn and Nielsen (1995) referred extraction up to 20% of the total biofilm protein in the Dowex-extractable fraction and a reduction of 28% of the number of viable cells in the batch culture after extraction. Also, most researchers find that most (if not all) quenching agents of microbial culture create large errors because of leakage of intracellular metabolites (Wu et al., 2010).

After EPS extraction, several methods are commonly applied to analyze the presence of carbohydrates, metals, proteins, DNA, and lipids in EPS samples. Attempts to generate quantitative descriptions of the biofilm matrix are usually limited to the soluble low-molecular-weight components by solution-based methods such as high-pressure liquid chromatography (HPLC) coupled to mass spectrometry (MS), Fourier transform infrared spectroscopy (FTIR), and solution NMR (Cegelski, 2015; Jiao et al., 2010). While MS methods have been used to analyze complex mixtures, the sample preparation and purification processes influence how the molecules of interest interact with other components in solution. The technological developments in the field of NMR spectroscopy have enabled the identification and quantitative measurement of many metabolites in a nontargeted and nondestructive manner (Smolinska et al., 2012). Solution NMR has been mostly applied to characterize metabolic contents on supernatants or cell extracts and also to characterize EPS-isolated components. The most industrially relevant topic is by far the NMR of proton (^1H) nucleus, ubiquitous in organic compounds, polymers, and natural materials. In addition to ^1H, other nuclei are also used. In the context of biofilms, ^{13}C, ^{15}N, and ^{31}P are examples of interesting heteronuclei that allow the in-depth insight into structural and chemical details beyond the geometrical properties of a biofilm. ^{13}C NMR can be used for the metabolic profiling of the carbohydrate cycle, whereas ^{31}P NMR is important for tissue metabolism. For example, Xu et al. (2017) analyzed phosphorus distribution by ^{31}P NMR to investigate the effect of cerium oxide nanoparticles on the process of phosphorus removal by EPS within a biofilm. Zhang et al. (2009) investigated through ^{31}P NMR the presence of phosphorus-containing species in the EPS, concluding that phosphorus is present as orthophosphate monoesters, DNA, pyrophosphate, and polyphosphate.

The limitation of NMR is, as previously discussed, its low sensitivity. As ^{1}H isotope enjoys a 99.989% natural abundance and the highest gyromagnetic constant, it is the most receptive isotope of all the periodic table. In the case of ^{13}C NMR measurements, the situation is significantly different; ^{13}C has a low natural abundance of approximately 1.07% and 3.9 times a lower gyromagnetic constant, making this nucleus less receptive than proton. Nevertheless, this problem may be solved by using polarization transfer techniques, inverse detection of heteronuclei through the more sensitive proton, or by using isotope selective labeled precursors (^{2}H, ^{13}C, ^{15}N) in the bacterial medium (Loquet et al., 2018; Mayer et al., 2001), e.g., by adding ^{13}C-labeled glycerol, which is predominantly used for the biosynthesis of alginate, the main polysaccharide component of EPS. In many cases, the problem of sensitivity has also been circumvented by the use of higher fields or cryoprobes. Generally, 500 or 600 MHz NMR instruments are used in most of the applications with complex mixtures, as these fields are cost-effective and easily accessed, although the use of 800 and 900 MHz fields has been reported (Bernini et al., 2009). As the interface between the sample and spectrometer, the NMR probe characteristics ultimately determine the sensitivity of the analytical method. Introduction of cryoprobes to cool down the probe electronics to temperatures close to the liquid helium (ca. 20 K) and to reduce the thermal noise and of miniaturized sample detection coils for measuring limited samples may have a large impact on sensitivity. The sensitivity enhancement obtainable from cryoprobes can be as high as four to fivefold and allows to measure metabolites at lower concentrations (Larive et al., 2015; Nagana Gowda and Raftery, 2015). Microcoil probes further enhance the ability of NMR to measure mass-limited biological samples. The signal-to-noise ratio (SNR) is increased by the use of small diameter coils since the coil efficiency is inversely proportional to the diameter of the coil. The use of microcoils with solenoidal geometry improves the SNR further, as they capture more magnetic flux than Helmholtz geometry coils (the ones used in standard probes and cryoprobes). Commercially available microcoil probes can analyze samples with volumes of a few microliters, and nanoliter detection volumes have also been reported (Olson et al., 1995; Bart et al., 2009; Gomez et al., 2010; Fratila et al., 2014). These methods are beneficial particularly when sample analytes can be concentrated into small volumes.

Conventional 2D techniques for molecular identification have been widely applied to verify ambiguous or overlapped signals that can be only unraveled by expanding them along the F1 dimension. *J*-resolved experiments help in the identification of the network of resonances associated with a specific metabolite via their multiplicity, whereas 2D ^{1}H-^{1}H correlation spectroscopy (COSY) and ^{1}H-^{1}H total correlation spectroscopy (TOCSY) allow to detect spin–spin coupling connectivities that identify chemically bonded pairs of protons (Simpson, 2012), etc. Also, 2D ^{1}H-^{13}C heteronuclear single-quantum correlation (HSQC) or 2D ^{1}H-^{13}C heteronuclear multiple-quantum correlation experiments have proven to be useful to follow carbon flow through the metabolome and identify specific metabolic pathways. Long-range 2D ^{1}H-^{13}C HMBC (heteronuclear multiple bond correlations) experiments are routinely employed as well to determine linkage positions in polysaccharides, as it can map correlations from anomeric atoms over the glycosidic linkage through long-range couplings between protons and carbons (Ståhle, 2017). In addition to these, multiple pulse sequences are available in the NMR arsenal, but the aforementioned experiments should give a hint at what sort of information can be extracted by NMR spectroscopy.

Several studies, some of them already described in Table 2.1, have demonstrated the power of NMR to identify the structures of novel polysaccharides that are within the biofilm and the structures of EPS. Bendaoud et al. (2011) applied ^{13}C NMR for the identification of galactofuranose residues in *Kingella kingae* exopolysaccharides. NMR, together with GC-MS and FTIR, were also applied to identify a polygalactan from the strain *Lactobacillus plantarum* 70810 (Wang et al., 2014). Izano et al. (2008) purified a poly-*N*-acetylglucosamine polysaccharide from a biofilm-producing clinical strain of *Actinobacillus actinomycetemcomitans* with LPS on a gel filtration column and analyzed its chemical structure by NMR spectroscopy. Fontana et al. (2015) reported the structural elucidation of the EPS produced by *L. plantarum* C88 using NMR and the computer program CASPER (computer-assisted spectrum evaluation of regular polysaccharides). The latter uses ^{1}H and ^{13}C chemical shifts of mono-to trisaccharides, stored on its database, for the prediction of chemicals shifts of ascertained polysaccharides. Säwén et al. (2010) investigated all aspects of the primary structure of the EPS polysaccharide obtained from *Streptococcus thermophilus* ST1, including component analysis and absolute configuration of the constituent monosaccharides, using an array of NMR spectroscopy techniques including, TOCSY, PANSY, HSQC, H2BC, HMBC, and ^{1}H-^{1}H-nuclear Overhauser effect spectroscopy (NOESY) tilted projections (tilt angles of +15 degrees and −15 degrees) obtained from the 3D NOESY-HSQC experiments (Säwén et al., 2010). The molecular mass of the polymers can be determined using pulsed-field gradient spin echo (PGSE) NMR diffusion experiments using the relationship developed for uncharged polysaccharides (Viel et al., 2003), together with dynamic light scattering (Säwén et al., 2010).

2.4.2 Solid-state nuclear magnetic resonance to determine the insoluble constituents of biofilms

The insoluble and complex nature of the EPS of most biofilms is a remarkable challenge for the plethora of current analytical techniques. As previously referred, estimates of relative quantities of these components in the intact matrix are in most of cases unreachable. In fact, the inability to completely solubilize EPS and the possible perturbations or degradation of the material during sample preparation can severely compromise the designed assays (Reichhardt et al., 2015a). Obtaining NMR spectra of such large and insoluble systems is not possible in solution as the influence of dipolar couplings and chemical shift anisotropy are not averaged out as they are in smaller, soluble, rapidly tumbling systems [94]. Solid-state NMR (SS-NMR) has emerged as the method of choice to achieve an adequate characterization of supramolecular assemblies in general, such as for investigation of insoluble noncrystalline biopolymers at atomic resolution. It can be used to quantify composition and to measure internuclear distances that help in determining key parameters in such macromolecular assemblies (Reichhardt et al., 2015a).

In solid or semisolid samples, the spectral acquisition is performed by using magic angle spinning (MAS-NMR) spectroscopy. With this technique, line broadening in solids can be reduced by spinning the sample rapidly about an axis inclined 54.7 degrees (the magic angle) relative to the external magnetic field. This angle averages out several anisotropic interactions that are orientation dependent (Sitter et al., 2008). Therefore, implementation of MAS-NMR experiments averages over the spatial coordinates in the dipolar coupling and chemical shift anisotropy Hamiltonians and permits the acquisition of high-resolution spectra in solids (Reichhardt and Cegelski, 2014), yielding a significant line narrowing and eventually increasing the resolution and the SNR. For ^{1}H SS-NMR, very fast MAS and/or multiple-pulse sequences may be used to suppress the strong dipolar couplings. In practice, the typical MAS rates range from a few kHz for large rotors containing spin-1/2 nuclei to over 100 kHz ("very fast" or "ultrafast") for quadrupolar nuclei ($I \geq 1$) or high-order spin systems (Bryce, 2017). In most biological solids applications, MAS is coupled with cross-polarization (CP) to increase the sensitivity usually in ^{13}C and ^{15}N (natural abundance of 0.368% and a gyromagnetic constant of -2.71 10^{7} rad s^{-1} T^{-1}, which is 9.9 times smaller than ^{1}H) detection, which necessitates long experimental times to provide suitable signal (Loquet et al., 2018). The resulting CP-MAS experiment is the experimental setup for most biological solid-state NMR studies (Reichhardt and Cegelski, 2014).

SS-NMR is thus uniquely suited to the study of complex and insoluble systems including bacterial whole cells and cell walls, amyloids, membrane proteins, biofilms, and their extracellular matrix (Reichhardt et al., 2015a; Reichhardt and Cegelski, 2014). It does not require homogeneous protein preparations or high-quality crystals (X-ray diffraction crystallography), high correlation times (high tumbling rates as in solution), chemical ionization (as in MS), or elaborated sample preparation (as in LC or GC). In comparison with diffraction methods, which benefit to a significant extent from a degree of long-range ordering of molecules in solids, NMR methods tend to provide much more local information. It has the sensitivity and resolution to profile the global carbon composition of the insoluble and intact biofilms, and it represents a valuable tool as new methods are needed to analyze and quantitatively define biofilm composition and architecture (Lim et al., 2012). Recent advances on ^{1}H-detected ultrafast MAS-NMR (Bryce, 2017) and dynamic nuclear polarization (DNP) for sensitivity enhancement in modern SS-NMR (Lilly Thankamony et al., 2017) will for sure give new insights on biofilms studies.

Multiple studies have been reported on the application of MAS-NMR to study biofilms and EPS. For instance, McCrate et al. (2013) integrated SS-NMR with electron microscopy and biochemical analysis to define the chemical composition of the intact and insoluble extracellular matrix of an *Escherichia coli* biofilm. Lim et al. (2012) evaluated the influence of DMSO and ethanol with respect to increases in curli production and biofilm formation by *E. coli* by SS-NMR. As an example, treatment with DMSO increased curli production, which was accompanied by a spectroscopic increase in carbons in the δ_C 50–75 ppm region, with respect to the carbonyl peak (δ_C 170 ppm). Except for the enhancement in peak intensity at 39 ppm (the isotropic carbon chemical shift of DMSO), growth in medium supplemented with labeled [$^{13}C_2$]-DMSO did not result in the increase of other carbon peak intensity as compared with treatment with unlabeled DMSO. ^{13}C CP-MAS and ^{13}C([^{15}N], ^{15}N[^{31}P], and ^{13}C [^{31}P] rotational-echo double-resonance (REDOR) were used to spectroscopically assign and quantify the carbon pools of the EPS matrix of *Aspergillus fumigatus* (Reichhardt et al., 2015a) and *Vibrio cholerae* (Reichhardt et al., 2015b). Tang et al. (2016) demonstrated the interaction between *S. mutans* cell surface–localized adhesin P1 protein (antigen I/II, PAc) and its C123 fragment also using SS-NMR. Thongsomboon et al. (2018) reported that ^{13}C, ^{31}P CP-MAS of the intact cellulosic material allowed to detect a zwitterionic phosphoethanolamine alteration in a chemically modified cellulose from one of the most commonly studied bacterial biofilm systems, which had evaded detection by conventional methods. SS-NMR can also offer excellent perspectives in the study of antibiotic–cell wall interaction and reveal how the biological functions of cell walls and biofilms depend on their chemical composition and architecture. For example, Medeiros-Silva et al. (2018) applied SS-NMR setups to study lipid II-binding antibiotics directly in cell membranes.

2.4.3 Imaging techniques to define biofilm structures and dynamics

The biofilm structure and its composition strongly depend on the chemical and hydrodynamic environmental parameters during biofilm growth and, thus, vary in time and space. As a direct consequence from environmental parameters, biofilms form smooth, flat, rough, fluffy, or filamentous morphologies (Herrling et al., 2019). For example, extensive shear stress at high flow velocities will lead to thin and more compact biofilms, whereas lower flow velocities to fluffier and open structured biofilms. Therefore, the main research questions in the field of biofilm structure concern internal heterogeneities, composition, porosity, spatial distribution of biochemical reactivity, and structural rearrangements.

Different imaging techniques have significantly contributed to the understanding of biofilm structure and functionality and their interplay (Ranzinger et al., 2016). Fluorescent microscopy allows applications with fluorescent in situ hybridization probes (FISH), fluorescent proteins to enhance visualization of bacterial cells in a biofilm, the use of stains for live and dead cells, for all microorganisms in a biofilm, for some components of biofilm matrix (Lewandowski and Beyenal, 2010), etc. Combined fluorescence in situ hybridization and microautoradiography provide species and substrate-uptake information at the single-cell level but are destructive and permit to assay only one substrate per sample (McLean et al., 2008). Electron microscopy, e.g., transmission electron microscopy (TEM) and scanning electron microscopy (SEM), can reveal physical structures and morphologies as well as chemical composition of biofilms on different length scales (Lewandowski and Beyenal, 2010). Electron microscopy can image biofilms at high spatial resolution but has limitations due to its invasiveness and destructive sample preparation (e.g., drying and application of vacuum). Geometric parameters as biofilm thickness and coverage can be obtained; however, resolution, contrast, and invasiveness of the named imaging techniques represent the major limitations.

Since biofilms are dynamic and not static entities, imaging approaches that maintain a temporal perspective are preferred to those that deliver only single-timepoint data (Palmer et al., 2006). The aforementioned imaging techniques for biofilm viewing rarely provide the real biofilm picture due to their invasiveness and destructiveness, i.e., they physically perforate the sample, thereby changing its permeability and, therefore, potentially its metabolism (Ranzinger et al., 2016). Invasive measurements can lead to inaccurate results and prohibit further (time-dependent) measurements, which are important for the mathematical modeling of biofilms. Few techniques exist that can continuously measure biofilm metabolite profiles in a truly noninvasive and nondestructive manner with adequate time and spatial resolution. Raman microscopy, and in particular, confocal microscopy, can reach down to submicrometer lateral spatial resolution. Confocal Raman microscopy is a label-free and nondestructive imaging technique. Confocal laser scanning microscopy (CLSM) has become an indispensable tool for studying in situ biofilm structure and composition and for a deeper understanding of structure and function of biofilms (Garny et al., 2010), because of its noninvasive nature and its 3D resolution capability. Although optical penetration depth is limited in CLSM due to optical absorption and scattering phenomena (Ranzinger et al., 2016), two-photon CLSM methods overcome depth penetration concerns; however, they typically require the addition of fluorescent tracers when detecting metabolic activity that could have undesirable effects on the cellular function (McLean et al., 2008). Optical coherence tomography can also noninvasively reveal the development of biofilms and their mechanical properties, for example, deformation, with a spatial resolution down to a few micrometers.

Moving on the NMR field, its improvements in electronic and computational specifications of NMR spectrometers and the use of more sophisticated signal processing methods have led to several new applications. Up to now, we have dealt with NMR spectra that furnish information on individual molecules, either in the liquid or in the solid state. In this sense, NMR spectra may be regarded as a technique that provides pictures or images of molecules after the intellectual transformation of the recorded data (chemical shifts, multiplicity, integral, spin—spin coupling constants, relaxation times, etc.) into an image of the molecule. A different kind of NMR imaging has been developed in the past decades, where structures well above the molecular level are studied and a real 2D or 3D representation of the object is derived from the spectral data. So, NMR, previously recognized as a powerful technique to provide information regarding composition, structure, configuration, and even conformation, is now widely used for microstructural investigations (Mariette, 2009). This field also known as spin mapping has shown incredible applications in biology and medicine, and it seems appropriate to discuss its applications in the biofilm characterization field.

The development of NMR imaging (MRI), in which the contrast in the image is governed by the NMR relaxation time, has opened up a completely new field. MRI is performed with an NMR instrument equipped with magnetic gradient coils that can spatially gather the NMR data, thus creating 2D and 3D images. These images display areas having different physicochemical properties of a specific molecule (e.g., water), creating different contrasts (Kirtil and Oztop, 2016). In other words, MRI provides spatial distribution of the signal due to the application of gradients in the three axes. Thus, MRI gives an overall noninvasively spatially resolved characterization of a biofilm system in vivo and in situ (Phoenix and Holmes, 2008) in the natural (wet) state, by allowing simultaneous imaging, diffusion, and flow velocity measurements as well as reaction monitoring and chemical analysis. ^{1}H MRI has a spatial resolution of a few 10th of micrometers with the advantage that various parameters beyond the ^{1}H spin density of the NMR experiment can be exploited to generate contrast in the images. Hoskins et al. (1999) reviewed several studies concerning selective imaging of biofilms in porous media by relaxation of MRI techniques. More information on NMR imaging and its application to study dynamic water transport phenomena will be discussed in the next section together with other methods usually applied for that purpose.

Due to the extreme structural heterogenicity in a biofilm, the combination of several techniques is the best way to try to get a representative view of the biofilm. For example, various methods for chemical structural analysis, such as FTIR spectroscopy and NMR spectroscopy, can be combined with CLSM to provide a comprehensive understanding of biofilm development and the molecular dynamics of the biofilm. Other promising combinations of two different analytical methods are, for example, CLSM and Raman microscopy as well as CLSM and MRI. Garny et al. (2010) combined CLSM and NMR spectroscopy to analyze the structure, composition, and molecular mobility of heterotrophic mixed-species biofilms cultivated in rotating annular reactors exposed to different flow conditions and glucose concentrations. Also, an integrated NMR and CLSM approach was used by McLean et al. (2008) for the noninvasive imaging, transport, and metabolites measurement of live biofilms of the dissimilatory metal-reducing bacterium *Shewanella oneidensis* strain MR-1 and the oral bacterium *S. mutans* strain UA159. Serra et al. (2013) used scanning electron and fluorescence microscopy to localize in situ cellulose filaments, sheets, and nanocomposites with curli fibers at cellular resolution within physiologically two-layered macrocolony biofilms. Yu et al. (2011) developed a novel method based on multiple fluorescence labeling and 2D FTIR—^{13}C NMR heterospectral correlation spectroscopy to gain insight on the composition, architecture, and function of EPS in biofilms during composting. However, the identification and quantification of specific EPS is limited by the number and type of fluorescently labeled probes available (Yu et al., 2011). SEM and matrix-assisted laser desorption ionization imaging high-resolution mass spectrometry (MALDI-imaging-HRMS) were combined by Eckelmann et al. (2018) for visualizing, in both high spatial and temporal resolution, the distribution and interplay of the compounds during microbial interactions.

2.4.4 Explore diffusion and mass transport within a biofilm

The metabolism of microorganisms in biofilms systems is strongly dictated by transport processes (Renslow et al., 2010). Diffusion of water and nutrients into the EPS matrix strongly varies between different biofilm systems, geometries, and growth conditions, thus affecting the substrate conversion. Also depending on its physical properties, the EPS matrix can change soil pore connectivity, effective pore size, and hydraulic conductivity, thereby affecting the hydrodynamic properties of the porous media (Kirkland et al., 2015b). Due to its important role in metabolic activity, the investigation and understanding of mass transfer and substrate consumption is essential to predict the activity of biofilms and the transport of nutrient and metabolic end products, to determine antibiotic penetration in biofilms, to model subsurface biofilms, and to improve design and operation of biofilm-based technical applications such as bioremediation strategies (Herrling et al., 2017). However, mass transport and diffusion in biofilms are extremely difficult to measure and gain access with the current technology because of the high complexity and heterogeneity of these biomass aggregates and of the temporally and spatially variable metabolic gradients that occur at the micrometer scale.

Since the 1990s, a variety of porous systems such as sand, glass bead packs, and flat plate channel reactors have been used to measure how the EPS production from bacterial cells changes a system's hydraulic conductivity, mass transport, and dynamic flow patterns (Codd et al., 2011). A range of transport measurement methods have been applied to determine the transport processes within porous biosystems. Most of them follow the local intensity in time-controlled sequential images of water or labeled molecules or map the effect of contrast agents. However, they might be difficult to apply to biofilms. For example, laser Doppler anemometry is only useful for measuring flow velocity distribution above the biofilm surface (Lewandowski and Beyenal, 2010). Also, flow velocity can be successfully measured by CSLM velocimetry and fluorescence recovery after photobleaching (Lorén et al., 2015) when the flow is steady and parallel to the bottom of the reactor, but it fails when the flow direction changes vertically (even slightly). Lawrence et al. (1994) used fluorescence recovery after photobleaching with CSLM to measure the effective diffusion coefficients of fluorescein and fluoroconjugated dextrans in *Pseudomonas* and mixed-species biofilms. Also, the effective diffusion coefficient of fluorescently tagged daptomycin was measured in *Staphylococcus epidermidis* biofilm cell clusters by CSLM (Stewart et al., 2009). Commonly applied techniques to investigate mass transport of substances into biofilms are microelectrodes (limited to certain substances as O_2, N_2O, NO_3), which offer a way to map a single chosen parameter. However, they are invasive and, as previously discussed, they might influence biofilm's structure and consequently its mass transfer (Ranzinger et al., 2016). Other methods include X-ray transmission or diffraction tomography (Van As and Lens, 2001), microwave heating, and ultrasonic tomography (Cravotto and Cintas, 2007). So, regardless of whether distinct cell cluster or surface-averaging methods are used, the position-dependent measurements of effective diffusion coefficients are commonly invasive to the biofilm, performed under unnatural conditions, lethal to cells, and/or spatially restricted to only certain regions of the biofilm (Renslow et al., 2010).

MRI provides an alternative method to monitor in situ biofilm development, allowing for the nondestructive examination of the relationship between biofilm development and porous media hydrodynamics and mass transport over various time and length scales (Kirkland et al., 2015b). The major attraction of MRI is that it overcomes the limitations of the competing methods for measuring effective diffusion coefficients in biofilms: Besides the fact that is noninvasive, so that no direct contact with the fluid is necessary, it uses naturally present isotopes such as ^{1}H, ^{13}C, ^{15}N, and ^{31}P, without the need of any ionizing radiation, in contrast to X-rays scattering flow methods. The quantitative measurement of the diffusion of water and metabolites can be performed on a microscopic scale and in any direction of displacement, whereas in X-ray methodologies, optical and ultrasound scattering flow methods only measure a net flow between the emitter and the detector. There are two main ways in which NMR may be used to study self-diffusion coefficients, which are also known as tracer diffusion or intradiffusion coefficients: (1) analysis of relaxation data and (2) application of pulsed-field gradients (PFGs). These methods report on motions in very different time scales, and thus, even though a translational diffusion coefficient can be derived in both cases, the two estimates will agree only under certain circumstances since the relaxation method is in fact sensitive to rotational diffusion, whereas the PFG method measures translational diffusion (Price, 1997).

2.4.4.1 Analysis of relaxation data

The T_2 relaxation time distribution in heterogeneous porous media is used to estimate pore size distribution in the formation and for fluid typing. Briefly, the NMR active nuclei can be oriented in a magnetic field and excited by an impulse of radiofrequency radiation. The strength of the free-induction decay (FID) signal is related to the density of protons in the sample volume. The process of returning to the equilibrium is called relaxation, and it is characterized by two parameters: the longitudinal relaxation time, T_I, which reflects the time needed for the magnetization to return to the equilibrium state, and the transverse relaxation time, T_2, which reflects the time needed for the FID pulse to decay. NMR relaxation times T_I and T_2 are affected by chemical and physical changes in a sample, e.g., pore size distributions, fluid viscosity, temperature, and chemical changes in the mineralogy of any solid matrix or dissolved ions. Thus, biofilms are known to reduce NMR relaxation times values of intracellular and extracellular water protons close to the film, enhancing sensitivity for biofilm detection. For example, whereas pure water exhibits relatively long T_2 relaxation times in the range of seconds, T_2 of water inside biofilms is reduced to values about 100 ms (Ranzinger et al., 2016). The shift of transverse relaxation to lower values in biofouled geometries such as tubes or porous media can thus be used to monitor biofilm growth. The T_I and T_2 values at different biofilm locations are also influenced by several factors, including composition, water content, and concentration of paramagnetic ions. So, they can be very revealing of biofilm properties (Phoenix and Holmes, 2008). T_2 measurements are generally considered the most robust low-field measurement considering acquisition times and SNR. Also, measuring T_2 is significantly faster than measuring T_1 (Kirkland et al., 2015a). But depending on the culture, T_1 contrast has also been explored, mainly to suppress bulk water with respect to biofilm. More recently, new 2D relaxation time distribution pulse sequences have been suggested, including T_IT_2, T_2-store-T_2, and T_2-diffusion coefficient D. Several of the applications published in the past decade include 2D relaxation/diffusion correlations in porous media (Berman et al., 2013).

Low-field benchtop ^{1}H nuclear magnetic resonance (^{1}H LF-NMR) relaxometry instruments have been increasingly popular as analytical tools for engineering research since they are less expensive and require less maintenance and relatively unspecialized personnel compared with high-field instruments (Berman et al., 2013). Kirkland et al. (2015a) used a small-diameter NMR well-logging tool using two NMR probes, operating at approximately 275 and 400 kHz, to detect biofilm accumulation in the subsurface using the change in T_2 relaxation behavior over time. The mean log T_2 relaxation times were reduced by 62% and 43%, respectively while biofilm was cultivated in the soil surrounding each well (Kirkland et al., 2015a). Similarly, Vista Clara Javelin NMR logging device, a slim down-the-borehole probe, detected changes in T_2 distribution over the course of 8 days while biofilm was cultivated in the sand-packed reactor. Measured NMR mean log T_2 relaxation times decreased from approximately 710 to 389 ms, indicating that the pore environment and bulk fluid properties were changing due to biofilm growth (Kirkland et al., 2015b). Fysun et al. (2019) investigated low-field ^{1}H NMR to measure transverse relaxation times T_2 obtained by inverse Laplace transform as well as diffusion coefficients D in deposit models of *Paenibacillus polymyxa* biofilm and dairy biofouling showed that with increasing biofouling thickness (up to 406.2 mg/cm^2), transverse relaxation times T_2 shift toward slower relaxation rates up to 111.9 ms. Also, the diffusion coefficient of water in this microbial biofilm matrix corresponded to approximately 65% and 75% of the value in pure water, respectively (Fysun et al., 2019).

2.4.4.2 Application of pulsed-field gradient nuclear magnetic resonance

We explained above how translational mobility can modify the relaxation decay curve from mono- to multiexponential behavior. Nevertheless, quantification of the diffusion coefficient from relaxation time experiments is still a challenge since the physical models used require several assumptions, which often cannot be verified [83]. Consequently, other NMR and MRI techniques based on the use of magnetic field gradient pulses are preferred because they do not require knowledge of the mechanism involved in the relaxation behavior. These techniques are all based on the use of well-defined linear magnetic field gradient pulses, which change the strength of the magnetic field probed by the molecule's protons locally. Consequently, if a molecule diffuses spatially in this magnetic field gradient, the NMR signal is reduced: the faster the diffusion rate, the higher the NMR signal attenuation. Thus, flow, mass transfer, and transport processes can be measured by mapping the (proton) intensity in a defined time interval directly in a so-called PFG experiment. A detailed description of this PFG methodology (PGSE-NMR for NMR liquid spectrometers, and diffusion-weighted MRI or diffusion tensor imaging [DTI] when using an MRI scanner) is provided by Price (1997).

PGSE-NMR is a sensitive tool that uses the nuclear magnetic spin properties of nuclei (typically ^{1}H in water) as a tracer for Brownian motion. Diffusion data are recorded using specific PFG pulse sequences. Depending on the use of either a spin echo (SE), a stimulated echo (STE), or a double stimulated echo (DSTE), the experiments are abbreviated as either PFG-SE, PFG-STE, or PFG-DSTE, respectively. When, after these sequences, a period of longitudinal eddy current delay (LED) is included, the LED abbreviation is included after their corresponding names. PGSE-NMR is widely recognized as a powerful method to study emulsions and porous materials (Mariette, 2009) and has been also used to measure effective diffusion coefficients in biofilms (Phoenix and Holmes, 2008). Potter et al. (1996) were the first to use PGSE NMR spectroscopy to detect bacterial cells in suspensions and porous media. Later, PGSE NMR was used by Beuling et al. (1998) to measure the diffusive properties of water in both natural and artificial biofilms, by Manz et al. (2003) to study the effects of biofilm structure on local fluid velocity, by Seymour et al. (2007) to study velocity and transport processes in a biofouled polystyrene-bead packed column, and by Phoenix and Holmes (2008) to measure local surface-averaged diffusion coefficients in a nonsustained phototrophic biofilm. Wieland et al. (2001) used combined PGSE and CPMG (Carr–Purcell–Meiboom–Gill) sequences together with a pulsed-field gradient turbo spin echo (PFG-TSE) sequence to map diffusion coefficients and water densities for natural microbial mats and to generate their diffusion coefficient depth profiles. Also, Renslow et al. (2010) obtained 2D effective diffusion coefficient maps in *S. oneidensis* MR-1 biofilms using PGSE-NMR, and from these maps, 1D surface-averaged effective diffusion coefficient profiles were generated to predict the mobility of heavy metals in subsurface biofilms.

Strong correlations between the diffusion of substrates and biofilm parameters have been reported. For example, Vogt et al. (2000) used PGSE NMR to study differences in metabolite diffusion within a biofilm of *P. aeruginosa* and found five groups of components including water, glycerol, and polysaccharides, with different ranges of diffusion coefficients that indicate locations in the biofilm pores or the EPS and an extreme heterogeneity of a biofilm. Herrling et al. (2017) compared water diffusion coefficients in multispecies biofilms with diverse geometries (sludge flocs, fluffy and compact biofilms grown on carriers, and aerobic and (an)aerobic granules) to identify correlations between biomass properties and water diffusion by different PGSE NMR data processing schemes, including (bi)exponential fit, **Γ** distribution, and 2D inverse Laplace transform. They reported that NMR diffusion was linked to biofilm structure (e.g., biomass density, organic and inorganic matter) and that diffusion was most restricted in granules with compact structures and was faster in heterotrophic biofilms with fluffy structures (Herrling et al., 2017). Kiamco et al. (2018) determined the effects of maltodextrin and vancomycin treatment on the metabolism and structure of *S. aureus* biofilms (biofilm porosity, thickness, biovolume, and relative diffusion coefficient depth profiles) through PGSE NMR measurements and high-resolution NMR spectroscopy, respectively.

2.4.5 Diffusion-ordered nuclear magnetic resonance spectroscopy applications to determine molecular size

The analysis of mixtures constitutes a challenge, because signals in regular NMR spectra cannot be assigned to the individual compounds, unless they are scalar coupled. A solution to this problem is the use of diffusion-ordered (DOSY) NMR spectroscopy, which is a 2D NMR technique to observe molecule diffusion by application of PGSE NMR (Li et al., 2009). This technique allows the separation of the NMR spectra of different compounds in a sample by exploiting their self-diffusion properties. Thus, it is called "chromatography by NMR" because NMR spectra of individual components of a complex mixture are easily resolved based upon their diffusion properties (Li et al., 2009). Also, as in all diffusion NMR experiments, it provides information on their intermolecular interactions as well as on their size and shape (Pagès et al., 2017). Recent improvements in DOSY NMR made it an increasingly valuable tool in complex mixture analysis (Gresley et al., 2012), enabling diffusion coefficients to be routinely measured and used to characterize chemical systems in solution (Le Gresley et al., 2015). DOSY has been applied for the analysis and characterization of mixtures and aggregates, for the study of intermolecular interactions, for the determination of molecular weight distributions for polymer mixtures and polymer blends, and for uncharged water-soluble oligo- and polysaccharides (Viel et al., 2003).

Due to their high gyromagnetic ratio and widespread occurrence, protons (^{1}H) are also the most commonly used nuclei for PGSE and DOSY NMR measurements; however, spectral overlap is a particularly severe problem for proton NMR that complicates the resolution of the mixture in terms of separated diffusion coefficients of each component of the mixture. Different active NMR nuclei can be used for diffusion NMR. Most of them present spin-quantum number equal to 1/2 (e.g., ^{13}C, ^{19}F, ^{29}Si, ^{31}P), but there are some other that are quadrupolar and therefore present spin-quantum number equal or higher than 1 (e.g., ^{2}H, $^{6/7}Li$, ^{11}B, ^{14}N, ^{35}Cl). The applicability of most of them is limited due to their low receptivity, low natural abundance, reduced relaxation times, high quadrupolar moments, or a combination of all these. Other possibility to obtain less overlapped DOSY spectra is by spreading the signals into a third dimension, by combining NOESY, TOCSY, HMQC, HSQC, or COSY experiments with diffusion-ordered spectra for obtaining better signal dispersion (Glanzer and Zangger, 2014). However, these 3D DOSY experiments require much longer measurement times and elaborate data processing. There are recent techniques to optimize DOSY experiments, which are described, for example, by Glanzer and Zangger (2014), such as the 2DJ/

IDOSY (a J-resolved DOSY experiment) that allows to reduce the experimental time by a factor of 4.

Another DOSY approach extremely useful in complex overlapped spectra is the one including a pure shift module (Aguilar et al., 2010; Foroozandeh et al., 2016) in which broadband homonuclear decoupling during acquisition is able to reduce or even remove spectral overlap. Further improvements, when looking for faster experiments, have been achieved in the one-shot sequence (Pelta et al., 2002) that allows the acquisition of high-resolution spectra with only one scan and provides good lineshapes, although quality can be improved with more scans. It has been shown that by including a gradient prepulse prior to the sequence, the repetition time can be shortened substantially, that is, the sequence run in a steady-state mode, without compromising the accuracy of the determined diffusion coefficient (Stait-Gardner et al., 2008; Zubkov et al., 2015). Numerous fast NMR approaches have been developed (Peled et al., 1999; Stamps et al., 2001; Thrippleton et al., 2003; Kittler et al., 2014), but these generally contain severe compromises, e.g., loss of spectral resolution, J-modulation effects, loss of a defined diffusion measuring time, etc. The best choice to avoid J-modulation effects in diffusion experiments would be to convert all magnetization into in-phase magnetization, rather than purging the antiphase magnetization. This can be done to a certain extent with a double spin echo with a 90 degrees pulse between the echoes in an experiment called J-compensated PGSE (Torres et al., 2010). The same approach could be used in bipolar pulse stimulated echo sequences, employing this double spin echo element instead of the spin echoes and adding the 45 degrees pulse at the end of the stimulated echo.

In biofilms, DOSY showed, for example, the presence of lower-molecular-weight saccharides (δ_H 2.5 and 2.75−3.5 ppm), as well as proteins in freeze-dried EPS from 96 h biofilms of nontypeable *Haemophilus influenzae* (Wu et al., 2014). Renslow et al. (2010) correlated the effective diffusion coefficient with biofilm depth by means of PFG-NMR and imaging in *S. oneidensis* and *Phormidium* biofilms (Ramanan et al., 2013). The results indicated heterogeneities in the biofilm, which represent local diffusion barriers. Other studies also confirm that D is biofilm specific and depth dependent with a linear decrease of D with biofilm depth (McLean et al., 2008). Compared with monocultures, multispecies biofilms form structures that are more complex. Diffusion coefficients are expected to show distributions rather than a single value. For example, internal heterogeneities significantly influenced the mass transport in methanogenic granular sludge. The diffusion coefficient of the biofilm was found to be about 65% lower than the self-diffusion coefficient of free water (Lens et al., 2003). In a recent study investigating different biomass geometries such as biofilms, granules, and sludge, the D value of water in these supramolecular architectures ranged from 36% to 96% of D_{water} (Herrling et al., 2017). In this sense, biomass organized in the form of granules shows the narrowest distribution centered in 4.71 10^{-10} m^2/s, together with the slowest

D_{mean} mainly due to the biofilm's compactness. Larger diffusion coefficients were found for sludge with a broader distribution mainly located at 8.55 10^{-10} m^2/s, even though the structure of the biomass appears to be completely different, whereas biofilm showed narrow distributions, but larger D_{mean}. The distributions give unique indications for the overall diffusion properties of diverse biofilm systems, which discrete, single parameters determined by, for example, a biexponential function could not express. This variability and distribution in diffusion highlights the necessity to combine biofilm structural details with diffusion for a more comprehensive view on biofilms.

2.5 Nuclear magnetic resonance–based metabolomics approach to study biofilms

Within the systems biology framework, functional analysis at all "omic" levels has seen an intense level of activity starting from the first decade of the 21st century (Ellis et al., 2007). Community proteomics and transcriptomics can provide insight into the potential function of coexisting microorganisms in situ. However, these analyses are blind to the flux of small-molecule metabolites that are foundational to the physiological or phenotypic state of an organism. Over the past decade, metabolomics offers some unique advantages over the other "omics" sciences. Metabolomics is the discipline where endogenous and exogenous metabolites are assessed, identified, and quantified within a biologic system, thus providing a chemical "snapshot" of an organism's metabolic state (Zhang et al., 2012). Metabolomic measurements can bring to light the key intra- and extracellular metabolites involved in cellular processes such as ion homeostasis, redox status, nutrient cycling, energetics, and cell–cell signaling. By capturing relative sizes of the metabolite pools, metabolomics is thus a reflection of the genetic regulation, which causes changes in protein expression (Mosier et al., 2013). Its approach is therefore analogous to the proteome and genome.

Nowadays, mass spectrometry, NMR, infrared (IR), and ultraviolet (UV) spectroscopy, with or without combination with chromatography (whether LC or GC), are well-established analytical methods for generating metabolomics profiles (Patel et al., 2010; Emwas et al., 2019). There are many reports that compare the advantages and limitations of the analytical platforms (Choi and Verpoorte, 2014). For instance, LC–MS and GC–MS are more time-consuming concerning the sample preparation. GC–MS requires sample derivatization (O'Gorman et al., 2013). On the other hand, GC–MS and LC–MS yield a higher sensitivity (10–100 times higher) than NMR and therefore may detect metabolites that are present in a concentration below the detection limit of ^1H NMR (Scalbert et al., 2009). This means that a typical NMR-based metabolomic study usually returns information on 50–200 identified metabolites with concentrations >1 μM, whereas a typical LC-MS study can return information on more than 1000 identified metabolites with concentration levels between 10 and 100 nM (Emwas et al., 2019). The advantages of NMR were already intensively described along this chapter (nondestructive, nonbiased, easily quantifiable, permits the identification of novel compounds, no chemical derivatization needed, etc.) (Wishart, 2008), but in terms of accessing to secondary metabolites of very low concentration, it cannot compete with MS spectrometry.

New methodologies on metabolomics coupled to multivariate data analysis (MVDA) techniques have been promising and open exciting perspectives in a number of fields including medicine, plant sciences, toxicology, and food sciences (Weljie et al., 2006; Abreu et al., 2018, 2019; Aguilera-Sáez et al., 2019), all with the ultimate goal of understanding cause and effect processes within biological systems (Gjersing et al., 2007). Also, it is a valuable tool for the discovery-oriented natural products chemistry (Kim et al., 2010). Through the statistical analysis of NMR spectra of complex mixtures of metabolites, unique spectral features can be identified from a determined biological system and correlated to a phenotype or biological property of interest, as illustrated in Fig. 2.2 (Larive et al., 2015).

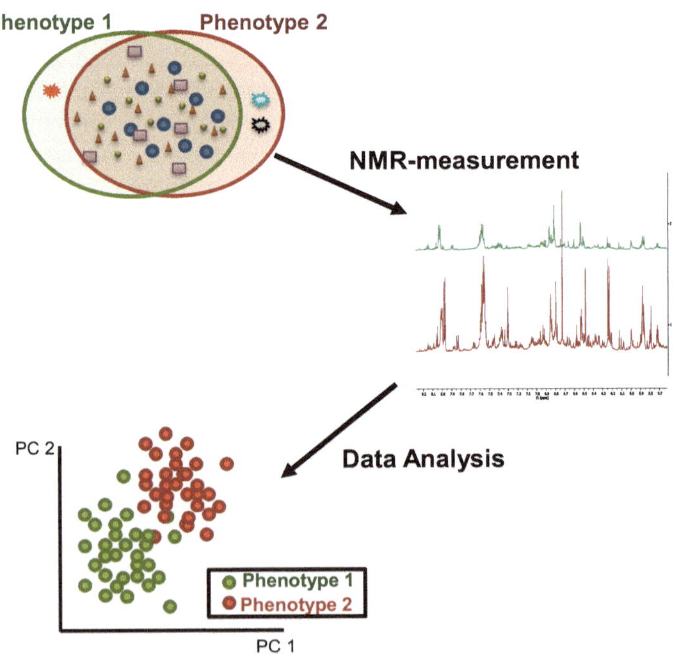

FIGURE 2.2 Scheme of NMR-based metabolomics used to identify metabolites in complex mixtures and correlate them to a phenotype or biological property of interest. Multivariate data analysis methods aim to differentiate between classes in highly complex data sets. *NMR*, nuclear magnetic resonance; *PC*, principal component.

NMR-based metabolomics is also finding use in biofilm research. Profiling the metabolome of bacteria, both extracellularly and intracellularly, using NMR spectroscopy or mass spectrometry, can provide mechanistic insights into the function and ecology of microbial communities, increase our understanding of the underlying biological processes related to the structure and formation of biofilms, their response to antimicrobial chemotherapy, virulence mechanisms (Stipetic et al., 2016), etc. Metabolite quantification and the pathway modeling of complex biological systems are also useful for exploring cell behavior in establishing a biofilm community. A systematic view of metabolic pathways or processes responsible for regulating this "social structure" of microorganisms may provide critical insights into biofilm-related drug resistance and lead to novel treatments.

Developing an organized approach is the most critical part of a metabolomic experiment, so some considerations might be considered before starting it. To ensure intra- and interlaboratory comparability, several efforts have been made to develop standardized procedures for (1) NMR sample preparation and (2) spectral acquisition. Some important considerations on an NMR metabolomic study will be discussed along this section. Also, recent applications on NMR coupled to MVDA techniques to study biofilms will be reviewed.

2.5.1 Designing a metabolomics experiment

2.5.1.1 Considerations for sample collection and preparation

First, the number of samples and/or size of the groups needed for a metabolomics experiment depends on the biological variability associated with the system being studied compared with the analytical variability of the analytical platform. It is important to avoid unintended bias. For instance, not controlling the effect of diet or the time of day of sample collection can lead to excessive variation and/or differences between groups that masquerade biologically relevant changes in metabolite levels. The sample preparation protocol should be minimal and relatively simple. The details of the procedure influence the accuracy, reliability, and reproducibility of the metabolomics data. Simple sample preparation has various advantages in terms of speed, capability, consistency, reproducibility, robustness, and efficiency. Sample integrity may alter the capacity of experimental design: If the samples have already been collected, it is important to know (1) how have they been collected and stored, (2) how can control and treatment samples be matched, and (3) if there is a clear phenotype between the control and experimental groups (Barnes et al., 2016).

A metabolite sample preparation usually includes cell quenching, cell harvesting, cell disruption, and metabolite extraction. A very critical issue in sample preparation is the need to rapidly and efficiently quench all enzymatic and biological activities to capture an accurate "snapshot" of the metabolome. This is because metabolites, such as pyruvate, fumarate, oxoglutarate, phosphoenolpyruvate, fructose-6-phosphate, and others, have a rapid turnover rate

(Zhang and Powers, 2012). In addition, it is important to avoid inducing a stress response or cell death that would completely invalidate the study. Thus, a quick quenching step that involves reducing the cell temperature has been shown to be a useful approach to slow down enzyme activity within a cell (Bolten et al., 2007). Cells can either be instantly frozen with liquid nitrogen to be extracted later or immediately extracted with the polar solvent of choice (e.g., methanol) precooled to dry ice temperature (−43°C) (Barnes et al., 2016).

A proper metabolite extraction technique is also critical. Intracellular metabolites are contained within a mechanical barrier, the cell membrane, or cell envelope. Therefore, to identify and quantify intracellular metabolites, it is necessary to extract metabolites from the intracellular compartment. Cell lysis and metabolite extraction can sometimes be carried out simultaneously. This is usually achieved using extracting solvents (organic, inorganic nonaqueous or a mixture of the two) that make the cell's walls porous, or "permeable," allowing the penetration of these solvents into the intracellular medium and greater recovery of intracellular metabolites (Pinu et al., 2017). Mechanical disruptions, such as the Fast-Prep^{R} system, are also widely used (Perry et al., 2008; Batzilla et al., 2006).

The choice of the extraction solvent is of utmost importance (Sapcariu et al., 2014). To obtain as much information as possible about metabolite levels associated with a biological sample, the optimal extraction system should extract the largest number of metabolites (Mushtaq et al., 2014), should be nonselective and not exclude molecules with particular physical or chemical properties, and should be nondestructive, as well as not modify metabolites through chemical or physical means. Other issues, such as solvent volume, sample:solvent ratio, and the conditioning of the sample for its introduction into the analytical instrument, can change the outcome of a metabolomic study completely because they affect the solubility of the metabolites. Protocols should thus be meticulously followed for reproducible results (Choi and Verpoorte, 2014). Mixtures of methanol and water are the most popular combinations for metabolomic studies because they have proven to be able to extract a wide range of metabolites (Mushtaq et al., 2014).

Variation in pH between aqueous samples can cause a significant difference in the chemical shifts of signals belonging to organic acids, amino acids, and other metabolites with acidic or basic functional groups. Reproducible chemical shifts can be obtained by using a buffered NMR solvent. A classic case of this occurs for the diastereotopic methylene hydrogens in citric acid. Changes in pH between samples will alter the ionization of the carboxylate groups in citric acid and thus affect the chemical shifts of the methylene hydrogens. In addition, it is well known that citric acid can chelate metal ions such as calcium, magnesium, and sodium. Thus, even if biofluid samples are buffered effectively to a constant pH, changes in metal ion concentrations between samples, which are not readily apparent by ^{1}H NMR, may have a significant effect on the chemical shifts and the half bandwidths of the signals of the methylene hydrogens of citric acid and also any other metabolites with similar properties (Dona et al., 2016).

Chemical shift variations introduce additional undesirable intersample variability that can distort, for instance, the results of multivariate analysis. Two different approaches to solve this problem can be applied: control of pH (buffered solution or pH adjustment) and specific data processing. Phosphate buffer is frequently used to stabilize pH in the range between 6 and 7, whereas oxalate buffer is used for more acidic solutions (pH 4). The concentration of the buffer must be sufficient to ensure a minimum pH variation. Additionally, the pH of a set of samples can be adjusted to a given value by adding small volumes of hydrochloric acid or sodium hydroxide to the solutions. The other approach consists in masking the pH-induced variation of chemical shifts by using exponential multiplication of free-induction decays with high values (10–50 Hz) of line broadening factor. A decrease in spectral resolution occurs, and small chemical shift differences between different compounds can be neglected. Alternatively, the effect of chemical shifts variation can be reduced by applying a bucketing procedure (Mannina et al., 2012).

2.5.1.2 Considerations for nuclear magnetic resonance acquisition

Care needs to be taken with respect to NMR analysis, especially when relying on databases for metabolite identification. These considerations rely especially on instrumental optimization, NMR pulse sequence selection, and choice of acquisition parameters. The selection of a "recommended" magnetic field facilitates comparison of spectra acquired in different laboratories and with available spectral databases (see Section 4.1). The use of 600 MHz spectrometers represents the best compromise between a good spectral sensitivity and resolution and affordable instrumental cost, and it is therefore considered the standard field for biofluid and tissue analyses (Vignoli et al., 2019).

The whole variety of NMR-influencing parameters needs be considered and can be explored to obtain a comprehensive description of the phenomena under investigation. Among them are temperature, pH value, concentration, and ionic conditions on the one hand (see previous section) and, on the other hand, chemical shift, multiplicity, magnitude and sign of the homo- and heteronuclear couplings, and the Overhauser effect [168]. There are a set of experiments routinely used for NMR-based metabolomics approaches, including PRESAT, 1D NOESY, PURGE, CPMG, T1 and T2 measurements, COSY, TOCSY, 2D J-resolved spectroscopy, gHSQC, gHMBC, etc. NMR spectra obtained using techniques such as HR-MAS can also be used in metabolomic studies (Gjersing et al., 2007). In this section, we will focus primarily on the description of proton NMR spectroscopy (^{1}H NMR), which is employed in most NMR-based metabolomics studies.

Because almost all 1D ^{1}H NMR spectra acquired for metabolomic studies are performed in water, solvent suppression is an important aspect of spectral acquisition that cannot be avoided. In effect, any resonance in the 1D ^{1}H NMR spectrum that does not originate from the bacterial metabolome will generate a "false feature" that needs to be removed. Depending on the nature of metabolites studied, different solvent suppression schemes or protocols may be used. In addition to solvent replacement methods, which often require lyophilization, the water suppression issue can be essentially avoided by the use of >99.9% deuterated solvents and by a variety of NMR pulse sequence techniques available for solvent suppression, such as the aforementioned NOESY experiment. This method provides a reproducible and easy-to-implement experiment for recording 1D ^{1}H spectra of biological samples with good water suppression. As a result, this pulse sequence has become the predominant approach used by NMR researchers in metabolomics (Emwas et al., 2019).

Broad resonances from proteins or other biomolecules may sometimes overlap or induce errors in the integration of relevant NMR resonances and interfering with the analysis. For instance, replicate samples may not cluster together because of a significant variation in the peak height and peak shape of the water resonance despite the overall similarity in all the metabolite NMR peaks. This will lead to an erroneous interpretation of the 1D ^{1}H NMR spectra and incorrect sample classification. These interferences can be readily removed by using a CPMG spin echo sequence. The CPMG pulse sequence takes advantage of the large difference in T_2 relaxation times between small-molecular-weight metabolites and large-molecular-weight biomolecules. The NMR resonances from the biomolecules rapidly decay during the CPMG pulse. Of course, such experiments require an initial optimization process that involves the correct tuning of the number of times you apply the transverse spin echo τ-180°-τ, and also the duration of this τ.

Other practical consideration includes the addition of an internal standard such as 3-(trimethylsilyl)-2,2′,3,3′-tetradeuteropropionic acid (usually abbreviated to TSP) or deuterated forms of 4,4-dimethyl-4-silapentane-1-sulfonic acid (DSS) or its sodium salt, for aqueous samples. For lipophilic samples, tetramethylsilane (TMS) is a good option (Dona et al., 2016). The chemical shift of the methyl resonances is defined to 0 ppm, and its line width should be less than 2 Hz (usually close to 1 Hz).

2.5.1.3 Considerations for nuclear magnetic resonance spectral analysis

A typical 1D ^{1}H NMR spectrum of a bacterial cell lysate may contain thousands of sharp lines from low-molecular-weight metabolites. The entire 1D ^{1}H NMR spectrum is used as a "fingerprint" to characterize the state of the bacterial cell. Then, a global metabolomic analysis is based on how similar or how different the 1D NMR spectra are between each class or group. Assigning a 1D ^{1}H NMR spectrum to identify the metabolites present in a sample is challenging. This is due to the diversity of metabolites and their range of concentrations, resulting in a large number and significant overlap of peaks, to the high chemical shift degeneracy (multiple metabolites have some chemical shifts in common), and to an incomplete database of NMR reference spectra for metabolites (Mosier et al., 2013). Chemical shifts will naturally be different between those of an authentic sample in water, D_2O, or phosphate buffer and those of the same metabolite in a biofluid such as urine or plasma, but generally ^{1}H NMR chemical shifts should agree to ~0.03 ppm and ^{13}C NMR chemical shifts to ~0.5 ppm (Dona et al., 2016).

The advantage of a metabolomics study is that assigning a complete 1D ^{1}H NMR spectrum is not necessary for a global analysis of the metabolome, but to identify the specific metabolites that are changing and are the main contributors to class distinction. The identification of such metabolites is extremely valuable for understanding the underlying biological differences. Statistical total correlation spectroscopy (STOCSY) can be used to associate multiple NMR peaks from the same molecule in a complex mixture. This significantly simplifies the assignment problem since most, if not all, of the NMR resonances for a given metabolite can be used together in a database search. A positive identification only occurs when all the observed chemical shifts match the metabolite's known chemical shifts in a database (Zhang and Powers, 2012). When the identity of the metabolites in a sample is known (or suspected), resonance assignments can be facilitated using libraries or databases. Public and commercial databases, such as COLMAR, HMDB, LipidMaps, and Metlin (Bartel et al., 2013; Bingol et al., 2014; Wishart et al., 2018), now contain experimental 1D ^{1}H, ^{13}C, and 2D ^{1}H-^{13}C spectra and extracted spectral parameters for over a 1000 compounds and theoretical data for thousands more (Ellinger et al., 2013). Also, as previously mentioned, the aforementioned multidimensional NMR experiments spectra can aid in the process of assigning resonances, despite this strategy can be time-consuming (Kim et al., 2010).

2.5.2 Multivariate data analysis in metabolomics

Scientific phenomena cannot be usually interpreted by a single variable but by multiple ones. So, a characteristic of metabolomics is the large amount of data generated (Brennan, 2013). Therefore, an important part of any metabolomics study is the analysis of the obtained data using data reduction, multicomponent statistics, and prediction methods (Berrueta et al., 2007). MVDA methods seek to capture changes of single metabolites between different groups whether by unsupervised or by supervised methods.

The unsupervised methods seek discriminating factors between the independent variables with the aim to obtain a graphical representation as the result of maximization of variances. For example, principal component analysis (PCA) is an unsupervised linear mixture mode and the most widely used multivariate analysis method for metabolic fingerprinting and in chemometrics in general. PCA is often used as pattern recognition technique and as a starting point for data analysis (Emwas et al., 2019) and attempts to gather useful information from the NMR spectra and to identify inherent grouping of samples as a result of the similarity of the metabolic composition by a smaller number of mutually decorrelated principal components (PCs) (Bartel et al., 2013; Brennan, 2013). Thus, PC regression analyzes X to obtain components, which can explain X in the best way. For studies on microorganisms, the application of PCA to liquid ^{1}H NMR spectra has been used to distinguish between different strains of *Bacillus cereus* (Gjersing et al., 2007).

Supervised methods find the best fitting relationship between independent and dependent variables. The most relevant examples of supervised techniques are partial least squares (PLS) and orthogonal projection to latent structures (OPLS) models (Brennan, 2013; Worley and Powers, 2013). PLS is a method for relating two data matrices of X and Y by a multivariate linear model. PLS regression finds components of X, which can predict Y in the best way. OPLS method is the improved form of PLS and removes X changes that have no correlation with Y. Among the results generated by multivariate approaches, four quantities are often analyzed first, which are scores and loadings plots, R^2 and Q^2. Scores are the coordinates of the new dimension-reduced coordinate space obtained by the PCA, PLS, or OPLS analyses, while loadings are the contributions of the original variables (NMR spectral bins or buckets or small frequency regions) to the new coordinates. For OPLS methods, the S-plot is proposed as a tool for visualization and interpretation of OPLS helping to identify statistically significant metabolites, based on both contributions to the model and their reliability (Sugimoto et al., 2012). R^2 indicates how well the model explains the dataset and Q^2 describes how good the model is able to predict.

MVDA analysis is usually performed by means of several software packages, including Mnova (Mestrelab Research, Santiago de Compostela, Spain), SIMCA-P+ (Umetrics), PLS Toolbox (Eigenvector Research, Wenatchee, WA, USA), and MetaboAnalyst (Xia Lab, McGill Univeristy), among many others.

2.5.3 Recent advances on nuclear magnetic resonance–based metabolomics applied to biofilms

After decades of extensive research into the morphology, physiology, and genomics of biofilm formation, studies on the metabolomics of biofilms are scarce. Attention has recently been directed toward the analysis of the cellular metabolome for a wide range of studies including genome annotation and pathway mapping, host–microbe interactions, infectious disease research, drug metabolism, heavy metal resistance, and transformation of a planktonic cell to a biofilm (Shommu et al., 2015; Booth et al., 2011). Table 2.2 reviews most studies applying comparative untargeted metabolomics using several analytical techniques to explain several biofilm-related aspects. These are mostly related to

TABLE 2.2 NMR-based metabolomics applied to biofilm studies.

Technique	Bacteria	Main results
Metabolic differences during biofilm formation		
^{1}H HR-MAS	*Pseudomonas aeruginosa*	PCA analysis of the ^{1}H HR-MAS NMR spectra shows differences between the planktonic and biofilm cells; no identification of the metabolites was made (Gjersing et al., 2007).
^{1}H NMR	*Acinetobacter baumannii* 1656-2	Clear separation between planktonic and biofilm cells. Acetate, pyruvate, succinate, UDP glucose, AMP, glutamate, and lysine and, particularly, the ratio of *N*-acetyl-D-glucosamine to D-glucosamine were increasingly involved in the energy metabolism of biofilm formation (Yeom et al., 2013).
UPLC/MS, 2D-PAGE, shotgun proteomics	*Vibrio fischeri* ETJB1H	Upregulated differences (twofold) were detected in biofilms for organic acids (e.g., carboxylic, phosphoric, aspartic, docosanoic, malonic, hydrobenzoic, and ketogluconic acids), sugars (e.g., fructose, mannose, and maltose), glycerol-derived components, and alcohols (mannitol and tetradecanol). Upregulated differences in the planktonic state (two- to threefold) include threonic acid, hydroxypyrimidine, tyramine, and cellobiose (Chavez-Dozal et al., 2015).
GC-MS	*Candida albicans*	Thirty-one differentially produced metabolites between the biofilm and planktonic states were identified and were involved in the tricarboxylic acid (TCA) cycle, lipid synthesis, amino acid metabolism, glycolysis, and oxidative stress. The lack of trehalose resulted in abnormal biofilm formation and increased sensitivity to amphotericin B and miconazole (Zhu et al., 2013).
GC-MS and NMR gene expression	Wild-type *Salmonella* and a CsgD deletion mutant	Metabolites associated with glucogenesis and major osmoprotectants were upregulated in wild-type *Salmonella,* whereas metabolites associated with the TCA cycle were upregulated in the mutant that does not produce EPS matrix. Common physiological properties of biofilms were induced independently of regulatory pathways that initiate biofilm formation (White et al., 2010).
^{1}H NMR	Methicillin-resistant (MRSA) and methicillin-susceptible (MSSA) *Staphylococcus aureus*	For both strains, MVDA analysis suggested key features distinguishing biofilm from planktonic growth, including selective amino acid uptake, lipid catabolism, butanediol fermentation, and a shift in metabolism from energy production to assembly of cell wall components and matrix deposition (Ammons et al., 2014).

Continued

TABLE 2.2 NMR-based metabolomics applied to biofilm studies.—cont'd

Technique	Bacteria	Main results
GC-MS and LC-MS	*Desulfovibrio vulgaris*	The overall metabolic level of the biofilm cells was downregulated for metabolites related to the central carbon metabolism, compared with planktonic cells, whereas fatty acid biosynthesis was upregulated, suggesting that these may be important for the formation, maintenance, and function of *D. vulgaris* biofilm (Zhang et al., 2016).
NMR	*Staphylococcus aureus*	Planktonic cells contained higher percentages of leucine, isoleucine, glutamate, glutamine, and proline, whereas biofilm cells contained higher percentages of lactate, citrulline, carnitine, choline, arginine, acetate, ornithine, and lysine (Wu et al., 2010).
LC-MS	*Helicobacter pylori* clinical strains	Low- and high-biofilm formers are presented as two distinct groups. Low-biofilm formers produced more metabolites than high-biofilm formers, especially lipids and metabolites involved in prostaglandin and folate metabolism (Wong et al., 2018).
Metabolic differences induced by environmental stress conditions		
1D 1H NMR, 2D ^{1}H-^{13}C HSQC and ^{1}H-^{1}H TOCSY	Wild-type *Staphylococcus epidermidis* 1457	The presence of 4% ethanol, 2% glucose, Fe-limitation, and sublethal dose of tetracycline perturbed the metabolome of *S. epidermidis* by inactivating TCA cycle, thus enabling metabolic precursors to flow into pathways linked to biofilm formation. A network of 37 metabolites affected by the stress factors was identified, which are the same set of metabolites affected by TCA cycle inactivation (Zhang et al., 2011).
^{1}H NMR and GC-MS	*Pseudomonas fluorescens*	The addition of copper led to changes on planktonic metabolism, showing an oxidative stress response characterized by changes in TCA cycle, glycolysis, pyruvate, and nicotinate/nicotinamide metabolism, which were not observed when copper was added to biofilms. Conversely, biofilms exhibited changes in exopolysaccharide-related metabolism suggestive of a protective response (Booth et al., 2011).
HPLC-MS	MSSA and MRSA	The sublethal dose of different antibiotics classes (β-lactams, aminoglycosides, and quinolones) on MRSA and MSSA strains induced similar and divergent metabolic perturbations after 6 h of coincubation, especially in important metabolic pathways such as pyrimidine, amino acid, and purine metabolisms (Schelli et al., 2017).

TABLE 2.2 NMR-based metabolomics applied to biofilm studies.—cont'd

Technique	Bacteria	Main results
Metabolic differences between susceptible and resistant strains		
^{1}H NMR	MRSA and MSSA	Differences were detected between the metabolic profiles of MRSA and MSSA strains on planktonic and biofilm states. MVDA suggested the two strains used distinguishably different metabolic strategies in planktonic state; however, when as biofilms, the metabolite profiles clustered identically (Ammons et al., 2014).
LC-MS	Polymyxin-susceptible and polymyxin-resistant *Acinetobacter baumannii*	Polymyxin-resistant strain showed perturbations in pentose phosphate pathway and TCA cycle intermediates (amino acids and carbohydrates) and in nucleotides and a shift in its glycerophospholipid profile toward increased abundance of short-chain lipids compared with the polymyxin-susceptible strain. Peptidoglycan biosynthesis metabolites were depleted in polymyxin-resistant strains (Maifiah et al., 2016).

GC-MS, gas chromatography–mass spectrometry; *HR-MAS*, high-resolution magic angle spinning; *HSQC*, heteronuclear single-quantum correlation; *LC-MS*, liquid chromatography–mass spectrometry; *MVDA*, multivariate data analysis; *NMR*, nuclear magnetic resonance; *PCA*, principal component analysis; *2D-PAGE*, two-dimensional polyacrylamide gel electrophoresis; *TOCSY*, total correlation spectroscopy; *UPLC/MS*, ultra-performance liquid chromatography–mass spectrometry.

the assessment of metabolic differences during biofilm formation, caused by stress-induced factors or between susceptible and resistance profiles.

Metabolomics is also gaining notable popularity in the studies of infectious pathogens as well as the resulting disease conditions (Shommu et al., 2015). The metabolic profiling of biofluids (blood, plasma, urine, etc.) of patients or model organisms has been performed to detect significant metabolites that could be used as an indicator of the infection. For instance, by applying a ^{1}H NMR metabolomics approach, researchers have been able to identify biomarkers that could be useful for early detection of sepsis, a life-threatening infectious disease (Mickiewicz et al., 2013, 2014). Moreover, metabolic profiling has successfully distinguished mice with gram-positive bacterial infection from those with gram-negative infection (Hoerr et al., 2012). Also, rapid metabolic changes can reflect drug mechanisms of action and reveal the active role of metabolism in mediating the first stress response to antimicrobials (Zampieri et al., 2017).

2.6 Conclusion

Bacteria within biofilms can rapidly acquire extensive phenotypic and genotypic diversity, which can enhance the ability of biofilm cells to persist and spread under diverse environmental stresses. This variation has implications in the adaptive evolution of bacterial communities, the metabolic capability of bacteria, and the ability of biofilm cells to establish chronic and antibiotic-resistant infections. To be able to fight them, it seems more urgent to understand first how they operate: the colonization process, communication process, pathogenicity and virulence process, biofilm formation and its recalcitrance, etc. Some key metabolites, nutrients, and autoinducers have been shown to significantly influence biofilm formation (Li and Tian, 2012). Thus, metabolite quantification and the pathway modeling of complex biological systems is useful for exploring cell behavior in establishing a biofilm community. Also, the identification of specific metabolites that are related to the formation and resistance in biofilms could allow us to anticipate biodegradation processes and to identify new drug targets and chemical leads fundamental for the drug discovery process. NMR methods and hardware and software advances open new perspectives for biofilm and EPS investigation, which have been shortly summarized in this chapter. Investing on new technical platforms or in the development of new pulse sequences or new combinations of the existing ones would allow the identification of key metabolites on the processes of biofilm development, communication, and resistance and set the basis for new therapeutic and diagnostic applications on our fight against MDR infections.

References

Abdel-Mawgoud, A.M., Lépine, F., Déziel, E., 2010. Rhamnolipids: diversity of structures, microbial origins and roles. Applied Microbiology and Biotechnology 86 (5), 1323–1336.

Abreu, A.C., Aguilera-Sáez, L.M., Peña, A., García-Valverde, M., Marín, P., Valera, D.L., et al., 2018. NMR-based metabolomics approach to study the influence of different conditions of water irrigation and greenhouse ventilation on zucchini crops. Journal of Agricultural and Food Chemistry 66 (31), 8422–8432.

Abreu, A.C., Molina-Miras, A., Aguilera-Sáez, L.M., López-Rosales, L., Cerón-García, M.d.C., Sánchez-Mirón, A., et al., 2019. Production of amphidinols and other bioproducts of interest by the marine microalga *Amphidinium carterae* unraveled by nuclear magnetic resonance metabolomics approach coupled to multivariate data analysis. Journal of Agricultural and Food Chemistry 67 (34), 9667–9682.

Aguilar, J.A., Faulkner, S., Nilsson, M., Morris, G.A., 2010. Pure shift ^{1}H NMR: a resolution of the resolution problem? Angewandte Chemie International Edition 49 (23), 3901–3903.

Aguilera-Sáez, L.M., Abreu, A.C., Camacho-Rodríguez, J., González-López, C.V., del Carmen Cerón-García, M., Fernández, I., 2019. NMR metabolomics as an effective tool to unravel the effect of light intensity and temperature on the composition of the marine microalgae *Isochrysis galbana*. Journal of Agricultural and Food Chemistry 67 (14), 3879–3889.

Al-Fattani, M.A., Douglas, L.J., 2006. Biofilm matrix of *Candida albicans* and *Candida tropicalis*: chemical composition and role in drug resistance. Journal of Medical Microbiology 55 (8), 999–1008.

Ammons, M.C.B., Tripet, B.P., Carlson, R.P., Kirker, K.R., Gross, M.A., Stanisich, J.J., et al., 2014. Quantitative NMR metabolite profiling of methicillin-resistant and methicillin-susceptible *Staphylococcus aureus* discriminates between biofilm and planktonic phenotypes. Journal of Proteome Research 13 (6), 2973–2985.

Baquero, F., Coque, T.M., Cantón, R., 2014. Counteracting antibiotic resistance: breaking barriers among antibacterial strategies. Expert Opinion on Therapeutic Targets 18 (8), 851–861.

Barnes, S., Benton, H.P., Casazza, K., Cooper, S.J., Cui, X., Du, X., et al., 2016. Training in metabolomics research. I. Designing the experiment, collecting and extracting samples and generating metabolomics data. Journal of Mass Spectrometry 51 (7), 461–475.

Bart, J., Kolkman, A.J., Oosthoek-de Vries, A.J., Koch, K., Nieuwland, P.J., Janssen, H., et al., 2009. A microfluidic high-resolution NMR flow probe. Journal of the American Chemical Society 131 (14), 5014–5015.

Bartel, J., Krumsiek, J., Theis, F.J., 2013. Statistical methods for the analysis of high-throughput metabolomics data. Computational and Structural Biotechnology Journal 4, e201301009.

Batzilla, C.F., Rachid, S., Engelmann, S., Hecker, M., Hacker, J., Ziebuhr, W., 2006. Impact of the accessory gene regulatory system (Agr) on extracellular proteins, codY expression and amino acid metabolism in *Staphylococcus epidermidis*. Proteomics 6 (12), 3602–3613.

Beitelshees, M., Hill, A., Jones, C.H., Pfeifer, B.A., 2018. Phenotypic variation during biofilm formation: implications for anti-biofilm therapeutic design. Materials (Basel) 11 (7), 1086.

Bendaoud, M., Vinogradov, E., Balashova, N.V., Kadouri, D.E., Kachlany, S.C., Kaplan, J.B., 2011. Broad-spectrum biofilm inhibition by *Kingella kingae* exopolysaccharide. Journal of Bacteriology 193 (15), 3879–3886.

Berman, P., Leshem, A., Etziony, O., Levi, O., Parmet, Y., Saunders, M., et al., 2013. Novel ^{1}H low field nuclear magnetic resonance applications for the field of biodiesel. Biotechnology for Biofuels 6 (1), 55.

Bernini, P., Bertini, I., Luchinat, C., Nepi, S., Saccenti, E., Schäfer, H., et al., 2009. Individual human phenotypes in metabolic space and time. Journal of Proteome Research 8 (9), 4264–4271.

Berrueta, L.A., Alonso-Salces, R.M., Héberger, K., 2007. Supervised pattern recognition in food analysis. Journal of Chromatography A 1158 (1–2), 196–214.

Beuling, E.E., van Dusschoten, D., Lens, P., van den Heuvel, J.C., Van As, H., Ottengraf, S.P.P., 1998. Characterization of the diffusive properties of biofilms using pulsed field gradient-nuclear magnetic resonance. Biotechnology and Bioengineering 60 (3), 283–291.

Bingol, K., Bruschweiler-Li, L., Li, D.-W., Brüschweiler, R., 2014. Customized metabolomics database for the analysis of NMR ^{1}H–^{1}H TOCSY and ^{13}C–^{1}H HSQC-TOCSY spectra of complex mixtures. Analytical Chemistry 86 (11), 5494–5501.

Bogino, P.C., de las Mercedes Oliva, M., Sorroche, F.G., Giordano, W., 2013. The role of bacterial biofilms and surface components in plant-bacterial associations. International Journal of Molecular Sciences 14 (8), 15838–15859.

Bolten, C.J., Kiefer, P., Letisse, F., Portais, J.-C., Wittmann, C., 2007. Sampling for metabolome analysis of microorganisms. Analytical Chemistry 79 (10), 3843–3849.

Booth, S.C., Workentine, M.L., Wen, J., Shaykhutdinov, R., Vogel, H.J., Ceri, H., et al., 2011. Differences in metabolism between the biofilm and planktonic response to metal stress. Journal of Proteome Research 10 (7), 3190–3199.

Branda, S.S., Chu, F., Kearns, D.B., Losick, R., Kolter, R., 2006. A major protein component of the *Bacillus subtilis* biofilm matrix. Molecular Microbiology 59 (4), 1229–1238.

Branda, S.S., Vik, Å., Friedman, L., Kolter, R., 2005. Biofilms: the matrix revisited. Trends in Microbiology 13 (1), 20–26.

Brennan, L., 2013. Metabolomics in nutrition research: current status and perspectives. Biochemical Society Transactions 41 (2), 670–673.

Brooun, A., Liu, S., Lewis, K., 2000. A dose-response study of antibiotic resistance in *Pseudomonas aeruginosa* biofilms. Antimicrobial Agents and Chemotherapy 44 (3), 640–646.

Bryce, D.L., 2017. NMR crystallography: structure and properties of materials from solid-state nuclear magnetic resonance observables. International Union of Crystallography Journal 4 (Pt 4), 350–359.

Burmølle, M., Webb, J.S., Rao, D., Hansen, L.H., Sørensen, S.J., Kjelleberg, S., 2006. Enhanced biofilm formation and increased resistance to antimicrobial agents and bacterial invasion are caused by synergistic interactions in multispecies biofilms. Applied and Environmental Microbiology 72 (6), 3916–3923.

Burnette-Curley, D., Wells, V., Viscount, H., Munro, C.L., Fenno, J.C., Fives-Taylor, P., et al., 1995. FimA, a major virulence factor associated with *Streptococcus parasanguis* endocarditis. Infection and Immunity 63 (12), 4669–4674.

Busetti, A., Crawford, D.E., Earle, M.J., Gilea, M.A., Gilmore, B.F., Gorman, S.P., et al., 2010. Antimicrobial and antibiofilm activities of 1-alkylquinolinium bromide ionic liquids. Green Chemistry 12 (3), 420–425.

Cegelski, L., 2015. Bottom-up and top-down solid-state NMR approaches for bacterial biofilm matrix composition. *Journal of Magnetic Resonance* (San Diego, California: 1997) 253, 91–97.

Chang, W.-S., van de Mortel, M., Nielsen, L., Nino de Guzman, G., Li, X., Halverson, L.J., 2007. Alginate production by *Pseudomonas putida* creates a hydrated microenvironment and contributes to biofilm architecture and stress tolerance under water-limiting conditions. Journal of Bacteriology 189 (22), 8290–8299.

Chavez-Dozal, A., Gorman, C., Nishiguchi, M.K., 2015. Proteomic and metabolomic profiles demonstrate variation among free-living and symbiotic *Vibrio fischeri* biofilms. BMC Microbiology 15, 226.

Choi, Y.H., Verpoorte, R., 2014. Metabolomics: what you see is what you extract. Phytochemical Analysis 25 (4), 289–290.

Chong, P.P., Chin, V.K., Wong, W.F., Madhavan, P., Yong, V.C., Looi, C.Y., 2018. Transcriptomic and genomic approaches for unravelling *Candida albicans* biofilm formation and drug resistance—an update. Genes 9 (11), 540.

Codd, S.L., Vogt, S.J., Hornemann, J.A., Phillips, A.J., Maneval, J.E., Romanenko, K.R., et al., 2011. NMR relaxation measurements of biofouling in model and geological porous media. Organic Geochemistry 42 (8), 965–971.

Colvin, K.M., Irie, Y., Tart, C.S., Urbano, R., Whitney, J.C., Ryder, C., et al., 2012. The Pel and Psl polysaccharides provide *Pseudomonas aeruginosa* structural redundancy within the biofilm matrix. Environmental Microbiology 14 (8), 1913–1928.

Costa, O.Y.A., Raaijmakers, J.M., Kuramae, E.E., 2018. Microbial extracellular polymeric substances: ecological function and impact on soil aggregation. Frontiers in Microbiology 9 (1636).

Costerton, J.W., Montanaro, L., Arciola, C.R., 2005. Biofilms in implant infections: its production and regulation. The International Journal of Artificial Organs 28 (11), 1062–1068.

Costerton, J.W., Stewart, P.S., Greenberg, E.P., 1999. Bacterial biofilms: a common cause of persistent infections. Science 284 (5418), 1318–1322.

Cozad, A., Jones, R.D., 2003. Disinfection and the prevention of infectious disease. American Journal of Infection Control 31 (4), 243–254.

Cravotto, G., Cintas, P., 2007. The combined use of microwaves and ultrasound: improved tools in process chemistry and organic synthesis. Chemistry - A European Journal 13 (7), 1902–1909.

Cugini, C., Shanmugam, M., Landge, N., Ramasubbu, N., 2019. The role of exopolysaccharides in oral biofilms. Journal of Dental Research 98 (7), 739–745.

Di Luca, M., Maccari, G., Nifosì, R., 2014. Treatment of microbial biofilms in the post-antibiotic era: prophylactic and therapeutic use of antimicrobial peptides and their design by bioinformatics tools. Pathogens and Disease 70 (3), 257–270.

Diggle, S.P., Stacey, R.E., Dodd, C., Cámara, M., Williams, P., Winzer, K., 2006. The galactophilic lectin, LecA, contributes to biofilm development in *Pseudomonas aeruginosa*. Environmental Microbiology 8 (6), 1095–1104.

Domenech, M., Pedrero-Vega, E., Prieto, A., García, E., 2016. Evidence of the presence of nucleic acids and β-glucan in the matrix of non-typeable *Haemophilus influenzae* in vitro biofilms. Scientific Reports 6, 36424.

Dona, A.C., Kyriakides, M., Scott, F., Shephard, E.A., Varshavi, D., Veselkov, K., et al., 2016. A guide to the identification of metabolites in NMR-based metabonomics/metabolomics experiments. Computational and Structural Biotechnology Journal 14, 135–153.

Donlan, R.M., 2001. Biofilm formation: a clinically relevant microbiological process. Clinical Infectious Diseases 33 (8), 1387–1392.

Duncan, K.D., Fyrestam, J., Lanekoff, I., 2019. Advances in mass spectrometry based single-cell metabolomics. Analyst 144 (3), 782–793.

Eckelmann, D., Spiteller, M., Kusari, S., 2018. Spatial-temporal profiling of prodiginines and serratamolides produced by endophytic *Serratia marcescens* harbored in *Maytenus serrata*. Scientific Reports 8 (1), 5283.

Ellinger, J.J., Chylla, R.A., Ulrich, E.L., Markley, J.L., 2013. Databases and software for NMR-based metabolomics. Current Metabolomics 1 (1), 1–22.

Ellis, D.I., Dunn, W.B., Griffin, J.L., Allwood, J.W., Goodacre, R., 2007. Metabolic fingerprinting as a diagnostic tool. Pharmacogenomics 8 (9), 1243–1266.

Emwas, A.-H., Roy, R., McKay, R.T., Tenori, L., Saccenti, E., Gowda, G.A.N., et al., 2019. NMR spectroscopy for metabolomics research. Metabolites 9 (7), 123.

Farrell, E.K., Tipton, P.A., 2012. Functional characterization of AlgL, an alginate lyase from *Pseudomonas aeruginosa*. Biochemistry 51 (51), 10259–10266.

Fernández, L., Breidenstein, E.B.M., Hancock, R.E.W., 2011. Creeping baselines and adaptive resistance to antibiotics. Drug Resistance Updates 14 (1), 1–21.

Fish, K.E., Osborn, A.M., Boxall, J., 2016. Characterising and understanding the impact of microbial biofilms and the extracellular polymeric substance (EPS) matrix in drinking water distribution systems. Environmental Sciences: Water Research, Technology 2 (4), 614–630.

Flemming, H.-C., Wingender, J., 2010. The biofilm matrix. Nature Reviews Microbiology 8 (9), 623–633.

Fontana, C., Li, S., Yang, Z., Widmalm, G., 2015. Structural studies of the exopolysaccharide from *Lactobacillus plantarum* C88 using NMR spectroscopy and the program CASPER. Carbohydrate Research 402, 87–94.

Foroozandeh, M., Castañar, L., Martins, L.G., Sinnaeve, D., Poggetto, G.D., Tormena, C.F., et al., 2016. Ultrahigh-resolution diffusion-ordered spectroscopy. Angewandte Chemie International Edition 55 (50), 15579–15582.

Franklin, M.J., Chang, C., Akiyama, T., Bothner, B., 2015. New technologies for studying biofilms. Microbiology Spectrum 3 (4), 1–41.

Fratila, R.M., Gomez, M.V., Sýkora, S., Velders, A.H., 2014. Multinuclear nanoliter one-dimensional and two-dimensional NMR spectroscopy with a single non-resonant microcoil. Nature Communications 5 (1), 3025.

Frieri, M., Kumar, K., Boutin, A., 2017. Antibiotic resistance. Journal of Infection and Public Health 10 (4), 369–378.

Fux, C.A., Costerton, J.W., Stewart, P.S., Stoodley, P., 2005. Survival strategies of infectious biofilms. Trends in Microbiology 13 (1), 34–40.

Fysun, O., Anzmann, T., Kleesattel, A., Gschwind, P., Rauschnabel, J., Kohlus, R., et al., 2019. Detection of *P. polymyxa* biofilm, dairy biofouling and CIP-cleaning agents using low-field NMR. European Food Research and Technology 245, 1719–1731.

Garny, K., Neu, T.R., Horn, H., Volke, F., Manz, B., 2010. Combined application of ^{13}C NMR spectroscopy and confocal laser scanning microscopy—investigation on biofilm structure and physico-chemical properties. Chemical Engineering Science 65 (16), 4691–4700.

Gilbert, P., Allison, D.G., McBain, A.J., 2002. Biofilms *in vitro* and *in vivo*: do singular mechanisms imply cross-resistance? Journal of Applied Microbiology 92, 98S–110S.

Gjersing, E.L., Herberg, J.L., Horn, J., Schaldach, C.M., Maxwell, R.S., 2007. NMR metabolomics of planktonic and biofilm modes of growth in *Pseudomonas aeruginosa*. Analytical Chemistry 79 (21), 8037–8045.

Glanzer, S., Zangger, K., 2014. Directly decoupled diffusion-ordered NMR spectroscopy for the analysis of compound mixtures. Chemistry 20 (35), 11171–11175.

Gomez, M.V., Verputten, H.H.J., Díaz-Ortíz, A., Moreno, A., de la Hoz, A., Velders, A.H., 2010. On-line monitoring of a microwave-assisted chemical reaction by nanolitre NMR-spectroscopy. Chemical Communications 46 (25), 4514–4516.

Gresley, A.L., Kenny, J., Cassar, C., Kelly, A., Sinclair, A., Fielder, M.D., 2012. The application of high resolution diffusion NMR to the analysis of manuka honey. Food Chemistry 135 (4), 2879–2886.

Griffith, C.J., Cooper, R.A., Gilmore, J., Davies, C., Lewis, M., 2000. An evaluation of hospital cleaning regimes and standards. Journal of Hospital Infection 45 (1), 19–28.

Hall, C.W., Mah, T.-F., 2017. Molecular mechanisms of biofilm-based antibiotic resistance and tolerance in pathogenic bacteria. FEMS Microbiology Reviews 41 (3), 276–301.

Herrling, M.P., Lackner, S., Nirschl, H., Horn, H., Guthausen, G., 2019. Chapter 4 – Recent NMR/MRI studies of biofilm structures and dynamics. In: WEBB, G.A. (Ed.), Annual Reports on NMR Spectroscopy. Academic Press.

Herrling, M.P., Weisbrodt, J., Kirkland, C.M., Williamson, N.H., Lackner, S., Codd, S.L., et al., 2017. NMR investigation of water diffusion in different biofilm structures. Biotechnology and Bioengineering 114 (12), 2857–2867.

Hesse, S.J.A., Ruijter, G.J.G., Dijkema, C., Visser, J., 2000. Measurement of intracellular (compartmental) pH by ^{31}P NMR in *Aspergillus niger*. Journal of Biotechnology 77 (1), 5–15.

Hoerr, V., Zbytnuik, L., Leger, C., Tam, P.P.C., Kubes, P., Vogel, H.J., 2012. Gram-negative and Gram-positive bacterial infections give rise to a different metabolic response in a mouse model. Journal of Proteome Research 11 (6), 3231–3245.

Hoskins, B.C., Fevang, L., Majors, P.D., Sharma, M.M., Georgiou, G., 1999. Selective imaging of biofilms in porous media by NMR relaxation. Journal of Magnetic Resonance 139 (1), 67–73.

Iwase, T., Uehara, Y., Shinji, H., Tajima, A., Seo, H., Takada, K., et al., 2010. *Staphylococcus epidermidis* Esp inhibits *Staphylococcus aureus* biofilm formation and nasal colonization. *Nature* 465 (7296), 346–349.

Izano, E.A., Sadovskaya, I., Wang, H., Vinogradov, E., Ragunath, C., Ramasubbu, N., et al., 2008. Poly-N-acetylglucosamine mediates biofilm formation and detergent resistance in *Aggregatibacter actinomycetemcomitans*. Microbial Pathogenesis 44 (1), 52–60.

Jahn, A., Nielsen, P.H., 1995. Extraction of extracellular polymeric substances (EPS) from biofilms using a cation exchange resin. Water Science and Technology 32 (8), 157–164.

Jiao, Y., Cody, G.D., Harding, A.K., Wilmes, P., Schrenk, M., Wheeler, K.E., et al., 2010. Characterization of extracellular polymeric substances from acidophilic microbial biofilms. Applied and Environmental Microbiology 76 (9), 2916–2922.

Kalpana, B.J., Aarthy, S., Pandian, S.K., 2012. Antibiofilm activity of α-amylase from *Bacillus subtilis* S8-18 against biofilm forming human bacterial pathogens. Applied Biochemistry and Biotechnology 167 (6), 1778–1794.

Kaper, J.B., Sperandio, V., 2005. Bacterial cell-to-cell signaling in the gastrointestinal tract. Infection and Immunity 73 (6), 3197–3209.

Kaplan, J.B., Ragunath, C., Ramasubbu, N., Fine, D.H., 2003. Detachment of *Actinobacillus actinomycetemcomitans* biofilm cells by an endogenous beta-hexosaminidase activity. Journal of Bacteriology 185 (16), 4693–4698.

Kiamco, M.M., Mohamed, A., Reardon, P.N., Marean-Reardon, C.L., Aframehr, W.M., Call, D.R., et al., 2018. Structural and metabolic responses of *Staphylococcus aureus* biofilms to hyperosmotic and antibiotic stress. *Biotechnology and Bioengineering* 115 (6), 1594–1603.

Kim, H.-K., Choi, Y.H., Verpoorte, R., 2010. NMR-based metabolomic analysis of plants. Nature Protocols 5, 536–549.

Kirkland, C.M., Herrling, M.P., Hiebert, R., Bender, A.T., Grunewald, E., Walsh, D.O., et al., 2015a. *In situ* detection of subsurface biofilm using low-field NMR: a field study. Environmental Science and Technology 49 (18), 11045–11052.

Kirkland, C.M., Hiebert, R., Phillips, A., Grunewald, E., Walsh, D.O., Seymour, J.D., et al., 2015b. Biofilm detection in a model well-bore environment using low-field NMR. Groundwater Monitoring, Remediation 35 (4), 36–44.

Kirtil, E., Oztop, M.H., 2016. ^{1}H nuclear magnetic resonance relaxometry and magnetic resonance imaging and applications in food science and processing. Food Engineering Reviews 8 (1), 1–22.

Kittler, W.C., Obruchkov, S., Galvosas, P., Hunter, M.W., 2014. Pulsed second order field NMR for real time PGSE and single-shot surface to volume ratio measurements. Journal of Magnetic Resonance 247, 42–49.

Kumar, A., Alam, A., Rani, M., Ehtesham, N.Z., Hasnain, S.E., 2017. Biofilms: survival and defense strategy for pathogens. International Journal of Medical Microbiology 307 (8), 481–489.

Lai, S., Tremblay, J., Déziel, E., 2009. Swarming motility: a multicellular behaviour conferring antimicrobial resistance. Environmental Microbiology 11 (1), 126–136.

Larive, C.K., Barding, G.A., Dinges, M.M., 2015. NMR spectroscopy for metabolomics and metabolic profiling. Analytical Chemistry 87 (1), 133–146.

Larsen, P., Nielsen, J.L., Dueholm, M.S., Wetzel, R., Otzen, D., Nielsen, P.H., 2007. Amyloid adhesins are abundant in natural biofilms. Environmental Microbiology 9 (12), 3077–3090.

Lawrence, J.R., Wolfaardt, G.M., Korber, D.R., 1994. Determination of diffusion coefficients in biofilms by confocal laser microscopy. Applied and Environmental Microbiology 60 (4), 1166–1173.

Le Gresley, A., Simpson, E., Sinclair, A.J., Williams, N., Burnett, G.R., Bradshaw, D.J., et al., 2015. The application of high resolution diffusion NMR for the characterisation and quantification of small molecules in saliva/dentifrice slurries. Analytical Methods 7 (6), 2323–2332.

Lebeaux, D., Ghigo, J.-M., Beloin, C., 2014. Biofilm-related infections: bridging the gap between clinical management and fundamental aspects of recalcitrance toward antibiotics. Microbiology and Molecular Biology Reviews 78 (3), 510–543.

Lens, P.N.L., Gastesi, R., Vergeldt, F., van Aelst, A.C., Pisabarro, A.G., Van As, H., 2003. Diffusional properties of methanogenic granular sludge: ^{1}H NMR characterization. Applied and Environmental Microbiology 69 (11), 6644–6649.

Lewandowski, Z., Beyenal, H., 2010. Chapter 2: Imaging and characterizing biofilm components. In: Fundamentals of Biofilm Research. CRC Press.

Lewis, K., 2008. Multidrug tolerance of biofilms and persister cells. In: Romeo, T. (Ed.), Bacterial Biofilms. Springer Berlin Heidelberg, Berlin, Heidelberg.

Li, B., Webster, T.J., 2018. Bacteria antibiotic resistance: new challenges and opportunities for implant-associated orthopedic infections. Journal of Orthopaedic Research 36 (1), 22–32.

Li, D., Kagan, G., Hopson, R., Williard, P.G., 2009. Formula weight prediction by internal reference diffusion-ordered NMR spectroscopy (DOSY). Journal of the American Chemical Society 131 (15), 5627–5634.

Li, Y.-H., Tian, X., 2012. Quorum sensing and bacterial social interactions in biofilms. Sensors (Basel) 12 (3), 2519–2538.

Lilly Thankamony, A.S., Wittmann, J.J., Kaushik, M., Corzilius, B., 2017. Dynamic nuclear polarization for sensitivity enhancement in modern solid-state NMR. Progress in Nuclear Magnetic Resonance Spectroscopy 102–103, 20–195.

Lim, J.Y., May, J.M., Cegelski, L., 2012. Dimethyl sulfoxide and ethanol elicit increased amyloid biogenesis and amyloid-integrated biofilm formation in *Escherichia coli*. Applied and Environmental Microbiology 78 (9), 3369–3378.

Liu, D., Momb, J., Thomas, P.W., Moulin, A., Petsko, G.A., Fast, W., et al., 2008. Mechanism of the quorum-quenching lactonase (AiiA) from *Bacillus thuringiensis*: 1. Product-bound structures. Biochemistry 47 (29), 7706–7714.

Loquet, A., El Mammeri, N., Jan, S., Berbon, M., Bardiaux, B., Pintacuda, G., et al., 2018. 3D structure determination of amyloid fibrils using solid-state NMR spectroscopy. Methods 138, 26–38.

Lorén, N., Hagman, J., Jonasson, J.K., Deschout, H., Bernin, D., Cella-Zanacchi, F., et al., 2015. Fluorescence recovery after photobleaching in material and life sciences: putting theory into practice. Quarterly Reviews of Biophysics 48 (3), 323–387.

Lynch, D.J., Fountain, T.L., Mazurkiewicz, J.E., Banas, J.A., 2007a. Glucan-binding proteins are essential for shaping *Streptococcus mutans* biofilm architecture. FEMS Microbiology Letters 268 (2), 158–165.

Lynch, S.V., Dixon, L., Benoit, M.R., Brodie, E.L., Keyhan, M., Hu, P., et al., 2007b. Role of the rapa gene in controlling antibiotic resistance of *Escherichia coli* biofilms. Antimicrobial Agents and Chemotherapy 51 (10), 3650–3658.

Mah, T.-F.C., O'Toole, G.A., 2001. Mechanisms of biofilm resistance to antimicrobial agents. Trends in Microbiology 9 (1), 34–39.

Maifiah, M.H.M., Cheah, S.-E., Johnson, M.D., Han, M.-L., Boyce, J.D., Thamlikitkul, V., et al., 2016. Global metabolic analyses identify key differences in metabolite levels between polymyxin-susceptible and polymyxin-resistant *Acinetobacter baumannii*. Scientific Reports 6, 22287.

Majors, P.D., McLean, J.S., Pinchuk, G.E., Fredrickson, J.K., Gorby, Y.A., Minard, K.R., et al., 2005. NMR methods for in situ biofilm metabolism studies. Journal of Microbiological Methods 62 (3), 337–344.

Mannina, L., Sobolev, A.P., Viel, S., 2012. Liquid state 1H high field NMR in food analysis. Progress in Nuclear Magnetic Resonance Spectroscopy 66, 1–39.

Manz, B., Volke, F., Goll, D., Horn, H., 2003. Measuring local flow velocities and biofilm structure in biofilm systems with Magnetic Resonance Imaging (MRI). Biotechnology and Bioengineering 84 (4), 424–432.

Mariette, F., 2009. Investigations of food colloids by NMR and MRI. Current Opinion in Colloid and Interface Science 13 (3), 203–211.

Matsuyama, T., Tanikawa, T., Nakagawa, Y., 2011. Serrawettins and other surfactants produced by *Serratia*. In: Soberón-Chávez, G. (Ed.), Biosurfactants: From Genes to Applications. Springer Berlin Heidelberg, Berlin, Heidelberg.

Mayer, C., Lattner, D., Schürks, N., 2001. ^{13}C Nuclear Magnetic Resonance studies on selectively labeled bacterial biofilms. Journal of Industrial Microbiology and Biotechnology 26 (1), 62–69.

McCrate, O.A., Zhou, X., Reichhardt, C., Cegelski, L., 2013. Sum of the parts: composition and architecture of the bacterial extracellular matrix. Journal of Molecular Biology 425 (22), 4286–4294.

McLean, J.S., Ona, O.N., Majors, P.D., 2008. Correlated biofilm imaging, transport and metabolism measurements via combined nuclear magnetic resonance and confocal microscopy. The ISME Journal 2 (2), 121–131.

Medeiros-Silva, J., Jekhmane, S., Paioni, A.L., Gawarecka, K., Baldus, M., Swiezewska, E., et al., 2018. High-resolution NMR studies of antibiotics in cellular membranes. Nature Communications 9 (1), 3963.

Metzger, U., Lankes, U., Fischpera, K., Frimmel, F.H., 2009. The concentration of polysaccharides and proteins in EPS of *Pseudomonas putida* and *Aureobasidum pullulans* as revealed by ^{13}C CPMAS NMR spectroscopy. Applied Microbiology and Biotechnology 85 (1), 197–206.

Mickiewicz, B., Duggan, G., Winston, B., Doig, C., Kubes, P., Vogel, H., 2014. Metabolic profiling of serum samples by ^{1}H nuclear magnetic resonance spectroscopy as a potential diagnostic approach for septic shock. Critical Care Medicine 42 (5), 1140–1149.

Mickiewicz, B., Vogel, H.J., Wong, H.R., Winston, B.W., 2013. Metabolomics as a novel approach for early diagnosis of pediatric septic shock and its mortality. American Journal of Respiratory and Critical Care Medicine 187 (9), 967–976.

Mora, P., Rosconi, F., Franco Fraguas, L., Castro-Sowinski, S., 2008. *Azospirillum brasilense* Sp7 produces an outer-membrane lectin that specifically binds to surface-exposed extracellular polysaccharide produced by the bacterium. Archives of Microbiology 189 (5), 519–524.

Mosier, A.C., Justice, N.B., Bowen, B.P., Baran, R., Thomas, B.C., Northen, T.R., et al., 2013. Metabolites associated with adaptation of microorganisms to an acidophilic, metal-rich environment identified by stable-isotope-enabled metabolomics. mBio 4 (2) e00484-12.

Mushtaq, M.Y., Choi, Y.H., Verpoorte, R., Wilson, E.G., 2014. Extraction for metabolomics: access to the metabolome. Phytochemical Analysis 25 (4), 291–306.

Nagana Gowda, G.A., Raftery, D., 2015. Can NMR solve some significant challenges in metabolomics? Journal of Magnetic Resonance 260, 144–160.

Nijampatnam, B., Zhang, H., Cai, X., Michalek, S.M., Wu, H., Velu, S.E., 2018. Inhibition of *Streptococcus mutans* biofilms by the natural stilbene piccatannol through the inhibition of glucosyltransferases. ACS Omega 3 (7), 8378–8385.

O'Gorman, A., Gibbons, H., Brennan, L., 2013. Metabolomics in the identification of biomarkers of dietary intake. Computational and Structural Biotechnology Journal 4 e201301004.

Olson, D.L., Peck, T.L., Webb, A.G., Magin, R.L., Sweedler, J.V., 1995. High-resolution microcoil ^{1}H-NMR for mass-limited, nanoliter-volume samples. Science 270 (5244), 1967–1970.

Olson, M., Ceri, H., Morck, D., Buret, A., Read, R., 2002. Biofilm bacteria: formation and comparative susceptibility to antibiotics. Canadian Journal of Veterinary Research 66 (1), 86–92.

Otter, J.A., Yezli, S., French, G.L., 2011. The role played by contaminated surfaces in the transmission of nosocomial pathogens. Infection Control and Hospital Epidemiology 32, 687–699.

Pagès, G., Gilard, V., Martino, R., Malet-Martino, M., 2017. Pulsed-field gradient nuclear magnetic resonance measurements (PFG NMR) for diffusion ordered spectroscopy (DOSY) mapping. Analyst 142 (20), 3771–3796.

Palmer, R.J., Haagensen, J.A.J., Neu, T.R., Sternberg, C., 2006. Confocal microscopy of biofilms — spatiotemporal approaches. In: Pawley, J.B. (Ed.), Handbook of Biological Confocal Microscopy. Springer US, Boston, MA.

Park, J.-H., Lee, J.-H., Cho, M.H., Herzberg, M., Lee, J., 2012. Acceleration of protease effect on *Staphylococcus aureus* biofilm dispersal. *FEMS Microbiology Letters* 335 (1), 31–38.

Patel, K.N., Patel, J.K., Patel, M.P., Rajput, G.C., Patel, H.A., 2010. Introduction to hyphenated techniques and their applications in pharmacy. Pharmaceutical Methods 1 (1), 2–13.

Peled, S., Tseng, C.H., Sodickson, A.A., Mair, R.W., Walsworth, R.L., Cory, D.G., 1999. Single-shot diffusion measurement in laser-polarized Gas. *Journal of Magnetic Resonance* (San Diego, California: 1997) 140 (2), 320–324.

Pelta, M., Morris, G., Stchedroff, M., Hammond, S., 2002. A one-shot sequence for high-resolution diffusion-ordered spectroscopy. Magnetic Resonance in Chemistry 40, S147–S152.

Perry, J.A., Lévesque, C.M., Suntharaligam, P., Mair, R.W., Bu, M., Cline, R.T., et al., 2008. Involvement of *Streptococcus mutans* regulator RR11 in oxidative stress response during biofilm growth and in the development of genetic competence. Letters in Applied Microbiology 47 (5), 439–444.

Peyyala, R., Kirakodu, S.S., Ebersole, J.L., Novak, K.F., 2011. Novel model for multispecies biofilms that uses rigid gas-permeable lenses. Applied and Environmental Microbiology 77 (10), 3413–3421.

Phoenix, V.R., Holmes, W.M., 2008. Magnetic Resonance Imaging of structure, diffusivity, and copper immobilization in a phototrophic biofilm. Applied and Environmental Microbiology 74 (15), 4934–4943.

Pinto, M.S., McGahan, M.E., Steiner, W.W., Priefer, R., 2011. The use of the pseudo-polyelectrolyte, poly(4-vinylphenol), in multilayered films as an antimicrobial surface coating. Colloids and Surfaces A 377 (1–3), 182–186.

Pinu, F.R., Villas-Boas, S.G., 2017. Extracellular microbial metabolomics: the state of the art. Metabolites 7 (3), 43.

Pinu, F.R., Villas-Boas, S.G., Aggio, R., 2017. Analysis of intracellular metabolites from microorganisms: quenching and extraction protocols. Metabolites 7 (4), 53.

Potter, K., Kleinberg, R.L., Brockman, F.J., McFarland, E.W., 1996. Assay for bacteria in porous media by diffusion-weighted NMR. Journal of Magnetic Resonance, Series B 113 (1), 9–15.

Price, W.S., 1997. Pulsed-field gradient nuclear magnetic resonance as a tool for studying translational diffusion: Part 1. Basic theory. Concepts in Magnetic Resonance 9 (5), 299–336.

Ramanan, B., Holmes, W.M., Sloan, W.T., Phoenix, V.R., 2013. Magnetic Resonance Imaging of mass transport and structure inside a phototrophic biofilm. Current Microbiology 66 (5), 456–461.

Ranzinger, F., Herrling, M.P., Lackner, S., Grande, V.W., Baniodeh, A., Powell, A.K., et al., 2016. Direct surface visualization of biofilms with high spin coordination clusters using Magnetic Resonance Imaging. Acta Biomaterialia 31, 167–177.

Reichhardt, C., Cegelski, L., 2014. Solid-state NMR for bacterial biofilms. Molecular Physics 112 (7), 887–894.

Reichhardt, C., Ferreira, J.A.G., Joubert, L.-M., Clemons, K.V., Stevens, D.A., Cegelski, L., 2015a. Analysis of the *Aspergillus fumigatus* biofilm extracellular matrix by solid-state Nuclear Magnetic Resonance spectroscopy. Eukaryotic Cell 14 (11), 1064–1072.

Reichhardt, C., Fong, J.C.N., Yildiz, F., Cegelski, L., 2015b. Characterization of the *Vibrio cholerae* extracellular matrix: a top-down solid-state NMR approach. Biochimica et Biophysica Acta (BBA) - Biomembranes 1848 (1, Part B), 378–383.

Renslow, R.S., Majors, P.D., McLean, J.S., Fredrickson, J.K., Ahmed, B., Beyenal, H., 2010. In situ effective diffusion coefficient profiles in live biofilms using pulsed-field gradient nuclear magnetic resonance. Biotechnology and Bioengineering 106 (6), 928–937.

Sapcariu, S.C., Kanashova, T., Weindl, D., Ghelfi, J., Dittmar, G., Hiller, K., 2014. Simultaneous extraction of proteins and metabolites from cells in culture. MethodsX 1, 74–80.

Säwén, E., Huttunen, E., Zhang, X., Yang, Z., Widmalm, G., 2010. Structural analysis of the exopolysaccharide produced by *Streptococcus thermophilus* ST1 solely by NMR spectroscopy. Journal of Biomolecular NMR 47 (2), 125–134.

Scalbert, A., Brennan, L., Fiehn, O., Hankemeier, T., Kristal, B.S., van Ommen, B., et al., 2009. Mass-spectrometry-based metabolomics: limitations and recommendations for future progress with particular focus on nutrition research. Metabolomics 5 (4), 435–458.

Schelli, K., Zhong, F., Zhu, J., 2017. Comparative metabolomics revealing *Staphylococcus aureus* metabolic response to different antibiotics. M*icrobial Biotechnology* 10 (6), 1764–1774.

Serra, D.O., Richter, A.M., Hengge, R., 2013. Cellulose as an architectural element in spatially structured *Escherichia coli* biofilms. Journal of Bacteriology 195 (24), 5540–5554.

Seviour, T., Derlon, N., Dueholm, M.S., Flemming, H.-C., Girbal-Neuhauser, E., Horn, H., et al., 2019. Extracellular polymeric substances of biofilms: suffering from an identity crisis. Water Research 151, 1–7.

Seymour, J.D., Gage, J.P., Codd, S.L., Gerlach, R., 2007. Magnetic resonance microscopy of biofouling induced scale dependent transport in porous media. Advances in Water Resources 30 (6), 1408–1420.

Shen, Y., Stojicic, S., Haapasalo, M., 2011. Antimicrobial efficacy of chlorhexidine against bacteria in biofilms at different stages of development. Journal of Endodontics 37 (5), 657–661.

Shin, J.-H., Lee, H.-W., Kim, S.-M., Kim, J., 2009. Proteomic analysis of *Acinetobacter baumannii* in biofilm and planktonic growth mode. Journal of Microbiology 47 (6), 728–735.

Shommu, N.S., Vogel, H.J., Storey, D.G., 2015. Potential of metabolomics to reveal *Burkholderia cepacia* complex pathogenesis and antibiotic resistance. Frontiers in Microbiology 6, 668.

Simões, L.C., Simões, M., Vieira, M.J., 2010. Influence of the diversity of bacterial isolates from drinking water on resistance of biofilms to disinfection. Applied and Environmental Microbiology 76 (19), 6673–6679.

Simões, M., 2005. Use of Biocides and Surfactants to Control Pseudomonas Fluorescens Biofilms: Role of the Hydrodynamic Conditions (Ph.D. thesis). University of Minho.

Simões, M., Rocha, R., Coimbra, M.A., Vieira, M., 2008. Enhancement of *Escherichia coli* and *Staphylococcus aureus* antibiotic susceptibility using sesquiterpenoids. Medicinal Chemistry 4, 616–623.

Simpson, J.H., 2012. Chapter 9 – Strategies for assigning resonances to atoms within a molecule. In: Simpson, J.H. (Ed.), Organic Structure Determination Using 2-D NMR Spectroscopy, second ed. Academic Press, Boston.

Singh, P.K., Schaefer, A.L., Parsek, M.R., Moninger, T.O., Welsh, M.J., Greenberg, E.P., 2000. Quorum-sensing signals indicate that cystic fibrosis lungs are infected with bacterial biofilms. Nature 407 (6805), 762–764.

Sitter, B., Bathen, T.F., Gribbestad, I.S., 2008. 5 - high-resolution magic angle spinning magnetic resonance spectroscopy. In: Hayat, M.A. (Ed.), Cancer Imaging. Academic Press, San Diego.

Smith, A.W., 2005. Biofilms and antibiotic therapy: is there a role for combating bacterial resistance by the use of novel drug delivery systems? Advanced Drug Delivery Reviews 57 (10), 1539–1550.

Smolinska, A., Blanchet, L., Buydens, L.M.C., Wijmenga, S.S., 2012. NMR and pattern recognition methods in metabolomics: from data acquisition to biomarker discovery: a review. Analytica Chimica Acta 750, 82–97.

Ståhle, J., 2017. Structure Elucidations of Bacterial Polysaccharides Using NMR Spectroscopy and Bioinformatics (Ph.D. dissertation). Stockholm University.

Stait-Gardner, T., Anil Kumar, P.G., Price, W.S., 2008. Steady state effects in PGSE NMR diffusion experiments. Chemical Physics Letters 462 (4), 331–336.

Stamps, J.P., Ottink, B., Visser, J.M., van Duynhoven, J.P.M., Hulst, R., 2001. Difftrain: a novel approach to a true spectroscopic single-scan diffusion measurement. Journal of Magnetic Resonance 151 (1), 28–31.

Stewart, P., 1996. Theoretical aspects of antibiotic diffusion into microbial biofilms. Antimicrobial Agents and Chemotherapy 40 (11), 2517–2522.

Stewart, P.S., 2002. Mechanisms of antibiotic resistance in bacterial biofilms. International Journal of Medical Microbiology 292 (2), 107–113.

Stewart, P.S., Davison, W.M., Steenbergen, J.N., 2009. Daptomycin rapidly penetrates a *Staphylococcus epidermidis* biofilm. Antimicrobial Agents and Chemotherapy 53 (8), 3505–3507.

Stewart, P.S., Franklin, M.J., 2008. Physiological heterogeneity in biofilms. Nature Reviews Microbiology 6, 199.

Stewart, P.S., William Costerton, J., 2001. Antibiotic resistance of bacteria in biofilms. The Lancet 358 (9276), 135–138.

Stipetic, L.H., Dalby, M.J., Davies, R.L., Morton, F.R., Ramage, G., Burgess, K.E.V., 2016. A novel metabolomic approach used for the comparison of *Staphylococcus aureus* planktonic cells and biofilm samples. Metabolomics 12 (4), 1–11.

Sugimoto, M., Kawakami, M., Robert, M., Soga, T., Tomita, M., 2012. Bioinformatics tools for mass spectroscopy-based metabolomic data processing and analysis. Current Bioinformatics 7 (1), 96–108.

Sun, J.-L., Zhang, S.-K., Chen, J.-Y., Han, B.-Z., 2012. Metabolic profiling of *Staphylococcus aureus* cultivated under aerobic and anaerobic conditions with 1^H NMR-based nontargeted analysis. *Canadian Journal of Microbiology* 58 (6), 709–718.

Tang, W., Bhatt, A., Smith, A.N., Crowley, P.J., Brady, L.J., Long, J.R., 2016. Specific binding of a naturally occurring amyloidogenic fragment of *Streptococcus mutans* adhesin P1 to intact P1 on the cell surface characterized by solid state NMR spectroscopy. Journal of Biomolecular NMR 64 (2), 153–164.

Teixeira, P.C., Leite, G.M., Domingues, R.J., Silva, J., Gibbs, P.A., Ferreira, J.P., 2007. Antimicrobial effects of a microemulsion and a nanoemulsion on enteric and other pathogens and biofilms. International Journal of Food Microbiology 118 (1), 15–19.

Thongsomboon, W., Serra, D.O., Possling, A., Hadjineophytou, C., Hengge, R., Cegelski, L., 2018. Phosphoethanolamine cellulose: a naturally produced chemically modified cellulose. Science 359 (6373), 334–338.

Thrippleton, M.J., Loening, N.M., Keeler, J., 2003. A fast method for the measurement of diffusion coefficients: one-dimensional DOSY. Magnetic Resonance in Chemistry 41 (6), 441–447.

Torino, M.I., Font de Valdez, G., Mozzi, F., 2015. Biopolymers from lactic acid bacteria. Novel applications in foods and beverages. Frontiers in Microbiology 6, 834.

Torres, A.M., Zheng, G., Price, W.S., 2010. J-compensated PGSE: an improved NMR diffusion experiment with fewer phase distortions. Magnetic Resonance in Chemistry 48 (2), 129–133.

Van As, H., Lens, P., 2001. Use of ^1H NMR to study transport processes in porous biosystems. Journal of Industrial Microbiology and Biotechnology 26 (1), 43–52.

Verstraeten, N., Braeken, K., Debkumari, B., Fauvart, M., Fransaer, J., Vermant, J., et al., 2008. Living on a surface: swarming and biofilm formation. Trends in Microbiology 16 (10), 496–506.

Viel, S., Capitani, D., Mannina, L., Segre, A., 2003. Diffusion-ordered NMR spectroscopy: a versatile tool for the molecular weight determination of uncharged polysaccharides. Biomacromolecules 4 (6), 1843–1847.

Vignoli, A., Ghini, V., Meoni, G., Licari, C., Takis, P.G., Tenori, L., et al., 2019. High-throughput metabolomics by [1]D NMR. Angewandte Chemie International Edition 58 (4), 968–994.

Vogt, M., Flemming, H.C., Veeman, W.S., 2000. Diffusion in *Pseudomonas aeruginosa* biofilms: a pulsed field gradient NMR study. Journal of Biotechnology 77 (1), 137–146.

Vogt, S.J., 2013. Nuclear Magnetic Resonance Studies of Biological and Biogeochemical Processes (Ph.D. dissertation in Philosophy in Engineering). Montana State University.

Wang, K., Li, W., Rui, X., Chen, X., Jiang, M., Dong, M., 2014. Characterization of a novel exopolysaccharide with antitumor activity from *Lactobacillus plantarum* 70810. International Journal of Biological Macromolecules 63, 133–139.

Weber, D.J., Rutala, W.A., Miller, M.B., Huslage, K., Sickbert-Bennett, E., 2010. Role of hospital surfaces in the transmission of emerging health care-associated pathogens: *Norovirus, Clostridium difficile*, and *Acinetobacter* species. American Journal of Infection Control 38 (5), S25–S33.

Weljie, A.M., Newton, J., Mercier, P., Carlson, E., Slupsky, C.M., 2006. Targeted profiling: Quantitative analysis of ^{1}H NMR metabolomics data. Analytical Chemistry 78 (13), 4430–4442.

White, A.P., Weljie, A.M., Apel, D., Zhang, P., Shaykhutdinov, R., Vogel, H.J., et al., 2010. A global metabolic shift is linked to salmonella multicellular development. PloS One 5 (7), e11814.

Wieland, A., de Beer, D., Damgaard, L.R., Kühl, M., van Dusschoten, D., Van As, H., 2001. Fine-scale measurement of diffusivity in a microbial mat with nuclear magnetic resonance imaging. Limnology and Oceanography 46 (2), 248–259.

Wishart, D.S., 2008. Quantitative metabolomics using NMR. TrAC Trends in Analytical Chemistry 27 (3), 228–237.

Wishart, D.S., Feunang, Y.D., Marcu, A., Guo, A.C., Liang, K., Vázquez-Fresno, R., et al., 2018. HMDB 4.0: the human metabolome database for 2018. Nucleic Acids Research 46 (D1), D608–D617.

Wong, E.H.J., Ng, C.G., Goh, K.L., Vadivelu, J., Ho, B., Loke, M.F., 2018. Metabolomic analysis of low and high biofilm-forming *Helicobacter pylori* strains. Scientific Reports 8 (1), 1409.

Worley, B., Powers, R., 2013. Multivariate analysis in metabolomics. Current Metabolomics 1 (1), 92–107.

Wu, H., Mintz, K.P., Ladha, M., Fives-Taylor, P.M., 1998. Isolation and characterization of Fap1, a fimbriae-associated adhesin of *Streptococcus parasanguis* FW213. Molecular Microbiology 28 (3), 487–500.

Wu, S., Baum, M.M., Kerwin, J., Guerrero, D., Webster, S., Schaudinn, C., et al., 2014. Biofilm-specific extracellular matrix proteins of nontypeable *Haemophilus influenzae*. Pathogens and Disease 72 (3), 143–160.

Wu, W., Jin, Y., Bai, F., Jin, S., 2015. Chapter 41 – *Pseudomonas aeruginosa*. In: Aang, Y.-W., Sussman, M., Liu, D., Poxton, I., Schwartzman, J. (Eds.), Molecular Medical Microbiology, second ed. Academic Press, Boston.

Wu, X.-H., Yu, H.-L., Ba, Z.-Y., Chen, J.-Y., Sun, H.-G., Han, B.-Z., 2010. Sampling methods for NMR-based metabolomics of *Staphylococcus aureus*. Biotechnology Journal 5 (1), 75–84.

Xu, K.D., McFeters, G.A., Stewart, P.S., 2000. Biofilm resistance to antimicrobial agents. Microbiology 146 (3), 547–549.

Xu, Y., Wang, C., Hou, J., Wang, P., You, G., Miao, L., et al., 2017. Effects of cerium oxide nanoparticles on the species and distribution of phosphorus in enhanced phosphorus removal sequencing batch biofilm reactor. Bioresource Technology 227, 393–397.

Yeom, J., Shin, J.-H., Yang, J.-Y., Kim, J., Hwang, G.-S., 2013. ^{1}H NMR-based metabolite profiling of planktonic and biofilm cells in *Acinetobacter baumannii* 1656-2. PloS One 8 (3), e57730.

Yu, G.-H., Tang, Z., Xu, Y.-C., Shen, Q.-R., 2011. Multiple fluorescence labeling and two dimensional FTIR–^{13}C NMR heterospectral correlation spectroscopy to characterize extracellular polymeric substances in biofilms produced during composting. Environmental Science and Technology 45 (21), 9224–9231.

Zampieri, M., Zimmermann, M., Claassen, M., Sauer, U., 2017. Nontargeted metabolomics reveals the multilevel response to antibiotic perturbations. Cell Reports 19 (6), 1214–1228.

Zhang, A., Sun, H., Wang, P., Han, Y., Wang, X., 2012. Recent and potential developments of biofluid analyses in metabolomics. Journal of Proteomics 75 (4), 1079–1088.

Zhang, B., Halouska, S., Schiaffo, C.E., Sadykov, M.R., Somerville, G.A., Powers, R., 2011. NMR analysis of a stress response metabolic signaling network. Journal of Proteome Research 10 (8), 3743–3754.

Zhang, B., Powers, R., 2012. Analysis of bacterial biofilms using NMR-based metabolomics. Future Medicinal Chemistry 4 (10), 1273–1306.

Zhang, Y., Pei, G., Chen, L., Zhang, W., 2016. Metabolic dynamics of *Desulfovibrio vulgaris* biofilm grown on a steel surface. Biofouling 32 (7), 725–736.

Zhang, Z.-C., Yang, H.-J., Xiao, K., Luo, X., Sha, H., Chen, Y.-M., 2009. Study on P forms in extracellular polymeric substances in enhanced biological phosphorus removal sludge by ^{31}P-NMR spectroscopy. Spectroscopy and Spectral Analysis 29, 536–539.

Zhu, Z., Wang, H., Shang, Q., Jiang, Y., Cao, Y., Chai, Y., 2013. Time course analysis of *Candida albicans* metabolites during biofilm development. Journal of Proteome Research 12 (6), 2375–2385.

Zubkov, M., Stait-Gardner, T., Price, W.S., Stilbs, P., 2015. Steady state effects in a two-pulse diffusion-weighted sequence. The Journal of Chemical Physics 142 (15), 154201.

Chapter 3

Design and fabrication of biofilm reactors

D.M. Goeres[1], S. Pedersen[2], B. Warwood[2], D.K. Walker[1], A.E. Parker[1,3], M. Mettler[1], P. Sturman[1]
[1]*Center for Biofilm Engineering, Montana State University, Bozeman, MT, United States;* [2]*BioSurface Technologies Corporation, Bozeman, MT, United States;* [3]*Department of Mathematical Sciences, Montana State University, Bozeman, MT, United States*

3.1 Definition of a biofilm reactor

A reactor is a vessel where controlled change occurs and is measured. A laboratory biofilm reactor is a tool that researchers use to generate a biofilm that exhibits particular qualities. Applied and fundamental biofilm research relies upon laboratory biofilm reactors to simulate environmental or medical conditions of interest. These reactors are used to grow biofilms to study inherent chemical and physiological characteristics and are commonly used to aid in the development of antimicrobial compounds for their control. Reactors can house a variety of surfaces on which to grow biofilm, including hard nonporous surfaces such as glass, metal, or plastics, or porous surfaces such as porcelain, cement, tile, textiles, or cell tissue. Historically, biofilm was associated with aqueous environments, which heavily influenced early biofilm reactor design and consequently much of the early published research (Characklis and Marshall, 1990; Costerton et al., 1978; Zobell, 1943). While biofilm may exist in what would be considered a dry environment (Vickery, 2019), for the purposes of this chapter, we will focus on biofilm reactor design for partially or fully hydrated environments.

A logical approach when designing a reactor is to look at the fluid dynamics present in an environment of interest and to create a system to mimic that environment. In the most general of terms, a system is considered to be open if there is an exchange of fluids and closed if no fluids are exchanged. In the case of aerobic research, a system may be closed with regard to liquid exchange but open with regard to gas exchange. Generally, with anaerobic systems, both liquids and gases are closely controlled. Batch reactors are commonly used closed systems. In a batch reactor, concentrations change over time—cell numbers increase, by-products increase, and nutrient components

Recent Trends in Biofilm Science and Technology. https://doi.org/10.1016/B978-0-12-819497-3.00003-9

decrease. A batch system may have mixing (fluid shear) present. An example of this is a microtiter plate in an environmental shaker or tubing configured to continuously recycle nutrients.

Continuously stirred tank reactors (CSTRs) and plug flow reactors (PFRs) are examples of open systems. A CSTR has flow into and out of the reactor so that there is no net accumulation of fluid. The fundamental assumption of a CSTR is that the bulk fluid in the reactor is perfectly mixed. This means a sample collected from any part of the reactor is representative of the entire system. As a note of caution, this assumption may not hold if the CSTR is growing biofilm and sections of the biofilm have randomly sloughed from the surface. The fluid dynamics in a CSTR are controlled by mixing (usually with a baffle, stir bar, internal rotating drum, etc.). An important feature of a CSTR is that the system will reach steady state with regard to bacterial growth, nutrient utilization, and product production. This is a desirable attribute because time is no longer a critical parameter once steady state is achieved.

In practice, CSTR biofilm reactors actually reach a "quasi" steady state wherein biofilm continually grows and randomly detaches from the growth surfaces, resulting in no net increase in biofilm over longer time periods, but often leading to substantial differences in biofilm at a particular location over short time spans.

The fluid dynamics in a PFR are defined by the flow rate of the fluid through the reactor. Unlike a CSTR, a PFR system is not perfectly mixed in the axial direction, but mixing may occur in the radial direction. The most classic example of a PFR is a pipe. In practical terms, this means that the biofilm changes along the length of the pipe or tubing. A PFR will also reach steady state, meaning that at a specific location in the pipe, the biofilm should be consistent over time. As with the CSTR, we must qualify the term "steady state" because biofilm growth is dynamic, and sections may detach and regrow.

A defining feature of open systems is the ability to control residence time, which is the time it takes for one entire fluid volume in a reactor to be completely exchanged. Mathematically, residence time is calculated by dividing the reactor volume by the fluid flow rate. For a CSTR operated specifically to grow biofilm, the residence time is less than the doubling time of the bacteria of interest. This means that the suspended bacteria wash out of the system and only attached cells (biofilm) remain. The residence time in many plug flow reactors is short, as the fluid quickly flows by a point on the pipe wall. In terms of biofilm, this is not of concern because those were the conditions under which the existing bacteria established, and they will continue to obtain the nutrients they need for growth. This feature becomes important, however, when testing antimicrobials, because a short residence time means a short contact time for the antimicrobial. It is important to note that closed systems do not have this feature, which means suspended bacteria, detached biofilm, and attached biofilm must all be considered when planning an experimental design and interpreting the results.

In the biofilm laboratory, typical reactors include microtiter plates (all configurations, including the MBEC 96-well plate reactor), CDC Biofilm Reactor, Drip Flow Biofilm Reactor, Rotating Disk Biofilm Reactor, Bio-inLine Biofilm Reactor (also known as the Modified Robbin's Device), Biofilm Annular Reactor, Constant Depth Film Fermenter, microfluidic devices, flow cells, agar plate methods, porous media columns, tubing or pipe reactors, and reactors that are modifications of the systems listed before. Multiple reviews describe these reactors and their use in detail (Azeredo et al., 2017; Coenye and Nelis, 2010; Gomes et al., 2014, 2018). ASTM standardized methods have been developed and validated based on the CDC Biofilm Reactor, Drip Flow Biofilm Reactor, Rotating Disk Reactor, and the MBEC Device (also known as the Calgary Device) (ASTM Methods E2647-13, E2196-17, E2562-17, and E2799-17). These standardized methods allow statistical comparison of biofilm-active compounds (antimicrobials) and, in the case of commercial products, marketing claims against biofilm. Both open and closed systems are represented in this list, and most of the reactor systems use submerged or hydrated coupons or surfaces for microbial growth. Although one reactor is not necessarily better than another, they all have associated advantages and limitations. Reactor choice does influence experimental outcomes (Buckingham-Meyer et al., 2007; Manner et al., 2017), which means results should always be put into the context of the system used to generate the biofilm and not extrapolated. The Biofilm Annular Reactor and Modified Robbins Device have been used extensively in drinking water and process water treatment systems to evaluate biofilm formation and corrosion control issues. The CDC, Rotating Disk, and Drip Flow Biofilm Reactors have been used extensively in the evaluation of various materials and chemistries for control of biofilms in a wide range of industries, including medical, dental, home products, industrial chemicals, and environmental, among others. The Constant Depth Film Fermenter has been used in the medical and dental biofilm research community, albeit more sparingly in the United States than in Europe. Microtiter plate methods are used extensively in fundamental biofilm research and for screening large chemical libraries for efficacy against biofilm.

Although a substantial number of biofilm researchers rely on the biofilm reactors listed before, in biofilm laboratories around the world, researchers are creating specialized reactors for specific applications. These specialized reactors rarely become commercial products, however. This chapter will focus on the process used to design and fabricate a biofilm reactor with the stated goal of generating a commercial product.

3.2 Design process

Biofilm is a collection of living microorganisms that are continuously growing and adapting to changes in the environment (Hall-Stoodley and Stoodley, 2005). A well-designed reactor must facilitate the growth, treatment, sampling, and analysis of biofilm that represents the environment that the researcher wants to study. Recreating key field or clinical test conditions in the laboratory enables the generation of relevant results that are predictive and useful for future decision-making. The biofilm reactor and associated protocol must also be affordable, practical, simple to operate, and relevant and provide repeatable, convincing results (Zelver et al., 2001). An example of this is the use of the Rotating Disk Biofilm Reactor for studying toilet bowl biofilms (Zelver et al., 1999).

BioSurface Technologies Corporation (BST) and the Standardized Biofilm Methods Laboratory (SBML) in the Center for Biofilm Engineering have collaborated on the design, fabrication, and/or commercialization of three biofilm reactors (CDC Biofilm Reactor, Drip Flow Biofilm Reactor, and Rotating Disk Biofilm Reactor), a tool known as the "splash guard" that is used when testing the effectiveness of biocides against biofilm following the Single Tube Method (Goeres et al., 2019) and the Treatment Imaging Flow Cell, a reactor used with a microscope to visualize the removal of biofilm from a surface or action of an applied antimicrobial (Ausbacher et al., 2018). The process of designing and bringing a reactor to market is depicted in Fig. 3.1.

The first step in the design and fabrication process of a commercial product is to define the environment, system, or condition that the laboratory reactor will model. This can be accomplished by performing a gap analysis to identify needs that are not currently being met, followed by a market analysis of what is available, and determining how the new system will compare. It seems obvious to note, but the most successful reactor will fill an unmet need. While the literature is packed with creative laboratory reactors that were designed to

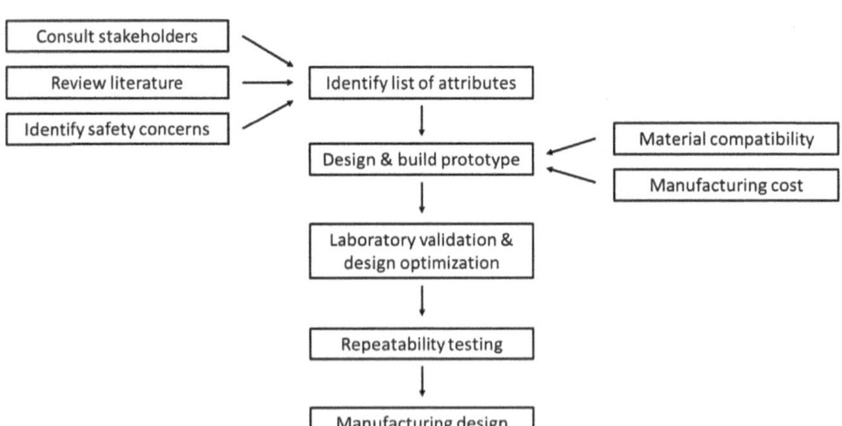

FIGURE 3.1 The step-by-step process BST and the SBML have developed and followed, which has led to successful product development. *BST*, BioSurface Technologies Corporation; *SBML*, Standardized Biofilm Methods Laboratory.

answer a specific research question, and many of these systems provide excellent results, the reactors are often so project specific or complex that any market would be limited.

Once the environment, system, or condition is identified, the next step is to identify the key attributes that the reactor must include to best model the environment of interest. Two basic strategies are employed to develop a relevant biofilm model. For example, in collaborative work with the US EPA, the focus was to measure efficacy of disinfectants against biofilms on hard, nonporous surfaces (Goeres et al., 2019). In this case, the strategy employed was to engineer a biofilm to have specific characteristics that emulate the biofilm in situ, matching, for example, the architecture, thickness, and strength of attachment. This was done by focusing on the fluid dynamics present while the biofilm was being formed. The second strategy incorporates the most important physiochemical and biological characteristics in the environment of interest. When evaluating medical devices, for instance, the second approach is more valid because it considers parameters such as the bodily fluids present and fluid dynamics the device will experience when placed in the body.

Soliciting stakeholder (market) input into the key attributes is an important step. Each category of stakeholder (industrial R&D researcher, academic researcher, regulator) will have requests for design attributes that will enable them to meet their research goals. Industry stakeholders may want to use the reactor to support a product marketing claim or to compare how their product performs against a competitor's product when tested under similar conditions. Regulators may want to reference the reactor in guidance documents and therefore require data that support product performance standards. Academics may want to conduct research that characterizes biofilm under a specific test condition or to measure how the biofilm responds to changes in an environmental condition. If possible, it is best to solicit input from multiple members of a stakeholder group. This can be done at conferences, by reaching people directly, or through a survey. It can become challenging when stakeholders have conflicting recommendations or if the request is simply not feasible due to manufacturing restraints or cost. In this case, the design team makes the decision whether to include the attribute or not, followed by a conversation with the stakeholders to explain the reasoning behind the decision.

Reviewing the literature, including patents, helps to identify reactors and methods other researchers have used that may be similar to the reactor under development. It is important that the new reactor does not duplicate a system already described in the literature. Although if this happens, a great option is to contact the authors and see if they want to collaborate on the commercialization of their system. This may present an intellectual property challenge, but this can generally be resolved. In addition, the literature may also reveal that a parameter that was perceived as critical to include may not actually have a large effect on the response of interest. Finally, when identifying the list of key attributes, it is best to address any potential or perceived safety concerns at the start of the process. Examples include potentially pathogenic bacteria or fungi or, if the system will be used for anaerobic biofilms, assessment of whether special containment is necessary for toxic gases such as hydrogen sulfide.

Up to this point, "attributes" has been used as a general term. Specific attributes to consider at the start of the design process include fluid dynamics, operational temperature, special containment, construction material of the reactor components and construction material of the surface used to support the biofilm growth, material compatibility (for sterilization or interaction with antimicrobials), number of necessary sampling opportunities, access for collecting samples, ability to conduct time course studies (collect one sample without disturbing the others), possibility to replace test surfaces, manufacturing expense, and ease of use.

For instance, when defining the fluid dynamics that the reactor will mimic, it is important to determine if the system will be closed or open. For open systems, the design must consider if the flow will be modeled as a CSTR or as plug flow. A biofilm reactor that is defined as a CSTR must be able to achieve and maintain a well-mixed environment. The reactor design must also consider if the flow is laminar (fluid particles follow smooth paths with little or no mixing) or turbulent (chaotic changes in pressure and flow velocity). The reactor design must also consider the doubling time of the bacteria used to grow the biofilm. Remembering that the residence time in the reactor must be less than the doubling time of the bacteria, a microbe with a fast doubling time in a reactor with a large volume will require a large volume of nutrients. For example, if the doubling time is 1 h and the reactor volume is 1 L, then the flow rate will need to be $\frac{1\mathrm{L}}{1\mathrm{hr}}$ or $\frac{1000\mathrm{mL}}{60\mathrm{min}}$ which equals 16.67 mL/min. This means 24 L of nutrients are necessary to grow the biofilm for 24 h with a continuous flow of nutrients. In addition, it is important to consider pump requirements so that the reactor may be used with commercially available pumps.

Growth surface (also known as coupon or carrier) selection is an important design choice. Stakeholders often want flexibility with regard to coupon material choice, which is based upon how they will assess biofilm growth. For instance, if the intent is to assess biofilm growth via microscopy, glass coupons are a logical growth surface. The reactor must therefore accommodate a variety of surfaces. From a statistical standpoint, it is important that the biofilm growth is repeatable regardless of location within the reactor. This means the fluid dynamics must be consistent across all coupons. Often, stakeholders request a reactor to hold as many coupons as possible and so a compromise must be reached. Time course studies can also be important in biofilm research, and being able to sample one coupon without disturbing the other coupons may also be a desirable attribute, although not always possible.

Once the key attributes have been identified, it is time to design and build a prototype of the reactor that incorporates the key attributes. Material selection is important. Tables 3.1 and 3.2 define the properties of common materials

TABLE 3.1 **Material properties of common metals used in biofilm reactor construction.**

Metal type	Corrosion resistance	Chemical compatibility	Autoclavable	Machinability	Overall cost
Anodized aluminum	High	Good	Yes	Very good	$
316 stainless steel	High	Good	Yes	Good	$$
304 stainless steel	Medium	Good	Yes	Good	$$
Titanium	High	Good	Yes	OK	$$$$$

Table courtesy of BST and reflects their experience with biofilm reactor fabrication.

used to construct biofilm reactors. There is not one perfect material that is consistently used for all biofilm reactors; rather, it is important to understand how the reactor will be used. If the reactor will be used to test the efficacy of strong biocides, then corrosion resistance and chemical compatibility are important to consider. It is often desirable to use off-the-shelf components for some of the pieces, which explains why one reactor may have parts constructed from multiple materials. Materials have different expansion coefficients, which is critical to consider, given that autoclaves will most likely be used to sterilize the reactor. Different expansion coefficients among materials can lead to cracks and breaks in the reactor. Stakeholders want a durable product that does not have to be replaced after only a few uses.

Part fabrication is considered at this stage. Machining parts is expensive and time-consuming. Injection molding has high upfront costs but is less expensive in the long term, and it is limited to certain materials. Laser cutting is inexpensive but limited to 2D parts.

Once a reactor prototype is built, the system is brought into the lab and tested. At this point, researchers provide feedback on how easy the system is to operate and if any of the parts fail to perform their designed function. This is an iterative process. It is not uncommon that several prototypes are designed and built before the final design is achieved. Once the design is deemed close to being finished, a set of experiments are conducted to determine if the biofilm growth is repeatable. Changes may still be made at this point in the process, but they are normally minimal. Once repeatable biofilm growth is achieved, then the reactor design is finalized, and the product is ready for market.

TABLE 3.2 Material properties of common plastics used in biofilm reactor construction.

Plastic type	Additional properties	Chemical compatibility	Autoclavable	Machinability	Overall cost
Polycarbonate	Clear, rigid	Good	Yes	Very good	$$
Polypropylene	Very soft	Good	Yes	OK	$
Polyethylene (HDPE, UHMW)	Soft	Good	Yes	Good	$
Polyethylene terephthalate	Rigid	Good	Yes	Very good	$$
Polysulfone	Very brittle	Very good	Yes	Very good (but easily cracked)	$$$
Polytetrafluoroethylene (Teflon)	Very soft, low friction	Very good	Yes	Poor	$$$
Nylon	Becomes brittle	Very good	Yes	Good	$$
Polystyrene			No		
Acrylic			No		
Polyvinyl chloride			No		

HDPE, high-density polyethylene; *UHMW*, ultrahigh-molecular-weight.
Table courtesy of BST and reflects their experience with biofilm reactor fabrication.

3.3 Implementing the design process: industrial surfaces biofilm reactor

The process described earlier was implemented during the design and fabrication of the industrial surfaces biofilm reactor (ISBR). After a market analysis of the currently available biofilm reactors, a need was identified for a bench-top biofilm reactor designed to provide a growth environment for biofilm under low shear, high gas transfer, and intermittent wetting. The goal was to model the environment present in a cooling tower, meaning the reactor will eventually need to accommodate the growth of biofilms that include *Legionella* spp. and tolerate the testing of strong biocides.

3.3.1 Step 1. Identify the list of key attributes for the biofilm reactor

A brief description of a cooling tower is provided to help explain the design choices made for the ISBR.

3.3.1.1 Description of cooling towers

Cooling towers are used to transfer heat from recirculating cooling water to air. Cooling water is used in virtually all industrial processes that generate waste heat. To accomplish the heat transfer, water flows down over packing material, whereas air travels upward to bring the water to ambient temperatures. Fig. 3.2

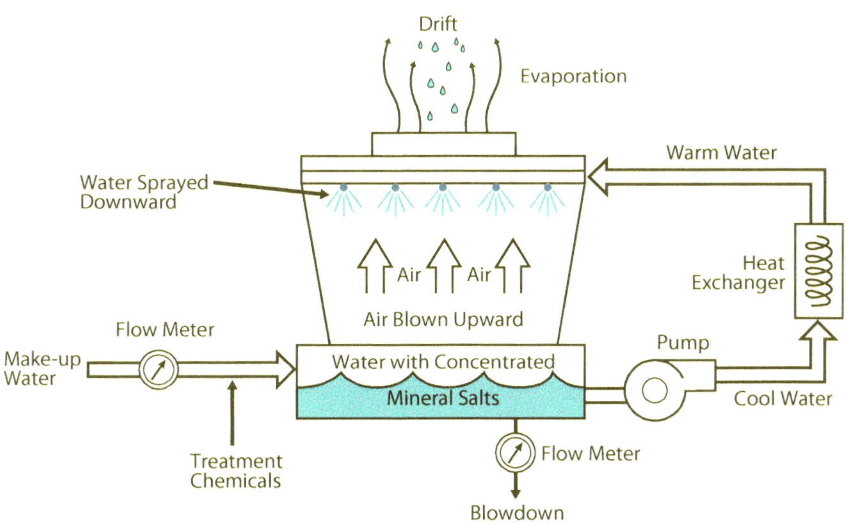

FIGURE 3.2 Schematic of a typical cooling tower design. *Image courtesy of Kolmetz, K., 2014. Cooling Tower Selection, Sizing and Troubleshooting (Engineering Design Guideline). KLM Technology Group.*

is a schematic of a typical cooling tower. Although different designs exist, the basic concept remains consistent. The challenge with cooling towers is that the water is not sterile and soon the packing material is covered in a thick biofilm, as shown in Fig. 3.3. Biofilm decreases the heat transfer efficiency, results in issues with equipment functionality including corrosion and plugging, and, most importantly, can cause potential health and safety concerns. Aerosols are generated as the water evaporates. These aerosols can contain pathogenic bacteria such as *Legionella*, whose route of transmission is through inhalation of aerosols. To control biofilm accumulation, biocides are added to the water. Biocides are costly, however, and associated with their own set of concerns including the aerosolization and release of toxic compounds to the environment and corrosion of equipment. Therefore, there is a need to optimize how and when the biocides are dosed before they are added to a cooling tower. A laboratory system allows for optimization of a biocide dosing strategy in a more cost-effective and timely manner. If a biocide fails in a laboratory test, it is not likely to work in an operating cooling tower.

In addition to modeling the key engineering parameters of a cooling tower, other important design attributes included:

- complete containment to prevent any escape of aerosolized biofilm,
- various surfaces (coupons) for biofilm growth to further model relevant conditions (for example, glass, polymers, stainless steel or mild steel, painted or coated material, etc.),
- a large number of sampling opportunities (coupons),
- operated with varying influent media conditions to mimic the range of industrial/environmental conditions to which surfaces could be exposed,
- in situ and ex situ testing possibilities (e.g., potential for use in either the laboratory or the field),

FIGURE 3.3 New high-efficiency cross-fluted cooling tower packing material (left). Fouled cross-fluted cooling tower packing material that is no longer highly efficient due to biofilm accumulation (right). *Images courtesy of D. McIlwaine, copyright 2013.*

- biofilm from reactor easily analyzed using various laboratory techniques.

Feedback from stakeholders was solicited and used to help refine the design. Comments included the following:

- Would like to see the spray rate of a cooling tower calculated and compared with the rate in the reactor.
- Would like to see a heat exchanger added to the flow path.
- Was interested in understanding why the coupons were placed at a 10 degrees angle (note: this angle helped to stabilize the coupons).
- Would like to include a heating element and monitor humidity in the reactor.
- Would like the reactor to accommodate a submerge-and-drain cycle.
- Would like the reactor to accommodate a wet/dry application for paint testing.
- Would like to see experiments run with bacteria sequenced from cooling towers.
- What are the safety concerns associated with growing *Legionella* biofilms in the reactor?

These comments were considered throughout the design and build process.

3.3.2 Step 2. Using the list of attributes identified in step 1, design a reactor and build a prototype

Three prototypes were fabricated during the design process (Fig. 3.4). Each design was critically evaluated to assess if it met the design criteria. Safety, cost, material compatibility and availability, and manufacturability were all considered during the design and decision-making process. We also considered biosafety concerns and ease of operation in the laboratory, as a complex design and high maintenance is not desirable.

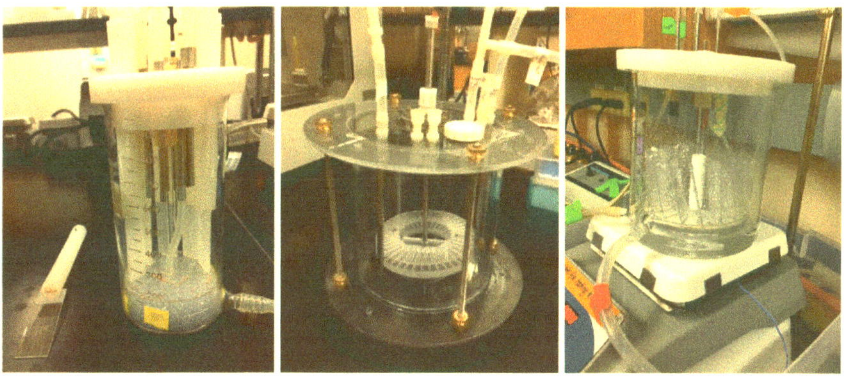

FIGURE 3.4 Early prototypes of the industrial surfaces biofilm reactor.

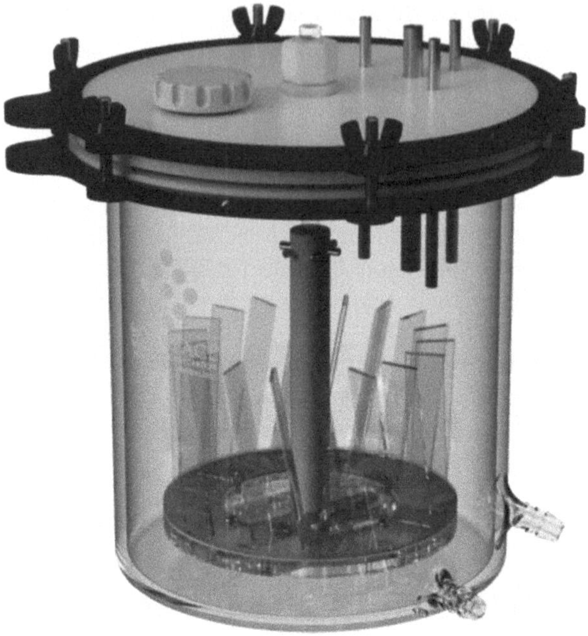

FIGURE 3.5 Final prototype of the industrial surfaces biofilm reactor.

The final prototype of the ISBR is shown in Fig. 3.5. The reactor consists of a glass vessel with two ports near the bottom. One of the ports maintains the water level within the bottom of the reactor, whereas the other is used to pump water from the bottom of the reactor to the top of the reactor. The recycled water is mixed with growth media before it goes back into the reactor through a port in the lid, which can be connected to a spray nozzle if desired. The lid contains three additional ports, one of which is used for air exchange, and the remaining two for adding other chemicals, such as biocides, to the reactor. The top also contains a removable port that is used to pull coupons from the reactor without disturbing the other coupons. On the bottom of the reactor rests a doughnut-shaped disk that is fabricated to hold 18 coupons at a 10 degree angle. The disk is connected to a shaft with a magnet at the base that is driven by a stir plate. The disk rotates while the recycled water is dripped or sprayed over the surface. The system was designed to be completely closed to the environment to prevent both cross-contamination of fresh influent water with the recycled feed and aerosolization out into the lab. In use, it is possible to connect the pump that recycles the water to a timer so that the coupons are wetted at intervals, to simulate a shower or rain, for example. The rotational speed may also be adjusted to allow for more active evaporation of the water off the coupon surfaces, if necessary. The stir plate has a heating mechanism if the water in the bottom of the reactor needs to be warmed above ambient temperature to facilitate evaporation, for instance.

A critical design decision was how to introduce water onto the coupons. It was decided that the final design would have the water drip into the reactor, with an option to attach a spray nozzle. Keeping in mind that bacteria are present as both suspended cells and biofilm in this reactor and that growth medium is added directly to the recycle loop, there is the potential for a spray nozzle to become fouled. Also, the recycle rate may change for different applications, which impacts the force of the spray coming out of the nozzle, which in turn impacts the biofilm growth on the coupons. Fig. 3.6 shows the different options considered for the spray nozzle. In the end, the nozzles that deflected the spray resulted in the best spray pattern, and the protocol developed for the reactor recommends using the nozzle pictured on the far left (not sold with the reactor).

3.3.3 Step 3. Laboratory validation and design optimization

During the laboratory phase of the testing, researchers conducted experiments using bacteria present in drinking water to determine if the reactor was relatively simple to operate in a safe condition, as compared with the other biofilm growth reactors. This testing resulted in numerous reactor design modifications including

1. a new glass vessel design,
2. coupon orientation in the bottom disk,
3. how the magnet was connected to the disk and the placement of the disk on the bottom of the reactor,
4. different spray nozzles versus none, and
5. reconfiguring nutrient addition to the recycle loop.

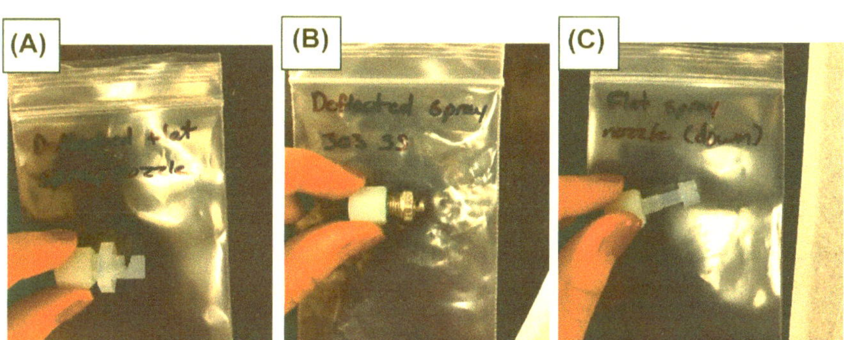

FIGURE 3.6 Four nozzles were considered during the design phase. (A) and (B) Nozzles that deflect the spray. (C) A nozzle that sprayed the water directly onto the coupons. The fourth nozzle considered was a large gauge needle (not pictured).

3.3.4 Step 4. Repeatability testing of the final design

The microbes present in Bozeman, MT, United States, drinking water were used throughout the development and final testing of the reactor prototype. Drinking water microbes are consistent with the type of microbes that may be growing in a cooling tower biofilm. The drinking water used in this testing was water that had run through two filters; the first filter removed any residual chlorine and the second filter enriched for the microbes present in the drinking water.

The laboratory protocol used in the final testing of this reactor started with assembling and sterilizing the reactor and all the parts. The reactor was then filled with the processed drinking water. The recycle water was supplemented with dissolved organic carbon to encourage biofilm growth. The water recycle rate was set to 55 mL/min. Tryptic Soy Broth (TSB) was added to the recycle tubing at a rate of 2 mL/min and concentration of 100 mg/L TSB. The experiment lasted 5 days with coupons being collected and analyzed for biofilm on days 1–5. The experiment was repeated twice. The data for each day were analyzed with a linear model with a random effect for experiment. The data for all days were analyzed with a linear mixed effects model with a random effect for experiment and a fixed effect for day (Pinheiro and Bates, 2000).

The results displayed in Table 3.3 show the repeatability standard deviations between runs for each day and averaged over all days to be well below the accepted repeatability standard deviation upper limit of 0.5 (Parker and Hamilton, 2012). The run-to-run variability decreased over time for the coupon log density (LD) of biofilm, dropping to zero after day 2. The plot in Fig. 3.7 compares the LDs of the three coupons across days and runs Fig. 3.8 shows that the broth LD measurements showed more variability between runs, with the run 2 LDs staying mostly steady across the 4 days and the run 1 LDs increasing and decreasing across time.

The final assessment is that the reactor shows promise that the repeatability standard deviations are within the range reported for other biofilm methods, but the protocol needs optimization that may result in additional small design changes.

3.3.5 Step 5. Incorporate a manufacturing approach to ready the reactor for commercialization

During the entire process, BST was evaluating if the design was sustainable, cost-efficient, and that the parts were either commercially available or if they could be manufactured locally. Important points to highlight include the following:

- Given that this reactor will be used in multiple experiments that may last several days, it was important for the reactor pieces to be interchangeable and durable, even after exposure to concentrated biocides and multiple sterilization cycles.

TABLE 3.3 Mean log density, coupon-to-coupon and run-to-run variability (with percent variance for each component), and repeatability standard deviation for log density measurements taken from coupons.

	Mean log density* (Log_{10}(CFU/ cm^2))	Coupon-to-coupon variability	Run-to-run variability	Repeatability standard deviation
Coupons—day 1	6.75	0.0022 (3.3%)	0.065 (96.7%)	0.260
Coupons—day 2	7.13	0.0064 (33.2%)	0.013 (66.8%)	0.139
Coupons—day 3	7.46	0.1292 (100%)	0 (0%)	0.359
Coupons—day 4	7.39	0.0564 (100%)	0 (0%)	0.237
Coupons—day 5	7.82	0.0076	NC	NC
Coupons—over all days	7.31	0.055 (100%)	0 (0%)	0.236
Broth—over all days	7.77 (Log_{10}(CFU/ mL)	0.0759 (30.5%)	0.173 (69.5%)	0.499

*Variability of the coupon log density (biofilm assessment) data is shown for each day separately and averaged across all 5 days over both runs. Only one run was performed on day 5, so no run-to-run variability or repeatability standard deviation could be calculated. The broth log density measurements only contained one value for each day of each run, so variability and repeatability could not be calculated for each individual day, only across all the days as seen in the last row of the table. *NC*, not calculable.

- Easy assembly was favored, and an important consideration when designing the mechanism to rotate the bottom disk, which is also why we switched to a single glass vessel in lieu of the reactor version shown in the center image in Fig. 3.4.
- Safety was considered during the entire design process.

3.4 Conclusions

Laboratory biofilm reactors are the tools that researchers use to generate biofilm with specific attributes. While no biofilm reactor is necessarily better than another, each has advantages and limitations; one reactor may better mimic the conditions present in the environment of interest than another.

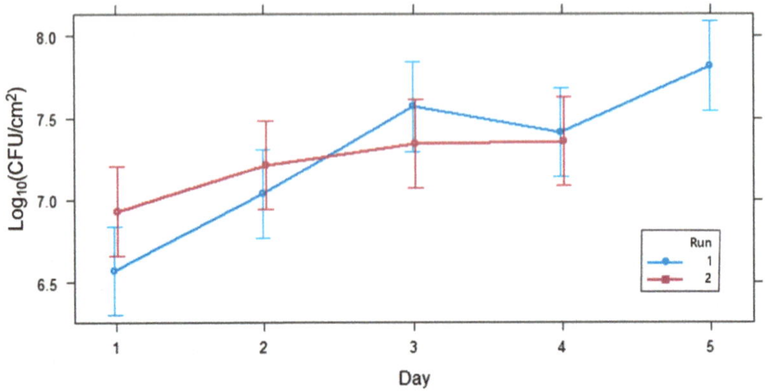

FIGURE 3.7 Log densities ($Log_{10}(CFU/cm^2)$) from coupons over time, separated by run. Each point for each day is a mean log density of three selected coupons from the reactor. Error bars for each point represent a 95% confidence interval for the average log density of that individual day and run.

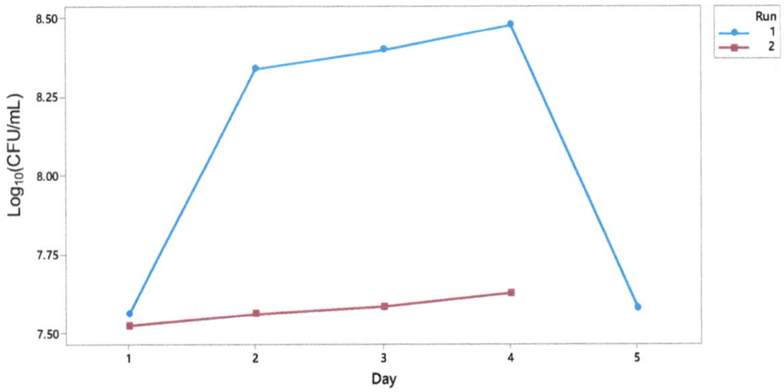

FIGURE 3.8 Log densities ($Log_{10}(CFU/mL)$) from broth samples over time, separated by run. Each point represents a single broth log density measurement for each individual day and run. Error bars could not be created for these points since they represent single values.

Commercially available biofilm reactors are developed following a traditional design and build process. This process was followed to design the new ISBR that was developed to model cooling tower biofilms but is suitable to the study of any biofilm grown under the conditions of low shear, high gas transfer, and intermittent wetting.

Acknowledgments

The design and fabrication of the industrial surfaces biofilm reactors was supported by a Montana Board of Research and Commercialization Technology grant from the State of Montana.

References

ASTM, 2013. Standard Test Method for Quantification of *Pseudomonas aeruginosa* Biofilm Grown Using a Drip Flow Biofilm Reactor with Low Shear and Continuous Flow. ASTM Standard E2647-13.

ASTM, 2017a. Standard Test Method for Quantification of *Pseudomonas aeruginosa* Biofilm Grown with Medium Shear and Continuous Flow Using Rotating Disk Reactor. ASTM Standard E2196-17.

ASTM, 2017b. Standard Test Method for Quantification of *Pseudomonas aeruginosa* Biofilm Grown with High Shear and Continuous Flow Using CDC Biofilm Reactor. ASTM Standard E2562-17.

ASTM, 2017c. Standard Test Method for Testing Disinfection Efficacy against *Pseudomonas aeruginosa* Biofilm Using MBEC Assay. ASTM Standard E2799-17.

Ausbacher, D., Lorenz, L., Pitts, B., Stewart, P.S., Goeres, D.M., 2018. Paired methods to measure biofilm killing and removal: a case study with Penicillin G treatment of *Staphylococcus aureus* biofilm. Letters in Applied Microbiology 66 (3), 231–237.

Azeredo, J., Azevedo, N.F., Briandet, R., Cerca, N., Coenye, T., Costa, A.R., Desvaux, M., Di Bonaventura, G., Hébraud, M., Jaglic, Z., Kačániová, M., 2017. Critical review on biofilm methods. Critical Reviews in Microbiology 43 (3), 313–351.

Buckingham-Meyer, K., Goeres, D.M., Hamilton, M.A., 2007. Comparative evaluation of biofilm disinfectant efficacy tests. Journal of Microbiological Methods 70 (2), 236–244.

Characklis, W.G., Marshall, K. (Eds.), 1990. Biofilms, vol. 5. Wiley-Interscience.

Coenye, T., Nelis, H.J., 2010. In vitro and in vivo model systems to study microbial biofilm formation. Journal of Microbiological Methods 83 (2), 89–105.

Costerton, J.W., Geesey, G.G., Cheng, K.J., 1978. How bacteria stick. Scientific American 238 (1), 86–95.

Goeres, D.M., Walker, D.K., Buckingham-Meyer, K., Lorenz, L., Summers, J., Fritz, B., Goveia, D., Dickerman, G., Schultz, J., Parker, A.E., 2019. Development, standardization, and validation of a biofilm efficacy test: the single tube method. Journal of Microbiological Methods 165, 105694.

Gomes, I.B., Meireles, A., Gonçalves, A.L., Goeres, D.M., Sjollema, J., Simões, L.C., Simões, M., 2018. Standardized reactors for the study of medical biofilms: a review of the principles and latest modifications. Critical Reviews in Biotechnology 38 (5), 657–670.

Gomes, I.B., Simões, M., Simões, L.C., 2014. An overview on the reactors to study drinking water biofilms. Water Research 62, 63–87.

Hall-Stoodley, L., Stoodley, P., 2005. Biofilm formation and dispersal and the transmission of human pathogens. Trends in Microbiology 13 (1), 7–10.

Kolmetz, K., 2014. Cooling Tower Selection, Sizing and Troubleshooting (Engineering Design Guideline). KLM Technology Group.

Manner, S., Goeres, D.M., Skogman, M., Vuorela, P., Fallarero, A., 2017. Prevention of *Staphylococcus aureus* biofilm formation by antibiotics in 96-microtiter well plates and drip flow reactors: critical factors influencing outcomes. Scientific Reports 7, 43854.

Parker, A.E., Hamilton, M.A., 2012. KSA-SM-10- assessing resemblance, repeatability, and reproducibility for quantitative methods. knowledge sharing articles. Revised January 23. Center for Biofilm Engineering at Montana State University, Bozeman, MT [E-publication].

Pinheiro, J., Bates, D., 2000. Mixed-effects Models in S and S-PLUS. Springer -Verlag, New York.

Vickery, K., 2019. Microbial biofilms in healthcare: formation, prevention and treatment. Materials 12. https://doi.org/10.3390/ma12122001.

Zelver, N., Hamilton, M., Pitts, B., Goeres, D., Walker, D., Sturman, P., Heersink, J., 1999. Measuring antimicrobial effects on biofilm bacteria: from laboratory to field. In: Methods in Enzymology, vol. 310. Academic Press, pp. 608–628.

Zelver, N., Hamilton, M., Goeres, D., Heersink, J., 2001. [24] Development of a standardized antibiofilm test. In: Methods in Enzymology, vol. 337. Academic Press, pp. 363–376.

Zobell, C.E., 1943. The effect of solid surfaces upon bacterial activity. Journal of Bacteriology 46 (1), 39.

Chapter 4

Oral biofilms

Marta Ribeiro, Manuel Simões
LEPABE – Laboratory for Process Engineering Environment, Biotechnology and Energy, Department of Chemical Engineering, Faculty of Engineering, University of Porto, Porto, Portugal

4.1 Defining the problematic: an introduction

The human oral cavity provides a multifaceted environment with numerous different surfaces, a constant salivary flow and a great variety of bacteria. Over the past decades, it has become increasingly clear that some oral bacteria represent a considerable clinical challenge, as they may be responsible for a range of diseases beyond the human oral cavity (Marsh, 2010). The development of dental plaque–mediated disease at a site may be regarded as a breakdown of the homeostatic mechanisms that in a normal way maintain a beneficial relationship between the resident oral microbiota and the host (Marsh, 2006; Colombo et al., 2015).

Dental plaque is a structurally and functionally organized multispecies microbial biofilm having a crucial role in the etiology of oral diseases with extensive implications for quality of life, systemic health, and economic costs (Marsh, 2010; Flemmig and Beikler, 2011). Oral biofilms are three-dimensional structured communities of multiple microbial species in close proximity with one another attached to a solid surface, like the tooth, root or dental implants surface, and embedded in a self-produced matrix of extracellular polymeric substances (EPS) (Zijnge et al., 2010; Maddi and Scannapieco, 2013). EPS are known to affect the features of the microbial communities, including surface properties, mass transfer, adsorption ability, and stability, being also used as a source of essential nutrients (Colombo et al., 2015). Biofilms protect the resident microbial community from potentially damaging factors, allowing the bacteria to develop resistance to antibiotic treatment, host immune defense system, and mechanical removal (Leid, 2009; Colombo et al., 2015; Rafii and Hart, 2015). In addition, the bacteria in biofilms may produce virulence factors that allow their invasion in the host tissues, being then responsible for the modulation of the host immune response as well as tissue destruction mediation (Colombo et al., 2015).

Recent Trends in Biofilm Science and Technology. https://doi.org/10.1016/B978-0-12-819497-3.00004-0

This chapter highlights the importance of understanding the oral cavity and its microbiota as well as the formation of dental plaque and its implications. The major mechanisms of oral bacterial interactions are emphasized. We build up previous research to outline the oral infectious diseases–associated oral biofilms, with particular reference to dental caries, periodontitis, and periimplant diseases as well as nonoral infections associated with oral bacteria.

4.2 The oral cavity and its microbiota

The oral cavity is estimated to host over 700 different bacterial species. This commensal microbiota in the mouth has a crucial role in maintaining oral and systemic health. The presence of oral microbiota inhibits the colonization by pathogens since all surfaces of the mouth are colonized by commensals, providing few binding sites available for pathogens (He and Shi, 2009; Zarco et al., 2012; Wade, 2013). Moreover, some health-associated bacteria can antagonize the growth of potentially harmful oral pathogens, producing bacteriocins and/or hydrogen peroxide (H_2O_2)—protecting the host against potential pathogen attack (Liu et al., 2018). For instance, in a work developed by Tong et al. (2007), it was observed that the oral bacterium *Streptococcus oligofermentans*, isolated from dental plaque of caries-free individuals, converted the lactic acid produced by *Streptococcus mutans* into hydrogen peroxide, inhibiting the growth of *S. mutans* (Tong et al., 2007). Furthermore, saliva is also crucial to maintain homeostasis and defend the oral cavity from disease. Saliva detaches layers of plaque and comprises several minerals, proteins, and antimicrobial enzymes that prevent biofilm build-up and activity. Moreover, saliva has nutrients that protect tooth enamel as well as antibodies that defend the oral cavity from infection (Filoche et al., 2010; Zarco et al., 2012).

4.3 Dental plaque

The dental plaque is polymicrobial in nature, and its formation has been described as a three-step process: pellicle formation, bacteria colonization, and biofilm maturation. In the early stage of development, a proteinaceous thin layer composed mainly of salivary proteins and bacterial enzyme, known as salivary pellicle, forms on the tooth surface, epithelium, and restorations. The bacterial adhesion to this salivary pellicle represents one of the most important factors in the colonization of oral surfaces (Jakubovics and Kolenbrander, 2010; Moreau et al., 2011). Some bacteria, known as early or primary colonizers, are capable to colonize surfaces coated by the salivary pellicle (Nabert-Georgi et al., 2018), providing binding sites for other bacteria that cannot directly bind to the pellicle (Fig. 4.1). During the initial colonization, oral streptococci and actinomyces bind to host-derived proteins in the salivary pellicle. Nevertheless, streptococci have been shown to be the main primary colonizers of tooth surfaces, constituting 60%–80% of dental plaque bacteria

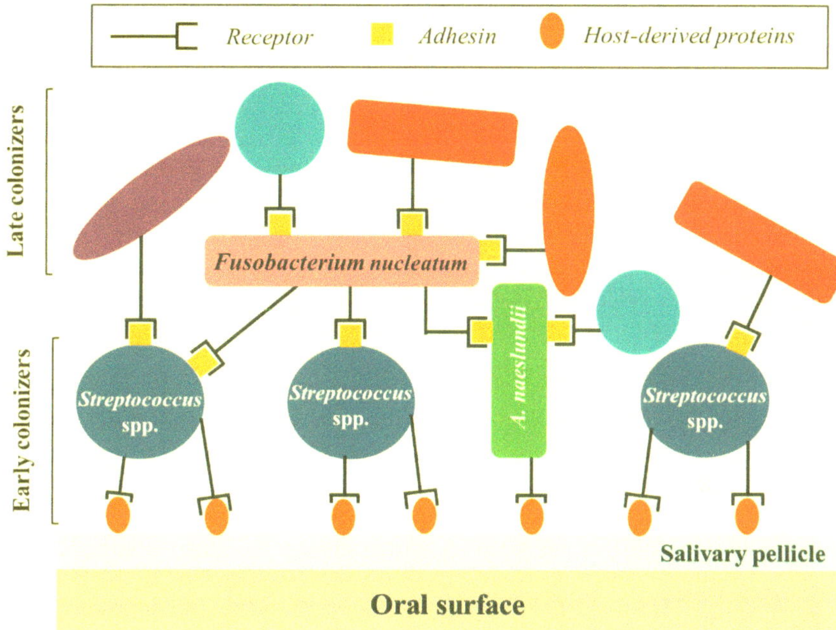

FIGURE 4.1 Bacterial colonization starts with the adhesion of early colonizers to host-derived proteins in the salivary pellicle, such as streptococci and some actinomyces bacteria. Then, the early adherent colonizers recruit other oral bacteria to produce a biofilm on the tooth surface. *Based on Jakubovics, N.S., Kolenbrander, P.E., 2010. The road to ruin: the formation of disease-associated oral biofilms. Oral Diseases 16, 729–739; Periasamy, S, Kolenbrander, P.E., 2010. Central role of the early colonizer Veillonella sp. in establishing multispecies biofilm communities with initial, middle, and late colonizers of enamel. Journal of Bacteriology 192, 2965–2972.*

within 4–8 h (Jakubovics and Kolenbrander, 2010; Periasamy and Kolenbrander, 2010). Then, the early adherent streptococci on oral surfaces recruit other bacteria to form a biofilm through an autoaggregation or coaggregation process, which corresponds to the attraction between same species or between different species, respectively (Jakubovics and Kolenbrander, 2010; Periasamy and Kolenbrander, 2010). The coaggregation phenomenon only happens between compatible partner microorganisms, being dependent on the presence of cell–surface adhesins on one cell type and cognate receptors on the other. Nevertheless, if one of the microorganisms is already adhered on the surface, the phenomenon is known as coadhesion (Jakubovics, 2015; Nabert-Georgi et al., 2018). Some microorganisms have coaggregation interactions with several different species, displaying a key central role in binding the biofilm together. This function has been commonly associated to the microorganism *Fusobacterium nucleatum*, as shown in Fig. 4.1. This bacterium has the ability to coaggregate strongly with many other different oral bacteria, being found in the middle layers of dental plaque where it can provide a bridge between the early and late colonizers (Benitez-Paez et al., 2014; Coppenhagen-Glazer et al., 2015; Jakubovics, 2015). Moreover, the coaggregation between *F. nucleatum* and gram-negative bacteria has been shown to be significantly distinct from *F. nucleatum* binding to gram-positive bacteria. For instance, the coaggregation between *F. nucleatum* and *Porphyromonas gingivalis* is regulated by an *F. nucleatum* lectin and a galactoside *P. gingivalis* moiety and inhibited by lactose, galactose, and associated monosaccharides (Rosen and Sela, 2006; Saito et al., 2008). On the contrary, the coaggregation of *F. nucleatum* and gram-positive bacteria is rarely inhibited by sugars (Kang et al., 2005; Nagaoka et al., 2008).

4.4 Disease-associated oral biofilms

Oral health is associated with an equilibrium between the host and its commensal microbiota. Shifts of microbial composition can lead to an imbalance that is responsible for the development of numerous oral pathologies (Frederic et al., 2018). Microbial biofilms in the oral cavity have been found to be involved in a wide variety of microbial infections including dental caries, periodontal and endodontic diseases, denture stomatitis, pulpal diseases, oral candidiasis, and periimplant diseases (Chandki et al., 2011; Soukos and Goodson, 2000; Nabert-Georgi et al., 2018). Nevertheless, the most common and widespread oral diseases are caries and periodontitis, as a consequence of an ecological imbalance of the resident microbiota—being responsible for more than 75% of all tooth extractions (Flemmig and Beikler, 2011; Gupta, 2011; Zarco et al., 2012). Moreover, nonoral infections such as brain or lung abscesses, endocarditis, cardiovascular disease, and septicemias have also been associated to oral bacteria that can access the bloodstream thorough untreated oral diseases (Parahitiyawa et al., 2009; Teles and Wang, 2011; Frederic et al., 2018). Therefore, understanding the driving forces behind the shifts in the ecological balance of the oral microbiota will become crucial for the future management and treatment of diseases associated with oral microorganisms. A summary of the microorganisms implicated in different oral infections is presented in Table 4.1.

4.4.1 Dental caries

Dental caries is an infectious communicable disease resulting in demineralization and destruction of tooth structure by acidic attack from cariogenic bacteria, such as *S. mutans* and *Lactobacillus* spp. However, other species with relevant traits can participate in demineralization as presented in Table 4.1 (Gupta, 2011; Melo et al., 2013; Singh et al., 2015). The worldwide prevalence of dental caries is between 60% and 90% for school children and affects the vast majority of the adult population (Saha et al., 2014; Chopra et al., 2015). The direct cause of this disease is related to the cariogenic plaque, which

TABLE 4.1 Summary of oral bacteria implicated in different oral infections.

Infection	Associated oral bacteria	References
Dental caries	*Streptococcus mutans* *Streptococcus sobrinus* *Lactobacillus* spp. *Actinomyces* spp. *Propionibacterium* spp. *Veillonella* spp.	Gupta (2011), Pepperney and Chikindas (2011), Melo et al. (2013)
Periodontitis	*Porphyromonas gingivalis* *Aggregatibacter actinomycetemcomitans* *Tannerella forsythia* *Treponema denticola* *Prevotella intermedia* *Fusobacterium nucleatum*	Cortelli et al. (2008), Stingu et al. (2008), Saha et al. (2014)
Peri-implantitis	*P. gingivalis* *A. actinomycetemcomitans* *T. denticola* *P. intermedia* *Prevotella nigrescens* *F. nucleatum* *T. forsythia*	Mouhyi et al. (2012), Vargas-Reus et al. (2012)

results when low populations of aciduric and acidogenic bacteria increase following high-frequency carbohydrate exposure. The metabolism of carbohydrate by these microorganisms has as consequence of causing acidification of plaque and acid-induced demineralization and destruction of the tooth hard tissues (Filoche et al., 2010).

S. mutans has been identified as a crucial modulator in dental caries since it is a major matrix producer, mainly EPS, which is recognized as an essential virulence mechanism associated with this oral disease. However, other constituents such as extracellular DNA and lipoteichoic acids have been also found in the matrix of cariogenic biofilms. The pathogens embedded in this EPS-rich matrix are then protected against physiological defense mechanisms of the mouth (Klein et al., 2015; Gao et al., 2016).

4.4.2 Periodontitis

Contrary to dental caries, which occurs on teeth above the gum line (supragingivally), periodontal disease occurs below (subgingivally) (Filoche et al., 2010). In periodontal diseases, caused predominantly by strictly anerobic gram-negative bacterial species, there is a cellular inflammatory response of the gingiva and surrounding tissues to the bacteria accumulation on the teeth, which can be grouped into gingivitis, characteristic of early-stage of periodontal disease, and periodontitis (Filoche et al., 2010; da Silva-Boghossian et al., 2011). It has been estimated that 50%–100% of adults are affected by gingivitis and 15%–20% of the middle-aged adult population are affected by severe periodontal diseases (Saha et al., 2014). While gingivitis is simply reversible with good oral hygiene, periodontitis is an irreversible and severe infection attacking soft tissues and bone that support the periodontium and teeth structures (Zarco et al., 2012). In chronic periodontitis, a periodontal destruction happens over a prolonged period of time, and it has been related with a prevalence of *P. gingivalis*, *Treponema denticola*, and *Tannerella forsythia*, known as the red complex bacteria. In contrast, in aggressive periodontitis, the periodontal destruction and bone loss occurs faster, and it has been linked with the bacterium *Aggregatibacter actinomycetemcomitans* (Maddi and Scannapieco, 2013). Periodontitis pathology is also dependent on the host inflammatory response. Pathogen recognition receptors (PRRs) from the host immune system are known to detect important molecular patterns on bacteria called pathogen-associated molecular patterns (PAMPs). In the presence of bacteria, the PRRs induce the activation and recruitment of downstream signaling pathways in inflammatory cells, which in turn lead to the production of inflammatory cytokines, aggravating the local and systemic inflammation (Maddi and Scannapieco, 2013).

4.4.3 Periimplantitis

Periimplant disease is characterized by host tissue inflammatory reactions that appear around dental implants. Periimplant mucositis and periimplantitis are the two main types of this disease. Periimplant mucositis is characterized by the appearance of a reversible inflammation reaction in soft tissues surrounding an implant. On the other hand, periimplantitis is defined as a destructive inflammatory process involving the soft and hard tissues surrounding an implant, leading to the loss of supporting bone (Norowski and Bumgardner, 2009; Mouhyi et al., 2012).

An interesting remark is that the pathogenic microorganisms involved in the colonization of dental implants (periimplantitis) are essentially the same of natural dentition (periodontitis) (Norowski and Bumgardner, 2009). Nevertheless, once a periimplant lesion occurs, opportunistic pathogens not directly involved in this disease are commonly found at these sites (i.e., *Staphylococcus* and *Candida* spp.) (Vargas-Reus et al., 2012).

4.5 Non-oral infections associated with oral bacteria

The oral cavity may also be an essential source of bacteria responsible to cause nonoral infections—which can access the bloodstream thorough untreated oral diseases and infect several organ systems, including brain, lungs, pancreas, heart valves, liver, and bone (Scannapieco, 2013). Therefore, oral bacteria have been associated to various systemic diseases, such as pneumonia (Scannapieco, 2013), cardiovascular disease (Meurman et al., 2004), and a number of cancers (Meyer et al., 2008; Fitzpatrick and Katz, 2010; Galvão-Moreira and da Cruz, 2016).

Dental plaque can influence the initiation and progression of pneumonia due to the relocalization of bacteria from the tooth-borne biofilm into the respiratory tract. The bacteria responsible for community-acquired pneumonia are species that frequently colonize the oropharynx, such as *Streptococcus pneumoniae*, *Haemophilus influenzae*, and *Mycoplasma pneumoniae* (Paju and Scannapieco, 2007). On the contrary, nosocomial pneumonia is normally instigated by bacteria that are not common members of the oropharyngeal microbiota, including *Staphylococcus aureus*, *Pseudomonas aeruginosa*, and enteric gram-negative bacteria (Scannapieco, 2013).

The association between periodontal and cardiovascular diseases has already been described (Mustapha et al., 2007; Blaizot et al., 2009). Three different mechanisms have been proposed to elucidate the relationship between periodontal and cardiovascular diseases. These mechanisms include the bacteria responsible for periodontal infection, the inflammatory mediators produced in response to periodontal infection, and the bacterial products known as lipopolysaccharides, which access the bloodstream and reach the heart and blood vessels (Masthan et al., 2016). Periodontal bacteria, including *P. gingivalis*, *A. actinomycetemcomitans*, *T. denticola*, and numerous others, have been shown to be able to invade the human cardiovascular system (Maddi and Scannapieco, 2013).

The association between oral health and cancer has also been described in several studies (Meyer et al., 2008; Fitzpatrick and Katz, 2010; Galvão-Moreira and da Cruz, 2016). The infection and inflammation have been recognized to play a crucial role in carcinogenesis, and highly prevalent oral diseases have been significantly associated to some types of cancer (Mantovani et al., 2008; Meurman, 2010; Meurman and Bascones-Martinez, 2011). Oral bacteria upregulate cytokines and other inflammatory mediators that affect the complex metabolic pathways and may consequently be involved in carcinogenesis (Meurman, 2010). A diversity of periodontal bacteria, including *P. gingivalis*, *Prevotella intermedia*, and *F. nucleatum*, are associated to oral squamous cell carcinoma (Mager et al., 2005; Gholizadeh et al., 2016). The bacteria *C. gingivalis*, *Prevotella melaninogenica*, and *Streptococcus mitis* could also be related to this type of oral carcinoma (Mager et al., 2005). Chronic periodontal disease could also be associated to head-and-neck squamous cell carcinoma, which could be correlated to direct toxic effects of bacteria and their products or to indirect effects of inflammation (Han et al., 2014). A correlation between oral microbiota and subsequent cancer risk has also been observed for pancreatic cancer (Fan et al., 2018), esophageal cancer (Peters et al., 2017), lung cancer (Maddi et al., 2019), gastrointestinal cancer (Zhang et al., 2019), and colorectal cancer (Yang et al., 2019).

4.6 Conclusions

The oral commensal microbiota in the mouth has an essential role in maintaining oral and systemic health. However, any shift in the oral microbiota composition is associated with the development of several oral infectious diseases, including dental caries, periodontitis, and periimplant disease. The oral microbiota, with hundreds of distinct microbial species typically residing in biofilms, also pose a warning scenario if the microorganisms are able to access blood circulation or spread locally to adjacent tissues. In this sense, oral biofilm-related diseases have wide implications on systemic health and oral health-associated quality of life and represent a substantial economic burden. Therefore, these threatening microbial profiles should be taken in consideration to develop novel strategies to control oral infections in the preclinical stages.

Acknowledgments

This work was financially supported by project UID/EQU/00511/2019—Laboratory for Process Engineering, Environment, Biotechnology and Energy (LEPABE) funded by national funds through FCT/MCTES (PIDDAC); project POCI-01-0145-FEDER-030219; POCI-01-145-FEDER-006939, funded by FEDER funds through COMPETE2020—Programa Operacional Competitividade e Internacionalização (POCI) and by national funds (PIDDAC) through FCT/MCTES. This work was also the result of the grant awarded to Manuel Simões (SFRH/BSAB/150379/2019) from national funds through FCT—Fundação para a Ciência e Tecnologia.

References

Benitez-Paez, A., Belda-Ferre, P., Simon-Soro, A., Mira, A., 2014. Microbiota diversity and gene expression dynamics in human oral biofilms. BMC Genomics 15, 311.

Blaizot, A., Vergnes, J.-N., Nuwwareh, S., Amar, J., Sixou, M., 2009. Periodontal diseases and cardiovascular events: meta-analysis of observational studies. International Dental Journal 59, 197—209.

Chandki, R., Banthia, P., Banthia, R., 2011. Biofilms: a microbial home. Journal of Indian Society of Periodontology 15, 111—114.

Chopra, A., Rao, N., Gupta, N., Vashisth, S., Lakhanpal, M., 2015. The predisposing factors between dental caries and deviations from normal weight. North American Journal of Medical Sciences 7, 151—159.

Colombo, A.P.V., do Souto, R.M., da Silva-Boghossian, C.M., Miranda, R., Lourenço, T.G.B., 2015. Microbiology of oral biofilm-dependent diseases: have we made significant progress to understand and treat these diseases? Current Oral Health Reports 2, 37—47.

Coppenhagen-Glazer, S., Sol, A., Abed, J., Naor, R., Zhang, X., Han, Y.W., Bachrach, G., 2015. Fap2 of *Fusobacterium nucleatum* is a galactose-inhibitable adhesin involved in coaggregation, cell adhesion, and preterm birth. Infection and Immunity 83, 1104—1113.

Cortelli, J.R., Aquino, D.R., Cortelli, S.C., Fernandes, C.B., de Carvalho-Filho, J., Franco, G.C.N., Costa, F.O., Kawai, T., 2008. Etiological analysis of initial colonization of periodontal pathogens in oral cavity. Journal of Clinical Microbiology 46, 1322—1329.

da Silva-Boghossian, C.M., do Souto, R.M., Luiz, R.R., Colombo, A.P., 2011. Association of red complex, *A. actinomycetemcomitans* and non-oral bacteria with periodontal diseases. Archives of Oral Biology 56, 899–906.

Fan, X., Alekseyenko, A.V., Wu, J., Peters, B.A., Jacobs, A.J., Gapstur, S.M., Purdue, M.P., Abnet, C.C., Stolzenberg-Solomon, R., Miller, G., Ravel, J., Hayes, R.B., Ahn, J., 2018. Human oral microbiome and prospective risk for pancreatic cancer: a population-based nested case-control study. Gut 67, 120–127.

Filoche, S., Wong, L., Sissons, C.H., 2010. Oral biofilms: emerging concepts in microbial ecology. Journal of Dental Research 89, 8–18.

Fitzpatrick, S.G., Katz, J., 2010. The association between periodontal disease and cancer: a review of the literature. Journal of Dentistry 38, 83–95.

Flemmig, T.F., Beikler, T., 2011. Control of oral biofilms. Periodontology 2000 55, 9–15.

Frederic, L.J., Michel, B., Selena, T., 2018. Oral microbes, biofilms and their role in periodontal and peri-implant diseases. Materials 11, 1802.

Galvão-Moreira, L.V., da Cruz, M.C.F.N., 2016. Oral microbiome, periodontitis and risk of head and neck cancer. Oral Oncology 53, 17–19.

Gao, L., Liu, Y., Kim, D., Li, Y., Hwang, G., Naha, P.C., Cormode, D.P., Koo, H., 2016. Nanocatalysts promote *Streptococcus mutans* biofilm matrix degradation and enhance bacterial killing to suppress dental caries in vivo. Biomaterials 101, 272–284.

Gholizadeh, P., Eslami, H., Yousefi, M., Asgharzadeh, M., Aghazadeh, M., Kafil, H.S., 2016. Role of oral microbiome on oral cancers, a review. Biomedicine and Pharmacotherapy 84, 552–558.

Gupta, G., 2011. Probiotics and periodontal health. Journal of medicine 4, 387–394.

Han, Y.W., Houcken, W., Loos, B.G., Schenkein, H.A., Tezal, M., 2014. Periodontal disease, atherosclerosis, adverse pregnancy outcomes, and head-and-neck cancer. Advances in Dental Research 26, 47–55.

He, X.-S., Shi, W.-Y., 2009. Oral microbiology: past, present and future. International Journal of Oral Science 1, 47–58.

Jakubovics, N.S., 2015. Intermicrobial interactions as a driver for community composition and stratification of oral biofilms. Journal of Molecular Biology 427, 3662–3675.

Jakubovics, N.S., Kolenbrander, P.E., 2010. The road to ruin: the formation of disease-associated oral biofilms. Oral Diseases 16, 729–739.

Kang, M.S., Na, H.S., Oh, J.S., 2005. Coaggregation ability of *Weissella cibaria* isolates with *Fusobacterium nucleatum* and their adhesiveness to epithelial cells. FEMS Microbiology Letters 253, 323–329.

Klein, M.I., Hwang, G., Santos, P.H.S., Campanella, O.H., Koo, H., 2015. *Streptococcus mutans*-derived extracellular matrix in cariogenic oral biofilms. Frontiers in Cellular and Infection Microbiology 5, 10.

Leid, J.G., 2009. Bacterial biofilms resist key host defenses. Microbe Magazine 4, 66–70.

Liu, Y., Palmer, S.R., Chang, H., Combs, A.N., Burne, R.A., Koo, H., 2018. Differential oxidative stress tolerance of *Streptococcus mutans* isolates affects competition in an ecological mixed-species biofilm model. Environmental Microbiology Reports 10, 12–22.

Maddi, A., Scannapieco, F.A., 2013. Oral biofilms, oral and periodontal infections, and systemic disease. American Journal of Dentistry 26, 249–254.

Maddi, A., Sabharwal, A., Violante, T., Manuballa, S., Genco, R., Patnaik, S., Yendamuri, S., 2019. The microbiome and lung cancer. Journal of Thoracic Disease 11, 280–291.

Mager, D.L., Haffajee, A.D., Devlin, P.M., Norris, C.M., Posner, M.R., Goodson, J.M., 2005. The salivary microbiota as a diagnostic indicator of oral cancer: a descriptive, non-randomized study of cancer-free and oral squamous cell carcinoma subjects. Journal of Translational Medicine 3, 27.

Mantovani, A., Allavena, P., Sica, A., Balkwill, F., 2008. Cancer-related inflammation. Nature 454, 436–444.

Marsh, P.D., 2010. Controlling the oral biofilm with antimicrobials. Journal of Dentistry 38 (Suppl. 1), S11–S15.

Marsh, P.D., 2006. Dental plaque as a biofilm and a microbial community – implications for health and disease. BMC Oral Health 6, S14.

Masthan, M.K., Anitha, N., Jacobina, J.J., Babu, N.A., 2016. Oral infections causing systemic diseases. Biomedical and Pharmacology Journal 9, 863–866.

Melo, M.A.S., Guedes, S.F.F., Xu, H.H.K., Rodrigues, L.K.A., 2013. Nanotechnology-based restorative materials for dental caries management. Trends in Biotechnology 31, 459–467.

Meurman, J.H., 2010. Oral microbiota and cancer. Journal of Oral Microbiology 2, 1–10.

Meurman, J.H., Bascones-Martinez, A., 2011. Are oral and dental diseases linked to cancer? Oral Diseases 17, 779–784.

Meurman, J.H., Sanz, M., Janket, S.-J., 2004. Oral health, atherosclerosis, and cardiovascular disease. Critical Reviews in Oral Biology and Medicine 15, 403–413.

Meyer, M.S., Joshipura, K., Giovannucci, E., Michaud, D.S., 2008. A review of the relationship between tooth loss, periodontal disease, and cancer. Cancer Causes and Control 9, 895–907.

Moreau, J.L., Sun, L., Chow, L.C., Xu, H.H.K., 2011. Mechanical and acid neutralizing properties and bacteria inhibition of amorphous calcium phosphate dental nanocomposite. Journal of Biomedical Materials Research 98, 80–88.

Mouhyi, J., Dohan Ehrenfest, D.M., Albrektsson, T., 2012. The peri-implantitis: implant surfaces, microstructure, and physicochemical aspects. Clinical Implant Dentistry and Related Research 14, 170–183.

Mustapha, I.Z., Debrey, S., Oladubu, M., Ugarte, R., 2007. Markers of systemic bacterial exposure in periodontal disease and cardiovascular disease risk: a systematic review and meta-analysis. Journal of Periodontology 78, 2289–2302.

Nabert-Georgi, C., Rodloff, A.C., Jentsch, H., Reissmann, D.R., Schaumann, R., Stingu, C.S., 2018. Influence of oral bacteria on adhesion of *Streptococcus mutans* and *Streptococcus sanguinis* to dental materials. Clinical and Experimental Dental Research 4, 72–77.

Nagaoka, S., Hojo, K., Murata, S., Mori, T., Ohshima, T., Maeda, N., 2008. Interactions between salivary *Bifidobacterium adolescentis* and other oral bacteria: *in vitro* coaggregation and coadhesion assays. FEMS Microbiology Letters 281, 183–189.

Norowski, P.A., Bumgardner, J.D., 2009. Biomaterial and antibiotic strategies for peri-implantitis: a review. Journal of Biomedical Materials Research Part B: Applied Biomaterials 88, 530–543.

Paju, S., Scannapieco, F.A., 2007. Oral biofilms, periodontitis, and pulmonary infections. Oral Diseases 13, 508–512.

Parahitiyawa, N.B., Jin, L.J., Leung, W.K., Yam, W.C., Samaranayake, L.P., 2009. Microbiology of odontogenic bacteremia: beyond endocarditis. Clinical Microbiology Reviews 22, 46–64.

Peppemey, A., Chikindas, M.L., 2011. Antibacterial peptides: opportunities for the prevention and treatment of dental caries. Probiotics and Antimicrobial Proteins 3, 68.

Periasamy, S., Kolenbrander, P.E., 2010. Central role of the early colonizer *Veillonella* sp. in establishing multispecies biofilm communities with initial, middle, and late colonizers of enamel. Journal of Bacteriology 192, 2965–2972.

Peters, B.A., Wu, J., Pei, Z., Yang, L., Purdue, M.P., Freedman, N.D., Jacobs, E.J., Gapstur, S.M., Hayes, R.B., Ahn, J., 2017. Oral microbiome composition reflects prospective risk for esophageal cancers. Cancer Research 77, 6777–6787.

Rafii, F., Hart, M.E., 2015. Antimicrobial resistance in clinically important biofilms. World Journal of Pharmacology 4, 31–46.

Rosen, G., Sela, M.N., 2006. Coaggregation of *Porphyromonas gingivalis* and *Fusobacterium nucleatum* PK 1594 is mediated by capsular polysaccharide and lipopolysaccharide. FEMS Microbiology Letters 256, 304–310.

Saha, S., Tomaro-Duchesneau, C., Rodes, L., Malhotra, M., Tabrizian, M., Prakash, S., 2014. Investigation of probiotic bacteria as dental caries and periodontal disease biotherapeutics. Beneficial Microbes 5, 447–460.

Saito, A., Inagaki, S., Kimizuka, R., Okuda, K., Hosaka, Y., Nakagawa, T., Ishihara, K., 2008. *Fusobacterium nucleatum* enhances invasion of human gingival epithelial and aortic endothelial cells by *Porphyromonas gingivalis*. FEMS Immunology and Medical Microbiology 54, 349–355.

Scannapieco, F.A., 2013. The oral microbiome: its role in health and in oral and systemic infections. Clinical Microbiology Newsletter 35, 163–169.

Singh, S., Sharma, A., Sood, P.B., Sood, A., Zaidi, I., Sinha, A., 2015. Saliva as a prediction tool for dental caries: an *in vivo* study. Journal of Oral Biology and Craniofacial Research 5, 59–64.

Soukos, N.S., Goodson, J.M., 2000. Photodynamic therapy in the control of oral biofilms. Periodontol 55, 143–166.

Stingu, C.S., Eschrich, K., Rodloff, A.C., Schaumann, R., Jentsch, H., 2008. Periodontitis is associated with a loss of colonization by *Streptococcus sanguinis*. Journal of Medical Microbiology 57, 495–499.

Teles, R., Wang, C.-Y., 2011. Mechanisms involved in the association between peridontal diseases and cardiovascular disease. Oral Diseases 17, 450–461.

Tong, H., Chen, W., Merritt, J., Qi, F., Shi, W., Dong, X., 2007. *Streptococcus oligofermentans* inhibits *Streptococcus mutans* through conversion of lactic acid into inhibitory H_2O_2: a possible counteroffensive strategy for interspecies competition. Molecular Microbiology 63, 872–880.

Vargas-Reus, M.A., Memarzadeh, K., Huang, J., Ren, G.G., Allaker, R.P., 2012. Antimicrobial activity of nanoparticulate metal oxides against peri-implantitis pathogens. International Journal of Antimicrobial Agents 40, 135–139.

Wade, W.G., 2013. The oral microbiome in health and disease. Pharmacological Research 69, 137–143.

Yang, Y., Cai, Q., Shu, X.O., Steinwandel, M.D., Blot, W.J., Zheng, W., Long, J., 2019. Prospective study of oral microbiome and colorectal cancer risk in low-income and African American populations. International Journal of Cancer 144, 2381–2389.

Zarco, M.F., Vess, T.J., Ginsburg, G.S., 2012. The oral microbiome in health and disease and the potential impact on personalized dental medicine. Oral Diseases 18, 109–120.

Zhang, Y., Niu, Q., Fan, W., Huang, F., He, H., 2019. Oral microbiota and gastrointestinal cancer. OncoTargets and Therapy 12, 4721–4728.

Zijnge, V., van Leeuwen, M.B.M., Degener, J.E., Abbas, F., Thurnheer, T., Gmür, R., Harmsen HJ, M., 2010. Oral biofilm architecture on natural teeth. PloS One 5, e9321.

Chapter 5

The role of filamentous fungi in drinking water biofilm formation

Ana F.A. Chaves[1], Lúcia Chaves Simões[2,3], Russell Paterson[2], Manuel Simões[3], Nelson Lima[2]

[1]*Faculty of Engineering, University of Porto, Porto, Portugal;* [2]*Centre of Biological Engineering, University of Minho, Braga, Portugal;* [3]*LEPABE — Laboratory for Process Engineering Environment, Biotechnology and Energy, Department of Chemical Engineering, Faculty of Engineering, University of Porto, Porto, Portugal*

5.1 Drinking water concerns

Water is fundamental to sustaining life, and access to safe drinking water (DW) is crucial to health, a basic human right and a very important issue for an effective policy in health protection. Safe DW is defined by the World Health Organization (WHO) as water that can be used without any risk of immediate- or long-term harm (WHO, 2011). The consumption of contaminated DW can cause a variety of health-related problems, especially in the more susceptible such as infants, children, the elderly, and immunocompromised people (WHO, 2011).

The supply of clean DW is a major public health milestone (Berry et al., 2006). Thus, there are many regulations with preestablished parameters (including microbiological ones) to ensure that the water provided is safe and free of hazards for human health. However, it is also well known that in 2012, about 748 million people remained without access to an improved source of DW, which is a very important reason for the many studies involving this subject (WHO, 2011) and to pursue United Nations Target 7.C of the millennium development goals that is to halve, by 2015, the proportion of the population without sustainable access to safe DW and basic sanitation.

Different concepts of DW exist around the world according to the stage of development of particular countries. Microbial contamination and chemical contamination are major concerns for developed countries, whereas for developing countries, where access to clean water and sanitation are not the rule, the main problems lie with microbial contamination (Gilbert, 2012).

Recent Trends in Biofilm Science and Technology. https://doi.org/10.1016/B978-0-12-819497-3.00005-2

Nevertheless, the risk from microbial pathogens remains present for developed countries, and the risk of outbreaks of waterborne diseases is not limited to developing countries (Beaudeau et al., 2008; Simões and Simões, 2013).

Two different chemical contamination sources can be attributed: anthropogenic activities (lead, nitrate, pesticides) and naturally occurring inorganic chemicals (arsenic, radon, fluoride) (WHO, 2011). Nonetheless, there are also emergent contaminants (antibiotics, human hormones, personal care products, and illicit drugs), which are appearing increasingly in surface waters (De Gusseme et al., 2011). The problem with these micropollutants lies with the fact that the conventional processes used for the treatment of DW are not capable of removing them (Kim et al., 2007; De Gusseme et al., 2011). Several chemical contaminants when ingested with DW are responsible for some human health problems such as vascular disease, dermal lesions, skin and lung cancer, and problems in neurodevelopment, among others (WHO, 2011).

Although chemical contamination is a crucial issue for developed countries, waterborne diseases are one of the most important health problems associated with water (Beaudeau et al., 2008). Waterborne diseases refer to any illness caused by the utilization of DW contaminated by pathogenic microorganisms such as bacteria, protozoa, virus, or helminths. Such pathogens are transmitted to the population when untreated or inadequately treated water is consumed (Simões and Simões, 2013). Several diseases are associated with waterborne outbreaks such as cholera, typhoid fever, meningitis, encephalitis, dysentery, hepatitis, legionellosis, pulmonary illness, poliomyelitis, leptospirosis, giardiasis, and salmonellosis, with gastroenteritis being the most notorious disease associated with waterborne outbreaks in developed countries (WHO, 2011).

It should be emphasized that adequate clean water supplies, combined with proper sanitation and improved hygiene standards, could significantly reduce the incidence of waterborne diseases (Simões and Simões, 2013). Also, DW quality is dependent on the specific composition and activity of microbial populations present during treatment, storage, and distribution phases, which means that microbial contamination is also a problem for drinking water distribution systems (DWDS) (Farkas et al., 2012). Thus, the creation of more robust and adequate disinfection treatments for DWDS is very important.

5.2 Microbiology of drinking water distribution systems

DWDS describes the system that distributes water from the original source such as a lake to the place where the water is accessible for human consumption (i.e., often a tap). It can be perceived as environments wherein bacteria, fungi (filamentous and yeast), protozoa, viruses, and algae interact, making them extremely complex systems (Berry et al., 2006; Hageskal et al., 2009). Therefore, and considering that each microorganism has a role in this environment, their significance should not be underestimated as a potential threat to human health and as contributors to a unique ecological niche (Paterson and Lima, 2005; Siqueira, 2011).

Approximately 95% of the total biomass in water is estimated as being attached to pipe walls in DWDS (Wingender and Flemming, 2004). Furthermore, several problems in DWDS have a microbial origin such as biofilm growth, nitrification, microbial-mediated corrosion, and the persistence of pathogens. Additionally, water microbiology can be affected by several factors, for example, pipe material, nutrient level, water flow, temperature, pH, and concentration of disinfectant (Berry et al., 2006).

Microorganisms in aqueous systems readily form biofilms, which offers them increased resistance to adverse factors such as high temperatures, dryness, poor oligotrophic conditions, and biocides when compared with their planktonic forms (Davey et al., 2000; Donlan, 2002). Alternatively in DW systems, these biofilms can work as a reservoir of pathogens and can also stimulate microbial-induced corrosion of system materials, leading to deterioration in potable water quality (Szewzyk et al., 2000). The presence of biofilm in the pipe walls can lead to poor water quality, because pathogenic microorganisms could be released by hydrodynamic forces from encrusting biofilms on system materials to consumable water, and chemical contaminants can be encapsulated in the biofilm matrix, leading to build-up of toxic substances. Furthermore, the presence of biofilms can alter the organoleptic quality, which is an indicative of poor DW (Gonçalves et al., 2006b). A conventional representation of a DW biofilm is shown in Fig. 5.1.

However, errors can occur during the study of biofilms in DW systems mainly due to variations in methodology (Berry et al., 2006; Hageskal et al., 2009). Also, the pipes in water networks are often inaccessible: in situ approaches are scarce and pilot experimental systems are used in laboratories for studies, or otherwise, it would be necessary to remove the pipes from the DWDS (Siqueira et al., 2011) for characterization of biofilms. Bench-top laboratory devices mimic DWDS, allowing to test different environmental conditions, and

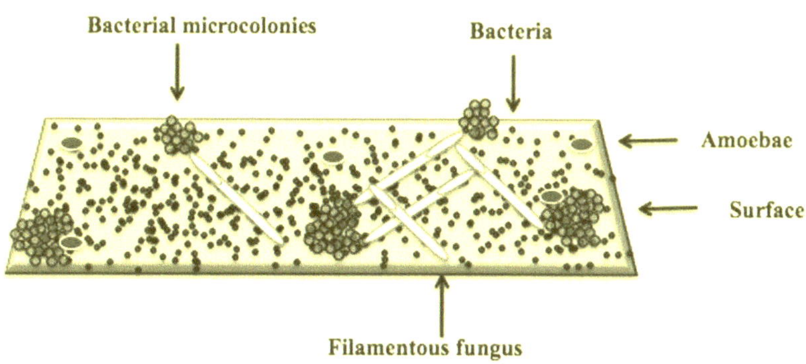

FIGURE 5.1 Microbial heterogeneity and architectural structure of water biofilms. *Based on Siqueira, V.M., 2011. Characterising Filamentous Fungal Biofilm in Drinking Water Distribution Systems Using Microscopic and Molecular Techniques (Ph.D. thesis). Universidade do Minho.*

can be used to study DW biofilm formation (Gomes et al., 2014). The development of such experimental devices aims to mimic DWDS and gather results that can be transposed to reality. Hence, the selection and use of an appropriate device is a crucial factor in obtaining reliable results and should be chosen, taking into account the goals of the study (Gomes et al., 2014).

5.2.1 Filamentous fungi in drinking water

Regarding DW problems, it is important to realize that the role played by filamentous fungi (ff) is underestimated and not fully understood (Doggett, 2000; Paterson and Lima, 2005; Gonçalves et al., 2006b; Hageskal et al., 2009). Some of the problems associated with fungal growth in DWDS are unsightly appearance, blocked pipes, odors and tastes, pigments, a source of potentially pathogenic and allergy-causing fungi, and mycotoxin production (Paterson and Lima, 2005; Hageskal et al., 2009; Simões et al., 2015). Fungal contamination of DW has been reported for decades, although investigations have been inadequate compared with those performed with bacteria (Paterson et al., 2009). The main reason for the infrequent discussion of ff as pathogenic microorganisms in water is that the presence of pathogenic bacteria, viruses, and parasites often leads to relatively acute symptoms in humans and the consumption of fungal-contaminated water has not (Hageskal et al., 2009). Additionally, this may be also explained by the reduced number of pathogenic ff reported in DW systems compared with those bacteria. However, the presence of ff in DW might cause organoleptic changes, which may lead to poor water quality (Paterson and Lima, 2005; Hageskal et al., 2009). Furthermore, it should be noticed that ff are especially adapted for growth on surfaces as evidenced by their absorptive nutrition mode, secretion of extracellular enzymes to digest complex molecules, and apical hyphal growth, which makes them the perfect candidates for biofilm formation (Jones, 1994). Furthermore, the ability of ff to form hyphal lattices in three dimensions makes them obvious candidates for biofilm formation.

Notwithstanding the fact that no direct association has been made linking ff with water contamination problems, several studies have been performed displaying that the recovery of ff in water samples varies between 7.5% and 89% positive samples (Hageskal et al., 2009). Filamentous fungi have been recovered from many types of water bodies such as raw, treated, heavily polluted, distilled or ultrapure, and bottled waters (Hageskal et al., 2009). They have been increasingly recognized as human pathogens, particularly for immunocompromised people. In fact, their significance in DW is underestimated. A range of attributable problems are changes in organoleptic properties of DW, infectious potential, allergic reactions, and toxin production (Paterson and Lima, 2005). Filamentous fungi can be implicated in three categories of disease in human and animals: infections (mycoses), allergies and toxicoses (Paterson et al., 2009).

Furthermore, considering that (1) microorganisms naturally interact with each other and (2) DWDS are complex environments full of different microorganisms such as bacteria, protozoa, fungi, and virus that compete with each other for nutrition and space, it is to be expected that bioactive secondary metabolites would be produced (Rateb et al., 2013), like patulin. Patulin is a mycotoxin produced by some ff and is recognized as bacterial quorum sensing inhibitor (Rasmussen et al., 2005). Mycotoxins are harmful secondary metabolites to humans and animals produced by a wide range of ff belonging mainly to *Fusarium*, *Aspergillus*, and *Penicillium* (Reverberi et al., 2010; Bayram and Braus, 2012), and many drinking water samples have been shown to contain *Aspergillus* and *Penicillium* (Table 5.1). Mycotoxins produced in water will be extremely diluted and represent a minor concern. However, their concentrations may increase and become a hazard to human health, especially when water is stored in cisterns, reservoirs, or even bottles, for prolonged periods of time (Hageskal et al., 2009). Mycotoxin-contaminated food and drink may have severe effects on human and animal health ranging from chronic to acute disease and death (Paterson et al., 2009). The bioactive actions include mutagenic, teratogenic, estrogenic, and carcinogenic effects. The frequency of such impacts and their severity depends on the type of mycotoxin, its concentration, the exposure pathway, and the duration of exposure. Mycotoxins may also damage major organs such as the liver, kidney, and lungs and the nervous, endocrine, and immune systems (Paterson et al., 2009).

Filamentous fungi in water may be aerosolized into the atmosphere, and introduced to immunocompromised patients via this route. Furthermore, tastes and odors changes have been associated with the occurrence of ff in DW systems (Paterson and Lima, 2005). Although not a health problem per se, many of the compounds produced by bacteria in relation to taste and odor are also produced by ff. In addition, they produce their own battery of compounds with distinctive odors and tastes. Some of the fungal isolates from DW are capable of transforming 2,4,6-tricholorophenol to 2,4,6-trichloroanisol, which causes taste and odor problems in distribution system (Gonçalves et al., 2006a). The genus *Acremonium*, commonly isolated from water (Table 5.1), produces octenol that is known as a major component of odors (Gonçalves et al., 2006a). *Penicillium expansum* strains are often isolated from DWDS in the north of Portugal, which can be patulin and geosmin producers. Geosmin, a volatile compound, is associated with earthy smells associated with water problems (Gonçalves et al., 2006a). According to these authors, during investigations of bad tasting water, the quantities of fungi present were found to be around 10^2-10^3 CFU/L, which represents a threshold level. Kelley et al. (2003) reported taste and odor problems associated with the presence of *Alternaria alternata*, *Aspergillus niger*, and *Cladosporium* sp. from samples of DW utilities indicated that these fungi might be involved. Additionally, some of the taxa were also present in swabs of pipes indicating that they may form parts of biofilms and contribute to tastes and odors in this manner.

TABLE 5.1 Studies performed in different countries where filamentous fungi were recovered from water supplies.

Country, place	Type of water	Most frequent fungi isolated	CFU/L	References
United Kingdom	Surface water and network	*Aspergillus, Cladosporium, Epicoccum, Penicillium,* and *Trichoderma*	0–9000 (surface water) 10–500 (tap water)	Kinsey et al. (1999)
Greece, Thessaloniki	Tap water	*Penicillium, Aspergillus,* and *Acremonium*	3.7×10^2	Arvanitidou et al. (1999)
Greece	Municipal water supplies of hemodialysis centers	*Aspergillus* and *Penicillium*	168–428 (tap water) 388–804 (treated water) 292–730 (dialysate)	Arvanitidou et al. (2000)
Germany, North Rhine-Westphalia	Drinking water	*Acremonium, Exophiala, Penicillium,* and *Phialophora*	1×10^3–2×10^3	Göttlich et al. (2002)
Norway, 14 different networks	Surface and underground drinking water	*Penicillium, Trichoderma,* and *Aspergillus* spp.	95 and 84	Hageskal et al. (2006, 2007)
Portugal, Braga	Tap water	*Penicillium* and *Acremonium* spp.	1–20	Gonçalves et al. (2006a)
Brazil, Maringá	Tap and mineral bottled water	Yeasts and filamentous fungi	2800 (yeasts) and 1000 (ff)	Yamaguchi et al.(2007)
Pakistan, Karachi	Drinking water	*Aspergillus niger* and *Aspergillus clavatus*		Nazim et al. (2008)
Poland, Warsaw	Municipal water supply system	*Aspergillus fumigatus* and *A. niger*	200–5000	Grabińska-Łoniewska et al. (2007)
Australia, Queensland	Municipal water	*Cladosporium, Penicillium, Aspergillus, and Fusarium*	1–50	Sammon et al. (2010)
Portugal, Lisbon	Groundwater, surface water, and spring water	*Penicillium, Aspergillus,* and *Cladosporium* spp.	660 (groundwater), 17,500 (surface water) and 10,250 (spring water)	Pereira et al. (2009, 2010)

Brazil, Recife	Water treatment plant and tap water	*Penicillium, Aspergillus,* and *Phoma*	1–28,667	Oliveira (2010)
Belgium, Liége	Tap water from the University Hospital of Liége	*A. fumigatus* and *Fusarium* spp.	7	Hayette et al. (2010)
Saudi Arabia, Jeddah	Drinking water distribution system	*Alternaria alternata*	40–150	Gashgari et al. (2013)
		Aspergillus niger	20–180	
		A. fumigatus	10–80	
		Acremonium strictum	40–320	
		Chaetomium globosum	30–120	
		Cladosporium cladosporioides	60–210	
		Cladosporium herbarum	20–40	
		Fusarium dimerum	80–430	
		Rhizopus azygosporus	10–30	
		Mucor plumbeus	20–50	
		Penicillium brevicompactum	30–80	
		Penicillium citrinum	10–40	
		Penicillium glabrum	0–20	
		Penicillium montanence	20–50	
		Penicillium melinii	0–20	
		Penicillium olsonii	0–20	
		Penicillium oxalium	20–60	
		Trichoderma viride	60–220	

Continued

TABLE 5.1 Studies performed in different countries where filamentous fungi were recovered from water supplies.—cont'd

Country, place	Type of water	Most frequent fungi isolated	CFU/L	References
China, Xiamen	Surface water, public system water, house water, and tank water	*Aspergillus* spp., *Fusarium* spp., *Penicillium* spp., *Trichoderma* spp., *Mucor* sp., and *Rhizopus* sp.	approx. 0–2400	Al-gabr et al. (2014)
Pittsburgh, PA, United States	Hospital hot water system	*Penicillium, Aspergillus, Peniophora, Cladosporium, and Rhodosporidium*		Ma et al. (2015)

In the future, if the microbiological quality of DW is to include ff, it is essential that standardization of ff methods in water regulations is implemented, including the methods for detection and quantification. ff are a difficult group to examine, but that should not mean that ff contamination of DW be ignored. As it has been mentioned, ff can affect the water quality in many ways and therefore should be considered when the microbiological safety and quality of DW is evaluated (Hageskal et al., 2009).

5.2.2 Biofilms

To maintain water quality and preserve human health, it becomes essential to control biofilm formation. As stated before, biofilms can act as a reservoir of pathogens and are the origin of several DWDS problems (Szewzyk et al., 2000). There are significant proportions of microorganisms that persist attached to surfaces in the form of biofilms (Thein et al., 2006). A biofilm is defined as an assemblage of microbial cells that are permanently attached to an abiotic or biotic surface and encircled in a matrix of primarily polysaccharide material (Davey et al., 2000; Donlan, 2002; Huq et al., 2008). Extracellular polymers (also known as extracellular polymeric substances [EPS]) are the key substances keeping biofilm organisms together, gluing them to the surface, and providing protection against stress (Simões et al., 2010).

The formation of biofilms may occur on a wide variety of surfaces including living tissues, indwelling medical devices, industrial or potable water system piping, or natural aquatic systems (Donlan, 2002). The solid–liquid interface between a surface and an aqueous medium, such as water, offers an ideal environment for the attachment and growth of microorganisms. Therefore, and to understand the attachment of the microorganisms, it is essential to consider the effects and features of the (1) substratum (roughness, hydrophobicity, surface charge, chemical composition), (2) characteristics of the aqueous medium (hydrodynamics, pH, temperature, cations, nutrient availability, presence of disinfectants), and (3) properties of the cell surface (hydrophobicity, extracellular appendages, EPS, signaling molecules) (Donlan, 2002; Yu et al., 2010). Thus, cell attachment to a surface and the succeeding biofilm formation is a very complex process, with many variables affecting the process. Furthermore, biofilms are a stable point in a biological cycle that includes four different steps: initiation, maturation, maintenance, and dissolution (O'Toole et al., 2000).

In natural habitats and in medical or industrial environments, there is a common concern regarding the biofilm mode of growth (Elvers et al., 1998). Consequently, the focus of biofilm research is understanding the mechanisms of biofilm formation, structure, detachment, and the methods for controlling its growth (Elvers et al., 1998; Siqueira and Lima, 2013). However, the great majority of these studies are performed using only pure bacterial cultures, which is not realistic when compared with nature (Elvers et al., 1998, 2001; Gonçalves et al., 2006b). Usually, in natural and industrial environments, multikingdom biofilm consortia are found, where these biofilms are complex and may contain many taxa (Elvers et al., 2001; Peters et al., 2010). Although studies using bacterial cultures are simpler to investigate, it is necessary to use multikingdom consortia for a better understanding of interactions and how they affect the biofilm (Elvers et al., 2001; Peters et al., 2010). These studies are more realistic.

5.2.2.1 Microbial biofilms in drinking water

Biofilms in nature can be composed of microorganisms from different kingdoms such as bacteria, viruses, protozoa, fungi, and algae (Momba et al., 2000). Nevertheless, bacteria are dominant generally in biofilms due to their high growth rates, small size, adaptation capacities, and the ability to produce EPS (Simões and Simões, 2013).

Microbes are able to come into closer contact with a surface, attach firmly, promote cell—cell interactions, and grow as a complex structure (Bryers and Ratner, 2004). Biofilm formation comprises of a sequence of steps (Bryers and Ratner, 2004), and biofilm formation mechanisms are described in diverse reports (Chmielewski and Frank, 2003; Verstraeten et al., 2008).

Currently, processes governing bacterial DW biofilm formation have been identified as follows: (1) preconditioning of the pipe surface either by macromolecules (organic and inorganic) present in the bulk water; (2) transport of planktonic cells from the water to the surface; (3) adsorption of cells at the pipe surface; (4) desorption of reversibly adsorbed cells; (5) irreversible adsorption of bacterial cells at a pipe surface; (6) production of cell—cell signaling molecules; (7) transport of substrates to and within the biofilm; (8) substrate metabolism by the biofilm-bound cells and transport of products out of the biofilm, accompanied by cell growth, replication, and EPS production; and (9) biofilm removal by detachment or sloughing (Bryers and Ratner, 2004; Simões and Simões, 2013).

The preconditioning of the pipe surface is considered the first step in DW biofilm formation and has to occur before any attachment to the surface by the bacteria. In this step, a thin layer of organic molecules and ions will cover the pipe surface due to physical or chemical adsorption. The strength of biofilm adhesion is largely dependent on the cohesion of the conditioning film (Busscher et al., 1995; Bos et al., 1999).

After the preconditioning of the pipe surface, it is possible to start the adhesion phase, which comprises steps 2 to 5 above. Initially, there is transport of bacterial cells to the pipe surface due to fluid dynamics, gravitational forces, and Brownian motion or by migration through active cell motility. The surface electrostatic charge and hydrophobic interactions affect this first surface—surface interaction and the adhesion process. Cells and the pipe surface can interact with each other by the establishment of long and short/intermediate distance forces. When these forces are in a favorable equilibrium, the end result will be the adhesion of the bacteria to the pipe surface (Simões, 2013).

Following cellular adhesion to pipe surfaces, growth and maturation are the following stages of DW biofilm formation (steps 6–9). Firstly, the attached bacteria start growing and form microcolonies, excrete organic polymers, and initiate the formation of the biofilm matrix (Allison and Sutherland, 1987). Transport of nutrients from the DW to the inner layers of biofilm and the transport of excreted metabolites in the opposite direction are important for biofilm maintenance and occur with the increase of thickness. Logically, the maturation phase is the next step. Here occurs the development of complex and organized consortia of bacteria, which are surrounded by an organic matrix that protects the bacteria from stress factors. Large amounts of EPS are produced in this stage. Also, in a mature biofilm, there is attachment of planktonic bacteria, bacterial detachment into the bulk water, growth, and death. Nonetheless, processes are at equilibrium, and the number of attached cells per unit surface area is considered constant in time, although fluctuations can occur throughout time (Bryers and Ratner, 2004).

Detachment of cells and other components from the biofilm is the last phase of biofilm formation (step 9). This phenomenon can happen due to different mechanisms such as sloughing (the rapid detachment of large portions of the biofilm), erosion (the continuous release of single cells or small clusters of cells), abrasion (collision of solid particles with the biofilm), and predator grazing by, for example, protozoa (Simões, 2013).

5.2.2.2 Filamentous fungi biofilms: can it happen?

Filamentous fungi are an excellent candidate species for studying biofilm formation. Nevertheless, the term "biofilm" is rarely used when talking about ff (Harding et al., 2009). Very few reports on ff biofilms can be found in the literature, which is probably due to the fact that ff do not fit completely or precisely within restrictive biofilm definitions, which are normally based on bacterial models (Harding et al., 2009). Fungi, as eukaryotic organisms, have peculiar characteristics such as nutrition by heterotrophic absorption, development of reproductive and vegetative structures (i.e., spores and hyphae), and sexual and asexual reproduction (Siqueira and Lima, 2013). Furthermore, ff biofilms are common in environments exposed to high moisture with a major air interface (i.e., unsaturated environments). In addition, ff often show invasive growth, by penetrating the substrata on which they grow. These differences in morphology and growth complicate the understanding on ff biofilm formation and behavior (Simões et al., 2015). Therefore, a direct comparison with the dynamics of bacterial biofilms can provide inaccurate conclusions. According to Harding et al. (2009), it is important to highlight that the unique features related to fungal biology that distinguish bacterial from fungal biofilm formation are the ability of fungi to have more than one planktonic form (i.e., sexual and asexual spores, sporangia, and hyphal fragments). These dispersive forms are not unicellular and often float in air rather than water, and the development of specialized reproductive tissues to produce dispersive forms occurs in fungi in response to specific environmental cues, biological stimuli, or stresses (Harding et al., 2009; Simões et al., 2015). Another interesting aspect of fungal biology that has no counterpart in bacterial biofilms is the aerial component of lifestyle in many species, with a strong dependency on aerial spore dissemination for dispersal (Harding et al., 2009). Fungal aerial morphogenesis involves the secretion of small proteins (hydrophobins). The hydrophobins are exclusive to ff and play a role in several processes in the growth and development of these microorganisms (formation of aerial structures, attachment of hyphae to hydrophobic surfaces, and changes in hyphal surface properties in response to environmental and developmental cues) (Harding et al., 2009; Simões et al., 2015). It needs to be remembered also that the diameter and length of fungal hyphae are much bigger than individual bacteria (Fig. 5.1). Sometimes, to describe the surface-associated growth of ff other terms such as "multicellular masses," "pellets," and "submerged/solid-state fermentation" are often used instead of the term "biofilm" (Gutiérrez-Correa and Villena, 2003; Mowat et al., 2009).

Although the reports about ff biofilms are sparse, there are many studies that describe the presence and growth of ff in several environments such as in medical, industrial, and environmental niches (Anaissie et al., 2003; Gutiérrez-Correa and Villena, 2003; Mowat et al., 2007). Harding et al. (2009) proposed a set of criteria for ff biofilm formation, and the studies aforementioned meet at least some of the criteria proposed. These criteria are concise into two groups: (1) structural features such as complex aggregated growth, surface-associated growth of cells, and secreted extracellular polymeric matrix and (2) altered gene expression resulting in phenotypic changes that include enhanced tolerance to antimicrobial compounds or biocides, changes in enzyme or metabolite production, and/or secretion and physiological changes (Harding et al., 2009).

Harding et al. (2009) proposed a preliminary model for ff biofilm formation, based on some of the already mentioned studies using bacterial and yeast models. This preliminary model is divided into six major stages: propagule adsorption, active attachment to a surface, microcolony formation, initial maturation, maturation or reproductive development, and dispersal or planktonic phase. Fig. 5.2 shows the main stages of this preliminary model. According to these authors, all the steps are comparable with those that occur for bacterial biofilm formation. However, it must be reinforced that filamentous fungal biofilms have morphological and growth features distinct from both bacterial and yeast biofilms that complicate the understanding of ff biofilms.

The propagule adsorption is comprised by the deposition of spores or other propagules such as hyphal fragments or sporangia. This phase involves mainly physical contact of the ff with a surface. When comparing it with the bacterial models, it represents the reversible attachment stage (Harding et al., 2009).

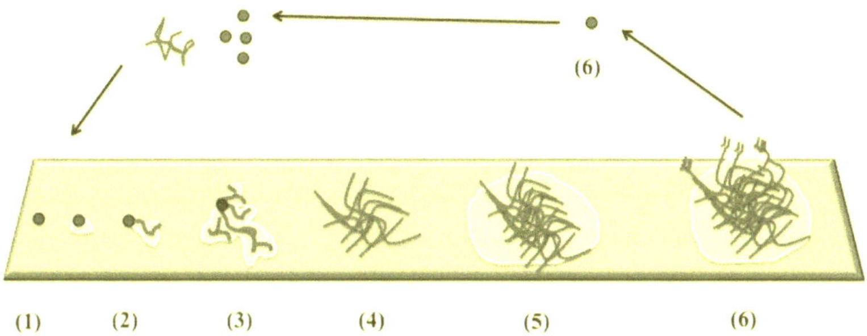

FIGURE 5.2 Harding et al. model for ff biofilm formation: (1) adsorption, (2) active attachment, (3) microcolony I (germling and/or monolayer), (4) microcolony II (mycelial development, hyphal layering, hyphal bundling), (5) development of the mature biofilm, and (6) dispersal or planktonic phase. *ff*, filamentous fungi. *Adapted from Harding, M.W., Marques, L.L.R., Howard, R.J., Olson, M.E., 2009. Can filamentous fungi form biofilms?. Trends in Microbiology 17, 475–480.*

Propagule adsorption is followed by the active attachment to the surface. In this phase, the ff secrete usually adhesive substances by germinating spores: It is comparable with the fixed attachment phase in bacteria (Harding et al., 2009).

The third phase is the microcolony formation where the initial stages of growth and surface colonization occur. To do so, the cells produce an extracellular polymeric matrix that allows the growing colony to adhere tenaciously to the substrate (Harding et al., 2009).

In the initial maturation phase, the formation of compacted hyphal or mycelial networks occurs, and hypha–hypha adhesion is achieved. Also, this includes the layer, formation of hyphal bundles bonded together by exopolymeric matrix, and formation of water channels via hydrophobic repulsion between hyphae or hyphal bundles (Harding et al., 2009). Lattices occur by the very nature of ff growth effectively forming a biofilm. A gossamer pellet was formed in shake flask growth in water of *Fusarium graminearum* (Paterson, 2007), which was similar to a floating biofilm, and filamentous fungal biofilms are notorious for fouling or blocking bioreactors used to grow ff for applied purposes (unpublished observations).

The maturation phase then occurs, which includes the formation of fruiting bodies, conidiophores, sporogenous cells, sclerotia, and other survival structures. Aerial growth is often a crucial feature of fungal fruiting and dispersal (Harding et al., 2009).

The dispersal or planktonic phase is the last stage of this preliminary model. This phase is characterized by the dispersal or release of spores or biofilm fragments. The detached cells can act as new propagules to reinitiate the cycle (Harding et al., 2009). It is important to mention that spores (or conidia), hyphal fragments, and other fungal propagules can be considered functional equivalents to planktonic bacterial cells (Harding et al., 2009). As stated above, fungi have more than one planktonic form involved in the dissemination phase of biofilm formation. The production of these dispersive forms occurs in response to specific environmental cues, biological stimuli, or stresses, for example, hydrodynamic conditions.

Studies performed by Beauvais et al. (2007) and Mowat et al. (2009) regarding *Aspergillus fumigatus* characterized the compact hyphal balls formed during growth (i.e., pellets) as biofilms. Pellet formation is well known when ff are grown in bioreactors and is good evidence of biofilm formation by ff. There was evidence of the presence of an extracellular polymeric matrix, differential gene expression, and also a variable sensitivity to antifungal drugs of these pellets. More recently, Simões et al. (2015) reported the dynamics of biofilm formation by *Penicillium expansum*, a strain isolated from DWDS, under controlled static conditions, mimicking the water behavior in stagnant regions of DWDS. This study demonstrated that *P. expansum* grows as a complex, multicellular biofilm in 48 h. Similar to bacterial biofilms, it was possible to identify the different phases of ff biofilm formation. The metabolic activity and biomass of fungal biofilms were found to increase over time. It was also possible to show a correlation between metabolism, biofilm mass, and hyphal development. Recently, Fernandes et al. (2019) evaluated the influence of several abiotic factors (substratum, hydrodynamic conditions, and nutrient availability) on the adhesion and biofilm formation by two ff (*Penicillium brevicompactum* and *P. expansum*) isolated from DW. These results allowed to refine a model for ff biofilm formation, and the overall results help to understand key aspects determining the presence of ff biofilms in DW, where stagnant conditions and the presence of nutrients should be avoided to prevent ff biofilm formation.

5.3 Drinking water distribution systems maintenance

Drinking water is probably the most important human resources, and it is essential to guarantee its quality. Therefore, treatment is a crucial measure to guarantee public health security. Also, DW is one of the most closely monitored resources and is subjected to several treatment processes before it can be distributed and consumed (WHO, 2011).

Avoiding biofilm formation would be a more logical and suitable option than treating it once formed. Nevertheless, there is currently no known technique that is able to effectively prevent or control the formation of unwanted biofilms without causing adverse side effects (Simões et al., 2010), but there are several measures that can be applied to limit DW biofilm formation. Such measures include (a) minimizing the concentration of organic matter entering the distribution system; (b) ensuring the material from which the pipework and fittings are made off is chemically and biologically stable; (c) prevention of water stagnation and sediment accumulation within the distribution systems; and (d) maintaining a sufficient disinfectant level throughout the system (WHO, 2011; Simões and Simões, 2013). However, chemical disinfection is the most widespread measure used to ensure biofilm control and microbiological DW quality.

5.3.1 Disinfection

Water disinfection means the removal, inactivation, or killing of microorganisms to guarantee safe water throughout drinking DWDS from a microbiological point of view. The most frequently used disinfectants are chlorine, chloramines, chlorine dioxide, ozone, and UV radiation (Chowdhury, 2012), of which chlorine is by far the most widely used.

Chlorination is a key step in the biofilm control process. Chlorine is a strong oxidizing agent, being the most commonly used disinfectant due to its effectiveness, stability, ease of use, and low cost (Rand et al., 2007; Fernandes et al., 2013). Chlorination can occur in two distinct points in the treatment process: pretreatment where a primary disinfection at the beginning of the treatment occurs, and a posttreatment or secondary disinfection aimed at maintaining a disinfectant residual in the distribution system. The posttreatment process plays an important role to control microbial regrowth (Deborde and von Gunten, 2008). Chemical disinfection and the maintenance of chlorine residual through the distribution systems are worldwide strategies to prevent bacterial regrowth during water transportation (LeChevallier et al., 1996; Momba et al., 1998; Codony et al., 2005). Even so, regrowth may occur when the residual chlorine decay further down in the distribution system (Chandy and Angles, 2001; Hallam et al., 2002). Some studies have demonstrated that chlorine is able to control biofilm formation by reducing the rate of biofilm growth, promoting biofilm detachment, and decreasing the activity of microorganisms (de Beer et al., 1994; Lund and Ormerod, 1995; Codony et al., 2002, 2005). However, the presence of residual chlorine is also one of the stress factors that leads to biofilm formation (Kokare et al., 2009). Nevertheless, some European countries notably the Netherlands, Germany, Austria, and Switzerland have taken the approach of distributing high-quality DW without the use of residual chlorine. The control of microbial growth in these countries is obtained through limitation of the nutrients essential for growth by more appropriate DW treatments (sedimentation, filtration, UV disinfection, ozone, peroxide), i.e., by the production of biologically stable DW (Simões and Simões, 2013).

Water disinfection is normally achieved by adding disinfectants in excess, and particularly chlorine, so that microbial accumulation in pipes and tanks can be controlled. Nevertheless, residual concentrations must be kept below guidelines to lower the potential to form disinfection by-products (DBPs) harmful to human health (Rand et al., 2007). Also, high chlorine concentrations can cause organoleptic problems (strong odor and tastes) and lead to the selection of resistant microorganisms (Nieuwenhuijsen et al., 2000).

According to the WHO (2011), 2–3 mg/L of chlorine should be added to water to achieve satisfactory disinfection and a residual concentration along DWDS. Nonetheless, the maximum amount of chlorine allowed is 5 mg/L. Also, the residual concentration of free chlorine leaving the treatment plant should be less than 1 mg/L and nearer 0.5 mg/L (WHO, 2011). However, this residual concentration seems insufficient to stop the growth and development of microbial biofilms (Zhou et al., 2009). Alternatively, disinfection with chlorine dioxide and chlorine can reduce the concentration of planktonic bacteria but has little or no effect on the density of biofilm bacteria (Simões et al., 2010). This resistance of biofilms to antimicrobials can be due to very low metabolic activity and severely downregulated rates of cell division of the deeply embedded microorganisms. Moreover, biofilms can function as a diffusion barrier to antimicrobial agents (Bridier et al., 2011). It should also be noticed that EPS matrix can reduce the disinfectant concentration and effectiveness since both compounds interact with each other (Bridier et al., 2011). In a previous study, conidial resistance of *Aspergillus niger*, *A. fumigatus*, *Cladosporium* spp., and *Penicillium oxalicum* was investigated (Rosenweig et al., 1983). *A. niger* exhibited the greatest resistance. Other study assessed the viability of spores of *P. brevicompactum* using three concentrations of sodium hypochlorite (1%, 5.7% and 10%) at different exposures times. Despite all concentrations being shown to be effective, increasing exposure times (2, 4, and 30 min) were required to cause total spore inactivation (Elbing, 2008). In another study, *P. brevicompactum* water biofilms were able to survive after 15 min in 2.38 mg/L of free chlorine, whereas free spores were susceptible to 1.83 mg/L (Siqueira and Lima, 2011).

Chloramines and chlorine dioxide are examples of other oxidant chlorine-based disinfectants used for DW disinfection. Chloramines are less effective than free chlorine and need longer contact times, or higher concentrations, to achieve the desired disinfection level but create smaller amount of DBPs. Yet, this compound is more effective in biofilm penetration than chlorine (Chandy and Angles, 2001). The use of chlorine dioxide is not a common practice due to factors such as high cost, onsite generation, lack of extensive information on it interaction with drinking water like breakdown products, and so on. Nonetheless, this disinfectant does not produce dangerous DBPs, such as trihalomethanes as in the case of chlorination (Simões and Simões, 2013). The proposed mechanism for antimicrobial activity of chlorine active compounds is the inhibition of key enzymatic reactions within the cell and protein denaturation (Fernandes et al., 2013).

Disinfection can also be performed by non–chlorine-based oxidants such as ozone. Ozone is a gas and a powerful oxidizing agent that has been used extensively for DW disinfection (Fernandes et al., 2013). This compound not only efficiently removes certain microorganisms, not easily neutralized by dissolved organics, effective in removing odor and improving taste. Comparatively, ozone produces low amount of DBPs. However, ozone is more expensive than chlorine, and its use does not allow a residual disinfection action in DWDS (Simões and Simões, 2013). Currently, ozone has been shown to possess greater bactericidal activity with less by-products for DW system; however, its cost of production seems to be a major factor hindering its widespread application. Furthermore, it was reported as the most effective water treatment method against fungal spores (Kelley et al., 2003). Ozone causes oxidation of cell components leading to protein inactivation, cell lysis, perturbation of adenosine triphosphate formation, and modification of purine and pyrimidine bases in nucleic acids (Fernandes et al., 2013).

UV radiation (electromagnetic energy in the range 250–265 nm) can be used as an alternative to chemical disinfection. UV is disinfected by destroying certain microorganisms, since it will alter their genetic material and render them unable to reproduce. This method is very effective against all bacteria, viruses, and protozoa cysts found in clarified waters. Nonetheless, it presents one major disadvantage, which is the fact that UV leaves no residual disinfectant in the water (Liberti et al., 2003). A possible solution for this disadvantage relies on the use of a second disinfectant to generate a residual amount of disinfecting power (Simões and Simões, 2013). Another limitation of this method is the contact time needed for large volumes of water. Therefore, its widespread application is limited. UV radiation can be used in combination with small domestic reverse osmosis systems. Recently, Al-Gabr et al. (2013) showed the efficacy of UV irradiation followed by chlorination on inactivation of *Aspergillus flavus* in DW. This study demonstrated that the combination of the two treatments was much more effective on inactivation of *A. flavus* than using either UV or chlorination alone. The efficiency of UV irradiation improves when followed by chlorination at low concentrations. Low concentration of chlorine and relatively short contact time with UV led to 4 log inactivation of *A. flavus*. Further UV irradiation can improve effectiveness of chlorination (at lower concentrations) in reducing *A. flavus*, thereby rendering more safe water, since lower amount of DBPs are produced (Al-Gabr et al., 2013).

The efficacy of chemical disinfectants against ff is variable between species (Kinsey et al., 2003). Chlorine disinfection is temperature dependent, and the inactivation of spores occurs less frequently at lower temperatures. The exposure time to free chlorine needed to inactivate ff is longer than for other chemical disinfectants, particularly ozone and chlorine dioxide (Paterson and Lima, 2005). Spores are more resistant than hyphal cells, with some being extremely chlorine resistant (Kelley et al., 1997). Such spores could thus allow the establishment of ff in the DWDS even if treatment processes have removed the vegetative cells. Once ff are established in the system, it can be difficult to maintain sufficient concentrations (i.e., 0.4–0.5 mg/L) (Rosenzweig et al., 1983) of free chlorine to prevent colonization and biofilm formation (Lund and Ormerod, 1995; Grabinska-Loniewska et al., 2007). This is because the chlorine demand of fungi is high (Rosenzweig et al., 1983; Kelley et al., 1997). It has been suggested that initial free chlorine concentrations of approximately 1 mg/L are sufficient for spore inactivation and to provide sufficient residual chlorine in the system to assist in prevention of new growth (Kelley et al., 1997; Kinsey et al., 2003) and development of biofilms (Lund and Ormerod, 1995; Momba et al., 2000).

Sisti et al. (2012) studied the efficacy of chlorine and peracetic acid alone or in combination against *Aspergillus* spp. in DW. These authors concluded that at the concentrations of chlorine that are usually present in DWDS, chlorine is ineffective against several *Aspergillus* spp. and peracetic acid cannot be considered an alternative to chlorine for disinfecting DW. Furthermore, the combination of the two disinfectants was not considered effective in eliminating ff at the concentrations permitted for DW disinfection.

5.4 Bacterial and fungal interactions

Previously, there has been a certain separation of microbiological research between especially bacteriologists and mycologists, which has led to the study of bacteria and fungi in exclusively axenic settings. However, this division is not realistic in vivo since bacteria and fungi often coexist and interact with each other in environmental biofilms (Wargo and Hogan, 2006; Frey-Klett et al., 2011). Current, studies show that fungi and bacteria frequently form physically and metabolically interdependent consortia that display properties distinct from them as individuals (Tarkka et al., 2009).

The bacteria–fungi interactions (BFIs) are of interest for several fields of study such as agriculture, forestry, environmental protection, food processing, biotechnology, and medicine (Kobayashi and Crouch, 2009; Frey-Klett et al., 2011). The most common reports about BFIs show that normally the bacterial partner exploits resources from the associated fungi through a parasitic or commensalism type of interaction (Kobayashi and Crouch, 2009). Nevertheless, there are examples where the fungi are able to take advantage of bacterial resources in mutualistic interactions (Kobayashi and Crouch, 2009).

The physical associations between the complexes formed by bacteria and fungi can range from apparently disordered polymicrobial communities to highly specific symbiotic associations of fungal hyphae and bacterial cells (Frey-Klett et al., 2011). Multispecies and multikingdom biofilms containing filamentous and nonfilamentous fungi (i.e., yeasts) and bacteria can be considered as a more intimate level of bacterial–fungal association. This kind of biofilm is different from multispecies communities from one kingdom. Once in the biofilm mode, the microorganisms form structured communities "glued" together by an extracellular matrix of macromolecules produced by the microorganisms, which have physical and physiological properties distinct from those of free-living cells (Donlan and Costerton, 2002). BFIs and adhesion are very important early events for the formation of multispecies bacterial–fungal biofilms (Frey-Klett et al., 2011) involving the two kingdoms.

The BFIs can be perceived as an equation (Fig. 5.3). In that equation, the combination of physical associations and molecular interactions between the two microorganisms can have a variety of different outcomes for each one. These changes may affect the influence of the bacterial–fungal complex on their biotic and abiotic environment (Frey-Klett et al., 2011).

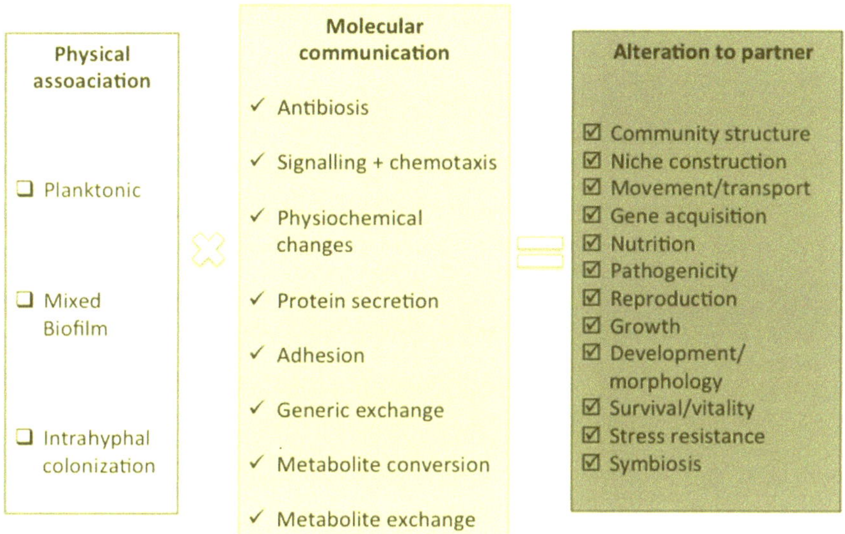

FIGURE 5.3 Bacteria—fungi interaction equation. *Adapted from Frey-Klett, P., Burlinson, P., Deveau, A., Barret, M., Tarkka, M., Sarniguet, A., 2011. Bacterial-fungal interactions: hyphens between agricultural, clinical, environmental, and food microbiologists. Microbiology and Molecular Biology Reviews 75, 583—609.*

Normally, the microbial communities that exist as a multispecies/bikingdom biofilm present an increased resistance to the treatment with antibiotics and antifungals. This resistance might be due not only to a more complex polymeric matrix composition but also to the fact that antimicrobial resistance profiles change in multispecies (and multikingdom) infections (Wargo and Hogan, 2006). Therefore, since the biofilms in DWDS are a complex environment, where the bacteria—fungi biofilms are present, the BFIs could play a significant role on biofilm formation and control in these systems. So, knowledge about multikingdom biofilms such as between bacteria and ff is essential to improve DW quality. Afonso et al. (2019) used ff and bacteria isolated from a DWDS to assess and characterize interkingdom biofilms. These authors found that interkingdom biofilm formation is microorganism dependent and that interkingdom biofilms may provide an advantage to the opportunistic bacteria to replicate and proliferate.

Acknowledgments

This study was supported by the Portuguese Foundation for Science and Technology (FCT) under the scope of the strategic funding of UID/BIO/04469/2019 unit and BioTecNorte operation (NORTE-01-0145-FEDER-000004) funded by the European Regional Development Fund under the scope of Norte2020—Programa Operacional Regional do Norte. Further support was obtained from the European Investment Funds by FEDER/COMPETE/

POCI—Operational Competitiveness and Internationalization Programme, under Projects: POCI-01-0145-FEDER-030219; POCI-01-0247-FEDER-035234; POCI-01-0247-FEDER-033298; POCI-01-0145-FEDER-028397; and POCI-01-0145-FEDER-006958 and National Funds by FCT—Portuguese Foundation for Science and Technology.

References

Afonso, T.B., Simões, L.C., Lima, N., 2019. In vitro assessment of inter-kingdom biofilm formation by bacteria and filamentous fungi isolated from a drinking water distribution system. Biofouling. https://doi.org/10.1080/08927014.2019.1688793.

Al-gabr, H.M., Zheng, T., Yu, X., 2013. Inactivation of Aspergillus flavus in drinking water after treatment with UV irradiation followed by chlorination. Science of the Total Environment 463–464, 525–529.

Al-gabr, H.M., Zheng, T., Yu, X., 2014. Occurrence and quantification of fungi and detection of mycotoxigenic fungi in drinking water in Xiamen City, China. Science of the Total Environment 466–467, 1103–1111.

Allison, D.G., Sutherland, I.W., 1987. The role of exopolysaccharides in adhesion of freshwater bacteria. Journal of General Microbiology 133, 1319–1327.

Anaissie, E.J., Stratton, S.L., Dignani, M.C., Lee, C., Summerbell, R.C., Rex, J.H., Monson, T.P., Walsh, T.J., 2003. Pathogenic molds (including *Aspergillus* species) in hospital water distribution systems: a 3-year prospective study and clinical implications for patients with hematologic malignancies. Blood 101, 2542–2546.

Arvanitidou, M., Kanellou, K., Constantinides, T.C., Katsouyannopoulos, V., 1999. The occurrence of fungi in hospital and community potable waters. Letters in Applied Microbiology 29, 81–84.

Arvanitidou, M., Spaia, S., Velegraki, A., Pezarloglou, M., Kanetidis, D., Pangidis, P., Askepidis, N., Katsinas, C.h., Vayonas, G., Katsouyannopoulos, V., 2000. High level of recovery of fungi from water and dialysate in haemodialysis units. Journal of Hospital Infection 45, 225–230.

Bayram, O., Braus, G.H., 2012. Coordination of secondary metabolism and development in fungi: the velvet family of regulatory proteins. FEMS Microbiology Reviews 36, 1–24.

Beaudeau, P., de Valk, H., Vaillant, V., Mannschott, C., Tillier, C., Mouly, D., Ledrans, M., 2008. Lessons learned from ten investigations of waterborne gastroenteritis outbreaks, France, 1998–2006. Journal of Water and Health 6, 491–503.

Beauvais, A., Schmidt, C., Guadagnini, S., Roux, P., Perret, E., Henry, C., Paris, S., Maleet, A., Prévost, M.C., Latgé, J.P., 2007. An extracellular matrix glues together the aerial-grown hyphae of *Aspergillus fumigatus*. Cellular Microbiology 9, 1588–1600.

Berry, D., Xi, C., Raskin, L., 2006. Microbial ecology of drinking water distribution systems. Current Opinion in Biotechnology 17, 297–302.

Bos, R., van der Mei, H.C., Busscher, H.J., 1999. Physico-chemistry of initial microbial adhesive interactions- its mechanisms and methods for study. FEMS Microbiology Reviews 23, 179–230.

Bridier, A., Briandet, R., Thomas, V., Dubois-Brissonnet, F., 2011. Resistance of bacterial biofilms to disinfectants: a review. Biofouling 27, 1017–1032.

Bryers, J.D., Ratner, B.D., 2004. Bioinspired implant materials befuddle bacteria. ASM News 70, 232–237.

Busscher, H.J., Bos, R., van der Mei, H.C., 1995. Initial microbial adhesion is a determinant for the strength of biofilm adhesion. FEMS Microbiology Letters 128, 229–234.

Chandy, J.P., Angles, M.L., 2001. Determination of nutrients limiting biofilm formation and the subsequent impact on disinfectant decay. Water Research 35, 2677–2682.

Chmielewski, R.A.N., Frank, J.F., 2003. Biofilm formation and control in food processing facilities. Comprehensive Reviews in Food Science and Food Safety 2, 22–32.

Chowdhury, S., 2012. Heterotrophic bacteria in drinking water distribution system: a review. Environmental Monitoring and Assessment 184, 6087–6137.

Codony, F., Morato, J., Ribas, F., Mas, J.D., 2002. Effect of chlorine, biodegradable dissolved organic carbon and suspended bacteria on biofilm development in drinking water systems. Journal of Basic Microbiology 42, 311–319.

Codony, F., Morato, J., Mas, J., 2005. Role of discontinuous chlorination on microbial production by drinking water biofilms. Water Research 39, 1896–1906.

Davey, M.E., ÓToole, G.A., 2000. Microbial biofilms : from ecology to molecular genetics. Microbiology and Molecular Biology Reviews 64, 847–867.

de Beer, D., Srinivasan, R., Stewart, P.S., 1994. Direct measurement of chlorine penetration into biofilms during disinfection. Applied and Environmental Microbiology 60, 4339–4344.

De Gusseme, B., Vanhaecke, L., Verstraete, W., Boon, N., 2011. Degradation of acetaminophen by *Delftia tsuruhatensis* and *Pseudomonas aeruginosa* in a membrane bioreactor. Water Research 45, 1829–1837.

Deborde, M., von Gunten, U., 2008. *R*eactions of chlorine with inorganic and organic compounds during water treatment- kinetics and mechanisms: a critical review. Water Research 42, 13–51.

Doggett, M.S., 2000. Characterization of fungal biofilms within a municipal water distribution system. Applied and Environmental Microbiology 66, 11–14.

Donlan, R., Costerton, J., 2002. Biofilms: survival mechanisms of clinically relevant microorganisms. Clinical Microbiology Reviews 15, 167–193.

Donlan, R.M., 2002. Biofilms: microbial life on surfaces. Emerging Infectious Diseases 8, 881–890.

Elbing, P.M., 2008. Effectiveness of Sodium Hypochlorite against Spores of *Penicillium brevicompactum* in an Insect Rearing Facility. Information Report 0832-7122; GLC-X-8E. Natural Resources Canada, Canadian Reserve Services, Great Lakes Forestry Centre, Sault Ste. Marie, Ontario.

Elvers, K.T., Leeming, K., Lappin-Scott, H.M., 2001. Binary culture biofilm formation by *Stenotrophomonas maltophilia* and *Fusarium oxysporum*. Journal of Industrial Microbiology & Biotechnology 26, 178–183.

Elvers, K.T., Leeming, K., Moore, C.P., Lappin-Scott, H.M., 1998. Bacterial-fungal biofilms in flowing water photo-processing tanks. Journal of Applied Microbiology 84, 607–618.

Farkas, A., Dragan-Bularda, M., Muntean, V., Ciataras, D., Tigan, S., 2012. Microbial activity in drinking water-associated biofilms. Central European Journal of Biology 8, 201–214.

Fernandes, S., Simões, M., Dias, N., Santos, C., Lima, N., 2013. Fungicidal activity of microbiocides. In: Fraise, A.P., Maillard, J.-Y., Sattar, S.A. (Eds.), Principles and Practice of Disinfection Preservation and Sterilization. Willey-Backwell Publishing Ltd.

Fernandes, S., Simões, L.C., Lima, N., Simões, M., 2019. Adhesion of filamentous fungi isolated from drinking water under different process conditions. Water Research 164, 114951.

Frey-Klett, P., Burlinson, P., Deveau, A., Barret, M., Tarkka, M., Sarniguet, A., 2011. Bacterial-fungal interactions: hyphens between agricultural, clinical, environmental, and food microbiologists. Microbiology and Molecular Biology Reviews 75, 583–609.

Gashgari, R.M., Elhariry, H.M., Gherbawy, Y.A., 2013. Molecular detection of mycobiota in drinking water at four different sampling points of water distribution system of Jeddah city (Saudi Arabia). Geomicrobiology Journal 30, 29–35.

Gilbert, N., 2012. Water under pressure. Nature 483, 256–257.
Gomes, I.B., Simões, M., Simões, L.C., 2014. An overview on the reactors to study drinking water biofilms. Water Research 62, 63–87.
Gonçalves, A.B., Paterson, R.R.M., Lima, N., 2006a. Survey and significance of filamentous fungi from tap water. International Journal of Hygiene and Environmental Health 209, 257–264.
Gonçalves, A.B., Santos, I.M., Paterson, R.R.M., Lima, N., 2006b. FISH and calcofluor staining techniques to detect in situ filamentous fungal biofilms in water. Revista Iberoamericana de Micologia 23, 194–198.
Göttlich, E., van der Lubbe, W., Lange, B., Fiedler, S., Melchert, I., Reifenrath, M., Flemming, H.-C., Hoog, S., 2002. Fungal flora in groundwater-derived public drinking water. International Journal of Hygiene and Environmental Health 205, 269–279.
Grabińska-Łoniewska, A., Koniłłowicz-Kowalska, T., Wardzyńska, G., Boryn, K., 2007. Occurrence of fungi in water distribution system. Polish Journal of Environmental Studies 16, 539–547.
Gutiérrez-Correa, M., Villena, G., 2003. Surface adhesion fermentation: a new fermentation category. Revista Peruana de Biología 10, 113–124.
Hageska, l, G., Knutsen, A.K., Gaustad, P., de Hoog, G.S., Skaar, I., 2006. Diversity and significance of mold species in Norwegian drinking water. Applied and Environmental Microbiology 72, 7586–7593.
Hageskal, G., Gaustad, P., Heier, B.T., Skaar, I., 2007. Occurrence of moulds in drinking water. Journal of Applied Microbiology 102, 774–780.
Hageskal, G., Lima, N., Skaar, I., 2009. The study of fungi in drinking water. Mycological Research 113, 165–172.
Hallam, N.B., West, J.R., Forster, C.F., Powell, J.C., Spencer, I., 2002. The decay of chlorine associated with the pipe wall in water distribution systems. Water Research 36, 3479–3488.
Harding, M.W., Marques, L.L.R., Howard, R.J., Olson, M.E., 2009. Can filamentous fungi form biofilms? Trends in Microbiology 17, 475–480.
Hayette, M.P., Christiaens, G., Mutsers, J., Barbier, C., Huynen, P., Melin, P., De Mol, P., 2010. Filamentous fungi recovered from the water distribution system of a Belgian university hospital. Medical Mycology 48, 969–974.
Huq, A., Whitehouse, C.A., Grim, C.J., Alam, M., Colwell, R.R., 2008. Biofilms in water, its role and impact in human disease transmission. Current Opinion in Biotechnology 19, 244–247.
Jones, E.B.G., 1994. Fungal adhesion. Mycological Research 98, 961–981.
Kelley, J., Paterson, R., Kinsey, G., Pitchers, R., Rossmoore, H., 1997. Identification, significance and control of fungi in water distribution systems. Water Technology Conference Proceedings. Public American Water Works Association, Denver, CO, US.
Kelley, J., Kinsey, G., Paterson, R., Brayford, D., Pitchers, R., Rossmore, H., 2003. Identification and Control of Fungi in Distribution Systems. AWWA Research Foundation and American Water Works Association, Denver.
Kim, S.D., Cho, J., Kim, I.S., Vanderford, B.J., Snyder, S.A., 2007. Occurrence and removal of pharmaceuticals and endocrine disruptors in South Korean surface, drinking, and waste waters. Water Research 41, 1013–1021.
Kinsey, G.C., Paterson, R.R.M., Kelley, J., 1999. Methods for the determination of filamentous fungi in treated and untreated waters. Journal of Applied Microbiology Symposium Supplement 85, 214S–224S.
Kinsey, G.C., Paterson, R.R.M., Kelley, J., 2003. Filamentous fungi in water systems. In: Mara, D., Horan, N. (Eds.), Handbook of Water and Wastewater Microbiology. Academic Press, London, UK.

Kobayashi, D.Y., Crouch, J.A., 2009. Bacterial/Fungal interactions: from pathogens to mutualistic endosymbionts. Annual Review of Phytopathology 47, 63–82.

Kokare, C.R., Chakraborty, S., Khopade, A.N., Mahadik, K.R., 2009. Biofilm: importance and applications. Indian Journal of Biotechnology 8, 159–168.

LeChevallier, M.W., Welch, N.J., Smith, D.B., 1996. Full-scale studies of factors related to coliform regrowth in drinking water. Applied and Environmental Microbiology 62, 2201–2211.

Liberti, L., Notarnicola, M., Petruzzelli, D., 2003. Advanced treatment for municipal wastewater reuse in agriculture. UV disinfection: parasite removal and by-product formation. Desalination 152, 315–324.

Lund, V., Ormerod, K., 1995. The influence of disinfection processes on biofilm formation in water distribution systems. Water Research 29, 1013–1021.

Ma, X., Baron, J.L., Vikram, A., Sout, J.E., Bibby, K., 2015. Fungal diversity and presence of potentially pathogenic fungi in a hospital hot water system treated with on-site monochloramine. Water Research 71, 197–206.

Momba, M.N.B., Cloete, T.E., Venter, S.N., Kfir, R., 1998. Evaluation of the impact of disinfection processes on the formation of biofilms in potable surface water distribution systems. Water Science and Technology 38, 283–289.

Momba, M., Kfir, R., Venter, S., Cloete, T., 2000. An overview of biofilm formation in distribution systems and its impact on the deterioration of water quality. Water S.A 26, 59–66.

Mowat, E., Butcher, J., Lang, S., Williams, C., Ramage, G., 2007. Development of a simple model for studying the effects of antifungal agents on multicellular communities of *Aspergillus fumigatus*. Journal of Medical Microbiology 56, 1205–1212.

Mowat, E., Williams, C., Jones, B., McChlery, S., Ramage, G., 2009. The characteristics of *Aspergillus fumigatus* mycetoma development: is this a biofilm? Medical Mycology: Official Publication of the International Society for Human and Animal Mycology 47, S120–S126.

Nazim, S., Dawar, S., Tariq, M., Zaki, M.J., 2008. Quantitative estimation of mycoflora in drinking water and fruit juices of Karachi. Pakistan Journal of Botany 240, 1263–1268.

Nieuwenhuijsen, M.J., Toledano, M.B., Eaton, N.E., Fawell, J., Elliott, P., 2000. Chlorination disinfection byproducts in water and their association with adverse reproductive outcomes: a review. Occupational and Environmental Medicine 57, 73–85.

Oliveira, H.M.B., 2010. Fungos filamentosos na água e em biofilmes na rede de distribuição de água potável do sistema Alto do Céu, Recife-PE. Masters thesis. In: Biologia de Fungos do Departamento de Micologia do Centro de Ciências Biológicas da Universidade Federal de Pernambuco (UFPE), Recife, Brazil.

O'Toole, G., Kaplan, H., Kolter, R., 2000. Biofilm formation as microbial development. Annual Review of Microbiology 54, 49–79.

Paterson, R.R.M., 2007. Zearalenone production and growth in drinking water inoculated with *Fusarium graminearum*. Mycological Progress 6, 109–113.

Paterson, R.R.M., Lima, N., 2005. Fungal contamination of drinking water. In: Lehr, J., Keeley, J., Lehr, J., Kingery, T.B. (Eds.), Water Encyclopedia. John Wiley & Sons, Inc.

Paterson, R.R.M., Hageskal, G., Skaar, I., Lima, N., 2009. Occurrence, problems, analysis and removal of filamentous fungi in drinking water. In: De Costa, P., Bezerra, P. (Eds.), Fungicides: Chemistry, Environmental Impact and Health Effects. Nova Science Publishers, Inc.

Pereira, V.J., Basílio, M.C., Fernandes, D., Domingues, M., Paiva, J.M., Benoliel, M.J., Crespo, M.T., San Romão, M.V., 2009. Occurrence of filamentous fungi and yeasts in three different drinking water sources. Water Research 43, 3813–3819.

Pereira, V.J., Fernandes, D., Carvalho, G., Benoliel, M.J., San Romão, M.V., Barreto Crespo, M.T., 2010. Assessment of the presence and dynamics of fungi in drinking water sources using cultural and molecular methods. Water Research 44, 4850–4859.

Peters, B.M., Jabra-Rizk, M.A., Scheper, M.A., Leid, J.G., Costerton, J.W., Shirtliff, M.E., 2010. Microbial interactions and differential protein expression in *Staphylococcus aureus* - *Candida albicans* dual-species biofilms. FEMS Immunology and Medical Microbiology 9, 493–503.

Rand, J.L., Hofmann, R., Alam, M.Z.B., Chauret, C., Cantwell, R., Andrews, R.C., Gagnon, G.A., 2007. A field study evaluation for mitigating biofouling with chlorine dioxide or chlorine integrated with UV disinfection. Water Research 41, 1939–1948.

Rasmussen, T.B., Skindersoe, M.E., Bjarnsholt, T., Phipps, R.K., Christensen, K.B., Jensen, P.O., Anderson, J.B., Koch, B., Larsen, T.O., Hentzer, M., Eberl, L., Hoiby, N., Givskov, M., 2005. Identity and effects of quorum-sensing inhibitors produced by *Penicillium* species. Microbiology 151, 1325–1340.

Rateb, M.E., Hallyburton, I., Houssen, W.E., Bull, A.T., Goodfellow, M., Santhanam, R., Jaspars, M., Ebel, R., 2013. Induction of diverse secondary metabolites in *Aspergillus fumigatus* by microbial co-culture. RSC Advances 3, 14444.

Reverberi, M., Ricelli, A., Zjalic, S., Fabbri, A.A., Fanelli, C., 2010. Natural functions of mycotoxins and control of their biosynthesis in fungi. Applied Microbiology and Biotechnology 87, 899–911.

Rosenzweig, W.D., Minnigh, H.A., Pipes, W.O., 1983. Chlorine demand and inactivation of fungal propagules. Applied and Environmental Microbiology 45, 182–186.

Sammon, N.B., Harrower, K.M., Fabbro, L.D., Reed, R.H., 2010. Incidence and distribution of microfungi in a treated 9 municipal water supply system in sub-tropical Australia. International Journal of Environmental Research and Public Health 7, 1597–1611.

Simões, L.C., 2013. Biofilms in drinking water. In: Simões, M., Mergulhão, F. (Eds.), Biofilms in Bioengineering. Nova Science Publishers, Inc.

Simões, L.C., Simões, M., 2013. Biofilms in drinking water: problems and solutions. *RSC Advances* 3, 2520.

Simões, L.C., Simões, M., Lima, N., 2015. Kinetics of biofilm formation by drinking water isolated *Penicillium expansum*. Biofouling 31, 349–362.

Simões, M., Simões, L.C., Vieira, M.J., 2010. A review of current and emergent biofilm control strategies. LWT - Food Science and Technology 43, 573–583.

Siqueira, V.M., 2011. Characterising Filamentous Fungal Biofilm in Drinking Water Distribution Systems Using Microscopic and Molecular Techniques. Ph.D. thesis. Universidade do Minho.

Siqueira, V.M., Lima, N., 2011. Efficacy of free chlorine against water biofilms and spores of *Penicillium brevicompactum*. In: Borchers, U., Thompson, K.C. (Eds.), Water Contamination Emergencies: Monitoring, Understanding and Acting. RSC Publishing, Cambridge.

Siqueira, V.M., Lima, N., 2013. Biofilm formation by filamentous fungi recovered from a water system. Journal of Mycology 2013, 1–9.

Siqueira, V.M., Oliveira, H.M.B., Santos, C., Paterson, R.R.M., Gusmão, N.B., Lima, N., 2011. Filamentous fungi in drinking water, particularly in relation to biofilm formation. International Journal of Environmental Research and Public Health 8, 456–469.

Sisti, M., Brandi, G., De Santi, M., Rinaldi, L., Schiavano, G.F., 2012. Disinfection efficacy of chlorine and peracetic acid alone or in combination against *Aspergillus spp.* and *Candida albicans* in drinking water. Journal of Water and Health 10 (1), 11–19.

Szewzyk, U., Szewzyk, R., Manz, W., Schleifer, K.H., 2000. Microbiological safety of drinking water. Annual Review of Microbiology 54, 81–127.

Tarkka, M.T., Sarniguet, A., Frey-Klett, P., 2009. Inter-kingdom encounters: recent advances in molecular bacterium-fungus interactions. Current Genetics 55, 233–243.

Thein, Z.M., Samaranayake, Y.H., Samaranayake, L.P., 2006. Effect of oral bacteria on growth and survival of *Candida albicans* biofilms. Archives of Oral Biology 51, 672–680.

Verstraeten, N., Braeken, K., Debkumari, B., Fauvart, M., Fransaer, J., Vermant, J., Michiels, J., 2008. Living on a surface: swarming and biofilm formation. Trends in Microbiology 16, 496–506.

Wargo, M.J., Hogan, D.A., 2006. Fungal-bacterial interactions: a mixed bag of mingling microbes. Current Opinion in Microbiology 9, 359–364.

WHO, 2011. Guidelines for Drinking-Water Quality, fourth ed. World Health Organization.

Wingender, J., Flemming, H.C., 2004. Contamination potential of drinking water distribution network biofilms. Water Science and Technology 49, 277–286.

Yamaguchi, M.U., Rampazzo, R.C.P., Ogatta, S.F.Y., Nakamura, C.V., Ueda-Nakamura, T., Filho, B.P.D., 2007. Yeasts and filamentous fungi in bottled mineral water and tap water from municipal supplies. Brazilian Archives of Biology and Technology 50, 1–9.

Yu, J., Kim, D., Lee, T., 2010. Microbial diversity in biofilms on water distribution pipes of different materials. Water Science and Technology 6, 163–171.

Zhou, L., Zhang, Y., Li, G., 2009. Effect of pipe material and low level disinfectants on biofilm development in a simulated drinking water distribution system. Journal of Zhejiang University - Science A 10, 725–731.

Chapter 6

Microalgal and cyanobacterial biofilms

Ana Cláudia Barros, A.L. Gonçalves, Manuel Simões
LEPABE – Laboratory for Process Engineering Environment, Biotechnology and Energy, Department of Chemical Engineering, Faculty of Engineering, University of Porto, Porto, Portugal

6.1 Microalgae and cyanobacteria

Microalgae and cyanobacteria are a diverse group of prokaryotic and eukaryotic photosynthetic microorganisms that can grow rapidly under hard conditions, which make them the most abundant organisms on earth (Li et al., 2008; Mata et al., 2010). They are present in both marine and freshwater environments, representing a huge variety of species (John et al., 2011). It is estimated that at least 50,000 species exist, but only a small amount have been studied (Richmond, 2004). Moreover, it is estimated that the biomass productivity of microalgae and cyanobacteria could be much higher than that of other microorganisms, such as switchgrass (Demirbaş, 2006).

These organisms can be classified into different phytoplankton taxonomic groups due to the wide range of structures, forms, and sizes already reported (Colares et al., 2013). They can be autotrophic, requiring inorganic compounds as carbon source and light as energy source or, otherwise, heterotrophic, requiring organic compounds as carbon and energy sources (Brennan and Owende, 2010). These microorganisms can also be mixotrophic, requiring both CO_2 and organic carbon (in this case, photosynthetic and respiratory metabolism occur simultaneously) (Aslan and Kapdan, 2006; Perez-Garcia et al., 2011). From now on, when we refer to microalgae, we include not only microalgae but also cyanobacteria.

6.2 Applications of microalgae and cyanobacteria

Over the past few years, the potential of microalgae and cyanobacteria in a wide range of applications has been ascertained, making them a valuable resource. Table 6.1 presents many applications already described for different microalgal/cyanobacterial species and the valuable products obtained by culturing these photosynthetic microorganisms.

Recent Trends in Biofilm Science and Technology. https://doi.org/10.1016/B978-0-12-819497-3.00006-4

TABLE 6.1 By-products and application areas described for different microalgal and cyanobacterial species.

Microorganisms	Applications	By-products	References
Chlorella sp.	Human nutrition, cosmetics, wastewater treatment		(Spolaore et al., 2006; Yamaguchi, 1996)
Chlorella minutissima	Food additive	Eicosapentaenoic acid	(Li et al., 2008)
Spirulina sp.	Human and animal nutrition, cosmetics	Nutraceuticals, phycobiliproteins, γ-linolenic acid	(Costa et al., 2003; Lee, 2001; Spolaore et al., 2006)
Dunaliella salina	Human and animal nutrition, cosmetics, health food, food additive	Carotenoids	(Becker, 1994; Brennan and Owende, 2010; Del Campo et al., 2007; Jin and Melis, 2003)
Haematococcus pluvialis	Human and animal nutrition, cosmetics, aquaculture	Carotenoids, astaxanthin, leutin	(Spolaore et al., 2006)
Porphyridium sp.	Clinical immunology	Phycobiliproteins	(Becker, 1994; Singh et al., 2005; Spolaore et al., 2006)
Spirulina platensis	Food additive	Proteins, polyunsaturated fatty acids, pigments, vitamins, phenolic compounds	(Colla et al., 2007; de Oliveira Rangel-Yagui et al., 2004; Madhyastha and Vatsala, 2007; Ogbonda et al., 2007; Soletto et al., 2005)
Schizochytrium sp.	Food additive, nutraceuticals	Docosahexaenoic acid	(Li et al., 2008)
Parietochloris incisa	Food additive, nutraceuticals	Arachidonic acid	(Li et al., 2008)
Euglena gracilis	Human nutrition	Biotin, α-tocopherol (vitamin E)	(Li et al., 2008)
Prototheca moriformis	Human nutrition	Ascorbic acid (vitamin C)	(Li et al., 2008)

Isochrysis galbana	Animal nutrition	Fatty acids	(Grima et al., 2003; Pulz and Gross, 2004)
Phaeodactylum tricornutum	Human nutrition, biofuels	Fatty acids	(Fernández et al., 1998; Priyadarshani and Rath, 2012; Yongmanitchai and Ward, 1991)
Odontella aurita	Pharmaceuticals, cosmetics, baby food	Fatty acids	(Del Campo et al., 2007; Priyadarshani and Rath, 2012)
Arthrospira sp.	Human and animal nutrition, cosmetics	Phycobiliproteins	(Delisle et al., 1991; Liang et al., 2004; Radmer, 1996; Stolz, 2005)
Porphyridium cruentum	Health food, pharmaceuticals, cosmetics	Arachidonic acid, phycocyanin, phycoerythrin, extracellular polysaccharides	(Borowitzka, 1997; Fuentes et al., 1999)
Chlorella vulgaris	Wastewater treatment, health food, food additive, feed surrogates	Fatty acids	(Aslan and Kapdan, 2006; Lee, 2001)
Aphanizomenon flos-aquae	Human nutrition		(Spolaore et al., 2006)
Crypthecodinium cohnii	Food additive, wastewater treatment	Docosahexaenoic acid	(Couto et al., 2010)
Schizochytrium sp.	Food additive, wastewater treatment	Docosahexaenoic acid	(Spolaore et al., 2006)
Lyngbya majuscula	Pharmaceuticals, human nutrition	Immune modulators	(Singh et al., 2005)
Muriellopsis sp.	Health food, food additive, animal nutrition	Carotenoids, lutein	(Blanco et al., 2007; Del Campo et al., 2007)

6.3 Microalgal/cyanobacterial cultivation

Microalgae and cyanobacteria can be cultured in closed systems, also called photobioreactors (PBRs), or in open systems (Borowitzka, 1999). The viability of each system depends on several factors, such as climatic conditions and intrinsic properties of the selected microalgal/cyanobacterial strain (Brennan and Owende, 2010).

In PBRs, the cultures are totally enclosed within a vessel and enable the control of all essential parameters and, therefore, the reproducibility of the cultivation conditions (Borowitzka, 1999; Pulz, 2001). The most commonly used PBRs are the tubular, flat plate, and column ones (Brennan and Owende, 2010; Posten, 2009; Ugwu et al., 2008; Xu et al., 2009). The use of closed PBRs is appropriated to overcome some problems related to the open pond systems (Brennan and Owende, 2010). Using these systems can be advantageous, since they have better light penetration, allowing higher biomass productivities and lower retention times. Also, PBRs allow the culture of single species of microalgae and/or cyanobacteria for longer periods of time with less contamination risks and no CO_2 losses (Pulz, 2001). However, due to their complexity, they require more energy and specialized people. Therefore, the investment and operation costs are much higher than those required by open ponds (Carvalho et al., 2006; Larsdotter, 2006; Ugwu et al., 2008). The scale-up of this kind of systems is also more difficult (Xu et al., 2009).

On the other hand, on open pond systems, the cultures are directly exposed to the environment, which makes them vulnerable to contamination by other microorganisms (Borowitzka and Moheimani, 2013). Shallow big ponds, circular ponds, raceway ponds, and tanks are the most commonly used systems (Lee, 2001; Ugwu et al., 2008). Some conditions, such as temperature and light, cannot be controlled for optimal microalgal/cyanobacterial growth. Therefore, the outdoor cultivation for commercial use is highly restricted to warm tropical locations, with low precipitation and cloud cover (Grobbelaar, 2013). Although they have larger production capacities, biomass productivities are lower than those achieved in PBRs, since they have poor mixing and oscillations in the culture conditions are very common (Christenson and Sims, 2011; Posten, 2009; Pulz, 2001; Ugwu et al., 2008). However, these systems are simpler than PBRs and, consequently, less expensive in terms of operation and production costs (Wolkers et al., 2011). The high evaporation rates can be seen as a disadvantage, but it also brings benefits to the culture system, since it helps with temperature control thanks to evaporative cooling (Christenson and Sims, 2011).

Both PBRs and open ponds are technologies based on microalgal/cyanobacterial growth in suspension. Indeed, the existing systems for both biomass production and wastewater treatment focus on microalgal/cyanobacterial growth in suspension on the culture medium. However, as biomass removal is essential for water recycling and commonly used harvesting procedures present some economic and technical limitations, new alternatives should be developed to optimize microalgal/cyanobacterial recovery. In this sense, microalgal/cyanobacterial immobilization has appeared as a valuable alternative for planktonic cultivation systems, since the cultured microorganisms grow in an immobilized form, making easy their removal at the end of cultivation time. Additionally, it has been reported that the application of an immobilized system in wastewater treatment processes allows more flexibility in terms of reactor design, comparatively to the planktonic ones (Mallick, 2002). Besides, immobilization provides higher cell densities and increased cell wall permeability and prevents washout and low leakage of cells from the matrix (Brouers and Hall, 1986; DOE, 1985; Mallick, 2002). The immobilization procedure begins with cell growth in a reactor, followed by harvesting and immobilization at a very high cell density. In 1966, for the first time in history, a study on immobilized algae was performed (Park et al., 1966). However, Prof. de la Noüe and his coworkers were the pioneers in using this technique applied to wastewater treatment. The promising results obtained have generated an increase in reports about microalgal/cyanobacterial immobilization for wastewater remediation applications (Mallick, 2002).

There are many methods of immobilization, such as adsorption, cross-linking, entrapment, covalent bonding, and encapsulation. Additionally, immobilization techniques comprise the immobilization in the surface of a substrate, in the form of biofilm, and the immobilization on the core of a matrix. An ideal substrate or matrix for immobilization should present the following properties: nontoxicity, stability, biomass retention, phototransparency, as well as resistance to disruption by cell growth (Mallick, 2002). These can be either a natural polymer, such as agar, agarose, collagen, cellulose, carrageen, and alginate, or a synthetic polymer, such as polyurethane, polyvinyl chloride, and acrylamide (Cai et al., 2013). A summary of the substrates and matrixes described for microalgal/cyanobacterial immobilization systems is presented in Table 6.2.

6.4 Microalgal harvesting techniques

The separation of microalgal biomass from the culture medium is called harvesting (Singh and Patidar, 2018). Although microalgae have a wide range of applications, their industrial production is still not economically viable (Christenson and Sims, 2011; Dassey and Theegala, 2013). Harvesting processes are responsible for 20%–30% of the overall production costs (Barros et al., 2015; Christenson and Sims, 2011; Rawat et al., 2011; Salim et al., 2011). An ideal scenario would be a harvesting process that achieves high amounts of biomass with minimum maintenance, energy, and operational costs (Danquah et al., 2009). Microalgae can be harvested by different methods: autoaggregation, gravitational sedimentation, centrifugation, chemical flocculation, bioflocculation, electrocoagulation–flocculation, or a combination of any of these (Grima et al., 2009; Mata et al., 2010).

TABLE 6.2 Summary of microalgae successfully immobilized onto different substrates and matrixes.

Microorganisms	Substrates/matrixes	Reference
Chlorella vulgaris	Glass, 3-aminopropyltriethoxysilane[a], propyltriethoxysilane[a]	(Sirmerova et al., 2013)
Scenedesmus obliquus *C. vulgaris* *Coccomyxa* sp. *Nannochloris* sp. *Nitzschia palea* *Oocystis* sp. *Oocystis polymorpha*	Polystyrene, polycarbonate, acrylic, glass	(Genin et al., 2014)
C. vulgaris *Nitzschia amphibia* *Chroococcus minutus*	Poly(methyl methacrylate), titanium, glass, copper[b], stainless steel, admiralty brass[b], aluminum brass[b]	(Sekar et al., 2004)
S. obliquus *C. vulgaris*	Borosilicate glass, polyethylene, polyurethane, poly(methyl methacrylate)	(Irving and Allen, 2011)
Botryococcus braunii *Botryococcus sudeticus* *Ankistrodesmus falcatus* *Cylindrotheca fusiformis* *Amphora coffeaeformis* *Nitzschia frustulum* *Scenedesmus dimorphus* *C. vulgaris* *Nannochloropsis oculata*	Glass, indium–tin oxide, stainless steel, polycarbonate, polyethylene, polystyrene	(Ozkan and Berberoglu, 2013)
Prototheca zopfii	Polyurethane	(Yamaguchi et al., 1999)
Aulosira fertilissima	Glass	(Banerjee et al., 2004)
Scenedesmus acutus *C. vulgaris*	Polyurethane	(Terry and Stone, 2002)
Phormidium laminosum *Phormidium uncinatum*	Polyvinyl chloride	(Garbisu et al., 1993; Gil and Serra, 1993)
Chlorella kessleri *C. vulgaris*	Polystyrene Polyurethane	(Travieso et al., 1996)
S. obliquus *C. vulgaris* *Coccomyxa* sp. *Nannochloris* sp. *N. palea* *Oocystis* sp. *O. polymorpha*	Cellulose acetate, silicone rubber	(Genin et al., 2014)

TABLE 6.2 Summary of microalgae successfully immobilized onto different substrates and matrixes.—cont'd

Microorganisms	Substrates/matrixes	Reference
Ascophyllum nodosum *C. vulgaris* *Anabaena doliolum* *Chlorella emersonii* *Chlorella homoshaera* *Chlorella salina* *Chlorella ellipsoidea* *Scenedesmus quadricauda* *Navicula canalis* *Nannochloropsis gaditana* *S. obliquus* *Tetraselmis chuii*	Alginate	(Abdel Hameed, 2002; Awasthi and Das, 2004; da Costa and Leite, 1991; Garnham et al., 1992; Ilangovan et al., 1998; Lau et al., 1998; Moreno-Garrido et al., 2002; Robinson and Wilkinson, 1994; Tam et al., 1998; Volesky and Prasetyo, 1994)
Chlorella emersonii	Agarose	(Robinson and Wilkinson, 1994)
Spirulina platensis *C. vulgaris*	Silica gel	(Rangsayatorn et al., 2004)
Scenedesmus acutus *C. vulgaris*	Carrageenan	(Travieso et al., 1999)
Chlorella sorokiniana	Loofa sponge	(Akhtar et al., 2004)
Phormidium sp. *Scenedesmus bicellularis*	Chitosan	(de la Noüe and Proulx, 1988; Kaya and Picard, 1996)
P. laminosum	Cellulose	(Sawayama et al., 1998)
C. kessleri *C. vulgaris*	Carrageenan, alginate	(Travieso et al., 1996)
C. vulgaris *S. dimorphus*	Acrylic, polypropylene, stainless steel, polycarbonate	(Roostaei et al., 2018)
Anabaena sp. *C. vulgaris*	Muslin, olefin, nylon, polyester	(Venable and Podbielski, 2019)
Nostoc muscorum *Mastigocladus laminosus*	Agar	(Rao and Hall, 1984)

[a]*Modified surfaces;*
[b]*Poor attachment.*

6.4.1 Chemical flocculation

Chemical flocculation is commonly used to a wide variety of applications, including those of nonalgae nature (Osborne, 2009). This method can be used to concentrate a suspension 200–300 times (Vandamme et al., 2013). Microalgae carry a negative charge that prevents the aggregation of suspended cells (Grima et al., 2003). The charge can be neutralized or reduced by adding chemical flocculants, such as alkaline compounds, metallic salts, or polyelectrolytes, resulting in the formation of metallic hydroxides that precipitate and take microalgal cells as they settle (González-Fernández and Ballesteros, 2013; Grima et al., 2003; Osborne, 2009). The most used flocculants are ferric chloride ($FeCl_3$), aluminum sulfate ($Al_2(SO_4)_3$), and ferric sulfate ($Fe_2(SO_4)_3$) (Gultom and Hu, 2013). However, the use of flocculant can increase the overall costs and affect negatively the downstream processing. So, inexpensive and nontoxic flocculants should be considered. Alternatively, nonchemical compounds can be used, such as potato starch or chitosan (Grima et al., 2003).

6.4.2 Autoflocculation

Autoflocculation occurs when there is a destabilization of microalgae particles, due to pH increasing to the point that the precipitation of carbonate salts with algal cells occurs, a consequence of photosynthetic CO_2 consumption. Increase in pH can occur naturally (photosynthetic cycles) or can be stimulated by the addition of NaOH (Gouveia, 2011).

6.4.3 Bioflocculation

In bioflocculation, flocculation is mediated by the presence of microorganisms or bioflocculants, which are biodegradable macromolecular flocculant secreted by microorganisms (Gao et al., 2006). This leads to changes in the surface charge that in turn causes the cells to aggregate (Lincoln and Koopman, 1986; Mohn, 1988). Microbial flocculation has been reported since *Pasteur* in 1876 for *Saccharomyces cerevisiae*. This process has applications in many areas such as wastewater treatment and fermentation (Lee et al., 2009). In their study, Shipin et al. (1999) reported that all known microalgae produce exopolysaccharides, which are indistinguishable from those produced by bacteria. These polymers will be responsible for the cell adhesion by bridging or charge neutralization, colloid entrapment, and double layer compression (Frølund et al., 1996; Lee et al., 2009; Salehizadeh and Shojaosadati, 2001). Besides, microalgae coated with polymer matrix did not show signs of stress or lysis (Shipin et al., 1999). Bacteria, actinomycetes, fungi, and algae are designed as producers of bioflocculants. These molecules are biodegradable and harmless (Abd-El-Haleem et al., 2008; Sanayei et al., 2010). But, there have been reported that some bioflocculants are not independent, and their dependency on cations to flocculate will bring pollution and more costs to the process

(Dermlim et al., 1999; Zhang et al., 2007). The bioflocculation process and secretion of flocculants are mostly notable in the wastewater treatment, in the treatment of activated sludge (Salehizadeh and Shojaosadati, 2001).

6.4.4 Electrocoagulation–flocculation

Electrocoagulation–flocculation (ECF) has arrived as an alternative for chemical flocculants (Vandamme et al., 2011). In ECF, microalgae are treated with an electrical field. Coagulants species, such as aluminum or iron ions, are generated in situ, due to electrolytic oxidation of an appropriate anode material (Harif et al., 2012; Vandamme et al., 2011). This technique has the advantage that no anions remain in the medium. Also, ECF is able to remove up to 95% of microalgae. Nevertheless, the electrolytic oxidation requires electricity, and there is a possibility of cell oxidation that leads to undesirable changes in lipid profile (Poelman et al., 1997). Furthermore, the harvesting efficiency was only tested for freshwater.

6.4.5 Gravitational sedimentation

Gravitational sedimentation allows a solid–liquid separation, because it separates a feed suspension into a slurry of higher concentration and an effluent with clear liquid (Svarovsky, 2000). This technique can be explained by Stoke's law, since it determines the characteristics of suspended solid by density, radius, and sedimentation velocity of algae cells. However, it is only applied to nonflocculating particles due to high complexity of the flocks (Schenk et al., 2008; Shelef et al., 1984). Despite the fact that this technology be considered as simple and low cost, this method is only suitable for larger particles (>70 μm) (Munoz and Guieysse, 2006). Sedimentation is a very slow process, which could cause deterioration of some biomass and become inapplicable to routine harvesting. However, it is a very simple process, with high energy efficiency and mostly used and water and wastewater treatment (Christenson and Sims, 2011; Pittman et al., 2011; Rawat et al., 2011).

6.4.6 Flotation

In flotation, gas bubbles fed to the broth provide the lifting force needed for particle transport and separation. This method can be compared with sedimentation for a large number of microalgal species, which explains why flotation is also designed as "inverted" sedimentation (Edzwald, 1993; Oswald, 1988; Singh et al., 2005). However, flotation can be more rapid and effective than sedimentation (Milledge and Heaven, 2013). Furthermore, this technique has relatively short operation times and requires low space equipment and lower initial equipment costs (Garg et al., 2012; Show et al., 2013). Although some microalgal strains naturally float to the surface when their lipid content increases, in most cases, flocculants need to be added to improve flotation efficiency (Gultom and Hu, 2013). Despite the many advantages of this technique, the high capital and operational costs, as well as the additional costs due to flocculants adding, make this technology not suitable for products with lower added value (Becker, 1994; Henderson et al., 2008). The possibility of contamination is a problem that raises in large-scale processes (Rawat et al., 2011).

6.4.7 Centrifugation

In centrifugation, a much greater force than gravity is used to make algae to settle to the bottom of the tank (Milledge and Heaven, 2013). This method can provide algae removal efficiencies higher than 95%, regardless of algae type, media, and other characteristics (Becker, 1994; Knuckey et al., 2006). Nevertheless, this method requires high capital and operational costs due to the energy consumption and the equipment maintenance (Milledge and Heaven, 2013). Although centrifugation is considered as the preferable and fastest harvesting method, the composition of the cell may change (González-Fernández and Ballesteros, 2013; Gultom and Hu, 2013).

6.4.8 Filtration

Filtration is used when large microalgal cells are present, but it cannot separate the biomass from the cultivation medium efficiently in microalgal cells with smaller dimensions (less than 10 μm). On the other hand, there are a wide variety of filter and membrane types available, which can be applied to more species. However, when membranes with small size pores are used, the pressure input and the operational energy required will be higher (Grima et al., 2003). Besides, in this process, clogging or fouling issues have been reported as a problem that is common (Milledge and Heaven, 2013). This is the cheapest harvesting method, since there are many costs associated with pumps and membrane maintenance or replacement (Bilad et al., 2012). Equipment operational energy requirements can be reduced by filter aids and flocculants, but additional materials need to be removed from the microalgal biomass, which increase costs (González-Fernández and Ballesteros, 2013).

6.5 Factors affecting microalgal/cyanobacterial biofilms

There are many factors influencing microalgal/cyanobacterial biofilm formation, such as environmental conditions (light, pH, nutrient composition, temperature), hydrodynamic conditions, the properties of the adhesion surface (hydrophobicity), and the characteristics of the microorganism (motility, surface hydrophobicity, chemotaxis, metabolic interactions, cell size, ability to produce EPS) (Choi et al., 2013; Simões, 2013).

6.5.1 Light

Light is one of the most important environmental factors that influences the development of microalgal/cyanobacterial biofilms (Hill et al., 1995). Light can affect the formation of algal biofilms through both photoinhibition (excessive light supply in the upper layers) or photolimitation (low light supply in the shaded layers) (Irving and Allen, 2011; Kesaano and Sims, 2014). Many researchers have found that photon flux densities can be critical in microalgal biofilm growth (Cheng et al., 2013; Schnurr et al., 2014; Wilkinson et al., 1990), but the specific value at which photoinhibition or light limitation occurs varies with species (Kesaano and Sims, 2014). Light intensity influences not only the adhesion of microalgae but also the EPS production (Katarzyna et al., 2015; Zippel et al., 2007).

Although light intensity is critical for microalgal biofilm growth, light quality is also extremely important as it can affect lipid composition and content and, therefore, biofilm growth (Bigogno et al., 2002; Hultberg et al., 2014; Khozin-Goldberg et al., 2002). Furthermore, in the study of Hultberg et al. (2014), they showed that different light colors (purple, blue, and white) could speed up biofilm formation by *Chlorella vulgaris* in comparison with green, yellow, and red light exposure. Additionally, light affects nutrients removal by influencing microalgal growth, whereby the selection of appropriate light conditions is essential for successful application of microalgal and cyanobacterial biofilms in wastewater treatment processes (Kesaano and Sims, 2014).

6.5.2 CO_2 concentration

During microalgal/cyanobacterial biofilm growth, the concentration of CO_2 is affected by the pH, and the pH is affected by the utilization of CO_2. More specifically, since photosynthesis requires CO_2, there is a rapid consumption of this compound, thus raising the pH of the culture medium. As CO_2 becomes limiting with distance into the biofilm due to the total inorganic carbon decrease, the HCO_3^- will form CO_2 and OH^-, leading to pH increase and the establishment of an equilibrium (Liehr et al., 1988, 1989).

6.5.3 pH

Microalgal biofilms are highly influenced by pH, as it is essential for cell establishment and biofilm growth (Katarzyna et al., 2015; Liehr et al., 1988). Additionally, depending on the microalgal species, more alkaline or acidic conditions can be preferred (Lavens and Sorgeloos, 1996; Starks et al., 1981). Usually, microalgal growth occurs for pH between 6 and 9 (Lavens and Sorgeloos, 1996). The same behavior was observed for microalgal attachment. In the study of Sekar et al. (2004), they observed that for pH ≥ 7, the microalga *Nitzschia amphibia* adhered more. The exception are microalgae from soil, which prefer acidic conditions.

6.5.4 Nutrients

Nutrients limitation can affect the biofilm growth, the type of biofilm produced, and the composition of species present on the biofilm (Hillebrand and Kahlert, 2001; Kesaano and Sims, 2014; Sekar et al., 2002). Heterotrophic biofilms can be favored by high contents of biodegradable organic matter (Olapade and Leff, 2006). Conversely, photoautotrophic biofilms can be enhanced by the presence of inorganic nutrients and light (Hillebrand et al., 2002). Carbon, nitrogen, and phosphorus are the most important nutrients for biofilm growth. So, C:N:P ratios are widely used to predict and minimize nutrient limitation in microalgae and cyanobacteria growth (Hillebrand and Sommer, 1999; Stelzer and Lamberti, 2001).

6.5.5 Temperature

Many metabolic reactions are influenced by temperature, meaning that microalgal/cyanobacterial biofilm formation is also affected by temperature oscillations (Babu, 2011). Some authors have found that warm temperatures can promote biofilm formation, whereas lower temperatures can inhibit biofilm growth (Babu, 2011; Donlan et al., 1994; Kesaano and Sims, 2014). So, it is expected that in tropical regions, where temperature is usually warm, biofilm formation is higher than in cold regions (Babu, 2011). It has been reported that diatoms are established at temperatures from 5 to 15°C, green algae at 15–30°C, and cyanobacteria above 30°C (DeNicola, 1996). However, the biofilm response to temperature changes is known to be species dependent (Kesaano and Sims, 2014).

6.5.6 Surface properties

The type of surface used can also influence biofilm formation. The most important surface properties are hydrophobicity, surface tension, roughness, and charge. Several studies have demonstrated that the extent of biofilm increases as the surface roughness increases, since the increased area provides more sites for colonization. According to the literature (Boyd et al., 2002; Taylor et al., 1998), a small increase in surface roughness leads to a significant adhesion of microbial cells. However, a larger increase in surface roughness can have no significant effect in microbial adhesion. Also, porous surfaces have been associated to higher cell attachment (Babu, 2011). Regarding hydrophobicity, some authors have demonstrated that cells adhere more to surfaces with low free energy of hydrophobic interaction (hydrophobic surfaces) than to hydrophilic surfaces (Ang, 2012; Becker, 1996; Becker and Wahl, 1991).

6.5.7 Hydrodynamic conditions

Biofilms can grow in laminar or turbulent flow (Kokare et al., 2009). Hydrodynamic conditions control drag forces and mass transfer (transport of cells, oxygen, and nutrients), playing an important role on biofilm formation (Simoes et al., 2007; Stoodley et al., 1998). Biofilms developed under laminar flow are different from those of turbulent flow. The first ones correspond to cell aggregates with interstitial voids, whereas the others consist of elongated streamers oscillating in the bulk fluid and forming a thicker biofilm (Kokare et al., 2009; Simões et al., 2004). Therefore, the hydrodynamic conditions imposed into the system will influence the nature of the biofilm, as well as the phenotype of microbial cells.

6.5.8 Extracellular polymeric substances

All benthic algae, diatoms, and cyanobacteria produce EPS (Evans, 2003). Diatoms are the most abundant EPS producers (Flemming and Wingender, 2010). The attachment of cells and biofilm formation is not possible without a matrix of EPS. EPS consist of a wide diversity of organic molecules, such as proteins, polysaccharides, amino acids, and uronic acid (Babu, 2011; Zhang and Fang, 2001). They have many advantages, such as protection of microorganisms from toxic substances and prevention of cells from desiccation (Babu, 2011). These substances are responsible for biofilms mechanical stability and cohesion and mediate their attachment to surfaces (Flemming and Wingender, 2010). Also, some authors found that a coating of EPS can protect microalgae and cyanobacteria against phagocytosis, dehydration, lysis, antibody recognition, and contact with toxic heavy metals (De Philippis and Vincenzini, 2003; De Vuyst and Degeest, 1999; Dudman, 1977; Hill et al., 1994; Scott et al., 1996; Tease and Walker, 1987). EPS represent 50%–90% of the organic carbon of the biofilm (Rossi et al., 2012). Both uronic acids and sulfate groups contribute to the negative charge (anionic nature) of the EPS, explaining the affinity between EPS and cations, such as metal ions. So, the presence of negatively charged polysaccharides around microalgae and cyanobacteria may help in metal cations sequestration, which can be favorable when there are low concentrations of metals that are essential for biofilm growth (Pereira et al., 2009).

6.5.9 Microalgal species

Species selection is the most important factor in biofilm formation. While some microalgal species grow better in suspension, others prefer the sessile state (Katarzyna et al., 2015). *Chlorella* and *Scenedesmus* are the two most studied species in which concerns biofilm experiments. Additionally, it was found that some microalgae are not able to produce EPS by themselves, which will affect biofilm structure and composition (Irving and Allen, 2011). It is well known that EPS protects the biofilm cells from the environment and provides the necessary nutrients for their growth (Shen et al., 2014). Additionally, the cell size or shape can influence biofilm formation. Filamentous species with large cells can favor biofilm thickening and biomass production, since they are less susceptible to external disturbances than unicellular or spherical ones (Wang et al., 2018). On the other hand, Shen et al. (2014) found that unicellular microalgae with spherical cells with 10–12 μm in size can promote a higher adhesion rate.

6.5.10 Presence of other microorganisms

In the past few years, the interactions between microalgae and other microorganisms have been studied. Although in the beginnings, the investigators looked at the presence of bacteria in microalgal medium as an undesirable contamination, more recently, it has been found that these interactions can be advantageous for microalgae (Mantzorou and Ververidis, 2018). Some authors reported that bacteria promote initial adhesion of microalgae through the establishment of strong interactions and production of EPS, in the surface conditioning phase (Lorite et al., 2011). There is a synergistic relationship between microalgae and bacteria. On the one hand, microalgae produce O_2 during photosynthesis, which is used by bacteria to oxidize organic matter and $NH4^+$. On the other hand, bacteria produce CO_2 that is then used my microalgae in their photosynthetic process (De Godos et al., 2009).

Most research studies focusing on the relationship between microalgae and bacteria were performed with members of the genus *Chlorella* (Lian et al., 2018). Several studies have demonstrated the beneficial effect of bacteria on microalgal growth, by increasing either the cell density (Amin et al., 2009; Cho et al., 2015; Hernandez et al., 2009; Park et al., 2008), the biomass productivity (Le Chevanton et al., 2013; Lépinay et al., 2018; Tanabe et al., 2015), or the growth rate (Kim et al., 2014; Lakaniemi et al., 2012; Watanabe et al., 2005). In the study of Boelee (2013), additional removal of nitrate, phosphate, and acetate was achieved in the presence of microalgae–bacterial biofilms.

On the opposite, some studies pointed out that bacteria have a negative influence on microalgal biofilms (Kim et al., 2008; Lee et al., 2010; Wang et al., 2010). In those cases, microalgal cells can be lysed to release intracellular nutrients for bacterial growth (Fuentes et al., 2016).

6.6 The role of microalgal/cyanobacterial biofilms in wastewater treatment processes

The recent increase of industrialization, agricultural practices, and urbanization has conducted to an excessive input of nutrients in water resources

(Aslan and Kapdan, 2006). Worldwide, water pollution has become a major problem, as it can lead to the degradation of natural ecosystems. Therefore, there is an urgent need for new wastewater treatment technologies that can effectively remove nutrients and, at the same time, be economically feasible (Shi et al., 2007). In fact, wastewaters are mainly composed by water (99.9%) and organic and inorganic compounds (0.1%) (Olutiola et al., 2010). Among all nutrients present in wastewater, nitrogen and phosphorus are the most abundant and, when in excess, can cause eutrophication (Kloc and Mendoza González, 2012). Wastewaters may also contain heavy metals, such as cadmium, copper, zinc, and lead, which can be dangerous for human health if not removed (Mendoza Gonzalez and Kloc, 2012). Recently, with the increasing concerns about wastewater remediation strategies for nutrient control, microalgal and cyanobacterial biofilms have emerged as a cost-effective and eco-friendly alternative to microalgal suspended-based technologies (Christenson and Sims, 2012; Kesaano and Sims, 2014). The microalgal/cyanobacterial remediation mechanism relies on nitrogen and phosphorus uptake from the effluent and the assimilation of these inorganic nutrients toward their growth (Pittman et al., 2011). Besides, biofilms have the ability to retain biomass that, after chemical and biological processing, can be transformed into high-valued products and fertilizers (Boelee et al., 2011; Rahman et al., 2012), resulting in the recycling of the nitrogen and phosphorus wastes. Their ability to retain microalgal/cyanobacterial biomass in a substrate or material makes microalgal biofilms a sustainable alternative for wastewater treatment, reducing the costs associated to the harvesting of biomass before effluent discharge into water bodies (Boelee et al., 2011; Hwang et al., 1998).

Several studies have focused on nutrients removal from wastewaters using microalgal and cyanobacterial biofilms (Chevalier et al., 2000; Doria et al., 2012; Renuka et al., 2013). Among all microalgae applied in wastewater treatment for nutrients removal, *Chlorella*, *Spirulina*, *Phormidium*, *Botryococcus*, *Chlamydomonas*, and *Scenedesmus* are the most widely used genera (Aslan and Kapdan, 2006; Rawat et al., 2011). The reactors/technologies used on some of those studies are described in Table 6.3, along with the operational conditions applied and nutrients removal efficiencies achieved.

The studies summarized in Table 6.3 show that microalgal and cyanobacterial biofilm systems can be used to effectively remove nitrogen and phosphorus from wastewaters. However, their efficacy will rely on many other factors apart from nutrients removal efficiencies, such as the biomass productivities, the area required, and the final concentration of wastewater, which will have to be in accordance with the limits for discharge effluents defined by European Union (EU) legislation (Boelee, 2013).

TABLE 6.3 Microalgal/cyanobacterial biofilm reactors, operation conditions, and nutrients removal efficiencies reported in the literature for the treatment of wastewaters from different sources.

Microorganisms	Reactor	Conditions	Wastewater source	Nutrients removal	Reference
Chlorella spp. + *Nitzschia* spp.	ND	OM=ND; V = 15 L; HRT = 10 h; T = ND; LI = 400 ft.c; L:D = ND	Settled domestic sewage	92% N 74% P	(McGriff and McKinney, 1972)
Microalgal consortia	Flow cell	OM = continuous; V = 0.36 L; HRT = ND; T = 22°C; LI = 230 μmol m^{-2} s^{-1}; L:D = ND	Domestic wastewater	ND	(Boelee et al., 2011)
Chlorella vulgaris	Two raceways	OM = continuous; V = 2200 L; HRT = 4.5 d; T = ND; LI = ND; L:D = ND	Diluted pig slurry	54%–98% N 42%–89% P	Fallowfield and Garrett, 1985)
C. vulgaris	Algal biofilm membrane photobioreactor	OM = continuous; V=ND; HRT = 2 d; T = 25–28°C; LI = 8000 lux; L:D = ND	Simulated secondary effluent	96% N 85% P	(Gao et al., 2015)
Scenedesmus sp.	PMMA reactor with cylindrical polypropylene fiber bundle carriers	OM = continuous; V = 96 L; HRT = 2 d; T = 20–22°C; LI = 2800 lx; L:D = 6:6	Municipal wastewater	24%–55% N	(He and Xue, 2010)
Microalgal consortia	Lab-scale ATS raceways with polyethylene mesh	OM=ND; V = 3700 L; HRT=ND; T = 19–24°C; LI = 270–390 μmol m^{-2} s^{-1}; L:D = 23:1	Anaerobically digested dairy manure	70%–90% N 70%–90% P	(Mulbry et al., 2008)
Scenedesmus obliquus *C. vulgaris*	PMMA flow cell with coupons	OM = fed-batch; V = 0.48 mL; HRT=ND; T = 23–27°C; LI = 100 μmol m^{-2} s^{-1}; L:D = 24:0	Municipal wastewater	ND	(Irving and Allen, 2011)

Trentepohlia aurea	Filter paper	OM = ND; V = ND; HRT = ND; T = 25°C; LI = 10 μmol m^{-2} s^{-1}; L:D = 24:0	Synthetic wastewater	75% N	(Abe et al., 2003)
Microalgal consortia	Polycarbonate flow lanes photobioreactor	OM = ND; V = ND; HRT = ND; T = 20–30°C; LI = 15–120 μmol m^{-2} s^{-1}; L:D = 16:8	Domestic sewage	100% P	(Guzzon et al., 2008)
Chlorella sp.	Polystyrene rocker system	OM = batch; V = ND; HRT = 6 d; T = 20°C; LI = 110–120 μmol m^{-2} s^{-1}; L:D = 24:0	Dairy manure	62% N 73% P	(Johnson and Wen, 2010)
Microalgal consortia	Lab-scale ATS units with polyethylene mesh	OM = continuous; V = ND; HRT = ND; T = 22°C; LI = 40–140 μmol m^{-2} s^{-1}; L:D = 16:8	Raw and anaerobically digested dairy manure	98% N 99% P	(Mulbry and Wilkie, 2001)
Microalgal consortia	Pilot-scale RABR	OM = continuous; V = 8000 L; HRT = 0.25 d; T = 9.6–19.2°C; LI = 170 μmol m^{-2} s^{-1}; L:D = 14:10	Municipal wastewater	76% N 24% P	(Christenson and Sims, 2012)
Chlorella sorokiniana Ralstonia basilensis	Flat plate photobioreactor	OM = continuous; V = 0.709 L; HRT = 3 d; T = 25°C; LI = 180 μE m^{-2} s^{-1}; L:D = 14:10	Mineral salt medium	ND	(Muñoz et al., 2009)
C. sorokiniana R. basilensis	Algal turf reactor shallow open pond	OM = continuous; V = 0.13 L; HRT = 1.3 d; T = 25°C; LI = 180 μE m^{-2} s^{-1}; L:D = 14:10	Mineral salt medium	ND	(Muñoz et al., 2009)
Botryococcus braunii	Multilayers photobioreactor	OM = continuous; V = 10°L; HRT = ND; T = 26°C; LI = 100 μmol m^{-2} s^{-1}; L:D = 12:12	Manure wastewater	ND	(Shen et al., 2015)

Continued

TABLE 6.3 Microalgal/cyanobacterial biofilm reactors, operation conditions, and nutrients removal efficiencies reported in the literature for the treatment of wastewaters from different sources.—cont'd

Microorganisms	Reactor	Conditions	Wastewater source	Nutrients removal	Reference
C. vulgaris Scenedesmus rubescens	Twin layers photobioreactor	OM = continuous; V = 2 L; HRT = ND; T = 30°C; LI = 20–120 μmol photons $m^{-2} s^{-1}$; L:D = ND	Municipal and synthetic secondary wastewater	94%–96% N 90% P	(Shi et al., 2007)
S. obliquus	Algem™ lab-scale photobioreactor	OM = continuous; V = 0.6 L; HRT = 3,6,12,20 h; T = 20°C; LI = ND; 200 μmol $m^{-2} s^{-1}$; L:D = ND	Secondary wastewater	70% N 33%–89% P	(Whitton et al., 2018)
Cyanobacteria and green algae consortia	Horizontal flat panel photobioreactor	OM = continuous; V = 100 L; HRT = 7 min; T = 19–24°C; LI = 90 ± 15 mmol photons $m^{-2} s^{-1}$; L:D = 24:0	Artificial wastewater	97% P	(Sukačová et al., 2015)
C. sorokiniana R. basilensis	Tubular photobioreactor	OM = continuous; V = 0.251 L; HRT = 2.3 d; T = 25°C; LI = 180 μE m^{-2} s^{-1}; L:D = 14:10	Mineral salt medium	ND	(Muñoz et al., 2009)

ATS, algal turf scrubber; *HRT*, hydraulic retention time; *L:D*, light:dark photoperiod; *LI*, light intensity; *N*, nitrogen; *ND*, not defined; *OM*, operation mode; *P*, phosphorus; *PMMA*, poly(methyl methacrylate); *RABR*, rotating algal biofilm reactor; *T*, temperature; *V*, volume.

6.7 Conclusions

The interest and research in microalgae have increased in the past years, due to their widespread range of applications. Microalgae can be used to produce biodiesel and other high-value products and, at the same time, treat wastewater by removing pollutants. Although the use of microalgae for industrial processes reveals to be quite promising, their application in large-scale systems in a cost-effective way remains a challenge. Microalgal harvesting in the end of the processes is hard and costly. In this sense, the growth of microalgae attached to surfaces emerged as a good alternative to suspended cultivation systems. There are many factors affecting microalgal biofilm, which includes light, CO_2 concentration, pH, hydrodynamic conditions, nutrients, temperature, EPS, surface properties, microalgal species, and the presence of other microorganisms. Therefore, such parameters must be in consideration when designing the reactor toward microalgal final application.

Acknowledgments

This work was financially supported by Project UID/EQU/00511/2019—Laboratory for Process Engineering, Environment, Biotechnology and Energy (LEPABE) funded by national funds through FCT/MCTES (PIDDAC); Project POCI-01-0247-FEDER-033298, funded by FEDER funds through COMPETE2020—Programa Operacional Competitividade e Internacionalização (POCI) and by national funds (PIDDAC) through FCT/MCTES; Project "LEPABE-2-ECO-INNOVATION"—NORTE-01-0145-FEDER-000005, funded by Norte Portugal Regional Operational Programme (NORTE 2020), under PORTUGAL 2020 Partnership Agreement, through the European Regional Development Fund (ERDF); Ph.D. grant awarded to Ana Barros (SFRH/BD/146028/2019/J653362bnpwq); grant attributed to Manuel Simões (SFRH/BSAB/150379/2019).

References

Abd-El-Haleem, D.A., Al-Thani, R.F., Al-Mokemy, T., Al-Marii, S., Hassan, F., 2008. Isolation and characterization of extracellular bioflocculants produced by bacteria isolated from Qatari ecosystems. Polish Journal of Microbiology 57 (3), 231–239.

Abdel Hameed, M., 2002. Effect of immobilization on growth and photosynthesis of the green alga *Chlorella vulgaris* and its efficiency in heavy metals removal. Bulletin of the Faculty of Science Assiut University 31 (1-D), 233–240.

Abe, K., Matsumura, I., Imamaki, A., Hirano, M., 2003. Removal of inorganic nitrogen sources from water by the algal biofilm of the aerial microalga *Trentepohlia aurea*. World Journal of Microbiology and Biotechnology 19 (3), 325–328.

Akhtar, N., Iqbal, J., Iqbal, M., 2004. Enhancement of lead (II) biosorption by microalgal biomass immobilized onto loofa (*Luffa cylindrica*) sponge. Engineering in Life Sciences 4 (2), 171–178.

Amin, S.A., Green, D.H., Hart, M.C., Küpper, F.C., Sunda, W.G., Carrano, C.J., 2009. Photolysis of iron–siderophore chelates promotes bacterial–algal mutualism. Proceedings of the National Academy of Sciences 106 (40), 17071–17076.

Ang Jr., P.O., 2012. Asian Pacific Phycology in the 21st Century: Prospects and Challenges: Proceeding of the Second Asian Pacific Phycological Forum, Held in Hong Kong, China, 21–25 June 1999. Springer Science & Business Media.

Aslan, S., Kapdan, I.K., 2006. Batch kinetics of nitrogen and phosphorus removal from synthetic wastewater by algae. Ecological Engineering 28 (1), 64–70.

Awasthi, M., Das, D.N., 2004. Heavy metal toxicity on nitrate reductase activity of free and immobilized algal cells. International Journal on Algae 6 (2).

Babu, M., 2011. Effect of Algal Biofilm and Operational Conditions on Nitrogen Removal in Waste Stabilization Ponds: UNESCO-IHE. Ph.D. thesis. CRC Press.

Banerjee, M., Mishra, S., Chatterjee, J., 2004. Scavenging of nickel and chromium toxicity in *Aulosira fertilissima* by immobilization: effect on nitrogen assimilating enzymes. Electronic Journal of Biotechnology 7, 13–14.

Barros, A.I., Gonçalves, A.L., Simões, M., Pires, J.C., 2015. Harvesting techniques applied to microalgae: a review. Renewable and Sustainable Energy Reviews 41, 1489–1500.

Becker, E.W., 1994. Microalgae: Biotechnology and Microbiology. Cambridge University Press.

Becker, K., 1996. Exopolysaccharide production and attachment strength of bacteria and diatoms on substrates with different surface tensions. Microbial Ecology 32 (1), 23–33.

Becker, K., Wahl, M., 1991. Influence of substratum surface tension on biofouling of artificial substrata in Kiel Bay (Western Baltic): *in situ* studies. Biofouling 4 (4), 275–291.

Bigogno, C., Khozin-Goldberg, I., Cohen, Z., 2002. Accumulation of arachidonic acid-rich triacylglycerols in the microalga *Parietochloris incisa* (Trebuxiophyceae, Chlorophyta). Phytochemistry 60 (2), 135–143.

Bilad, M.R., Vandamme, D., Foubert, I., Muylaert, K., Vankelecom, I.F., 2012. Harvesting microalgal biomass using submerged microfiltration membranes. Bioresource Technology 111, 343–352.

Blanco, A.M., Moreno, J., Del Campo, J.A., Rivas, J., Guerrero, M.G., 2007. Outdoor cultivation of lutein-rich cells of *Muriellopsis* sp. in open ponds. Applied Microbiology and Biotechnology 73 (6), 1259–1266.

Boelee, N., Temmink, H., Janssen, M., Buisman, C., Wijffels, R., 2011. Nitrogen and phosphorus removal from municipal wastewater effluent using microalgal biofilms. Water Research 45 (18), 5925–5933.

Boelee, N.C., 2013. Microalgal Biofilms for Wastewater Treatment.

Borowitzka, M.A., 1997. Microalgae for aquaculture: opportunities and constraints. Journal of Applied Phycology 9 (5), 393.

Borowitzka, M.A., 1999. Commercial production of microalgae: ponds, tanks, and fermenters. In: Progress in Industrial Microbiology. Elsevier.

Borowitzka, M.A., Moheimani, N.R., 2013. Algae for Biofuels and Energy. Springer.

Boyd, R.D., Verran, J., Jones, M., Bhakoo, M., 2002. Use of the atomic force microscope to determine the effect of substratum surface topography on bacterial adhesion. Langmuir 18 (6), 2343–2346.

Brennan, L., Owende, P., 2010. Biofuels from microalgae—a review of technologies for production, processing, and extractions of biofuels and co-products. Renewable and Sustainable Energy Reviews 14 (2), 557–577.

Brouers, M., Hall, D.O., 1986. Ammonia and hydrogen production by immobilized cyanobacteria. Journal of Biotechnology 3 (5), 307–321.

Cai, T., Park, S.Y., Li, Y., 2013. Nutrient recovery from wastewater streams by microalgae: status and prospects. Renewable and Sustainable Energy Reviews 19, 360–369.

Carvalho, A.P., Meireles, L.A., Malcata, F.X., 2006. Microalgal reactors: a review of enclosed system designs and performances. Biotechnology Progress 22 (6), 1490–1506.

Cheng, P., Ji, B., Gao, L., Zhang, W., Wang, J., Liu, T., 2013. The growth, lipid and hydrocarbon production of *Botryococcus braunii* with attached cultivation. Bioresource Technology 138, 95–100.

Chevalier, P., Proulx, D., Lessard, P., Vincent, W., De la Noüe, J., 2000. Nitrogen and phosphorus removal by high latitude mat-forming cyanobacteria for potential use in tertiary wastewater treatment. Journal of Applied Phycology 12 (2), 105–112.

Cho, D.-H., Ramanan, R., Heo, J., Lee, J., Kim, B.-H., Oh, H.-M., Kim, H.-S., 2015. Enhancing microalgal biomass productivity by engineering a microalgal–bacterial community. Bioresource Technology 175, 578–585.

Choi, N.-Y., Kim, B.-R., Bae, Y.-M., Lee, S.-Y., 2013. Biofilm formation, attachment, and cell hydrophobicity of foodborne pathogens under varied environmental conditions. Journal of the Korean Society for Applied Biological Chemistry 56 (2), 207–220.

Christenson, L., Sims, R., 2011. Production and harvesting of microalgae for wastewater treatment, biofuels, and bioproducts. Biotechnology Advances 29 (6), 686–702.

Christenson, L.B., Sims, R.C., 2012. Rotating algal biofilm reactor and spool harvester for wastewater treatment with biofuels by-products. Biotechnology and Bioengineering 109 (7), 1674–1684.

Colares, R.G., Machado, P., de Faria, M., Detoni, A., Tavano, V., 2013. Microalgae classification using semi-supervised and active learning based on Gaussian mixture models. Journal of the Brazilian Computer Society 19 (4), 411.

Colla, L.M., Reinehr, C.O., Reichert, C., Costa, J.A.V., 2007. Production of biomass and nutraceutical compounds by *Spirulina platensis* under different temperature and nitrogen regimes. Bioresource Technology 98 (7), 1489–1493.

Costa, J.A.V., Colla, L.M., Duarte Filho, P., 2003. *Spirulina platensis* growth in open raceway ponds using fresh water supplemented with carbon, nitrogen and metal ions. Zeitschrift für Naturforschung C 58 (1–2), 76–80.

Couto, R.M., Simoes, P.C., Reis, A., Da Silva, T.L., Martins, V.H., Sánchez-Vicente, Y., 2010. Supercritical fluid extraction of lipids from the heterotrophic microalga *Crypthecodinium cohnii*. Engineering in Life Science 10 (2), 158–164.

da Costa, A.C.A., Leite, S.G.F., 1991. Metals biosorption by sodium alginate immobilized *Chlorella homosphaera* cells. Biotechnology Letters 13 (8), 559–562.

Danquah, M.K., Gladman, B., Moheimani, N., Forde, G.M., 2009. Microalgal growth characteristics and subsequent influence on dewatering efficiency. Chemical Engineering Journal 151 (1–3), 73–78.

Dassey, A.J., Theegala, C.S., 2013. Harvesting economics and strategies using centrifugation for cost effective separation of microalgae cells for biodiesel applications. Bioresource Technology 128, 241–245.

De Godos, I., González, C., Becares, E., García-Encina, P.A., Muñoz, R., 2009. Simultaneous nutrients and carbon removal during pretreated swine slurry degradation in a tubular biofilm photobioreactor. Applied Microbiology and Biotechnology 82 (1), 187–194.

de la Noüe, J., Proulx, D., 1988. Biological tertiary treatment of urban wastewaters with chitosan-immobilized Phormidium. Applied Microbiology and Biotechnology 29 (2–3), 292–297.

de Oliveira Rangel-Yagui, C., Danesi, E.D.G., de Carvalho, J.C.M., Sato, S., 2004. Chlorophyll production from *Spirulina platensis*: cultivation with urea addition by fed-batch process. Bioresource Technology 92 (2), 133–141.

De Philippis, R., Vincenzini, M., 2003. Outermost polysaccharidic investments of cyanobacteria: nature, significance and possible applications. Recent Research Developments in Microbiology 7 (17), 13–22.

De Vuyst, L., Degeest, B., 1999. Heteropolysaccharides from lactic acid bacteria. FEMS Microbiology Reviews 23 (2), 153–177.

Del Campo, J.A., García-González, M., Guerrero, M.G., 2007. Outdoor cultivation of microalgae for carotenoid production: current state and perspectives. Applied Microbiology and Biotechnology 74 (6), 1163–1174.

Delisle, H., Alladoumgué, M., Begin, F., Nandjingar, K., Lasorsa, C., 1991. Household food consumption and nutritional adequacy in wadi zones of Chad, Central Africa. Ecology of Food and Nutrition 25 (3), 229–248.

Demirbaş, A., 2006. Oily products from mosses and algae via pyrolysis. Energy Sources, Part A 28 (10), 933–940.

DeNicola, D.M., 1996. Periphyton responses to temperature at different ecological levels. Algal Ecology: Feshwater Benthic Ecosystems 149–181.

Dermlim, W., Prasertsan, P., Doelle, H., 1999. Screening and characterization of bioflocculant produced by isolated *Klebsiella* sp. Applied Microbiology and Biotechnology 52 (5), 698–703.

DOE, U., November 1985. Review and Evaluation of Immobilized Algae Systems for the Production of Fuels from Microalgae. Solar Energy Research Institute, Alexandria (VA).

Donlan, R., Pipes, W., Yohe, T., 1994. Biofilm formation on cast iron substrata in water distribution systems. Water Research 28 (6), 1497–1503.

Doria, E., Longoni, P., Scibilia, L., Iazzi, N., Cella, R., Nielsen, E., 2012. Isolation and characterization of a *Scenedesmus acutus* strain to be used for bioremediation of urban wastewater. Journal of Applied Phycology 24 (3), 375–383.

Dudman, W., 1977. The role of surface polysaccharides in natural environments. In: Sutherland, I. (Ed.), Surface Carbohydrates of the Prokaryotic Cell. Academic Press, New York, pp. 357–414.

Edzwald, J., 1993. Algae, bubbles, coagulants, and dissolved air flotation. Water Science and Technology 27 (10), 67–81.

Evans, L.V., 2003. Biofilms: Recent Advances in Their Study and Control. CRC press.

Fallowfield, H., Garrett, M., 1985. The photosynthetic treatment of pig slurry in temperate climatic conditions: a pilot-plant study. Agricultural Wastes 12 (2), 111–136.

Fernández, F.A., Camacho, F.G., Pérez, J.S., Sevilla, J.F., Grima, E.M., 1998. Modeling of biomass productivity in tubular photobioreactors for microalgal cultures: effects of dilution rate, tube diameter, and solar irradiance. Biotechnology and Bioengineering 58 (6), 605–616.

Flemming, H.-C., Wingender, J., 2010. The biofilm matrix. Nature Reviews Microbiology 8 (9), 623.

Frølund, B., Palmgren, R., Keiding, K., Nielsen, P.H., 1996. Extraction of extracellular polymers from activated sludge using a cation exchange resin. Water Research 30 (8), 1749–1758.

Fuentes, J.L., Garbayo, I., Cuaresma, M., Montero, Z., González-del-Valle, M., Vílchez, C., 2016. Impact of microalgae-bacteria interactions on the production of algal biomass and associated compounds. Marine Drugs 14 (5), 100.

Fuentes, M.R., Sánchez, J.G., Sevilla, J.F., Fernández, F.A., Pérez, J.S., Grima, E.M., 1999. Outdoor continuous culture of *Porphyridium cruentum* in a tubular photobioreactor: quantitative analysis of the daily cyclic variation of culture parameters. Journal of Biotechnology 70 (1–3), 271–288.

Gao, F., Yang, Z.-H., Li, C., Zeng, G.-M., Ma, D.-H., Zhou, L., 2015. A novel algal biofilm membrane photobioreactor for attached microalgae growth and nutrients removal from secondary effluent. Bioresource Technology 179, 8–12.

Gao, J., Bao, H.-y., Xin, M.-x., Liu, Y.-x., Li, Q., Zhang, Y.-f., 2006. Characterization of a bioflocculant from a newly isolated *Vagococcus* sp. W31. Journal of Zhejiang University Science B 7 (3), 186–192.

Garbisu, C., Hall, D.O., Serra, J.L., 1993. Removal of phosphate by foam-immobilized *Phormidium laminosum* in batch and continuous-flow bioreactors. Journal of Chemical Technology and Biotechnology 57 (2), 181–189.

Garg, S., Li, Y., Wang, L., Schenk, P.M., 2012. Flotation of marine microalgae: effect of algal hydrophobicity. Bioresource Technology 121, 471–474.

Garnham, G.W., Codd, G.A., Gadd, G.M., 1992. Accumulation of cobalt, zinc and manganese by the estuarine green microalga *Chlorella salina* immobilized in alginate microbeads. Environmental Science & Technology 26 (9), 1764–1770.

Genin, S.N., Stewart Aitchison, J., Grant Allen, D., 2014. Design of algal film photobioreactors: material surface energy effects on algal film productivity, colonization and lipid content. Bioresource Technology 155, 136–143.

Gil, J.M., Serra, J.L., 1993. Nitrate removal by immobilized cells of *Phormidium uncinatum* in batch culture and a continuous-flow photobioreactor. Applied Microbiology and Biotechnology 39 (6), 782–787.

González-Fernández, C., Ballesteros, M., 2013. Microalgae autoflocculation: an alternative to high-energy consuming harvesting methods. Journal of Applied Phycology 25 (4), 991–999.

Gouveia, L., 2011. Microalgae as a feedstock for biofuels. In: Microalgae as a Feedstock for Biofuels. Springer.

Grima, E.M., Belarbi, E.-H., Fernández, F.A., Medina, A.R., Chisti, Y., 2003. Recovery of microalgal biomass and metabolites: process options and economics. Biotechnology Advances 20 (7–8), 491–515.

Grima, E.M., Fernández Sevilla, J.M., Acién Fernández, F.G., 2009. Microalgae, mass culture methods. Encyclopedia of Industrial Biotechnology: Bioprocess, Bioseparation, and Cell Technology 1–24.

Grobbelaar, J.U., 2013. Mass production of microalgae at optimal photosynthetic rates. Photosynthesis. IntechOpen 357–371.

Gultom, S., Hu, B., 2013. Review of microalgae harvesting via co-pelletization with filamentous fungus. Energies 6 (11), 5921–5939.

Guzzon, A., Bohn, A., Diociaiuti, M., Albertano, P., 2008. Cultured phototrophic biofilms for phosphorus removal in wastewater treatment. Water Research 42 (16), 4357–4367.

Harif, T., Khai, M., Adin, A., 2012. Electrocoagulation *versus* chemical coagulation: coagulation/flocculation mechanisms and resulting floc characteristics. Water Research 46 (10), 3177–3188.

He, S., Xue, G., 2010. Algal-based immobilization process to treat the effluent from a secondary wastewater treatment plant (WWTP). Journal of Hazardous Materials 178 (1–3), 895–899.

Henderson, R.K., Parsons, S.A., Jefferson, B., 2008. Surfactants as bubble surface modifiers in the flotation of algae: dissolved air flotation that utilizes a chemically modified bubble surface. Environmental Science & Technology 42 (13), 4883–4888.

Hernandez, J.-P., de-Bashan, L.E., Rodriguez, D.J., Rodriguez, Y., Bashan, Y., 2009. Growth promotion of the freshwater microalga *Chlorella vulgaris* by the nitrogen-fixing, plant growth-promoting bacterium *Bacillus pumilus* from arid zone soils. European Journal of Soil Biology 45 (1), 88–93.

Hill, D.R., Peat, A., Potts, M., 1994. Biochemistry and structure of the glycan secreted by desiccation-tolerant *Nostoc commune* (Cyanobacteria). Protoplasma 182 (3–4), 126–148.

Hill, W.R., Ryon, M.G., Schilling, E.M., 1995. Light limitation in a stream ecosystem: responses by primary producers and consumers. Ecology 76 (4), 1297–1309.

Hillebrand, H., Kahlert, M., 2001. Effect of grazing and nutrient supply on periphyton biomass and nutrient stoichiometry in habitats of different productivity. Limnology and Oceanography 46 (8), 1881–1898.

Hillebrand, H., Kahlert, M., Haglund, A.-L., Berninger, U.-G., Nagel, S., Wickham, S., 2002. Control of microbenthic communities by grazing and nutrient supply. Ecology 83 (8), 2205–2219.

Hillebrand, H., Sommer, U., 1999. The nutrient stoichiometry of benthic microalgal growth: redfield proportions are optimal. Limnology and Oceanography 44 (2), 440–446.

Hultberg, M., Asp, H., Marttila, S., Bergstrand, K.-J., Gustafsson, S., 2014. Biofilm formation by *Chlorella vulgaris* is affected by light quality. Current Microbiology 69 (5), 699–702.

Hwang, S.J., Havens, K.E., Steinman, A.D., 1998. Phosphorus kinetics of planktonic and benthic assemblages in a shallow subtropical lake. Freshwater Biology 40 (4), 729–745.

Ilangovan, K., Cañizares-Villanueva, R., González Moreno, S., Voltolina, D., 1998. Effect of cadmium and zinc on respiration and photosynthesis in suspended and immobilized cultures of *Chlorella vulgaris* and *Scenedesmus acutus*. Bulletin of Environmental Contamination and Toxicology 60 (6), 936–943.

Irving, T.E., Allen, D.G., 2011. Species and material considerations in the formation and development of microalgal biofilms. Applied Microbiology and Biotechnology 92 (2), 283–294.

Jin, E.S., Melis, A., 2003. Microalgal biotechnology: carotenoid production by the green algae *Dunaliella salina*. Biotechnology and Bioprocess Engineering 8 (6), 331.

John, R.P., Anisha, G.S., Nampoothiri, K.M., Pandey, A., 2011. Micro and macroalgal biomass: a renewable source for bioethanol. Bioresource Technology 102 (1), 186–193.

Johnson, M.B., Wen, Z., 2010. Development of an attached microalgal growth system for biofuel production. Applied Microbiology and Biotechnology 85 (3), 525–534.

Katarzyna, L., Sai, G., Singh, O.A., 2015. Non-enclosure methods for non-suspended microalgae cultivation: literature review and research needs. Renewable and Sustainable Energy Reviews 42, 1418–1427.

Kaya, V.M., Picard, G., 1996. Stability of chitosan gel as entrapment matrix of viable *Scenedesmus bicellularis* cells immobilized on screens for tertiary treatment of wastewater. Bioresource Technology 56 (2–3), 147–155.

Kesaano, M., Sims, R.C., 2014. Algal biofilm based technology for wastewater treatment. Algal Research 5, 231–240.

Khozin-Goldberg, I., Didi-Cohen, S., Shayakhmetova, I., Cohen, Z., 2002. Biosynthesis of eicosapentaenoic acid (EPA) in the freshwater eustigmatophyte *Monodus subterraneus* (Eustigmatophyceae). Journal of Phycology 38 (4), 745–756.

Kim, B.-H., Ramanan, R., Cho, D.-H., Oh, H.-M., Kim, H.-S., 2014. Role of *Rhizobium*, a plant growth promoting bacterium, in enhancing algal biomass through mutualistic interaction. Biomass and Bioenergy 69, 95–105.

Kim, M.-J., Jeong, S.-Y., Lee, S.-J., 2008. Isolation, identification, and algicidal activity of marine bacteria against *Cochlodinium polykrikoides*. Journal of Applied Phycology 20 (6), 1069–1078.

Kloc, J., Mendoza González, I., 2012. The Study of Biological Wastewater Treatment through Biofilm Development on Synthetic Material vs. Membranes. Worcester Polytechnic Institute Report, 1–72.

Knuckey, R.M., Brown, M.R., Robert, R., Frampton, D.M., 2006. Production of microalgal concentrates by flocculation and their assessment as aquaculture feeds. Aquacultural Engineering 35 (3), 300–313.

Kokare, C., Chakraborty, S., Khopade, A., Mahadik, K.R., 2009. Biofilm: Importance and Applications.

Lakaniemi, A.M., Intihar, V.M., Tuovinen, O.H., Puhakka, J.A., 2012. Growth of *Chlorella vulgaris* and associated bacteria in photobioreactors. Microbial biotechnology 5 (1), 69–78.

Larsdotter, K., 2006. Wastewater treatment with microalgae-a literature review. Vatten 62 (1), 31.

Lau, P., Tam, N., Wong, Y., 1998. Operational optimization of batchwise nutrient removal from wastewater by carrageenan immobilized *Chlorella vulgaris*. Water Science and Technology 38 (1), 185–192.

Lavens, P., Sorgeloos, P., 1996. Manual on the Production and Use of Live Food for Aquaculture: Food and Agriculture Organization (FAO).

Le Chevanton, M., Garnier, M., Bougaran, G., Schreiber, N., Lukomska, E., Bérard, J.-B., Fouilland, E., Bernard, O., Cadoret, J.-P., 2013. Screening and selection of growth-promoting bacteria for *Dunaliella cultures*. Algal Research 2 (3), 212–222.

Lee, A.K., Lewis, D.M., Ashman, P.J., 2009. Microbial flocculation, a potentially low-cost harvesting technique for marine microalgae for the production of biodiesel. Journal of Applied Phycology 21 (5), 559–567.

Lee, Y.-K., 2001. Microalgal mass culture systems and methods: their limitation and potential. Journal of Applied Phycology 13 (4), 307–315.

Lee, Y.-K., Ahn, C.-Y., Kim, H.-S., Oh, H.-M., 2010. Cyanobactericidal effect of *Rhodococcus* sp. isolated from eutrophic lake on Microcystis sp. Biotechnology Letters 32 (11), 1673–1678.

Lépinay, A., Turpin, V., Mondeguer, F., Grandet-Marchant, Q., Capiaux, H., Baron, R., Lebeau, T., 2018. First insight on interactions between bacteria and the marine diatom *Haslea ostrearia*: algal growth and metabolomic fingerprinting. Algal research 31, 395–405.

Li, Y., Horsman, M., Wu, N., Lan, C.Q., Dubois-Calero, N., 2008. Biofuels from microalgae. Biotechnology Progress 24 (4), 815–820.

Lian, J., Wijffels, R.H., Smidt, H., Sipkema, D., 2018. The effect of the algal microbiome on industrial production of microalgae. Microbial Biotechnology 11 (5), 806–818.

Liang, S., Liu, X., Chen, F., Chen, Z., 2004. Current microalgal health food R & D activities in China. In: Asian Pacific Phycology in the 21st Century: Prospects and Challenges. Springer.

Liehr, S.K., Eheart, J.W., Suidan, M.T., 1988. A modeling study of the effect of pH on carbon limited algal biofilms. Water Research 22 (8), 1033–1041.

Liehr, S.K., Suidan, M.T., Eheart, J.W., 1989. Effect of concentration boundary layer on carbon limited algal biofilms. Journal of Environmental Engineering 115 (2), 320–335.

Lincoln, E., Koopman, B., 1986. Bioflocculation of microalgae in mass cultures. Beihefte zur Nova Hedwigia 83, 207–211.

Lorite, G.S., Rodrigues, C.M., De Souza, A.A., Kranz, C., Mizaikoff, B., Cotta, M.A., 2011. The role of conditioning film formation and surface chemical changes on *Xylella fastidiosa* adhesion and biofilm evolution. Journal of Colloid and Interface Science 359 (1), 289–295.

Madhyastha, H., Vatsala, T., 2007. Pigment production in *Spirulina fussiformis* in different photophysical conditions. Biomolecular Engineering 24 (3), 301–305.

Mallick, N., 2002. Biotechnological potential of immobilized algae for wastewater N, P and metal removal: a review. Biometals 15 (4), 377–390.

Mantzorou, A., Ververidis, F., 2018. Microalgal biofilms: a further step over current microalgal cultivation techniques. Science of the Total Environment 651, 3187–3201.

Mata, T.M., Martins, A.A., Caetano, N.S., 2010. Microalgae for biodiesel production and other applications: a review. Renewable and Sustainable Energy Reviews 14 (1), 217–232.

McGriff Jr., E.C., McKinney, R.E., 1972. The removal of nutrients and organics by activated algae. Water Research 6 (10), 1155–1164.

Mendoza Gonzalez, I.M., Kloc, J.J., 2012. The Study of Biological Wastewater Treatment through Biofilm Development on Synthetic Material vs. Membranes.

Milledge, J.J., Heaven, S., 2013. A review of the harvesting of micro-algae for biofuel production. Reviews in Environmental Science and Biotechnology 12 (2), 165–178.

Mohn, F., 1988. Harvesting of Micro-algal Biomass. Cambridge University Press, Cambridge.

Moreno-Garrido, I., Codd, G., Gadd, G., Lubián, L.M., 2002. Cu and Zn accumulation by calcium alginate immobilized marine microalgal cells of *Nannochloropsis gaditana* (Eustigmatophyceae). Ciencias Marinas 28 (1), 107–119.

Mulbry, W., Kondrad, S., Pizarro, C., Kebede-Westhead, E., 2008. Treatment of dairy manure effluent using freshwater algae: algal productivity and recovery of manure nutrients using pilot-scale algal turf scrubbers. Bioresource Technology 99 (17), 8137–8142.

Mulbry, W.W., Wilkie, A.C., 2001. Growth of benthic freshwater algae on dairy manures. Journal of Applied Phycology 13 (4), 301–306.

Munoz, R., Guieysse, B., 2006. Algal–bacterial processes for the treatment of hazardous contaminants: a review. Water Research 40 (15), 2799–2815.

Muñoz, R., Köllner, C., Guieysse, B., 2009. Biofilm photobioreactors for the treatment of industrial wastewaters. Journal of Hazardous Materials 161 (1), 29–34.

Ogbonda, K.H., Aminigo, R.E., Abu, G.O., 2007. Influence of temperature and pH on biomass production and protein biosynthesis in a putative *Spirulina* sp. Bioresource Technology 98 (11), 2207–2211.

Olapade, O.A., Leff, L.G., 2006. Influence of dissolved organic matter and inorganic nutrients on the biofilm bacterial community on artificial substrates in a northeastern Ohio, USA, stream. Canadian Journal of Microbiology 52 (6), 540–549.

Olutiola, P., Awojobi, K., Oyedeji, O., Ayansina, A., Cole, O., 2010. Relationship between bacterial density and chemical composition of a tropical sewage oxidation pond. African Journal of Environmental Science and Technology 4 (9), 595–602.

Osborne, A.L., 2009. Harvesting Microalgae for Biofuel: Processes and Mechanisms.

Oswald, W.J., 1988. Role of microalgae in liquid waste treatment and reclamation. In: Lembi, C.A., Robert Waaland, J. (Eds.), Algae and Human Affairs. sponsored by the Phycological Society of America, Inc.

Ozkan, A., Berberoglu, H., 2013. Cell to substratum and cell to cell interactions of microalgae. Colloids and Surfaces B: Biointerfaces 112, 302–309.

Park, R.B., Kelly, J., Drury, S., Sauer, K., 1966. The Hill reaction of chloroplasts isolated from glutaraldehyde-fixed spinach leaves. Proceedings of the National Academy of Sciences of the United States of America 55 (5), 1056–1062.

Park, Y., Je, K.-W., Lee, K., Jung, S.-E., Choi, T.-J., 2008. Growth promotion of *Chlorella ellipsoidea* by co-inoculation with *Brevundimonas* sp. isolated from the microalga. Hydrobiologia 598 (1), 219–228.

Pereira, S., Zille, A., Micheletti, E., Moradas-Ferreira, P., De Philippis, R., Tamagnini, P., 2009. Complexity of cyanobacterial exopolysaccharides: composition, structures, inducing factors and putative genes involved in their biosynthesis and assembly. FEMS Microbiology Reviews 33 (5), 917–941.

Perez-Garcia, O., Escalante, F.M., de-Bashan, L.E., Bashan, Y., 2011. Heterotrophic cultures of microalgae: metabolism and potential products. Water Research 45 (1), 11–36.

Pittman, J.K., Dean, A.P., Osundeko, O., 2011. The potential of sustainable algal biofuel production using wastewater resources. Bioresource Technology 102 (1), 17–25.

Poelman, E., De Pauw, N., Jeurissen, B., 1997. Potential of electrolytic flocculation for recovery of micro-algae. Resources, Conservation and Recycling 19 (1), 1–10.

Posten, C., 2009. Design principles of photo-bioreactors for cultivation of microalgae. Engineering in Life Science 9 (3), 165–177.

Priyadarshani, I., Rath, B., 2012. Commercial and industrial applications of micro algae – A review. Journal of Algal Biomass Utilization 3 (4), 89–100.

Pulz, O., 2001. Photobioreactors: production systems for phototrophic microorganisms. Applied Microbiology and Biotechnology 57 (3), 287–293.

Pulz, O., Gross, W., 2004. Valuable products from biotechnology of microalgae. Applied Microbiology and Biotechnology 65 (6), 635–648.

Radmer, R.J., 1996. Algal diversity and commercial algal products. BioScience 46 (4), 263–270.

Rahman, A., Ellis, J.T., Miller, C.D., 2012. Bioremediation of domestic wastewater and production of bioproducts from microalgae using waste stabilization ponds. Journal of Bioremediation and Biodegradation 3 (06).

Rangsayatorn, N., Pokethitiyook, P., Upatham, E., Lanza, G., 2004. Cadmium biosorption by cells of *Spirulina platensis* TISTR 8217 immobilized in alginate and silica gel. Environment International 30 (1), 57–63.

Rao, K., Hall, D., 1984. Photosynthetic production of fuels and chemicals in immobilized systems. Trends in Biotechnology 2 (5), 124–129.

Rawat, I., Kumar, R.R., Mutanda, T., Bux, F., 2011. Dual role of microalgae: phycoremediation of domestic wastewater and biomass production for sustainable biofuels production. Applied Energy 88 (10), 3411–3424.

Renuka, N., Sood, A., Ratha, S.K., Prasanna, R., Ahluwalia, A.S., 2013. Evaluation of microalgal consortia for treatment of primary treated sewage effluent and biomass production. Journal of Applied Phycology 25 (5), 1529–1537.

Richmond, A., 2004. Handbook of Microalgal Culture: Biotechnology and Applied Phycology. Wiley Online Library.

Robinson, P.K., Wilkinson, S.C., 1994. Removal of aqueous mercury and phosphate by gel-entrapped *Chlorella* in packed-bed reactors. Enzyme and Microbial Technology 16 (9), 802–807.

Roostaei, J., Zhang, Y., Gopalakrishnan, K., Ochocki, A.J., 2018. Mixotrophic microalgae biofilm: a novel algae cultivation strategy for improved productivity and cost-efficiency of biofuel feedstock production. Scientific Reports 8 (1), 12528.

Rossi, F., Micheletti, E., Bruno, L., Adhikary, S.P., Albertano, P., De Philippis, R., 2012. Characteristics and role of the exocellular polysaccharides produced by five cyanobacteria isolated from phototrophic biofilms growing on stone monuments. Biofouling 28 (2), 215–224.

Salehizadeh, H., Shojaosadati, S., 2001. Extracellular biopolymeric flocculants: recent trends and biotechnological importance. Biotechnology Advances 19 (5), 371–385.

Salim, S., Bosma, R., Vermuë, M.H., Wijffels, R.H., 2011. Harvesting of microalgae by bio-flocculation. Journal of Applied Phycology 23 (5), 849–855.

Sanayei, Y., Ismail, N., Teng, T.T., Morad, N., 2010. Studies on flocculating activity of bio-flocculant from closed drainage system (CDS) and its application in reactive dye removal. International Journal of Chemistry 2 (1), 168.

Sawayama, S., Rao, K., Hall, D., 1998. Nitrate and phosphate ion removal from water by *Phormidium laminosum* immobilized on hollow fibres in a photobioreactor. Applied Microbiology and Biotechnology 49 (4), 463–468.

Schenk, P.M., Thomas-Hall, S.R., Stephens, E., Marx, U.C., Mussgnug, J.H., Posten, C., Kruse, O., Hankamer, B., 2008. Second generation biofuels: high-efficiency microalgae for biodiesel production. Bioenergy Research 1 (1), 20–43.

Schnurr, P.J., Espie, G.S., Allen, D.G., 2014. The effect of light direction and suspended cell concentrations on algal biofilm growth rates. Applied Microbiology and Biotechnology 98 (20), 8553–8562.

Scott, C., Fletcher, R., Bremer, G., 1996. Observations on the mechanisms of attachment of some marine fouling blue-green algae. Biofouling 10 (1–3), 161–173.

Sekar, R., Nair, K., Rao, V., Venugopalan, V., 2002. Nutrient dynamics and successional changes in a lentic freshwater biofilm. Freshwater Biology 47 (10), 1893–1907.

Sekar, R., Venugopalan, V.P., Satpathy, K.K., Nair, K.V.K., Rao, V.N.R., 2004. Laboratory Studies on Adhesion of Microalgae to Hard Substrates. Springer, Dordrecht, Netherlands, pp. 109–116.

Shelef, G., Sukenik, A., Green, M., 1984. Microalgae Harvesting and Processing: A Literature Review (No. SERI/STR-231-2396). Technion Research and Development Foundation Ltd., Haifa (Israel).

Shen, Y., Xu, X., Zhao, Y., Lin, X., 2014. Influence of algae species, substrata and culture conditions on attached microalgal culture. Bioprocess and Biosystems Engineering 37 (3), 441–450.

Shen, Y., Zhang, H., Xu, X., Lin, X., 2015. Biofilm formation and lipid accumulation of attached culture of *Botryococcus braunii*. Bioprocess and Biosystems Engineering 38 (3), 481–488.

Shi, J., Podola, B., Melkonian, M., 2007. Removal of nitrogen and phosphorus from wastewater using microalgae immobilized on twin layers: an experimental study. Journal of Applied Phycology 19 (5), 417–423.

Shipin, O., Meiring, P., Phaswana, R., Kluever, H., 1999. Integrating ponds and activated sludge process in the PETRO concept. Water Research 33 (8), 1767–1774.

Show, K.-Y., Lee, D.-J., Chang, J.-S., 2013. Algal biomass dehydration. Bioresource Technology 135, 720–729.

Simões, L.C., 2013. Biofilms in Drinking Water.

Simoes, M., Pereira, M.O., Sillankorva, S., Azeredo, J., Vieira, M.J., 2007. The effect of hydrodynamic conditions on the phenotype of *Pseudomonas fluorescens* biofilms. Biofouling 23 (4), 249–258.

Simões, M., Pereira, M.O., Vieira, M., 2004. Cellular Changes Due to Biofilm Formation: Influence of Flow Regime.

Singh, G., Patidar, S., 2018. Microalgae harvesting techniques: a review. Journal of Environmental Management 217, 499–508.

Singh, S., Kate, B.N., Banerjee, U., 2005. Bioactive compounds from cyanobacteria and microalgae: an overview. Critical Reviews in Biotechnology 25 (3), 73–95.

Sirmerova, M., Prochazkova, G., Siristova, L., Kolska, Z., Branyik, T., 2013. Adhesion of *Chlorella vulgaris* to solid surfaces, as mediated by physicochemical interactions. Journal of Applied Phycology 25 (6), 1687–1695.

Soletto, D., Binaghi, L., Lodi, A., Carvalho, J., Converti, A., 2005. Batch and fed-batch cultivations of *Spirulina platensis* using ammonium sulphate and urea as nitrogen sources. Aquaculture 243 (1–4), 217–224.

Spolaore, P., Joannis-Cassan, C., Duran, E., Isambert, A., 2006. Commercial applications of microalgae. Journal of Bioscience and Bioengineering 101 (2), 87–96.

Starks, T.L., Shubert, L.E., Trainor, F., 1981. Ecology of soil algae: a review. Phycologia 20 (1), 65–80.

Stelzer, R.S., Lamberti, G.A., 2001. Effects of N: P ratio and total nutrient concentration on stream periphyton community structure, biomass, and elemental composition. Limnology and Oceanography 46 (2), 356–367.

Stolz, P., 2005. Manufacturing microalgae for skin care. Cosmetics & Toiletries 120, 99–106.

Stoodley, P., Dodds, I., Boyle, J.D., Lappin-Scott, H., 1998. Influence of hydrodynamics and nutrients on biofilm structure. Journal of Applied Microbiology 85 (S1), 19S–28S.

Sukačová, K., Trtílek, M., Rataj, T., 2015. Phosphorus removal using a microalgal biofilm in a new biofilm photobioreactor for tertiary wastewater treatment. Water Research 71, 55–63.

Svarovsky, L., 2000. Solid-liquid Separation. Elsevier.

Tam, N.F., Wong, Y.-S., Simpson, C.G., 1998. Removal of copper by free and immobilized microalga, *Chlorella vulgaris*. In: Wastewater Treatment with Algae. Springer.

Tanabe, Y., Okazaki, Y., Yoshida, M., Matsuura, H., Kai, A., Shiratori, T., Ishida, K.-i., Nakano, S.-i., Watanabe, M.M., 2015. A novel alphaproteobacterial ectosymbiont promotes the growth of the hydrocarbon-rich green alga *Botryococcus braunii*. Scientific Reports 5, 10467.

Taylor, R.L., Verran, J., Lees, G.C., Ward, A.P., 1998. The influence of substratum topography on bacterial adhesion to polymethyl methacrylate. Journal of Materials Science: Materials in Medicine 9 (1), 17–22.

Tease, B.E., Walker, R.W., 1987. Comparative composition of the sheath of the cyanobacterium *Gloeothece* ATCC 27152 cultured with and without combined nitrogen. Microbiology 133 (12), 3331–3339.

Terry, P.A., Stone, W., 2002. Biosorption of cadmium and copper contaminated water by *Scenedesmus abundans*. Chemosphere 47 (3), 249–255.

Travieso, L., Benitez, F., Weiland, P., Sanchez, E., Dupeyron, R., Dominguez, A., 1996. Experiments on immobilization of microalgae for nutrient removal in wastewater treatments. Bioresource Technology 55 (3), 181–186.

Travieso, L., Canizares, R., Borja, R., Benitez, F., Dominguez, A., Dupeyrón y, R., Valiente, V., 1999. Heavy metal removal by microalgae. Bulletin of Environmental Contamination and Toxicology 62 (2), 144–151.

Ugwu, C., Aoyagi, H., Uchiyama, H., 2008. Photobioreactors for mass cultivation of algae. Bioresource Technology 99 (10), 4021–4028.

Vandamme, D., Foubert, I., Muylaert, K., 2013. Flocculation as a low-cost method for harvesting microalgae for bulk biomass production. Trends in Biotechnology 31 (4), 233–239.

Vandamme, D., Pontes, S.C.V., Goiris, K., Foubert, I., Pinoy, L.J.J., Muylaert, K., 2011. Evaluation of electro-coagulation–flocculation for harvesting marine and freshwater microalgae. Biotechnology and Bioengineering 108 (10), 2320–2329.

Venable, M.E., Podbielski, M.R., 2019. Impact of substrate material on algal biofilm biomass growth. Environmental Science and Pollution Research 26 (7), 7256–7262.

Volesky, B., Prasetyo, I., 1994. Cadmium removal in a biosorption column. Biotechnology and Bioengineering 43 (11), 1010–1015.

Wang, J.-H., Zhuang, L.-L., Xu, X.-Q., Deantes-Espinosa, V.M., Wang, X.-X., Hu, H.-Y., 2018. Microalgal attachment and attached systems for biomass production and wastewater treatment. Renewable and Sustainable Energy Reviews 92, 331–342.

Wang, X., Li, Z., Su, J., Tian, Y., Ning, X., Hong, H., Zheng, T., 2010. Lysis of a red-tide causing alga, *Alexandrium tamarense*, caused by bacteria from its phycosphere. Biological Control 52 (2), 123–130.

Watanabe, K., Takihana, N., Aoyagi, H., Hanada, S., Watanabe, Y., Ohmura, N., Saiki, H., Tanaka, H., 2005. Symbiotic association in *Chlorella* culture. FEMS Microbiology Ecology 51 (2), 187–196.

Whitton, R., Santinelli, M., Pidou, M., Ometto, F., Henderson, R., Roddick, F., Jarvis, P., Villa, R., Jefferson, B., 2018. Tertiary nutrient removal from wastewater by immobilised microalgae: impact of wastewater nutrient characteristics and hydraulic retention time (HRT). H2Open Journal 1 (1), 12–25.

Wilkinson, S., Goulding, K., Robinson, P., 1990. Mercury removal by immobilized algae in batch culture systems. Journal of Applied Phycology 2 (3), 223–230.

Wolkers, H., Barbosa, M., Kleinegris, D., Bosma, R., Wijffels, R., Harmsen, P., 2011. Microalgae: The Green Gold of the Future?: Large-Scale Sustainable Cultivation of Microalgae for the Production of Bulk Commodities.

Xu, L., Weathers, P.J., Xiong, X.R., Liu, C.Z., 2009. Microalgal bioreactors: challenges and opportunities. Engineering in Life Science 9 (3), 178–189.

Yamaguchi, K., 1996. Recent advances in microalgal bioscience in Japan, with special reference to utilization of biomass and metabolites: a review. Journal of Applied Phycology 8 (6), 487–502.

Yamaguchi, T., Ishida, M., Suzuki, T., 1999. An immobilized cell system in polyurethane foam for the lipophilic micro-alga *Prototheca zopfii*. Process Biochemistry 34 (2), 167–171.

Yongmanitchai, W., Ward, O.P., 1991. Growth of and omega-3 fatty acid production by *Phaeodactylum tricornutum* under different culture conditions. Applied and Environmental Microbiology 57 (2), 419–425.

Zhang, T., Fang, H.H., 2001. Quantification of extracellular polymeric substances in biofilms by confocal laser scanning microscopy. Biotechnology Letters 23 (5), 405–409.

Zhang, Z.-q., Bo, L., Xia, S.-q., Wang, X.-j., Yang, A.-m., 2007. Production and application of a novel bioflocculant by multiple-microorganism consortia using brewery wastewater as carbon source. Journal of Environmental Sciences 19 (6), 667–673.

Zippel, B., Rijstenbil, J., Neu, T.R., 2007. A flow-lane incubator for studying freshwater and marine phototrophic biofilms. Journal of Microbiological Methods 70 (2), 336–345.

Chapter 7

Biofilms in membrane systems for drinking water production

Nadia M. Farhat, Szilard S. Bucs, Johannes S. Vrouwenvelder
Water Desalination and Reuse Center (WDRC), Division of Biological and Environmental Science and Engineering (BESE), King Abdullah University of Science and Technology (KAUST), Thuwal, Saudi Arabia

7.1 Introduction

The demand for clean freshwater is rising due to the expanding needs of agriculture, food production, and energy consumption. However, the available clean freshwater resources are declining due to increased water pollution and the weaknesses of water management. Problems related to freshwater availability are expected to become larger in the coming decades. Water scarcity and inadequacy of water available for potable use have pushed for investments for improvement of water treatment technologies. The only methods to increase drinking water supply beyond what is available from the hydrological cycle are desalination and water reuse (Shannon et al., 2008). Seawater desalination offers a seemingly unlimited, steady supply of high-quality fresh water, without impairing natural freshwater ecosystems (Elimelech and Phillip, 2011). Desalination substantially removes dissolved solutes from the brackish or seawater sources to create the desired product water (Service, 2006). Constant enhancements to existing desalination technologies, especially in the past decade, lead to the reliable use of these technologies to desalinate seawater as well as brackish water to produce high-quality drinking water. The most commonly used technologies to achieve this objective include membrane-based and thermal desalination processes (Schiermeier, 2008).

Membrane filtration processes can produce high-quality drinking water, free of pathogenic microorganisms and (in)organic contaminants. Membrane filtration is a process in which a membrane is used as a selective physical barrier to separate compounds by applying a driving force across the membrane. In a membrane system, a feedwater stream is separated in two streams, the product or permeate, containing solutes that passed the membrane and the concentrate containing solutes and particles rejected by the membrane (Amjad, 1993; Mallevialle et al., 1996). Membrane operations can be

Recent Trends in Biofilm Science and Technology. https://doi.org/10.1016/B978-0-12-819497-3.00007-6

classified considering the parameters driving force, separation mechanism, and rejection properties. In case of key industrially applied membrane processes, the driving force is a hydraulic pressure difference across the membrane. Four pressure-driven membrane filtration processes can be discriminated based on differences in feed pressures and membrane rejection capacities: microfiltration (MF), ultrafiltration (UF), nanofiltration (NF), and reverse osmosis (RO), ranked by increasing feed pressure (Fig. 7.1).

A classification generally made is low-pressure membranes for MF and UF and high-pressure membranes for NF and RO. MF screens particles from 0.1 to 0.5 microns, and UF screens particles from 0.005 to 0.05 microns. NF is applied for removal of divalent ions, e.g., sulfate and hardness, natural color (humic acids), and partial removal of monovalent ions, e.g., sodium and chloride. RO membranes are able to remove mono- and divalent ions for more than 99%. The pores in NF and RO membranes are smaller than 1 nm. RO is the global leading technology to desalinate water. NF and RO membranes are used to produce high-quality drinking water from abundantly available brackish and seawater sources (Elimelech and Phillip, 2011). While NF/RO technology has been widely used for seawater desalination for more than 4 decades, NF/RO advanced membrane technology has rapidly grown and is constantly improving, reaching lower levels of salt passage and becoming more energy efficient. These improvements eventually led to a decrease in the total water cost of desalinated water production. Improvement of NF/RO desalination technology resulted also in an increase in the share of membrane-based desalination up to presently about 80% of the total global desalination capacity (Greenlee et al., 2009). Different types of NF/RO membrane modules for water desalination exist; however, the most

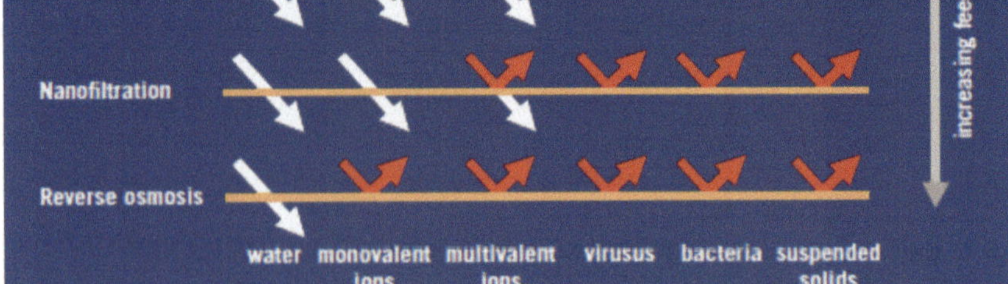

FIGURE 7.1 Scheme of different hydraulic pressure-driven membrane filtration processes and rejection capacities (Vrouwenvelder et al., 2011a).

commonly used type is the spiral-wound membrane module. Spiral-wound membrane modules use sheets of membrane, feed spacer, and permeate spacer wrapped around a hollow tube called the permeate tube (Fig. 7.2). Spiral-wound membrane elements have a surface to volume ratio of 300–1000 m^2/m^3 enabling small footprint systems and relative low prices per m^2 membrane area (Amjad, 1993; Mallevialle et al., 1996).

Water after passing the membrane must follow a tortuous pathway through the membrane envelope to reach the permeate collection tube. Applying pressure across the membrane forces the water contained in the feed flow to permeate through the membrane. To overcome the feed side osmotic pressure, relatively high feed pressure is required to produce clean water. In seawater desalination, the feed pressure commonly ranges from 40 to 70 bar, and the permeate water recovery from each membrane module is approximately 10% the feed flow, reaching a water recovery of 45% for the whole pressure vessel. Feed spacers create intermembrane space in the spiral-wound membrane elements necessary for water transport. The spacers improve water mixing, thus reducing the overall diffusive boundary layer thickness as well as reducing concentration polarization, defined as the increase in the concentration of salts in the feedwater closer to the membrane surface (Herzberg and Elimelech, 2007), thus improving the overall treatment efficiency. Feed spacers can have different designs (e.g., geometries) but, generally, have a meshlike structure with mesh sizes of 2–3 mm and thicknesses varying from 28 to 34 mil (1 mil equals 25.4 μm) (Araujo et al., 2012a; Siddiqui et al., 2016).

A drawback in membrane filtration applications is membrane fouling restricting membrane performance. Fouling can be defined as the accumulation of material on the surface of the membrane or feed spacer in the feed

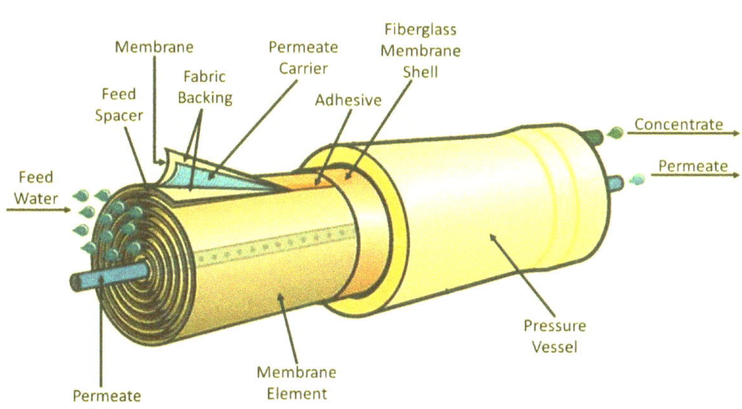

FIGURE 7.2 Spiral-wound reverse osmosis (RO) membrane module showing the RO membrane sandwiched between a feed spacer and permeate carrier (Vrouwenvelder et al., 2011a).

channel of the membrane element. Four categories of membrane fouling can be distinguished: (1) particle/colloid fouling, (2) inorganic fouling, (3) organic fouling, and (4) biological fouling (biofouling) (Amiri and Samiei, 2007). Fouling can develop unevenly through a membrane module and can occur between the membrane sheets of a module, where spacers are located to create space for the concentrate stream (Tran et al., 2007). These different types of fouling may occur simultaneously and can influence each other (Flemming, 2003). Scaling and particulate fouling can be relatively easy to control by physical and chemical pretreatment, but biofouling and organic fouling seem to be less controllable (Vrouwenvelder et al., 2008).

Biofouling is defined as the unwanted accumulation of biofilms (Flemming, 2002). Biofouling can be operationally defined as the accumulation of a biofilm, resulting in an unacceptable degree of system performance loss in industrial membrane systems (Roe et al., 1994). Different phases of biofilm development can be distinguished: attachment phase with reversible adsorption followed by irreversible adsorption of bacteria on a surface (Palmer et al., 2007), exponential bacterial cell growth and division phase followed by production of extracellular polymeric substances, and detachment phase (Characklis, 1981; Kostakioti et al., 2013). Before the attachment phase, a conditioning layer comprising organics present in the feedwater starts to form on the pristine surface, allowing the subsequent attachment of bacterial cells. The biomass present at the last stage when the biofilm becomes mature is characterized by a balance between bacterial cell growth, cell degradation, and cell detachment.

Fouling results in reduced membrane performance and increased costs. The reduction in performance is characterized by an increase of feed channel pressure (FCP) drop and/or decline of flux and/or increase of salt passage (Flemming, 2003; Vrouwenvelder et al., 2008). The FCP drop is the earliest and strongest impacted performance indicator followed sequentially by permeability and salt rejection decline (Siebdrath et al., 2019) (Fig. 7.3). Major consequences of fouling occurrence on membrane system operation are (1) deterioration of produced permeate quality and quantity, (2) higher energy consumption, (3) higher chemical use for cleaning, (4) more plant downtime (less water production), and (5) shorter membrane lifetime.

7.2 Methods to evaluate biofilm growth potential of feedwater

Biofouling occurrence in membrane systems is considered inevitable. Recently, the use of growth potential methods of water to predict biofouling propensity of the membrane system has gained interest among researchers. Several studies could directly link the feedwater growth potential to biofilm formation on RO membrane (Jeong et al., 2013; Wang et al., 2014; Abushaban et al., 2019a). Moreover, the frequency of membrane cleaning primarily depends on the biofouling potential of the feedwater and the operational conditions.

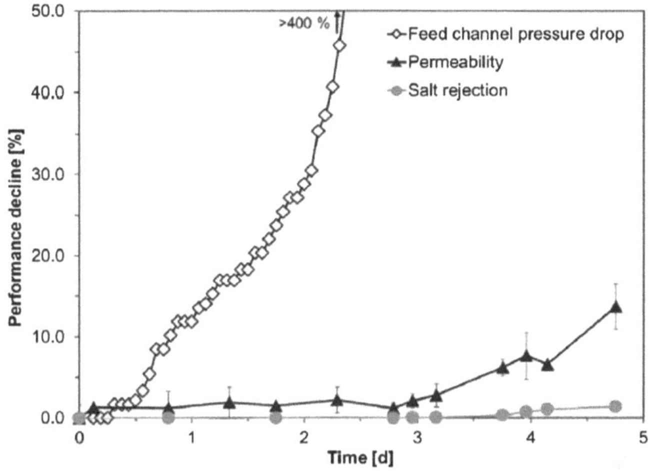

FIGURE 7.3 Performance decline of FCP drop, permeability and salt rejection over four 1-m-long membrane test cells, illustrating that the FCP drop was impacted earliest and strongest by biofouling (Siebdrath et al., 2019). *FCP*, feed channel pressure.

Bacterial growth potential is the quantification of the extent of bacterial growth that can occur under defined conditions (Farhat et al., 2018a). Nutrient availability, mainly organic carbon and other growth-promoting/limiting compounds (e.g., nitrogen, phosphorus and iron), governs the bacterial growth potential (Prest et al., 2016; Nescerecka et al., 2018). Numerous methods to determine the bacterial growth potential and growth-promoting properties of water have been developed throughout the past 3 decades (Van der Kooij et al., 1982; Servais et al., 1989; Ross et al., 2013; Farhat et al., 2018a). Various studies contributed to constantly optimize the bacterial growth potential bioassays with a main focus on three aspects: the selection of inoculum, the optimization of inoculation and incubation, and the evolution of bacterial growth measurements (Le chevallier et al., 1993; Sathasivan and Ohgaki, 1999; Wang et al., 2014; Van der Kooij et al., 2017). The conventional bioassays to measure the bacterial growth potential use selected pure cultures highly abundant in water distribution systems as test strains. A major drawback of using pure cultures is the inability of some pure strains to universally grow in different water types and to assimilate all the AOC (assimilable organic carbon) present in the water. The inoculum selection has been a point of focus in many studies and a principal alteration to the initial bacterial growth potential methods (Kemmy et al., 1989; Sathasivan and Ohgaki, 1999; Haddix et al., 2004; Weinrich et al., 2011), from using a single bacterial strain to a mix of two or more bacterial strains or the indigenous bacteria. Indigenous microbial consortium has been used to further broaden and diversify the substrate utilization range in comparison with a single pure culture. Shifting from cultivation-dependent quantification methods to cultivation-independent methods was another primary variation to the initial growth potential methods (Le chevallier et al., 1993; Hammes and Egli, 2005; Abushaban et al., 2017). Bacterial growth measurements evolved from the use of plate counting (Van der Kooij et al., 1982; Escobar and Randall, 2000) and turbidity measurements (Werner and Hambsch, 1986) to the use of adenosine triphosphate (ATP) luminescence method (Le chevallier et al., 1993; Van der Kooij et al., 2017; Abushaban et al., 2019b), bioluminescence method (Weinrich et al., 2011), and total cell count with fluorescence staining and flow cytometry method (Farhat et al., 2018a).

Although studies such as Abushaban et al. (2019a) observed that a higher bacterial growth potential in seawater RO feed corresponded to a higher chemical cleaning frequency, more research is required to confirm if a correlation exists between bacterial growth potential and prevention of biofouling in seawater RO plants by early cleaning.

7.3 Conventional biofouling control strategies

7.3.1 Preventive biofouling control

Biodegradable nutrients in the water, passing RO pretreatment, can be converted into bacteria biomass causing biofouling of the RO membrane installation. Is essence, the presence of a single bacterium in the feedwater or in the membrane system itself can convert nutrients into biomass-causing biofouling. Very low nutrient concentrations in the order of 1 μg C/L in the feedwater can cause biofouling of RO membranes. Therefore, biofouling prevention strategies targeting bacteria removal or inactivation requires that bacteria are totally removed from the water or completely inactivated. Practically, such an approach is difficult to achieve as RO systems have large surfaces in contact with water and the installation requires maintenance, such as membrane installation and replacement, which cannot be done under full sterile conditions. Hence, it is a challenge to have and maintain all the time a fully sterile RO membrane installation.

To prevent RO membrane biofouling, a suite of strategies has been studied, such as, (1) biomass removal in pretreatment, (2) biomass inactivation during pretreatment, (3) biomass inactivation using dosage of biocides, and (4) nutrient removal from water in pretreatment to restrict bacterial growth. For bacterial removal from the water, a commonly applied pretreatment technology is low-pressure filtration such as UF with a smaller pore size than bacteria.

Low-pressure membrane filtration systems such as MF and UF have gained importance in the past years as pretreatment for RO systems. Although MF and UF pretreatment removes very well the bacterial cells from the feedwater, biodegradable nutrients can pass UF membranes enabling eventual microbial growth in the subsequent RO installation. In some cases, activated carbon or biofiltration is used to remove dissolved organic matter from the feedwater.

Activated carbon adsorption, either in granular or in powder form, has also been considered as a feasible mean for reducing membrane fouling, either alone or in combination with other pretreatment processes (Meier and Melin, 2005; Gur-Reznik et al., 2008; Zhao and Yu, 2015). Chinu et al. (2009) delayed fouling development in a lab-scale setup by using biofiltration as pretreatment using seawater.

Biodegradable nutrients in the water will pass (partly) most-if-not-all conventional water treatment processes (e.g., coagulation and sedimentation, rapid/slow sand filtration, activated carbon filtration), as well as membrane-based filtration processes such as MF and UF. So far, no biological pretreatment has demonstrated to completely prevent downstream biofilm development even minimal amounts of nutrient (microgram per liter level) in the feedwater lead to a significant organic substrate supply for biofilm growth, which will occur over days, weeks or months of operation (Fig. 7.4).

For bacteria inactivation, "chlorination and downstream dechlorination" can be considered. Chlorine is typically dosed at the intake point of a seawater desalination installation to avoid microbial growth (mussels) in the intake pipe. However, thin-film composite RO membranes are damaged when in short contact with residual chlorine; therefore, before supplying water to the RO installation, residual chlorine is removed by dosing, e.g., sodium bisulfate (Prihasto et al., 2009; Farhat et al., 2018b). The rationale is that all bacteria are killed by the chlorine before entering the RO. It has been found that RO desalination plants with pretreatment involving "chlorination and dechlorination" suffer from more severe biofouling problems than plants without "chlorination and dechlorination" pretreatment, which is most likely

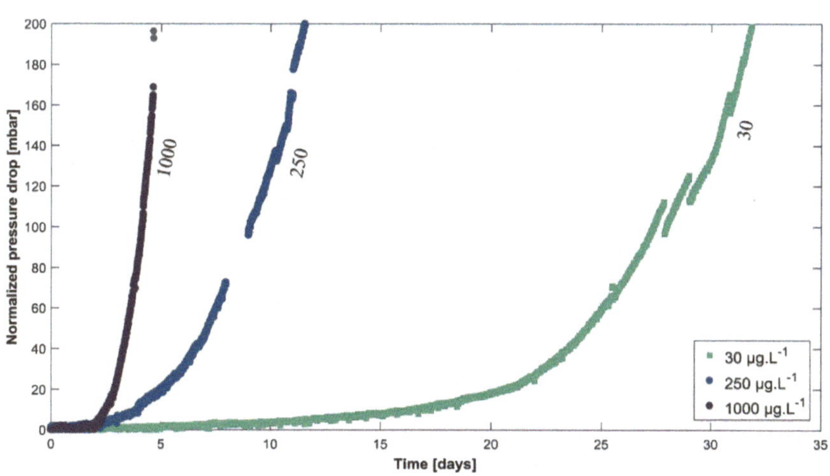

FIGURE 7.4 Feed channel pressure drop increase in time in lab-scale setup at three different biodegradable nutrient concentrations. With decreasing biodegradable nutrient concentration, biofilm formation and related pressure drop increase can be delayed but not prevented.

explained by the oxidation of the water organic material by the chlorine resulting in an increased concentration of easily biodegradable nutrients, leading to more excessive biofilm growth.

Ultraviolet (UV) irradiation of the water to inactivate the bacteria has been studied as strategy for biofouling control. UV treatment has demonstrated to be effective in bacteria inactivation, initially slowing down the biofilm growth, however, not preventing biofouling for the long run.

Nonoxidizing biocides can be used as an effective biofouling control strategy. An alternative nonoxidizing biocide, monochloramine 2,2-dibromo-3-nitrilopropionamide (DBNPA) has been applied in limited number of water treatment plants. DBNPA is rapidly degrading in alkaline aqueous solutions and is compatible with polyamide-based membranes and shows high rejection rates for RO membranes (Bertheas et al., 2009). The antimicrobial effect is due to the fast reaction between DBNPA and sulfur-containing organic molecules in microorganisms such as glutathione or cysteine. Inactivation of bacteria has been demonstrated using a biocide such as DBNPA (Siddiqui et al. 2017b). However, continuous dosage is needed for DBNPA to be effective. DBNPA dosage did not demonstrate curative biofouling control. The use of such biocides raises a lot of concern regarding the potential contamination of the produced water and the discharged concentrate. Governmental regulation prohibits, e.g., DBNPA use in RO systems for drinking water production due to environmental and health risks (Bertheas et al., 2009; Kim and Park, 2016).

In summary, no technology for biomass removal, biomass inactivation, and nutrient removal has demonstrated to be effective in all cases for biofouling prevention.

7.3.2 Risk of chemical dosage: antiscalants, acids, and biocides

Chemicals are dosed to the feedwater of an RO installation to avoid scaling (acid and/or antiscalants) or biofouling (biocides). Studies of 14 different antiscalants used in practice installations showed that antiscalants contain biodegradable compounds that can lead to rapid and severe biofouling (Vrouwenvelder et al., 2000). The biodegradable compounds could originate from, e.g., (impure) polymers and additions such as biodegradable conserving chemicals. The total organic carbon concentration was not a suitable measurement to predict the growth potential of the antiscalant (Fig. 7.5).

Dosage of only one batch of impure acid has caused, in unrelated cases, biofouling at two plants. The impact of the short use of one batch of impure acid caused prolonged performance decline and early membrane replacement. In other words, a short period of nutrient dosage can impact performance for prolonged periods. Research has elucidated that the acid was transported by not dedicated transport trucks, possibly leading to contamination with nutrients. Subsequently, a guideline was made for dedicated transport of acid and for a rapid testing method for the growth potential of acids before dosage to the RO feedwater.

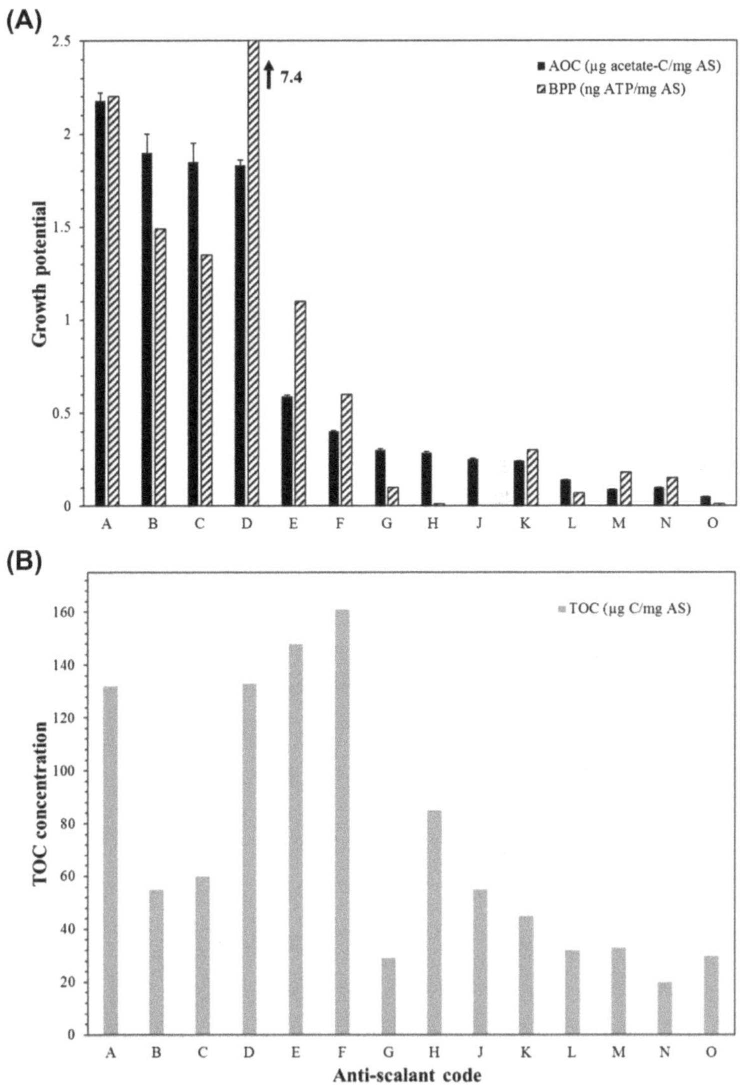

FIGURE 7.5 Bacterial growth potential and TOC content of 14 commercially available antiscalants used in practice (coded A to O). (A) Growth potential (AOC concentration per mg and ATP yield (BPP) per mg) and (B) dissolved organic carbon (DOC) concentration of antiscalants determined a concentration of 50-mg antiscalant per liter slow sand filtrate (Vrouwenvelder et al., 2000). *AOC*, assimilable organic carbon; *ATP*, adenosine triphosphate; *TOC*, total organic carbon.

A biofouling inhibitor can contain biodegradable compounds enhancing biofilm formation. It is recommended to avoid dosages of chemicals as much as possible to reduce the membrane fouling risk and environmental impact of concentrate discharge containing such chemicals (Sweity et al., 2013).

7.3.3 Conventional curative biofouling control: cleanings

To maintain a constant clean water recovery without compromising energy input, it is important to control fouling in the system. When performance declines, membrane modules need to be chemically and/or physically cleaned to maintain the plant operation. Curative chemical cleaning is the conventional approach practiced for quite a few decades where chemicals are fed into the system when there is a performance decline. Chemical cleaning broadly includes weakening of the biofilm matrix mostly by chemicals such chelating agents, detergents, bases, acids, and enzymes and removal of the biofilm by mechanical forces such as rinsing with water (Cornelissen et al., 2007; Wibisono et al., 2015). Most commonly, a combination of acidic and/or basic (alkaline) chemicals is used to clean the membrane modules. Common acidic solutions (pH 2) include hydrochloric acid, phosphoric acid, sodium hydrosulfate, and sulfamic acid, whereas alkaline (pH 12) chemicals include sodium lauryl sulfate, sodium hydroxide, and sodium ethylene diamine tetra acetic acid. Commercial cleaning agents are found to be more effective than conventional cleaning agents recommended by the membrane manufacturers in some cases (DOW, 2014). In general, all the used cleaning protocols lead to mainly biomass inactivation and very limited removal of biofilm (Fig. 7.6).

Hydraulic cleaning is the most commonly used physical cleaning method wherein water is flushed through the system in forward or backward direction

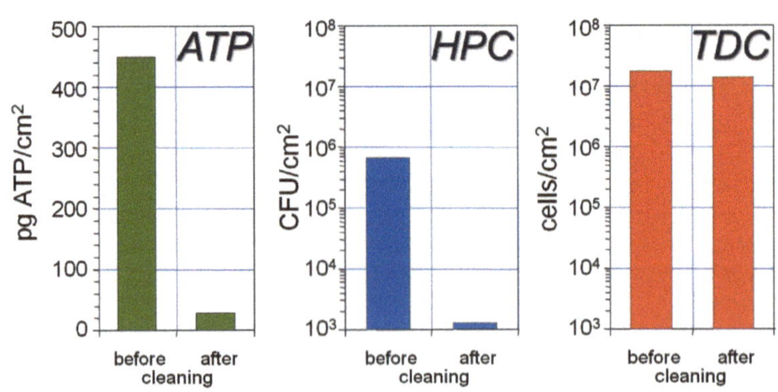

FIGURE 7.6 Biomass parameters obtained during autopsy membrane module from the plant with and without conventional cleaning showed that biomass was partly inactivated (adenosine triphosphate [ATP] and heterotrophic place count [HPC]) but not removed (total direct cell count [TDC]). Membrane installation performance decline was not controlled by enhanced frequency of a conventional cleaning strategy (Vrouwenvelder et al., 2008).

to remove the accumulated biomass and organic foulants. Forward flushing in fact can cause further biofouling, as the biomass accumulated in the lead membrane is pushed to the ones downstream, where they form new biofilms almost instantly. Due to this reason, some plants perform a backwash by reversing the flow direction in the module, thereby reducing the chance of spreading the biomass to all the adjacent membrane modules during cleaning (Vrouwenvelder et al., 2011b).

7.4 New control strategies

7.4.1 Membrane modification

Physical properties (e.g., hydrophilicity, surface charge, roughness) of the membrane surface impact membrane biofouling (Louie et al. 2006, 2011). It is generally accepted that hydrophilic membranes are more resistant to fouling (Kang and Cao, 2012). Deposition of foulant is less likely on neutrally or close to neutrally charged membrane surfaces (Kim and Lee, 2006; Kang and Cao, 2012). Also surface morphology plays a significant role on membrane biofouling, because foulants are more likely to be trapped by rougher topologies than by the smoother ones (Louie et al., 2006; Sagle et al., 2009). Membranes and feed spacers with antifouling coatings to prevent attachment and growth of bacteria have been produced and tested (Araujo et al., 2012a; Miller et al., 2012; Ronen et al., 2015; Bucs et al., 2017). Many studies showed that, compared with uncoated membranes, less fouling developed when membranes surfaces had a hydrophilic coating (Tang et al., 2009; Araujo et al., 2012b). The surface modification of membranes and spacers, however, in general does not last in the long term either due to detachment of antifouling coating/material in time or due to formation of a conditioning biofilm layer on the membrane and feed spacer (Suwarno et al., 2016).

Most research focusing on membrane modification for biofouling prevention demonstrated inability of modified membranes to control biofouling for the long run. Nevertheless, modified membranes might have the potential for more effective cleanings. This is an underexplored research field, which should be explored more.

Instead of trying to reduce membrane fouling by chemically modifying the membrane, another novel approach involving a sacrificial coating on the membrane has been investigated by Son et al. (2018) and by Nava-Ocampo et al. (under review). The objective of their studies was to apply a re-applicable sacrificial multilayer polyelectrolyte coating on the membrane surface to enhance the system cleanability. The membrane coating was based on a layer-by-layer assembly approach, which avoids the use of toxic linkers using poly(diallyl-dimethyl ammonium chloride) and poly(sodium-4-styrene sulfonate). The polyelectrolyte coating could be effectively applied on the membrane surface under cross-flow condition, and it was stable on the membrane surface with continuous operation. The coating layer and any biofilm on top of the sacrificial layer was successfully removed, leaving a clean membrane surface (Nava-Ocampo et al. under review). The use of a re-applicable sacrificial coating could be successful for biofouling control of spiral-wound RO membranes.

Graphene materials as antimicrobial agents have received quite many contradicting results and conclusions. A recent study by Alayande et al. (2019) on the antimicrobial properties of graphene-based membranes such as single-layer graphene oxide (GO) and modified graphene oxide (rGO) on top of a cellulose ester membrane showed that the found antimicrobial effect is neither caused by the graphene nor the membrane support. The antimicrobial effect is found to be clearly and conclusively linked to the hydriodic acid vapor treatment used for preparing the rGO membrane, emphasizing the importance of caution in the reporting of antimicrobial properties of graphene-based surfaces (Alayande et al., 2019).

7.4.2 Feed spacer modification

Feed spacers are used in spiral-wound membrane installations for two reasons: to separate the adjacent membranes and to allow proper mixing of feedwater, thus decreasing concentration polarization. Concentration polarization is the accumulation of solute close to the membrane surface, causing a higher salt concentration of feedwater near the membrane surface compared with the bulk feedwater (Radu et al., 2014; Bucs et al., 2016; Zargar et al., 2020). This can cause reduced permeate flux, increased salt-passage, and/or scaling. Commonly used feed spacers in spiral-wound membrane modules have a diamond-shaped geometry with varying thickness ranging from 26 to 34 mil (1 mil is 25.4 μm). Feed spacers are also the main contributors to biofouling in NF/RO spiral-wound membranes (Vrouwenvelder et al., 2009). Spacers provide support to the microbes to attach and grow. Biofouling on the spacer rather than on the membrane itself causes most of the pressure drop increase in the feed channel in practice. A plethora of research in the past has focused on modifying feed spacers in terms of design and/or material, mainly to improve mass transfer for a better permeate production (Cao et al., 2001; Subramani et al., 2006). A recent approach is to produce geometrically modified feed spacers to obtain a lower pressure drop increase caused by biofouling compared with feed spacers from practice (Li et al., 2016; Kerdi et al., 2018; Ali et al., 2019; Wu et al., 2019). The thickness, material, and geometry of feed spacers have an impact also on membrane modules cleanability. Geometrically modified feed spacers with alternating thick and thin strands demonstrated the ability to reduce the biofouling impact on feed channel pressure drop by 80% with the same amount of biomass present (Siddiqui et al., 2017a).

7.4.3 Advanced cleaning strategies

Conventionally applied acid/alkali based cleaning treatments often fail to fully restore membrane performance (Vrouwenvelder et al., 1998; Huiting et al., 2001; Beyer et al., 2014). A recent study examining chemical cleaning in three full-scale RO plants in the Netherlands underlined the need for novel cleaning approaches, as none of the standard cleaning chemicals resulted in highly effective membrane regeneration (Beyer et al., 2017). Moreover, rapid biofilm regrowth is observed, inducing more frequent and rigorous cleaning, resulting in increased costs and chemical waste (Vrouwenvelder et al., 1998). There is an apparent need for novel chemical cleaning strategies to enhance biofilm solubilization and removal from membrane modules, achieve higher recoveries of membrane performance parameters, and decrease the required cleaning frequency. Sanawar et al. (2018) tested urea, a chaotropic agent, as a novel cleaning chemical for RO polyamide membranes. Results from that study demonstrated that urea cleaning was as effective as the conventional chemical cleaning in terms of restoring the normalized feed channel pressure drop and more effective in terms of (1) restoring membrane permeability; (2) solubilizing organic foulants, and (3) removing the surface fouling layer.

The use of formulated cleaning products with multiple cleaning mechanisms offers enhanced cleaning performance by targeting the composite layers of biofilms on a membrane surface. Cleaning chemicals that incorporate effervescent reagents produce a range of microbubble sizes (5−500 m), which agitate the biofilm depositing on the membrane surface assisting their removal, therefore performing as chemical and physical cleaning agents. Some novel cleaning procedures also employ additional cleaning mechanisms such as high-ionic-strength compounds causing normal osmosis to occur during periods of soaking, from permeate to the feed side breaking up layered deposits. Physically increasing the amount of generated microbubbles is also done by using a specially designed microbubble generator, which inducts air into the cleaning solution being circulated through the RO plant. The shear effect that the bubble creates has shown to better enhance biofilm removal in several studies (Chesters et al., 2015; Fazel and Chesters, 2015; Wibisono et al., 2015). Pneumatic cleaning refers to the use of air or gas mixed with water for flushing (air−water flushing) (Vrouwenvelder et al., 2010). A series of experiments done by Cornelissen et al. shows promising results on pilot scale for the use of air/water flushing (Cornelissen et al., 2007; Wibisono et al., 2014, 2015). CO_2 dissolved in water has proved to restore initial hydraulic resistance as well as visible reduction in biofouling (Ngene et al., 2010).

An approach that combines both bubbling and cleaning developed by Yang et al. (2019) on the basis of intermittently adding a low concentration hydrogen peroxide (H_2O_2) to the feed water. With periodic dosing (every 12 h) of 0.3% (w/w) H_2O_2, no detectable biofouling occurred, while biofouling was evident without H_2O_2 dosing. The control of biofouling mainly resulted from bubble production when H_2O_2 dissociated to shear. The use of H_2O_2 dosage had the combined advantages of physically shearing off biofilms from spacers and chemically killing bacteria, while providing a low cost approach for biofouling control in membrane-based desalination processes (Yang et al., 2019). The cleaning efficiency can be evaluated with novel non-destructive techniques such as e.g. MRI (Ujihara et al., 2018) and optodes and the use of sensitive feed channel pressure drop measurements.

7.5 Future perspectives

Biofouling is defined as the biofilm accumulation causing an unacceptable performance decline (Characklis and Marshall, 1990). So, the presence of a biofilm is not necessarily a problem. An effective strategy to control biofouling should most likely involve the combination of (1) delaying the built-up of biofilms by pretreatment removing extensively the easily biodegradable nutrients before supply to the membrane installation, (2) restricting the impact of accumulated biomass in the membrane system on the system performance by, e.g., modified spacer/module/plant design and operation, and (3) removing the accumulated biofilm from the membrane system using advanced cleaning strategies (Bucs et al. 2014, 2018).

Water pretreatment processes such as cartridge filters, located directly before the RO membranes, are usually not studied because of the relative "low cost" of these filters. Surprisingly, a very recent study showed that selection of the proper cartridge filter and the right time of filter replacement could lead to cost savings in the order of 0.48 Eurocent per m^3, which add up to an annual saving of 1 million Euros at a 600,000 m^3/day plant (1,400,000 m^3/day passes the cartridge filters) (Farhat et al., 2020). This underlines that significant cost savings and better membrane performance may be achieved at treatments steps or chemical dosages not expected.

For restricting the impact of accumulated biomass on reduced membrane performance, it has been demonstrated that modified feed spacer designs and modified hydraulics in membrane modules can reduce biofilm impact. Most likely, the adaptation of the design and operation of membrane modules and installations also may lead to a reduction of biofilm impact. This research field is still relatively unexplored and therefore has perspective.

Extensive research has been done on conventional chemical cleaning and some research on air/water flush cleaning. However, so far, no cleaning has been developed that is effective for biofouling control in all cases. In general, conventional chemical cleanings inactivate a large part of the biomass but does not remove the biomass from the membrane system. Cleanings should include solubilization of the biofilm and subsequent removal from the membrane system. Modified (more open) spacers are expected to ease the biofilm removal, and novel strategies such as re-applicable sacrificial coatings (Son et al., 2018; Nava-Ocampo et al. under review) and the use of reusable solubilizers for biofilms and organics such as urea (Sanawar et al., 2018) are expected to lead to more effective and more environmentally friendly cleaning.

Novel modified membrane and feed spacer materials should not be studied in short-term adhesion experiments with proteins and/or pure bacteria strains (Miller et al., 2012), but with longer-term biofilm experiments (Miller et al., 2012; Siddiqui et al., 2017a). There are extensive research efforts on developing new membranes and spacer materials and coatings. Proposed is to study, in addition to the biofouling prevention properties of new materials and coatings, the cleaning efficiency of such material. It could be that a material does not prevent biofilm accumulation but enables very effective biofilm removal.

Clearly, this chapter identifies and describes a selection of highly new and promising directions for research aiming at biofouling control.

Acknowledgment

The research reported in this publication was supported by funding from King Abdullah University of Science and Technology (KAUST).

References

Abushaban, A., Mangal, M.N., Salinas-Rodriguez, S.G., Nnebuo, C., Mondal, S., Goueli, S.A., Schippers, J.C., Kennedy, M.D., 2017. Direct measurement of ATP in seawater and application of ATP to monitor bacterial growth potential in SWRO pre-treatment systems. Desalination and Water Treatment 99, 91–101.

Abushaban, A., Salinas-Rodriguez, S.G., Dhakal, N., Schippers, J.C., Kennedy, M.D., 2019a. Assessing pretreatment and seawater reverse osmosis performance using an ATP-based bacterial growth potential method. Desalination 467, 210–218.

Abushaban, A., Salinas-Rodriguez, S.G., Mangal, M.N., Mondal, S., Goueli, S.A., Knezev, A., Vrouwenvelder, J.S., Schippers, J.C., Kennedy, M.D., 2019b. ATP measurement in seawater reverse osmosis systems: eliminating seawater matrix effects using a filtration-based method. Desalination 453, 1–9. https://doi.org/10.1016/j.desal.2018.11.020.

Alayande, A.B., Park, H.-D., Vrouwenvelder, J.S., Kim, I.S., 2019. Implications of chemical reduction using hydriodic acid on the antimicrobial properties of graphene oxide and reduced graphene oxide membranes. Small 15 (28). https://doi.org/10.1002/smll.201901023 art. no. 1901023.

Ali, S.M., Qamar, A., Kerdi, S., Phuntsho, S., Vrouwenvelder, J.S., Ghaffour, N., Shon, H.K., 2019. Energy efficient 3D printed column type feed spacer for membrane filtration. Water Research 164.

Amiri, M.C., Samiei, M., 2007. Enhancing permeate flux in a RO plant by controlling membrane fouling. Desalination 207 (1–3), 361–369.

Amjad, Z., 1993. Reverse Osmosis : Membrane Technology, Water Chemistry & Industrial Applications. Van Nostrand Reinhold, New York.

Araujo, P.A., Kruithof, J.C., Van Loosdrecht, M.C.M., Vrouwenvelder, J.S., 2012a. The potential of standard and modified feed spacers for biofouling control. Journal of Membrane Science 403, 58—70.

Araujo, P.A., Miller, D.J., Correia, P.B., van Loosdrecht, M.C.M., Kruithof, J.C., Freeman, B.D., Paul, D.R., Vrouwenvelder, J.S., 2012b. Impact of feed spacer and membrane modification by hydrophilic, bactericidal and biocidal coating on biofouling control. Desalination 295, 1—10.

Bertheas, U., Majamaa, K., Arzu, A., Pahnke, R., 2009. Use of DBNPA to control biofouling in RO systems. Desalination and Water Treatment 3 (1—3), 175—178.

Beyer, F., Laurinonyte, J., Zwijnenburg, A., Stams, A.J.M., Plugge, C.M., 2017. Membrane fouling and chemical cleaning in three full-scale reverse osmosis plants producing demineralized water. Journal of Engineering 2017, 14.

Beyer, F., Rietman, B.M., Zwijnenburg, A., van den Brink, P., Vrouwenvelder, J.S., Jarzembowska, M., Laurinonyte, J., Stams, A.J.M., Plugge, C.M., 2014. Long-term performance and fouling analysis of full-scale direct nanofiltration (NF) installations treating anoxic groundwater. Journal of Membrane Science 468, 339—348.

Bucs, S.S., Farhat, N., Kruithof, J.C., Picioreanu, C., van Loosdrecht, M.C.M., Vrouwenvelder, J.S., 2018. Review on strategies for biofouling mitigation in spiral wound membrane systems. Desalination 434, 189—197.

Bucs, S.S., Linares, R.V., Farhat, N., Matin, A., Khan, Z., Van Loosdrecht, M.C.M., Yang, R., Wang, M., Gleason, K.K., Kruithof, J.C., Vrouwenvelder, J.S., 2017. Coating of reverse osmosis membranes with amphiphilic copolymers for biofouling control. Desalination and Water Treatment 68, 1—11.

Bucs, S.S., Linares, R.V., van Loosdrecht, M.C.M., Kruithof, J.C., Vrouwenvelder, J.S., 2014. Impact of organic nutrient load on biomass accumulation, feed channel pressure drop increase and permeate flux decline in membrane systems. Water Research 67, 227—242.

Bucs, S.S., Valladares Linares, R., Vrouwenvelder, J.S., Picioreanu, C., 2016. Biofouling in forward osmosis systems: an experimental and numerical study. Water Research 106, 86—97.

Cao, Z., Wiley, D.E., Fane, A.G., 2001. CFD simulations of net-type turbulence promoters in a narrow channel. Journal of Membrane Science 185 (2), 157—176.

Characklis, W.G., 1981. Fouling biofilm development — a process analysis. Biotechnology and Bioengineering 23 (9), 1923—1960.

Characklis, W.G., Marshall, K.C., 1990. Biofilms. Wiley, New York.

Chesters, S.P., Armstrong, M.W., Fazel, M., 2015. Microbubble RO membrane cleaning reduces fouling on WWRO plant. Desalination and Water Treatment 55 (11), 2900—2908.

Chinu, K.J., Johir, A.H., Vigneswaran, S., Shon, H.K., Kandasamy, J., 2009. Biofilter as pretreatment to membrane based desalination: evaluation in terms of fouling index. Desalination 247 (1—3), 77—84.

Cornelissen, E.R., Vrouwenvelder, J.S., Heijman, S.G.J., Viallefont, X.D., Van Der Kooij, D., Wessels, L.P., 2007. Periodic air/water cleaning for control of biofouling in spiral wound membrane elements. Journal of Membrane Science 287 (1), 94—101.

DOW, 2014. Membranes Cleaning Procedures for DOW FILMTEC FT30 Elements.

Elimelech, M., Phillip, W.A., 2011. The future of seawater desalination: energy, technology, and the environment. Science 333 (6043), 712—717.

Escobar, I.C., Randall, A.A., 2000. Sample storage impact on the assimilable organic carbon (AOC) bioassay. Water Research 34 (5), 1680—1686.

Farhat, N., Hammes, F., Prest, E., Vrouwenvelder, J., 2018a. A uniform bacterial growth potential assay for different water types. Water Research 142, 227—235.

Farhat, N.M., Christodoulou, C., Placotas, P., Blankert, B., Sallangos, O., Vrouwenvelder, J.S., 2020. Cartridge filter selection and replacement: optimization of produced water quantity, quality, and cost. Desalination 473, 114172.

Farhat, N.M., Loubineaud, E., Prest, E.I.E.C., El-Chakhtoura, J., Salles, C., Bucs, S.S., Trampé, J., Van den Broek, W.B.P., Van Agtmaal, J.M.C., Van Loosdrecht, M.C.M., Kruithof, J.C., Vrouwenvelder, J.S., 2018b. Application of monochloramine for wastewater reuse: effect on biostability during transport and biofouling in RO membranes. Journal of Membrane Science 551, 243–253.

Fazel, M., Chesters, S., 2015. RO membrane cleaning using microbubbles at 6,800 m3/d wastewater RO plant in UAE. Desalination and Water Treatment 55 (12), 3358–3366.

Flemming, H.C., 2002. Biofouling in water systems – cases, causes and countermeasures. Applied Microbiology and Biotechnology 59 (6), 629–640.

Flemming, H.C., 2003. Role and levels of real-time monitoring for successful anti-fouling strategies – an overview. Water Science and Technology 47 (5), 1–8.

Greenlee, L.F., Lawler, D.F., Freeman, B.D., Marrot, B., Moulin, P., 2009. Reverse osmosis desalination: water sources, technology, and today's challenges. Water Research 43 (9), 2317–2348.

Gur-Reznik, S., Katz, I., Dosoretz, C.G., 2008. Removal of dissolved organic matter by granular-activated carbon adsorption as a pretreatment to reverse osmosis of membrane bioreactor effluents. Water Research 42 (6–7), 1595–1605.

Haddix, P.L., Shaw, N.J., LeChevallier, M.W., 2004. Characterization of bioluminescent derivatives of assimilable organic carbon test bacteria. Applied and Environmental Microbiology 70 (2), 850–854.

Hammes, F.A., Egli, T., 2005. New method for assimilable organic carbon determination using flow-cytometric enumeration and a natural microbial consortium as inoculum. Environmental Science and Technology 39 (9), 3289–3294.

Herzberg, M., Elimelech, M., 2007. Biofouling of reverse osmosis membranes: role of biofilm-enhanced osmotic pressure. Journal of Membrane Science 295 (1–2), 11–20.

Huiting, H., Kappelhof, J.W.N.M., Bosklopper, T.G.J., 2001. Operation of NF/RO plants: from reactive to proactive. Desalination 139 (1–3), 183–189.

Jeong, S., Naidu, G., Vigneswaran, S., Ma, C.H., Rice, S.A., 2013. A rapid bioluminescence-based test of assimilable organic carbon for seawater. Desalination 317, 160–165.

Kang, G.D., Cao, Y.M., 2012. Development of antifouling reverse osmosis membranes for water treatment: a review. Water Research 46 (3), 584–600.

Kemmy, F.A., Fry, J.C., Breach, R.A., 1989. Development and operational implementation of a modified and simplified method for determination of assimilable organic-carbon (Aoc) in drinking-water. Water Science and Technology 21 (3), 155–159.

Kerdi, S., Qamar, A., Vrouwenvelder, J.S., Ghaffour, N., 2018. Fouling resilient perforated feed spacers for membrane filtration. Water Research 140, 211–219.

Kim, I.C., Lee, K.H., 2006. Dyeing process wastewater treatment using fouling resistant nanofiltration and reverse osmosis membranes. Desalination 192 (1–3), 246–251.

Kim, T.S., Park, H.D., 2016. Lauroyl arginate ethyl: an effective antibiofouling agent applicable for reverse osmosis processes producing potable water. Journal of Membrane Science 507, 24–33.

Kostakioti, M., Hadjifrangiskou, M., Hultgren, S.J., 2013. Bacterial biofilms: development, dispersal, and therapeutic strategies in the dawn of the postantibiotic era. Cold Spring Harbor Perspectives in Medicine 3 (4).

Le chevallier, M.W., Shaw, N.E., Kaplan, L.A., Bott, T.L., 1993. Development of a rapid assimilable organic-carbon method for water. In: Water Quality Technology Conference 1992, Pts I and Ii, pp. 1135–1157.

Li, W.Y., Chen, K.K., Wang, Y.N., Krantz, W.B., Fane, A.G., Tang, C.Y., 2016. A conceptual design of spacers with hairy structures for membrane processes. Journal of Membrane Science 510, 314–325.

Louie, J.S., Pinnau, I., Ciobanu, I., Ishida, K.P., Ng, A., Reinhard, M., 2006. Effects of polyether–polyamide block copolymer coating on performance and fouling of reverse osmosis membranes. Journal of Membrane Science 280 (1), 762–770.

Louie, J.S., Pinnau, I., Reinhard, M., 2011. Effects of surface coating process conditions on the water permeation and salt rejection properties of composite polyamide reverse osmosis membranes. Journal of Membrane Science 367 (1–2), 249–255.

Mallevialle, J., Odendaal, P.E., Wiesner, M.R., American Water Works Association Research Foundation, Lyonnaise des eaux-Dumez (Firm), Water Research Commision of South America, AWWA Research Foundation, 1996. Water Treatment Membrane Processes. McGraw-Hill, New York, NY.

Meier, J., Melin, T., 2005. Wastewater reclamation by the PAC-NF process. Desalination 178 (1–3), 27–40.

Miller, D.J., Araujo, P.A., Correia, P.B., Ramsey, M.M., Kruithof, J.C., van Loosdrecht, M.C.M., Freeman, B.D., Paul, D.R., Whiteley, M., Vrouwenvelder, J.S., 2012. Short-term adhesion and long-term biofouling testing of polydopamine and poly(ethylene glycol) surface modifications of membranes and feed spacers for biofouling control. Water Research 46 (12), 3737–3753.

Nava-Ocampo, M.F., Bucs, S.S., Farinha, A., Son, M., Logan, B.E., Vrouwenvelder, J.S., under review. Sacrificial Coating Development for Biofouling Control in Membrane Systems.

Nescerecka, A., Juhna, T., Hammes, F., 2018. Identifying the underlying causes of biological instability in a full-scale drinking water supply system. Water Research 135, 11–21.

Ngene, I.S., Lammertink, R.G.H., Kemperman, A.J.B., van de Ven, W.J.C., Wessels, L.P., Wessling, M., Van der Meer, W.G.J., 2010. CO_2 nucleation in membrane spacer channels remove biofilms and fouling deposits. Industrial and Engineering Chemistry Research 49 (20), 10034–10039.

Palmer, J., Flint, S., Brooks, J., 2007. Bacterial cell attachment, the beginning of a biofilm. Journal of Industrial Microbiology and Biotechnology 34 (9), 577–588.

Prest, E.I., Hammes, F., Kotzsch, S., van Loosdrecht, M.C.M., Vrouwenvelder, J.S., 2016. A systematic approach for the assessment of bacterial growth-controlling factors linked to biological stability of drinking water in distribution systems. Water Science and Technology: Water Supply 16 (4), 865–880.

Prihasto, N., Liu, Q.F., Kim, S.H., 2009. Pre-treatment strategies for seawater desalination by reverse osmosis system. Desalination 249 (1), 308–316.

Radu, A.I., Bergwerff, L., van Loosdrecht, M.C.M., Picioreanu, C., 2014. A two-dimensional mechanistic model for scaling in spiral wound membrane systems. Chemical Engineering Journal 241, 77–91.

Roe, F.L., Wentland, E., Zelver, N., Warwood, B., Waters, R., Characklis, W.G., 1994. Online side-stream monitoring of biofouling. Biofouling and Biocorrosion in Industrial Water Systems 137–150.

Ronen, A., Lerman, S., Ramon, G.Z., Dosoretz, C.G., 2015. Experimental characterization and numerical simulation of the anti-biofuling activity of nanosilver-modified feed spacers in membrane filtration. Journal of Membrane Science 475, 320–329.

Ross, P.S., Hammes, F., Dignum, M., Magic-Knezev, A., Hambsch, B., Rietveld, L.C., 2013. A comparative study of three different assimilable organic carbon (AOC) methods: results of a round-robin test. Water Science and Technology: Water Supply 13 (4), 1024–1033.

Sagle, A.C., Van Wagner, E.M., Ju, H., McCloskey, B.D., Freeman, B.D., Sharma, M.M., 2009. PEG-coated reverse osmosis membranes: desalination properties and fouling resistance. Journal of Membrane Science 340 (1–2), 92–108.

Sanawar, H., Pinel, I., Farhat, N.M., Bucs, S.S., Zlopasa, J., Kruithof, J.C., Witkamp, G.J., van Loosdrecht, M.C.M., Vrouwenvelder, J.S., 2018. Enhanced biofilm solubilization by urea in reverse osmosis membrane systems. Water Research X 1, 100004.

Sathasivan, A., Ohgaki, S., 1999. Application of new bacterial regrowth potential method for water distribution system – a clear evidence of phosphorus limitation. Water Research 33 (1), 137–144.

Schiermeier, Q., 2008. Purification with a pinch of salt. Nature 452 (7185), 260–261.

Servais, P., Anzil, A., Ventresque, C., 1989. Simple method for determination of biodegradable dissolved organic-carbon in water. Applied and Environmental Microbiology 55 (10), 2732–2734.

Service, R.F., 2006. Desalination freshens up. Science 313 (5790), 1088–1090.

Shannon, M.A., Bohn, P.W., Elimelech, M., Georgiadis, J.G., Marinas, B.J., Mayes, A.M., 2008. Science and technology for water purification in the coming decades. Nature 452 (7185), 301–310.

Siddiqui, A., Farhat, N., Bucs, S.S., Linares, R.V., Picioreanu, C., Kruithof, J.C., van Loosdrecht, M.C.M., Kidwell, J., Vrouwenvelder, J.S., 2016. Development and characterization of 3D-printed feed spacers for spiral wound membrane systems. Water Research 91, 55–67.

Siddiqui, A., Lehmann, S., Bucs, S.S., Fresquet, M., Fel, L., Prest, E.I.E.C., Ogier, J., Schellenberg, C., van Loosdrecht, M.C.M., Kruithof, J.C., Vrouwenvelder, J.S., 2017a. Predicting the impact of feed spacer modification on biofouling by hydraulic characterization and biofouling studies in membrane fouling simulators. Water Research 110, 281–287.

Siddiqui, A., Pinel, I., Prest, E.I., Bucs, S.S., van Loosdrecht, M.C.M., Kruithof, J.C., Vrouwenvelder, J.S., 2017b. Application of DBNPA dosage for biofouling control in spiral wound membrane systems. Desalination and Water Treatment 68, 12–22.

Siebdrath, N., Farhat, N., Ding, W., Kruithof, J., Vrouwenvelder, J.S., 2019. Impact of membrane biofouling in the sequential development of performance indicators: feed channel pressure drop, permeability, and salt rejection. Journal of Membrane Science 585, 199–207.

Son, M., Yang, W., Bucs, S.S., Nava-Ocampo, M.F., Vrouwenvelder, J.S., Logan, B.E., 2018. Polyelectrolyte-based sacrificial protective layer for fouling control in reverse osmosis desalination. Environmental Science and Technology Letters 5 (9), 584–590.

Subramani, A., Kim, S., Hoek, E.M.V., 2006. Pressure, flow, and concentration profiles in open and spacer-filled membrane channels. Journal of Membrane Science 277 (1–2), 7–17.

Suwarno, S.R., Hanada, S., Chong, T.H., Goto, S., Henmi, M., Fane, A.G., 2016. The effect of different surface conditioning layers on bacterial adhesion on reverse osmosis membranes. Desalination 387, 1–13.

Sweity, A., Oren, Y., Ronen, Z., Herzberg, M., 2013. The influence of antiscalants on biofouling of RO membranes in seawater desalination. Water Research 47 (10), 3389–3398.

Tang, C.Y.Y., Kwon, Y.N., Leckie, J.O., 2009. Effect of membrane chemistry and coating layer on physiochemical properties of thin film composite polyamide RO and NF membranes I. FTIR and XPS characterization of polyamide and coating layer chemistry. Desalination 242 (1–3), 149–167.

Tran, T., Bolto, B., Gray, S., Hoang, M., Ostarcevic, E., 2007. An autopsy study of a fouled reverse osmosis membrane element used in a brackish water treatment plant. Water Research 41 (17), 3915–3923.

Ujihara, R., Fridjonsson, E.O., Bristow, N.W., Vogt, S.J., Bucs, S.S., Vrouwenvelder, J.S., Johns, M.L., 2018. Earth's field MRI for the non-invasive detection of fouling in spiral-wound membrane modules in pressure vessels during operation. Desalination and Water Treatment 135, 16–24. https://doi.org/10.5004/dwt.2018.23156.

Van der Kooij, D., Veenendaal, H.R., van der Mark, E.J., Dignum, M., 2017. Assessment of the microbial growth potential of slow sand filtrate with the biomass production potential test in comparison with the assimilable organic carbon method. Water Research 125, 270–279.

Van der kooij, D., Visser, A., Hijnen, W.A.M., 1982. Determining the concentration of easily assimilable organic-carbon in drinking-water. Journal American Water Works Association 74 (10), 540–545.

Vrouwenvelder, H.S., van Paassen, J.A.M., Folmer, H.C., Hofman, J.A.M.H., Nederlof, M.M., van der Kooij, D., 1998. Biofouling of membranes for drinking water production. Desalination 118 (1–3), 157–166.

Vrouwenvelder, J., Kruithof, J., van Loosdrecht, M., 2011a. Biofouling of Spiral Wound Membrane Systems. IWA Publishing.

Vrouwenvelder, J.S., Buiter, J., Riviere, M., van der Meer, W.G.J., van Loosdrecht, M.C.M., Kruithof, J.C., 2010. Impact of flow regime on pressure drop increase and biomass accumulation and morphology in membrane systems. Water Research 44 (3), 689–702.

Vrouwenvelder, J.S., Manolarakis, S.A., van der Hoek, J.P., van Paassen, J.A.M., van der Meer, W.G.J., van Agtmaal, J.M.C., Prummel, H.D.M., Kruithof, J.C., van Loosdrecht, M.C.M., 2008. Quantitative biofouling diagnosis in full scale nanofiltration and reverse osmosis installations. Water Research 42 (19), 4856–4868.

Vrouwenvelder, J.S., Manolarakis, S.A., Veenendaal, H.R., van der Kooij, D., 2000. Biofouling potential of chemicals used for scale control in RO and NF membranes. Desalination 132 (1), 1–10.

Vrouwenvelder, J.S., Van Loosdrecht, M.C.M., Kruithof, J.C., 2011b. A novel scenario for biofouling control of spiral wound membrane systems. Water Research 45 (13), 3890–3898.

Vrouwenvelder, J.S., von der Schulenburg, D.A.G., Kruithof, J.C., Johns, M.L., van Loosdrecht, M.C.M., 2009. Biofouling of spiral-wound nanofiltration and reverse osmosis membranes: a feed spacer problem. Water Research 43 (3), 583–594.

Wang, Q.H., Tao, T., Xin, K.L., Li, S.P., Zhang, W.F., 2014. A review research of assimilable organic carbon bioassay. Desalination and Water Treatment 52 (13–15), 2734–2740.

Weinrich, L.A., Schneider, O.D., LeChevallier, M.W., 2011. Bioluminescence-based method for measuring assimilable organic carbon in pretreatment water for reverse osmosis membrane desalination. Applied and Environmental Microbiology 77 (3), 1148–1150.

Werner, P., Hambsch, B., 1986. Investigations on the growth of bacteria in drinking water. Water Supply 4.

Wibisono, Y., Cornelissen, E.R., Kemperman, A.J.B., van der Meer, W.G.J., Nijmeijer, K., 2014. Two-phase flow in membrane processes: a technology with a future. Journal of Membrane Science 453, 566–602.

Wibisono, Y., El Obied, K.E., Cornelissen, E.R., Kemperman, A.J.B., Nijmeijer, K., 2015. Biofouling removal in spiral-wound nanofiltration elements using two-phase flow cleaning. Journal of Membrane Science 475, 131–146.

Wu, B., Zhang, Y., Mao, Z., Tan, W.S., Tan, Y.Z., Chew, J.W., Chong, T.H., Fane, A.G., 2019. Spacer vibration for fouling control of submerged flat sheet membranes. Separation and Purification Technology 210, 719–728.

Yang, W., Son, M., Xiong, B., Kumar, M., Bucs, S., Vrouwenvelder, J.S., Logan, B.E., 2019. Effective biofouling control using periodic H_2O_2 cleaning with CuO modified and polypropylene spacers. ACS Sustainable Chemistry and Engineering 7 (10), 9582–9587. https://doi.org/10.1021/acssuschemeng.9b01086.

Zargar, M., Ujihara, R., Vogt, S.J., Vrouwenvelder, J.S., Fridjonsson, E.O., Johns, M.L., 2020. Imaging of membrane concentration polarization by NaCl using 23Na nuclear magnetic resonance. Journal of Membrane Science 600, 117868.

Zhao, D.S., Yu, S.L., 2015. A review of recent advance in fouling mitigation of NF/RO membranes in water treatment: pretreatment, membrane modification, and chemical cleaning. Desalination and Water Treatment 55 (4), 870–891.

Chapter 8

Biofilm fuel cells

Marcela N. Gatti, Facundo Quiñones, Ruben H. Milocco
Grupo de Control Automático y Sistemas, Facultad de Ingeniería. Universidad Nacional del Comahue, Neuquén, Argentina

8.1 Processes involved in the biofilm of a microbial fuel cell

Microbial fuel cell (MFC) is a field of science and technology that combines biological catalytic redox activity with abiotic electrochemical reactions and capacitive charging accumulation. The biofilm plays an important role in an MFC since the processes mentioned occur on it. In this chapter, the principal processes linked to the biofilm in an MFC are revised and applied by using mathematical interpretation.

8.1.1 Bacteria-catalyzed reactions

MFCs are bioelectrochemical reactors where the electroactive bacteria grow as biofilm attached on the anode surface and releasing electrons and protons. Current releasing results from the electron flow through an external load connected between the anode and the cathode while protons reach the cathode surface crossing a proton exchange membrane (PEM). The negative electrode is the anode immersed in a solution at a given substrate concentration. The electrons are generated during oxidation of the organic substrate by the electroactive bacteria. Considering acetate as the carbon source, the main reaction in the anode is the following:

$$(CH_2O)_2 + 2H_2O \rightarrow 2CO_2 + 8H^+ + 8e^- \qquad 8.1$$

Bacteria are capable of transferring the electrons to the anode surface by means of different mechanisms. The anode transfers electrons to the cathode via the external circuit, which results in the production of electric energy under load. Simultaneously, protons are transferred via the membrane to the cathode to maintain charge balance. The cathode is the positive electrode where reduction is carried on given by the following reaction:

Recent Trends in Biofilm Science and Technology. https://doi.org/10.1016/B978-0-12-819497-3.00008-8

$$O_2 + 4H^+ + 4e^- \rightarrow 2H_2O \qquad 8.2$$

MFC allows converting the chemical energy stored in organic material to electrical energy by using microorganisms as biocatalysts. Organic material comes from urban or industrial wastewater, giving a valuable alternative for treating wastewater while simultaneously generating renewable energy. MFCs are clean, effective, and recyclable without producing toxic by-products (Reguera et al., 2005; Schröder, 2007). The use of bacteria as catalysts in electrochemical reactions provides potential applications for this technology.

8.1.2 Mechanisms involved in releasing electrons on the anode surface

Electron transference can be mediated by two mechanisms: direct electron transfer (DET) and mediated electron transfer (MET).

Direct electron transfer—This mechanism takes place via a physical contact of the bacterial cell membrane with the fuel cell anode involving outer membrane redox proteins. The DET mechanism needs physical contact between the microorganism and the anode surface. Therefore, the bacterium growing on the first monolayer over the anode surface is electronically active. Moreover, electron transfer can evolve by conducting molecular pili (nanowires). These nanowires allow the microorganisms to use an electrode that is not in direct cell contact. The nanowires are connected to the membrane-bound cytochromes via which the electron transfer to the outside of the cell is accomplished (Schröder et al., 2015). This allows the development of thicker biofilms and thus higher anode performances.

Mediated electron transfer—This mechanism can evolve via exogenous mediators or metabolites produced by cells. Several compounds have been studied as exogenous mediators, the majority based on phenazines, phenothiazines, phenoxazines, and quinones. However, this approach presents the greatest disadvantage of adding exogenous compounds, which is not a sustainable procedure. In some conditions, there is not a direct contact between microorganisms and the anode surface, and there is a lack of soluble electron acceptors, which occurs in thick biofilms. Under these conditions, microorganisms can produce electron shuttling compounds via secondary metabolic pathways. Secondary metabolites make the electron transfer independent of the presence of exogenous redox shuttles.

The importance of mediators lies in the fact that they can transfer electrons to the MFC anode or to the biofilm aerobic layers, where mediators are reoxidized and become available for further redox processes. Then, one molecule is reduced and oxidized many times favoring the current production. The electrochemical oxidation can be described by the following reaction (Picioreanu et al., 2007):

$$M_{red} \leftrightarrow M_{ox} + 2H^+ + 2e^- \qquad 8.3$$

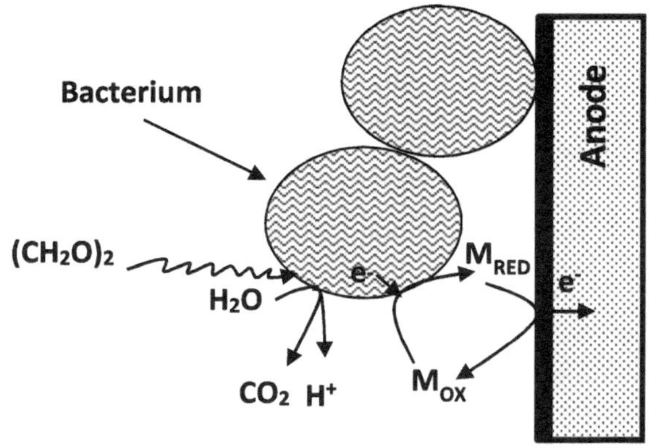

FIGURE 8.1 Biological oxidation of acetic acid and redox reactions of soluble mediator.

where M_{red} and M_{ox} are the reduced and oxidized form of the mediator, respectively. Fig. 8.1 shows biological oxidation of acetic acid and redox reactions of soluble mediators.

8.1.3 The faradaic charge transfer processes

The electrochemical oxidation and reduction reactions taking place at the anode—and simultaneously at the cathode to keep the electrochemical balance—give rise to the faradaic current, $I_f(t)$. This can be mathematically interpreted by means of a Butler—Volmer-type equation. This expression links the faradaic current with its driving force, which results from the anode potential and the substrate concentration (Bockris and Reddy, 2015):

$$I_f = \overset{\sim r}{k} c^r e^{\alpha^r E_a} - \overset{\sim o}{k} c^o e^{-\alpha^o E_a} \qquad 8.4$$

where c^r and c^o, $\overset{\sim r}{k}$ and $\overset{\sim o}{k}$ are the concentrations and constant rates of reduced and oxidized species at the electrochemical interface, respectively; $\alpha^r = a^r F/_{RT}$ and $\alpha^o = a^o F/_{RT}$ are constants where a^r with values in the interval (0,1) and $a^o = 1 - a^r$ are symmetry factors. The values assumed in this chapter are $a_r = a_o = 0.5$ that means $\alpha^r = \alpha^0 = \alpha$; F is the Faraday constant; R is the universal gas constant; T is the absolute temperature; and E_a is the anode potential.

The importance of taking into account the oxidation reactions in the anode surface is clear when only the reduction term on the right side of Eq. (8.4) is considered. Under zero current $I_f = 0$, if the term corresponding to the reduction reactions is missing, the anode potential should be $E_a = -\infty$, which is not possible. By multiplying and dividing the first term on the right side of Eq. (8.4) by the maximum acetate concentration—which is the reduced state of the active species ($\overline{c}^r$), and taking into account that the saturated concentration of the oxidized species (c^o) remains almost constant, the following direct equation is obtained:

$$I_f = k^r S e^{\alpha E_a} - k^o e^{-\alpha E_a} \qquad 8.5$$

where $k^r = \tilde{k}^r \bar{c}^r$ and $k^o = \tilde{k}^o c^o$ are constants, and $S = c^r/\bar{c}^r$ is the fractional substrate or acetate concentration at the interface, which lies in the interval (0,1). This is a Butler–Volmer equation expressing the functional relationship of current with respect to the potential and concentration.

The MFC potential relationship is given by the following equation:

$$\mathrm{V} = E_c - E_a - IR = E_c^o + \eta - E_a - IR' \qquad 8.6$$

where V is the potential between terminals of both electrodes, E_c^o is the reversal cathodic potential given by the Nernst equation (Bockris and Reddy, 2004), η is the sum of the concentration and activation cathodic overpotentials, I is the current that flows through the MFC, which is the difference between the Faradaic current (I_f) and the double-layer current (I_c) as will be described later, and R is an ohmic resistance—including both the resistance to the flow of electrons through the electrodes and interconnections, and the resistance to the flow of ions through the proton exchange membrane. By replacing E_a from Eq. (8.6) to Eq. (8.5), the following expression for the current is obtained:

$$I_f = K^r S e^{-\alpha(V+IR_i)} - K^o e^{\alpha(V+IR_i)} \qquad 8.7$$

where $K^r = k^r e^{\alpha\left(E_c^o\right)}$, $K^o = k^o e^{-\alpha\left(E_c^o\right)}$, and the internal resistance (R_i) that depends on the current (I) is given by $R_i = \frac{R-\eta}{I}$. From Eq. (8.7), it is possible to obtain the substrate concentration at the electrode surface (S) explicitly, which is a function of V, I, and I_f:

$$S = \frac{K^o e^{\alpha(V+IR_i)} + I_f}{K^r e^{-\alpha(V+IR_i)}} \qquad 8.8$$

Note that, when the total current is zero, the open-circuit voltage (OCV)(OCV) can be written explicitly as:

$$\mathrm{OCV} = A + B\ln(S_{I=0}) \qquad 8.9$$

where $A = B\ln\left(\frac{K^r}{K^o}\right)$ and $\mathrm{B} = 1/(2\ \mathrm{alpha}) = \mathrm{RT/F}$ and $S_{I=0}$ is the substrate concentration at zero current. This expression that relates the potential with the substrate concentration is called electromotive force (EMF).

8.1.4 The double-layer capacitance

Double-layer capacitance appears at the interface between a conductive electrode and an electrolyte. This phenomenon is characterized by the presence of two layers with opposite charge and separated by a single layer of solvent molecules adhered on the surface of the electrode (Bockris and Reddy, 2015). The formation of a biofilm on the electrode surface can have a pronounced effect on the thickness of the double layer, which in turn influences the double-layer capacitance of the system (Malvankar et al., 2012; Sekar and

Ramasamy, 2013; Sekar and Ramasamy, 2013, 2013). According to Kim et al. (2011), several bacteria and biofilm-associated components are able to instantaneously affect the double-layer capacitance, such as physical appendages of bacteria, extracellular components, and various metabolites. Moreover, the double-layer capacitance depends on the electrode polarization, ionic concentration, temperature, type of ions, oxide layers, roughness of the electrode, etc. The charging accumulation in the interface electrode/electrolyte origins a capacitive current— $I_c(t)$. This can be represented by Eq. (8.10):

$$I_c(t) = C_{dl}\frac{\mathrm{d}(E_c - E_a)}{\mathrm{d}t} \quad 8.10$$

where C_{dl} is the double-layer capacity. The capacitive current is modeled in parallel with the charge transfer process. Then, the total measured current, $I(t)$, can be modeled by Eq. (8.11):

$$I_f(t) = I(t) + I_c(t) \quad 8.11$$

8.1.5 Mass transfer

In the biofilm, the substrate moves by diffusion from high to low substrate concentration zones. In MFCs, the zone with low substrate concentration is close to the anode surface, where electroactive bacteria degrade organic compounds. The substrate diffusional transport may be expressed by Fick's first and second laws, which in the case of planar geometry can be expressed according to Eqs. (8.12 and (8.13), as can be seen in Crank (1979): corresponds to Crank (1979), according to Eqs. (8.12) and (8.13)

$$J(z,t) = -D\overline{c}_r\frac{\partial S(z,t)}{\partial z}, \quad 8.12$$

$$\frac{\partial S(z,t)}{\partial t} = D\frac{\partial^2 S(z,t)}{\partial z^2}, \quad 8.13$$

where D is the effective diffusion coefficient, $S(z,t)$ and $J(z,t)$ are the substrate concentration and flux of substrate in the biofilm at time t and spatial position z, respectively. Using Eq. (8.12) in Eq. (8.13), the following equation is obtained:

$$\frac{\partial S(z,t)}{\partial t} = -\frac{1}{\overline{c}_r}\frac{\partial J(z,t)}{\partial z} \quad 8.14$$

The analytical solution of Eq. (8.14) is complex—see Crank (1979) for details. To simplify, it can be approximated into a set of ordinary differential equations by using a spatial discretization. Spatial discretization is a very well-known method to approximate partial differential equations in ordinary differential equations. For details, see Lasia and Grégoire (1995) and Britz

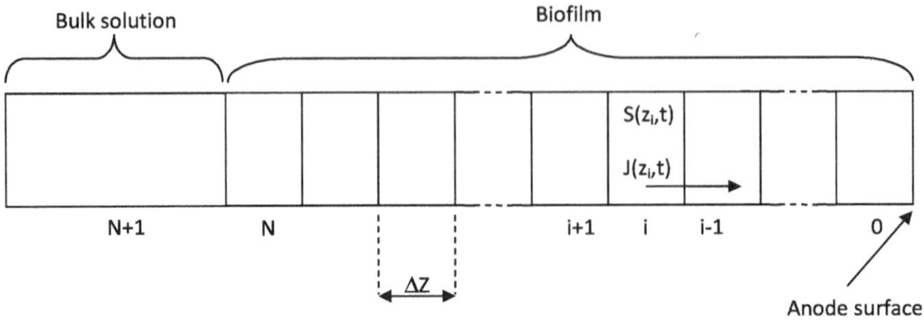

FIGURE 8.2 Spatial discretization.

(1988). Eq. (8.14) can be discretized along the space variable z by considering $N+1$ slices of the biofilm with thickness Δz, as illustrated in Fig. 8.2. In this figure, the slice $N+1$ corresponds to the bulk solution, and the slice 0 is the closest to the anode surface.

If each slice is small enough, the concentration $S(z_i, t)$ in the ith slice, where $(0 \le i \le N)$, can be considered constant with input and output substrate flux given by $J(z_{i+1}, t)$ and $J(z_i, t)$, respectively. Using this approximation, Eqs. (8.12) and (8.14) can be written as:

$$J(z_i, t) = \frac{D\overline{c}_r}{\Delta z}(S(z_i, t) - S(z_{i-1}, t)), \tag{8.15}$$

$$\frac{dS(z_i, t)}{dt} = -\frac{1}{\overline{c}_r \Delta z}(J(z_i, t) - J(z_{i+1}, t)), \tag{8.16}$$

By replacing Eq. (8.15) in Eq. (8.16), and considering a boundary condition for the flux $J(z_{N+1}, t) = 0$, the following set of ordinary differential equations that describe the acetate concentration profile is fulfilled:

$$\begin{aligned}
\frac{dS(z_0, t)}{dt} &= \frac{-aS(z_0, t) + aS(z_1, t) - J(z_0, t)}{(\overline{c}\Delta z)}, \\
\frac{dS(z_1, t)}{dt} &= aS(z_0, t) - 2aS(z_1, t) + aS(z_2, t), \\
\vdots &= \vdots, \\
\frac{dS(z_i, t)}{dt} &= aS(z_{i-1}, t) - 2aS(z_i, t) + aS(z_{i+1}, t), \\
\vdots &= \vdots, \\
\frac{dS(z_{N-1}, t)}{dt} &= aS(z_{N-2}, t) - 2aS(z_{N-1}, t) + aS(z_N, t),
\end{aligned} \tag{8.17}$$

where $a = \frac{D}{\Delta z^2}$. Notice that different boundary conditions can be used to model several operation modes. In this case, the concentration at the slice in contact whit the bulk, $S(z_N, t)$, is considered constant, according to Eq. (8.18):

$$\frac{dS(z_N, t)}{dt} = 0 \tag{8.18}$$

The flux and the faradaic current are given by the following relationship (Eq. 8.19):

$$J(z_0, t) = \frac{I_f(t)}{\gamma F N_e}, \tag{8.19}$$

where γ is the biofilm area and N_e is the electron-mol/acetate-mol ratio involved in the electrochemical reaction. Using Eqs. (8.18) and (8.19) in Eq. (8.17), a set of ordinary differential equations is obtained, which can be written in a vector−matrix formulation as follows:

$$\begin{aligned} \dot{\mathbf{S}}(t) &= \Phi\mathbf{S}(t) + \Gamma I_f(t) \\ S(z_0, t) &= CS \end{aligned} \tag{8.20}$$

where $S = [S(z_0, t), S(z_1, t), \ldots, S(z_i, t), \ldots, S(z_{N-1}, t), S(z_N, t)]^T$, and with

$$\Phi = a\begin{bmatrix} -1 & 1 & 0 & 0 & \cdots & 0 \\ 1 & -2 & 1 & 0 & \cdots & 0 \\ 0 & 1 & -2 & 1 & \cdots & 0 \\ \vdots & & & & & \\ 0 & 0 & \cdots & 1 & -2 & 1 \\ 0 & 0 & 0 & \cdots & 0 & 0 \end{bmatrix}; \Gamma = \begin{bmatrix} -b \\ 0 \\ 0 \\ \vdots \\ 0 \end{bmatrix}; C = [10, \ldots, 0];$$

$$b = \frac{1}{F\gamma\bar{c}\Delta z N_e}$$

8.2 Microbial fuel cell structures

On the anode surface, biofilm grows forming a thick layer of microorganisms. To favor this, several structures of MFCs are implemented. Some of them are introduced in this section.

Single-chamber MFC: This consists of an anode placed in an anode chamber coupled with a porous air−cathode that is exposed directly to the air. Protons travel throughout the anolyte solution to the porous air−cathode (Liu et al., 2005; Kim et al., 2007; Ren et al., 2011; Zhang et al., 2013; Sirinutsomboon, 2014). Fig. 8.3 shows a scheme of a single-chamber MFC.

Two-chamber MFC: This MFC has an anodic and a cathodic chamber connected by the PEM. The cathode chamber is permanently aerated, and the anodic chamber operates in anaerobic conditions (Min et al., 2005; Kim et al., 2007; Lee et al., 2016). Fig. 8.4 depicts the two-chamber MFC.

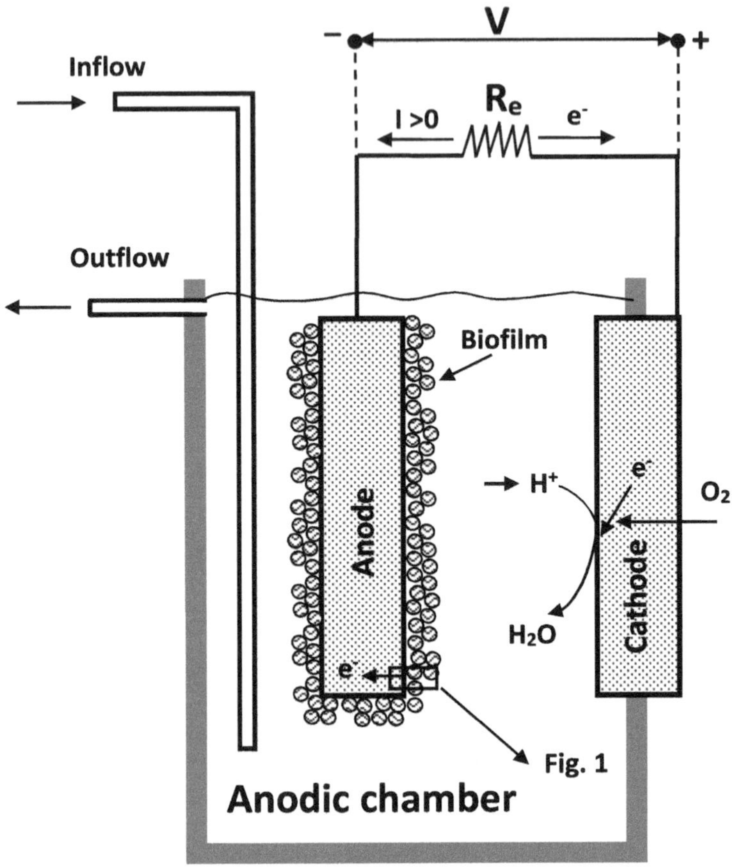

FIGURE 8.3 Single-chamber MFC. *MFC*, microbial fuel cell.

Up-flow mode MFC: This consists on a tube where the anode and cathode are placed at the bottom and the top of the reactor, respectively. Air is fed between both electrodes allowing a proper oxygen gradient (Jang et al., 2004).

Stacked MFC: This is an assembled of several MFCs connected in series and in parallel (Aelterman et al., 2006).

Flat configuration MFC: This has a flat configuration (Fig. 8.5) where the anode is located in the deeper zone of the cell and the cathode is partially submerged. In this way, the proton produced by substrate oxidation diffuses to the cathode zone to be consumed on the oxygen reduction. In the cathode neighborhood, there is oxygen because of its proximity to the air. Flat configuration has the advantages of being easily implemented and scalable in a nonaerated stage of a wastewater treatment plant.

There exists a great diversity of materials used for manufacturing the anode and cathode electrodes and the PEM (Oliveira et al., 2013a,b): For the anode,

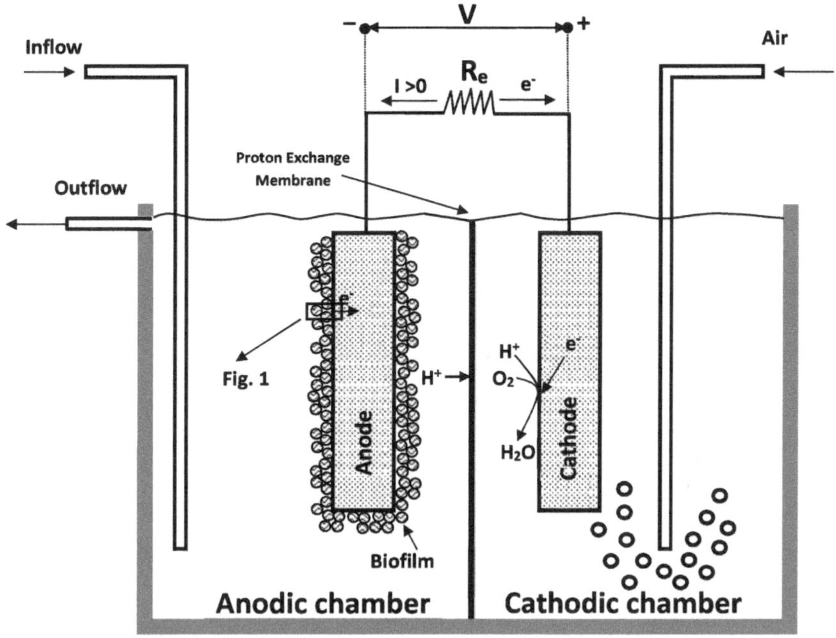

FIGURE 8.4 Two-chamber MFC. *MFC*, microbial fuel cell.

carbon cloth, carbon fibber veil, graphite felt, graphite plates, graphite road, graphite brush, titanium plates, etc., and for the cathode, graphite road, carbon fibber veil, carbon cloth/platinum, titanium plates, MnO_2, etc. Regarding the proton exchange system, the following materials have been used: Nafion, a sulfonated tetrafluoroethylene copolymer that consists of a hydrophobic fluorocarbon backbone to which hydrophilic sulfonate groups (SO_3) are attached (Min et al., 2005); salt bridge, a saturated solution of KCl and agar (Min et al., 2005); Ultrex CMI 7000, a product of Membranes Inc., United States, is a cation exchange membrane consisting in gel polystyrene and divinylbenzene cross-link structure containing large amounts of sulfuric acid groups, among others (Rahimnejad et al., 2014). For an existing MFC, it is possible to enhance MFC power yield by optimization of the operating conditions, such as organic loading, feed rate and shear stress, pH, and temperature. An exhaustive revision of the operating conditions of an MFC can be found in Oliveira et al. (2013a,b). This great variety of structures, materials, and experimental conditions results in a wide range of power delivered by the MFC. Implemented size range from microliters to liters results in delivered powers falling in the interval of 8−2500 $\frac{mW}{m^2}$, as reported in many studies (Liu et al., 2005; Kim et al., 2007; Park et al., 2007; Ren et al., 2011; Zhang et al., 2013; Tota-Maharaj and Parneet, 2015; Lee et al., 2016; Khater et al., 2017; Nasirahmadi and Akbari-Adergani, 2018).

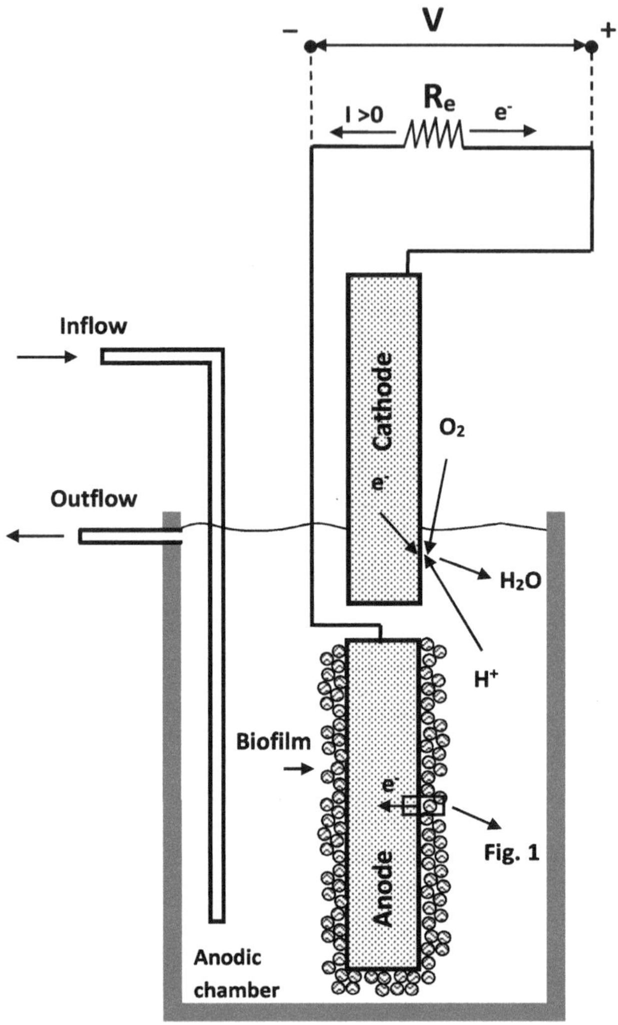

FIGURE 8.5 Flat configuration MFC. *MFC*, microbial fuel cell.

8.3 Integration of main processes in a microbial fuel cell model

To represent the temporal evolution of the biofilm growing on the anode surface, the principal processes must be considered. From a previous work (Gatti and Milocco, 2017), these processes are (1) the charge transfer due to the electrochemical reactions at the electrode interfaces, modeled by a Butler–Volmer-type equation, (2) a mass transport process governed by the Fick's laws, and (3) the biofilm charge capacity. The complete model is given

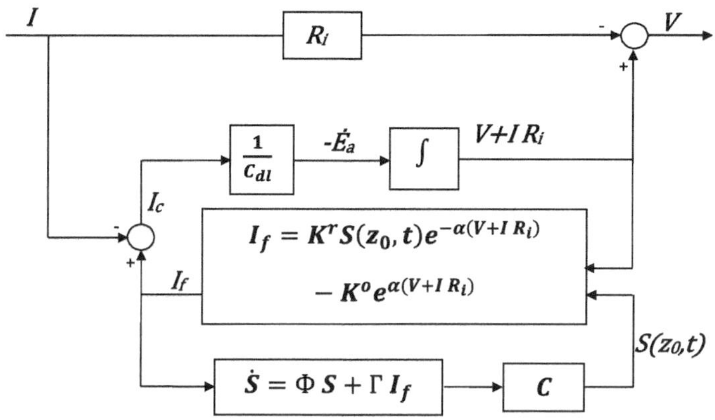

FIGURE 8.6 Schematic representation of the MFC model which can be used for simulating MFC temporal evolution. *MFC*, microbial fuel cell.

by Eqs. (8.7), (8.20), and (8.10). Considering that $S(t)$ in Eq. (8.7) is equal to $S(z_0, t)$, and given initial conditions of $V(0)$ and S(0), the complete model is given by the following equations and is presented in Fig. 8.6:

$$I_f(t) = K^r S(z_0, t)e^{-\alpha(V(t)+I(t)R_i)} - K^o e^{\alpha(V(t)+I(t)R_i)} \tag{8.21}$$

$$\frac{d\mathbf{S}(t)}{dt} = \Phi\mathbf{S}(t) + \Gamma I_f(t) \tag{8.22}$$

$$S(z_0, t) = CS \tag{8.23}$$

$$\frac{d(V(t) + I(t)R_i)}{dt} = \frac{I_f(t) - I(t)}{C_{dl}} \tag{8.24}$$

8.3.1 Parameter identification

The unknown vector parameters of the model are $\theta = [K^r, R_i, C_{dl}, a, b]$. K^o is obtained from the definition of the parameter A in Eq. (8.9) as follows:

$$K^o = K^r e^{-A/B} \tag{8.25}$$

For a given number of slices N, the five parameters in θ are obtained by optimizing numerically the following quadratic problem:

$$\underset{\theta}{\operatorname{argmin}}\{Q(\theta)\}, \tag{8.26}$$

where the quadratic cost function is given by:

$$Q(\theta) = \frac{1}{M}\sum_{i=1}^{M}\frac{\left(V(i) - \widehat{V}(i,\theta)\right)^2}{V^2(i)}, \tag{8.27}$$

where V is the M-data record obtained by regularly sampling the MFC voltage and $\widehat{V}(i, \theta)$ is the voltage estimated by the model with parameter θ at the same sampling-time. The minimization can be carried out using the SIMPLEX numerical optimization method (Venkataraman, 2009). The amount of slices (N) is chosen using the following strategy: starting with a few slices for the diffusional state space discretization, the minimization procedure is repeated by increasing the number of slices until the cost does not vary significantly. To avoid local minimum, several runs can be carried out by starting from different initial values and verifying that the minimum is always the same in the constrained set of admissible values.

To apply the procedure, a flat-configuration MFC was used. The chamber has a volume of 0.12 L, and the anode and cathode are built of plain carbon cloth (C), 0.356 mm thickness (The Fuel Cell Store, Texas, United States). Sediment from the shore of the Negro river (Patagonia-Argentina) was used as inoculum. A 0.2 m depth was used to assure anaerobic conditions (Sacco et al., 2012). A feed solution containing 0.5 gL^{-1} of sodium acetate dissolved in 50 mM phosphorus buffer solution (Na_2HPO_4, 4.58 gL^{-1}; $Na_2PO_4.H_2O$ 2.45 gL^{-1}; NH_4Cl 0.31 gL^{-1}; KCl 0.13 gL^{-1}; trace minerals; and vitamins) (Zhang, 2010) was used in the anodic chamber fed at a continuous rate of 60 $\pm 2\ \frac{\text{mL}}{\text{h}}$. The effluent outflows through an overflow system. The initial substrate concentration on the anode surface was equal to 0.5 gL^{-1}, which corresponds to a fractional concentration of $S(Z_0, 0) = \overline{S} = 1$. Different current amplitudes were obtained by connecting an external varying resistor load, R_e, between the electrode terminals. The ohmic resistance ranged in the interval of $R_e \in [50, 45000]\ \Omega$. By using the optimization method, the following values were obtained: $K^r = 1.72\ \ 10^{-2}$ (mA), $R_i = 1175$ (Ω), $C_{dl} = 1.15\ \ 10^{-1}\ F$, $a = 4.75\ \ 10^{-2}\text{sec}^{-1}$, $b = -6.69\ \ 10^{-1}(\text{mAsec})^{-1}$, with $N = 10$ slices. Fig. 8.7 shows two measured experiments, where external load changed to force a variation on total current between terminals of the MFC. Mathematical description of the processes occurring in the biofilm is adequate to represent experimental data in a suitable way. Both the transient behavior and the steady state are well estimated. Fig. 8.7B and D depict the estimated substrate concentrations on each slice from experiments of Fig. 8.7A and C, respectively. The anode surface concentration is represented by the curve with the lowest values. An important substrate depletion occurs on anode surface driving organic matter diffusion across the biofilm.

8.3.2 Electrical analogy for small-signal operation

The output voltage V can be expressed as a sum of two components, one as a function of the substrate at zero current, $S_{I=0}$, according to the EMF in Eq. (8.9) and the other taking into account the contribution of the consumed current I. To obtain such decomposition, Eq. (8.8) can be rewritten as follows:

$$S = \frac{K^o}{K^r} e^{2\alpha(V + IR_x)}, \qquad 8.28$$

where R_x is a dummy resistance obtained by combining Eqs. (8.8) and (8.28),

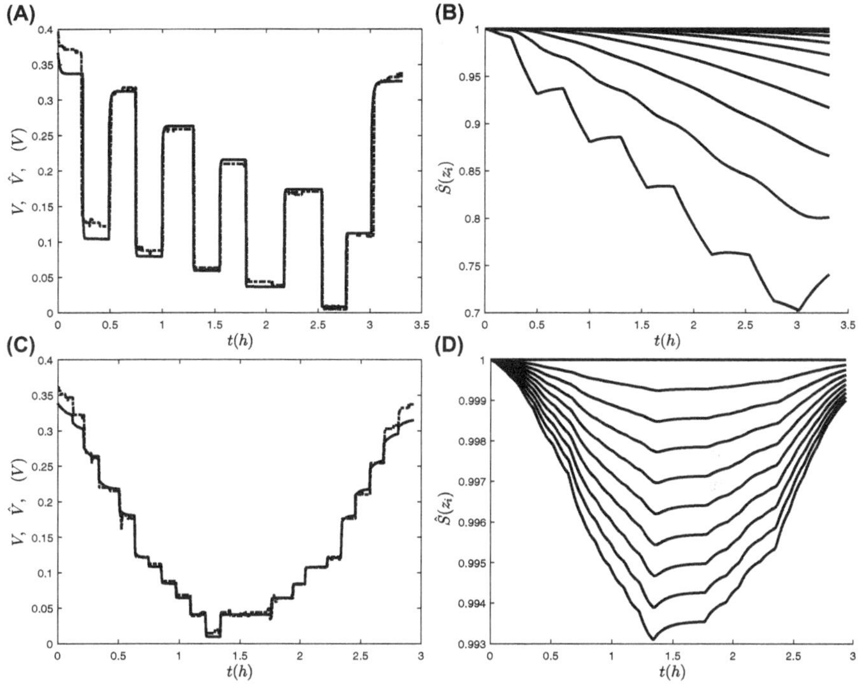

FIGURE 8.7 (A and C) Measured voltage: $V(-.)$, estimated voltage: $\widehat{V}(-)$, (B and D) Estimated substrate concentration on the slices: $\widehat{S}(z_i), i = 0:10$.

and taking logarithms in both. After some algebra, the following relationship is obtained:

$$IR_x = IR_i + I_f R_z, \tag{8.29}$$

where

$$R_z = \frac{1}{2I_f \alpha} \ln\left(1 + \frac{I_f}{K^o} e^{-\alpha(V+IR_i)}\right). \tag{8.30}$$

Note that resistances R_x and R_z are functions of the currents and the potential. Moreover, R_x and R_z are both positive definite values since the currents are always positive. Taking logarithms in both sides of Eq. (8.28), the following equality is obtained:

$$V = A + B\ln(S) - IR_x, \tag{8.31}$$

Then, it follows that the potential V, for arbitrary values of current I and substrate concentration S, is obtained by subtracting the factor IR_x to the output of the EMF function, where $S = S(z_0, t)$.

Considering now the special case in which the substrate has small deviations from the constant bulk concentration $S = S^o$ and the current has small variations around $I = \bar{I} = 0$. Then, the potential can be well approximated by a pseudolinear function as follows:

$$V \approx V(S^o, I^o) + \left.\frac{\partial V}{\partial S}\right|_{S^o, I^o}(S - S^o) + \left.\frac{\partial V}{\partial I}\right|_{S^o, I^o}(I - I^o)$$

$$= A + B\ln(S^o) + \frac{B}{S^o}(S - S^o) - IR_x - I^2\frac{\partial R_x}{\partial I}$$

$$\approx A + B\ln(S^o) + \frac{B}{S^o}(S - S^o) - IR_x. \quad 8.32$$

where the quadratic and the higher terms were neglected.

Using a first-order approximation for the diffusion Eqs. (8.23 and 8.24), the expression of the time derivative for substrate is obtained as:

$$\frac{\mathrm{d}S(z_0, t)}{\mathrm{d}t} = -aS(z_0, t) + aS^o + bI_f(t) \quad 8.33$$

Thus, the complete linear model equations are given by Eqs. (8.24, 8.29, 8.32, and 8.33), which can be rewritten in a compact form as:

$$\frac{\mathrm{d}X_1}{\mathrm{d}t} = -aX_1 + \frac{bB}{S^o}I_f, \quad 8.34$$

$$\frac{\mathrm{d}X_2}{\mathrm{d}t} = \frac{1}{C_{dl}}I_c, \quad 8.35$$

$$V = X_2 - IR_i, \quad 8.36$$

$$\mathrm{OCV} = X_1 + X_2 + I_f R_z. \quad 8.37$$

where

$$X_1 = \frac{B}{S^o}(S^o - S);\ X_2 = V + IR_i;\ \text{and OCV} = A + B\ln(S^o).$$

which has the electrical analogy shown in Fig. 8.8.

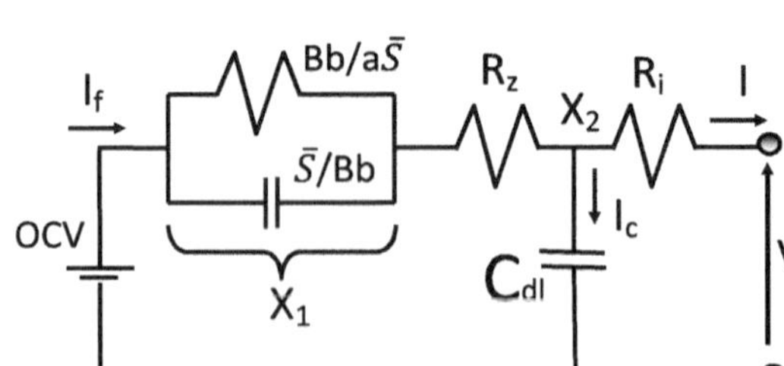

FIGURE 8.8 Equivalent circuit.

8.3.3 Maximum steady-state power delivered by microbial fuel cell

A system is in the steady state when the time derivatives are equal to zero. For stable systems, it is equivalent to consider a constant input and $t \rightarrow \infty$, when the transient state is vanished. In MFC, the steady state of the substrate when the current is a constant different from zero, at the electrode surface, $z = z_0$ and $t \rightarrow \infty$, is $S = S(z_0, \infty)$, and it is obtained by considering the derivatives equal to zero in the diffusion Eqs. (8.23) and (8.24):

$$0 = a\Phi\mathbf{S} + b\Gamma I, \qquad 8.38$$

$$S = C\mathbf{S}, \qquad 8.39$$

with the boundary condition $S(z_N, t) = 1$.

It is important to note that the different types of diffusion governed by Fick's laws are dynamic linear processes where only the entries in the matrix Φ and vector Γ change. Then, in the steady state, there is always a linear relationship between the current and the substrate, which is obtained by solving Eqs. (8.38) and (8.39). In general, for any given matrix and vector, the substrate fulfills

$$I = \mathrm{d}(1 - S), \qquad 8.40$$

where $\mathrm{d} = \frac{a}{bN} = \frac{DF\gamma\bar{c}N_e}{L}$ with $L = N\Delta_z$ the thickness of the biofilm. Note at this point that by replacing Eq. (8.40) in Eq. (8.30), a convenient description is obtained showing the dependency with diffusion. Thus, it is reasonable to interpret such term as a resistance due to the diffusion process.

Now, the goal is to obtain the expression of the maximum delivered power at steady state. To do that, note that Eq. (8.31) can be represented using an equivalent Thévenin electric circuit as it is shown in Fig. 8.9.

The maximum power transfer theorem states that the maximum amount of power, $\overline{P}$, will be dissipated in the load resistance if it is equal in value to the Thévenin source resistance of the network supplying the power. The maximum power $\overline{P}$ transferred to the external load occurs when $R_L = R_x$, which is equivalent to say that the voltage is half of the source, therefore:

$$\overline{P} = \overline{VI} \qquad 8.41$$

where the bar means steady state at the maximum power delivered and

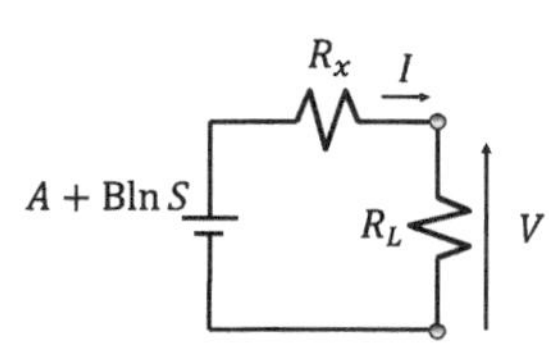

FIGURE 8.9 Thévenin circuit for obtaining the maximum power of the MFC in steady-state condition. *MFC*, microbial fuel cell.

$$\overline{V} = \frac{A + B\ln(\overline{S})}{2} \tag{8.42}$$

$$\overline{I} = d(1 - \overline{S}) \tag{8.43}$$

By replacing Eqs. (8.42) and (8.43) in Eq. (8.41), the maximum delivered power at any steady state is:

$$\overline{P} = \overline{I}\overline{V} = d(1 - \overline{S})\frac{A + B\ln(\overline{S})}{2}. \tag{8.44}$$

For a given pair (d, A), the maximum power is a function of $\overline{S}$. Then, to know $\overline{S}$, a second equation is needed. To this end, using Eqs. (8.42) and (8.43), $\overline{\mu}$ can be obtained:

$$\overline{\mu} = \overline{V} + \overline{I}R_i = \frac{A + B\ln(\overline{S})}{2} + dR_i(1 - \overline{S}) \tag{8.45}$$

By replacing Eqs. (8.25) and (8.45) in the Butler−Volmer Eq. (8.21), the following important relationship for $\overline{S}$ is obtained:

$$d(1 - \overline{S}) = K^r e^{-\overline{\mu}/(2B)}\left(\overline{S} - e^{(\overline{\mu}-A)/B}\right) \tag{8.46}$$

Then, given the parameter vector $\theta = (A, K^r, d, R_i)$, it is possible to obtain the corresponding $\overline{S}$ by replacing Eq. (8.45) in Eq. 8.46 and finally $\overline{P}$ in Eq. (8.44). Unfortunately, there is no explicit expression to obtain $\overline{S}$. Instead, a numerical search procedure based on some 1D optimization method can be carried on to obtain $\overline{S} \in (0, \ 1)$ using Eq. (8.45) in Eq. (8.46). Note that the expression of R_x is not needed to obtain the maximum delivered power because the maximum power transfer theorem states that the terminal voltage is equal to the half of the source. Moreover, the resistance R_x is given by the quotient $\overline{V}/\overline{I}$ as follows:

$$R_x = \frac{A + B\ln(\overline{S})}{2d(1 - \overline{S})} \tag{8.47}$$

Using the parameter vector, it is possible to carry out a sensitivity analysis with respect to the maximum power obtainable within a range of bounded variation of the parameters. This allows to know how much it is possible to improve a certain MFC to increase the power delivered as well as to compare the potential improvement between different MFCs.

8.3.4 Sensitivity analysis of the maximum power

To apply the sensitivity analysis, a single-chamber (MFC1) and a two-chamber (MFC2) MFC were studied. Both MFCs are constructed of glass sheets with 5 mm thickness. In MFC1, the anode and cathode electrodes were located on the opposite sides of the chamber; thereby the MFC lacks the PEM. The MFC2 was composed of an anodic and cathodic chamber separated by a PEM made of NAFION 117, 0.183 mm thickness (The Fuel Cell Store, Texas, United States).

Anodic and cathodic chambers had equal volume of 0.12 L. MFCs used the same type of anode and cathode. The anodes were built of plain carbon cloth, 0.356 mm thickness (The Fuel Cell Store, Texas, United States), and the cathodes were built of 0.5 mg cm^{-2} Pt catalyst on vulcan carbon cloth (C_{PT}), 0.410 mm thickness, with 30 % PTFE treatment and a microporous layer on catalyzed side. The cathodes included a Nafion postcoat to improve water management and adhesion on the membrane. Anodes and cathodes had 4 × 4 cm, mounted on a Teflon frame—resulting in a projected area of 0.0016 m^2. Three copper wires surrounded the Teflon frame in contact with the electrode. PEM were of 4 × 5 cm mounted on a Teflon frame, resulting in a projected area of 0.0020 m^2. The cathodic chamber on the MFC2 was continuously aerated, and the MFCs were maintained at a temperature of 25 25± 0.5°C in a temperature-controlled room.

MFC1 and MFC2 were inoculated with sediment from the shore of the Negro river (Patagonia-Argentina). A feed solution containing 0.5 gL^{-1} of sodium acetate dissolved in 50 mM phosphorus buffer solution (Na_2HPO_4, 4.58 gL^{-1}; $Na_2PO_4.H_2O$ 2.45 gL^{-1}; NH_4Cl 0.31 gL^{-1}; KCl 0.13 gL^{-1}; trace minerals, and vitamins) was used in the anodic chamber fed at a continuous rate of 60 ±2 $\frac{mL}{h}$. The effluent outflows through an overflow system. The cathodic chamber was filled with a phosphorus buffer solution.

Table 8.1 shows parameter values obtained by means of an optimization procedure, and Table 8.2 depicts steady-state values of the MFCs at the maximum power condition.

TABLE 8.1 Model parameter values and their standard deviations.

Parameter	MFC1	MFC2
$K^r(mA)$	$1.45\ 10^4 \pm 0.17\ 10^4$	$3.54\ 10^5 \pm 0.26\ 10^5$
$R_i(\Omega)$	143 ± 23	53 ± 4
$a(s^{-1})$	$9.21\ 10^{-2} \pm 2.11\ 10^{-2}$	$1.93\ 10^{-1} \pm 0.32\ 10^{-1}$
$b\ (mAs^{-1})$	$-6.60\ 10^{-3} \pm 1.33\ 10^{-3}$	$-8.71\ 10^{-3} \pm 1.24\ 10^{-3}$
$d\ (mA)$	$9.91\ 10^{-1} \pm 3\ 10^{-1}$	$1.58\ 10^{-1} \pm 0.41\ 10^{-1}$

MFC, microbial fuel cell.

TABLE 8.2 Steady-state values at the maximum power condition obtained by simulation of MFC1 and MFC2.

	MFC1	MFC2
$\overline{S}$	0.33	0.19
$\overline{V}(mV)$	340	468
$\overline{I}(mA)$	0.72	1.38
$R_i(\Omega)$	143	56
$\overline{P}\left(mW/m^2\right)$	154	402

MFC, microbial fuel cell.

8.3.5 Achievable maximum power

The parameters K^r and K^o (and consequently A) depends on the type of material, R_i of the distance between electrodes, and d of the type of solution homogenization. The goal now is to analyze the parameter sensitivity with respect to the maximum achievable delivered power. To this end, the maximum power in a given admissible interval of possible parameter variations can be reached. Then, given the parameter vector $\theta = (A, K^r, d, R_i)$, it is possible to obtain the corresponding $\overline{S}$ replacing Eq. (8.45) in Eq. (8.46) and finally $\overline{P}$ in Eq. (8.44) by means of a numerical 1D optimization procedure. For the identified parameter vector θ_o, the size of the intervals is $[\theta] = [\theta_o(1-g), \theta_o(1+g)]$ where $g \in (0,1)$. The maximum power obtained in the parameter interval together with the parameter values and also the steady state of substrate corresponding to the maximum power are depicted with respect to the amplitude of intervals g in Figs. 8.10–8.12. From the experimental results, it can be seen that MFC2 delivers more power than MFC1. Even the slope of the delivered power grows faster for MFC2. In both structures, power increases if the internal resistance R_i decreases and the diffusion constant d increases. The last two parameters A and K^r also need to increase for both structures to reach more power. However, MFC1 is insensitive for these last two parameters. It is observed from the figures that the two-chamber MFC allows to achieve a greater power than the single-chamber MFC. That fundamental parameter for the increase of the power is the reduction of the internal resistance and the increase of the substrate diffusion in the liquid medium. One way to achieve a diffusion enhance is by conveniently stirring the bulk solution and choosing cell structures that keep the electrodes clean of impurities.

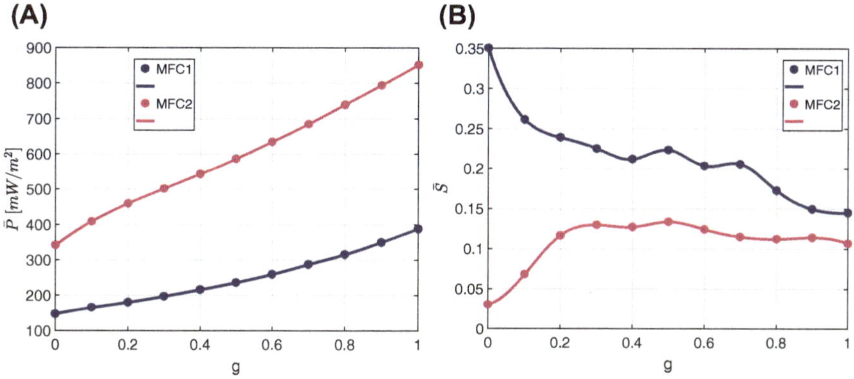

FIGURE 8.10 Maximum delivered power $\overline{P}$ (A) and the corresponding substrate at the electrode surface $\overline{S}$ (B) as function of the parameter intervals in both MFCs. MFC, microbial fuel cell.

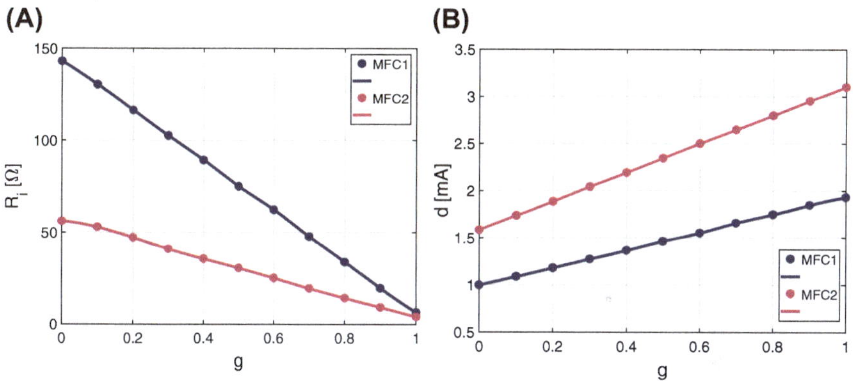

FIGURE 8.11 Internal resistance R_i (A) and diffusion parameter d (B) as function of the parameter intervals in both MFCs. MFC, microbial fuel cell.

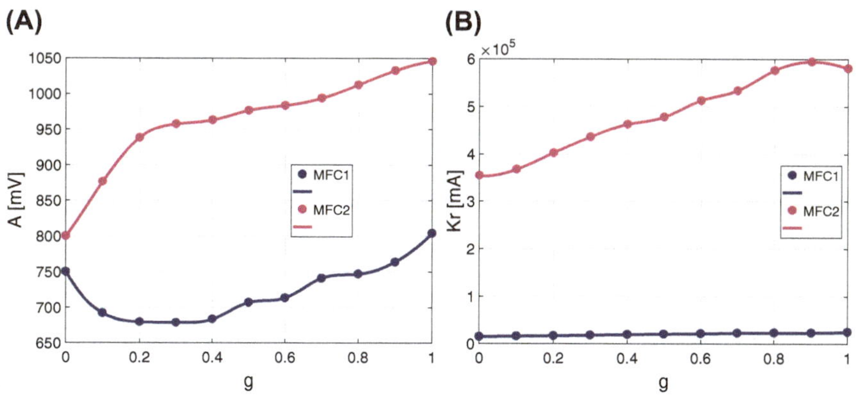

FIGURE 8.12 Open-circuit voltage A (A) and constant K^r (B) as function of the parameter intervals in both MFCs. MFC, microbial fuel cell

8.4 Dimensional electrodes

8.4.1 Problem formulation

Three-dimensional electrodes are a new research area in which biofilm can develop. As an example, it is of interest a mathematical representation of an electrode formed by a porous cylinder. Bacteria are located inside the pores which are represented by spheres of radius r. The substrate concentration C is expressed in fractional form $S = \frac{C}{\mathrm{Cm}}$, where Cm is the maximum concentration within the electrode. The electrode is submerged in a cell where the fractional concentration of the bulk is constant, $S = 1$. The two processes considered to obtain the spatial and temporal variations are (1) the substrate diffusion and (2) the charge transference within the sphere walls induced by the bacteria.

8.4.2 Substrate diffusion

To represent the substrate diffusion, the cylinder is divided into several layers of thickness L. In turn, each layer is subdivided into concentric rings of thickness $L_j - L_{j-1}$. Each ring contains a different number of spherical porous in which two types of substrate diffusion coexist: a planar diffusion in vertical direction and a radial diffusion toward the center of the cylinder. The scheme is shown in Fig. 8.13. To obtain the equivalences between the areas of the rings and the volumes of n spheres within each one, the subscript i for vertical divisions (planar diffusion) and j for radial divisions (cylindrical diffusion) are defined, giving place to the following set of equalities:

$$\text{Equivalences of areas: } n_j 4\pi r^2 = \pi\left(L_j^2 - L_{j-1}^2\right) + 2\pi L_j L. \qquad 8.48$$

$$\text{Equivalences of volumes: } n_j \frac{4}{3}\pi r^3 = \pi\left(L_j^2 - L_{j-1}^2\right)L. \qquad 8.49$$

Calling $\sigma = \frac{L}{r}$ and $\sigma_j = \frac{L_j}{r}$, both conditions lead to

$$0 = \sigma_j^2(3\sigma - 4) - \sigma_j 8\sigma - \sigma_{j-1}^2(3\sigma - 4), \qquad 8.50$$

$$n_i = \left(\sigma_j^2 - \sigma_{j-1}^2\right)\sigma\frac{3}{4}. \qquad 8.51$$

Given a value of $\sigma$and starting with $\sigma_0 = 0$, σ_j is obtained by solving the second polynomial degree. Repeating the process, the widths of each ring and the number of spheres are obtained. The total volume of each ring matches that of the included spheres. The area of the outer edge plus the lower one of each ring is equal to the area of the included spheres. A value of L is search such that the partitions obtained maintain the proportion of the original cylinder, that is, the radius with respect to its length.

For planar diffusion, the following holds:

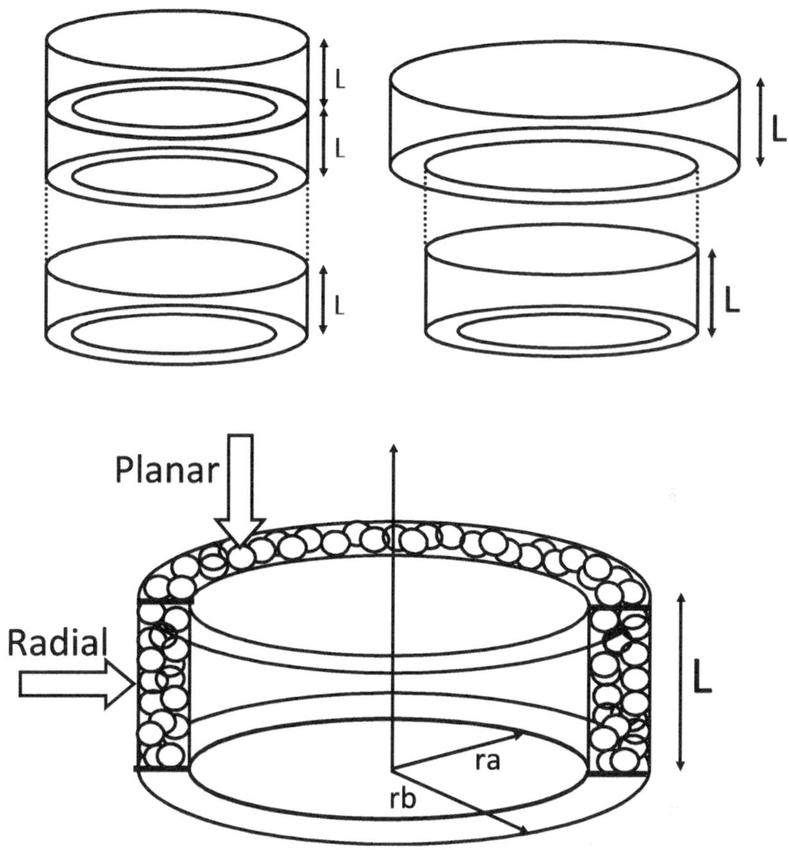

FIGURE 8.13 Porous electrode structure.

$$\frac{\partial S^p}{\partial t} = D\frac{\partial^2 S}{\partial y^2} + q^p(t) \qquad 8.52$$

where D is the diffusion coefficient that considers the dynamics of mass transport between spheres. S^p is the concentration of planar diffusion, and q^p is the flux. The discrete version in space and time, on vertical slides, every T seconds, is

$$\frac{S^p_{i,j,t+1} - S^p_{i,j,t}}{T} = \frac{D}{L^2}\left(S_{i-1,j,t} - 2S_{i,j,t} + S_{i+1,j,t}\right) + q^p(t) \qquad 8.53$$

The spherical substrate diffusion follows the following law:

$$\frac{\partial S^r}{\partial t} = \frac{D}{x}\left(\frac{\partial S}{\partial x} + x\frac{\partial^2 S}{\partial x^2}\right) + q^r(t) \qquad 8.54$$

The discrete sequence in space and time, on vertical subdivisions, every T seconds is

$$\frac{S^r_{i,j,t+1} - S^r_{i,j,t}}{T} = \left(S_{i,j-1,t}\delta_j - S_{i,j,t}\beta_j + S_{i,j+1,t}\lambda_j\right) + q^r(t) \qquad 8.55$$

$$\delta_j = \frac{D}{AC} - \frac{D}{L_jA} \qquad 8.56$$

$$\beta_j = \frac{D}{AC} + \frac{D}{AB} \qquad 8.57$$

$$\lambda_j = \frac{D}{L_jA} + \frac{D}{AB} \qquad 8.58$$

The total diffusion is the sum of the planar plus the radial:

$$\frac{\partial S}{\partial t} = \frac{\partial S^p}{\partial t} + \frac{\partial S^r}{\partial t} \qquad 8.59$$

from which the discrete time—space value of $S$is

$$S_{i,j,t} = S^p_{i,j,t} + S^r_{i,j,t} \qquad 8.60$$

The boundary conditions to solve the differential equation are given by a constant concentration $S = 1$ on the total electrode surface. The inflows in each ring are given by the volumetric flow of moles consumed by the bacteria in its surface. The relationship between the current and the consumed moles is given by the Faraday law, and the volumes of each subdivision, V_j, are as follows:

$$q^p(i,j,t) + q^r(i,j,t) = q(i,j,t) = \frac{I_f(i,j,t)}{C_mF \;\; V(i,j,t)}; \;\; \left[\frac{1}{s.}\right] \qquad 8.61$$

$$\text{units } I_f[A];\, C_m\left[\text{mol}/m^3\right];\, F\left[A\ s/\text{mol}\right];\, V\left[m^3\right]$$

8.4.3 Charge transfer

The faradaic current generated at the interface for each of the diffusion components and per ring is determined by the Butler—Volmer equation:

$$If^p(i,j,t) = A^p(i,j)\left(K^rS^p(i,j,t)e^{-\eta(t)} - K^oe^{\eta(t)}\right), \qquad 8.62$$

$$If^r(i,j,t) = A^r(i,j)\left(K^rS^r(i,j,t)e^{-\eta(t)} - K^oe^{\eta(t)}\right), \qquad 8.63$$

$$\eta(t) = \frac{F}{2RT}(E(t) + I(t)R_i) \quad 8.64$$

$$A^p(i,j) = \left(L_j^2 - L_{j-1}^2\right)\pi, \quad 8.65$$

$$A^r(i,j) = 2\pi L_j L. \quad 8.66$$

$$I_f(t) = \sum_{i,j} If^p(i,j,t) + If^r(i,j,t) \quad 8.67$$

where K^r and K^o are constants, E is the measured potential and I is the measured current given by $I = I_f - I_c$. The faradaic current, $I_f = If^p + If^r$, is the sum of both components, whereas I_c is the current throughout the double layer of the interface. The model is completed by closing the circuit with an external resistor R_e giving rise to the current $I = E/R_e$.

$$\dot{E} = \frac{I_f - E/R_e}{C_d\left(1 + R_i/R_e\right)}, \quad 8.68$$

The model parameters are the diffusion constant $D\left[m^2/s\right]$, which determines the speed of the process and the spatial distribution of S(it can be obtained approximately by using the voltagram); the constants $K^r$and $K^o\left[A/m^2\right]$, which determines, along with the internal resistance $R_i[\Omega]$ and the current $I[A]$; the double-layer capacity $C_d[F]$, which determines the energy available immediately; and the maximum concentration C_m. Rest of parameters depend on the geometry in [m]. Finally, the complete model is given by Eqs. (8.53, 8.55, 8.60, 8.62, 8.64, 8.67, 8.61, 8.68), in this order.

8.4.4 Simulation results

To simulate the porous anode, the following parameter values have been adopted: electrode height 2*cm*, $D = 6\ 10^{-8}\ \mathrm{m^2/s}$; $C_m = 0.020\ \mathrm{mol/m^3}$; $K^r = 1\ 10^5\ \mathrm{A/m^2}$; $K^o = 1\ 10^{-10}\ \mathrm{A/m^2}$; $C_d = 2500\ \mathrm{F/m^2}$; $R_i = 25\ \Omega$; $R_e = 5\ \Omega$. The steady-state concentration profile is shown in Fig. 8.14 for two

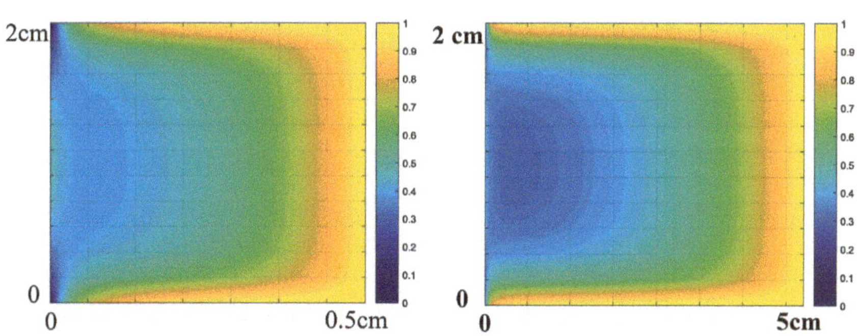

FIGURE 8.14 Cross section of the porous electrode seen from the cylinder axis outward; 0.166 cm per division for height and 0.0125 cm for width.

cases: (case 1) left side of Fig. 8.14 for a sphere radius of $r = 10\ \mu m$ and a cylinder radio of 0.5 cm reaching a potential of $E = 1.6$ V and a power of 1.6 W; (case 2) right side of Fig. 8.14 for a sphere radius of $r = 10\ \mu m$ and a cylinder radio of 5 cm achieving a potential of $E = 0.106$ V and a power of 2.27 W.

8.5 Conclusions

In this chapter, several processes occurring in the biofilm of an MFC were presented: substrate diffusional process, substrate oxidation catalyzed by bacteria that produces the faradaic current, and charge accumulation on double layer, which results in the capacitive current. Mathematical expressions widely used to represent these processes were detailed with the purpose to adequately describe the temporal evolution. Several MFC configurations were introduced, even a novelty flat configuration easily scalable. Moreover, a method was presented to analyze the sensitivity of the maximum delivered power of the MFC with respect to five important parameters. Finally, a 3D porous anode, where biofilm can develop, was exhaustively described to introduce such complex system.

References

Aelterman, P., Rabaey, K., Pham, H.T., Boon, N., Verstraete, W., 2006. Continuous electricity generation at high voltages and currents using stacked microbial fuel cells. Environmental Science and Technology 40 (10), 3388–3394.

Bockris, J.O.M., Reddy, A.,K.R., 2015. Modern Electrochemistry: An Introduction to an Interdisciplinary Area, vol. 2. Springer US.

Bockris, J.O.M., Reddy, A.,K.R., 2004. Modern Electrochemistry 2B: Electrodics in Chemistry, Engineering, Biology and Environmental Science, vol. 2. Springer Science & Business Media.

Britz, D., 1988. Digital Simulation in Electrochemistry, second ed. Springer, Berlin.

Crank, J., 1979. The Mathematics of Diffusion, second ed. Oxford university press.

Gatti, M.N., Milocco, R.H., 2017. A biofilm model of microbial fuel cells for engineering applications. International Journal of Energy and Environmental Engineering 8 (4), 303–315.

Jang, J.K., Pham, T.H., Chang, I.S., Kang, K.H., Moon, H., Cho, K.S., Kim, B.H., 2004. Construction and operation of a novel mediator-and membrane-less microbial fuel cell. Process Biochemistry 39 (8), 1007–1012.

Khater, D.Z., El-Khatib, K.M., Hassan, H.M., 2017. Microbial diversity structure in acetate single chamber microbial fuel cell for electricity generation. Journal of Genetic Engineering and Biotechnology 15 (1), 127–137.

Kim, J.R., Jung, S.H., Regan, J.M., Logan, B.E., 2007. Electricity generation and microbial community analysis of alcohol powered microbial fuel cells. Bioresource Technology 98 (13), 2568–2577.

Kim, T., Kang, J., Lee, J.H., Yoon, J., 2011. Influence of attached bacteria and biofilm on double-layer capacitance during biofilm monitoring by electrochemical impedance spectroscopy. Water Research 45 (15), 4615–4622.

Lasia, A., Grégoire, D., 1995. General model of electrochemical hydrogen absorption into metals. Journal of the Electrochemical Society 142 (10), 3393–3399.

Lee, Y.Y., Kim, T.G., Cho, K.S., 2016. Characterization of the COD removal, electricity generation, and bacterial communities in microbial fuel cells treating molasses wastewater. Journal of Environmental Science and Health, Part A 51 (13), 1131–1138.

Liu, H., Cheng, S., Logan, B.E., 2005. Production of electricity from acetate or butyrate using a single-chamber microbial fuel cell. Environmental Science and Technology 39 (2), 658–662.

Malvankar, N.S., Mester, T., Tuominen, M.T., Lovley, D.R., 2012. Supercapacitors based on c-type cytochromes using conductive nanostructured networks of living bacteria. ChemPhysChem 13 (2), 463–468.

Min, B., Kim, J., Oh, S., Regan, J.M., Logan, B.E., 2005. Electricity generation from swine wastewater using microbial fuel cells. Water Research 39 (20), 4961–4968.

Nasirahmadi, S., Akbari-adergani, B., 2018. Microbial fuel cell with a nano-membrane and two expired medicinal drug-feeding cathode: a novel strategy. International Journal of Energy and Environmental Engineering 9 (3), 249–256.

Oliveira, V.B., Simões, M., Melo, L.F., Pinto, A.M.F.R., 2013a. Overview on the developments of microbial fuel cells. Biochemical Engineering Journal 73, 53–64.

Oliveira, V.B., Simões, M., Melo, L.F., Pinto, A.M.F.R., 2013b. A 1D mathematical model for a microbial fuel cell. Energy 61, 463–471.

Park, H.I., Mushtaq, U., Perello, D., Lee, I., Cho, S.K., Star, A., Yun, M., 2007. Effective and low-cost platinum electrodes for microbial fuel cells deposited by electron beam evaporation. Energy and Fuels 21 (5), 2984–2990.

Picioreanu, C., Head, I.M., Katuri, K.P., van Loosdrecht, M.C., Scott, K., 2007. A computational model for biofilm-based microbial fuel cells. Water Research 41 (13), 2921–2940.

Rahimnejad, M., Bakeri, G., Ghasemi, M., Zirepour, A., 2014. A review on the role of proton exchange membrane on the performance of microbial fuel cell. Polymers for Advanced Technologies 25 (12), 1426–1432.

Reguera, G., McCarthy, K.D., Mehta, T., Nicoll, J.S., Tuominen, M.T., Lovley, D.R., 2005. Extracellular electron transfer via microbial nanowires. Nature 435 (7045), 1098.

Ren, Z., Ramasamy, R.P., Cloud-Owen, S.R., Yan, H., Mench, M.M., Regan, J.M., 2011. Time-course correlation of biofilm properties and electrochemical performance in single-chamber microbial fuel cells. Bioresource Technology 102 (1), 416–421.

Sacco, N.J., Figuerola, E.L., Pataccini, G., Bonetto, M.C., Erijman, L., Cortón, E., 2012. Performance of planar and cylindrical carbon electrodes at sedimentary microbial fuel cells. Bioresource Technology 126, 328–335.

Schröder, U., Harnisch, F., Angenent, L.T., 2015. Microbial electrochemistry and technology: terminology and classification. Energy and Environmental Science 8 (2), 513–519.

Schröder, U., 2007. Anodic electron transfer mechanisms in microbial fuel cells and their energy efficiency. Physical Chemistry Chemical Physics 9 (21), 2619–2629.

Sekar, N., Ramasamy, R.P., 2013. Electrochemical impedance spectroscopy for microbial fuel cell characterization. Journal of Microbial & Biochemical Technology 6 (2).

Sirinutsomboon, B., 2014. Modeling of a membraneless single-chamber microbial fuel cell with molasses as an energy source. International Journal of Energy and Environmental Engineering 5 (2–3), 93.

Tota-Maharaj, K., Paul, P., 2015. Performance of pilot-scale microbial fuel cells treating wastewater with associated bioenergy production in the Caribbean context. International Journal of Energy and Environmental Engineering 6 (3), 213–220.

Venkataraman, P., 2009. Applied Optimization with MATLAB Programming. John Wiley & Sons.

Zhang, F., Xia, X., Luo, Y., Sun, D., Call, D.F., Logan, B.E., 2013. Improving startup performance with carbon mesh anodes in separator electrode assembly microbial fuel cells. Bioresource Technology 133, 74–81.

Zhang, Y.H.P., 2010. Production of biocommodities and bioelectricity by cell-free synthetic enzymatic pathway biotransformations: challenges and opportunities. Biotechnology and Bioengineering 105 (4), 663–677.

Chapter 9

Application of lactic acid bacteria and their metabolites against foodborne pathogenic bacterial biofilms

E. Giaouris
Department of Food Science and Nutrition, School of the Environment, University of the Aegean, Ierou Lochou 10 and Makrygianni, Lemnos, Greece

9.1 Introduction

Foodborne pathogenic bacteria, among other microorganisms as well, can easily attach to various surfaces encountered within food processing and create biofilms on them, resisting this way the antimicrobial action of common sanitizing agents and other harsh environmental conditions, such as desiccation and nutrients deprivation (Bridier et al., 2015). As thus, new efficient antibiofilm approaches are needed to combat these detrimental biofilms, ensuring the safety of our food supply. In the past years, the bioprotective action of LAB and some of their purified metabolites, such as bacteriocins, exopolysaccharides (EPS), and biosurfactants, contained in their cell-free culture supernatants (CFSs), are included among those approaches that have been successfully tested (Camargo et al., 2018; Castellano et al., 2017). As an example, Fig. 9.1 presents some of the main proposed mechanisms of LAB action against biofilm development by *Listeria monocytogenes* and other pathogens as well.

Indeed, in recent years, numerous in vitro studies have shown not only the protective role of LAB against pathogenic biofilm formation (e.g., by leaving them to form biofilms on surfaces, which subsequently inhibit their colonization by pathogenic cells, through competition for nutrients and space, production of antimicrobials, etc.), but also their significant eradicating action on preformed pathogenic biofilms, mainly through the application of LAB metabolites (either in pure form or as crude extracts in the form CFSs). LAB genera that have been mostly employed in these studies comprise *Lactococcus*,

Recent Trends in Biofilm Science and Technology. https://doi.org/10.1016/B978-0-12-819497-3.00009-X

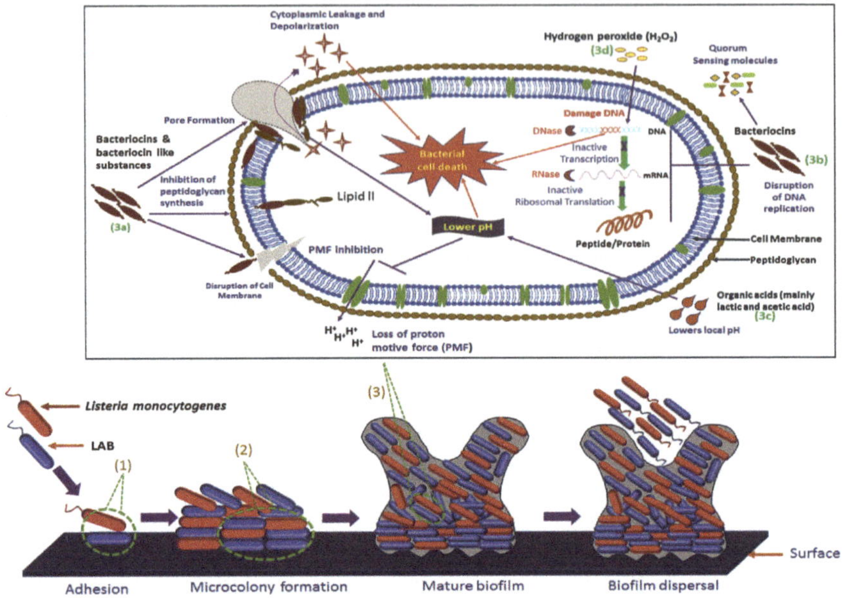

FIGURE 9.1 Proposed mechanisms of LAB action against biofilm development by *Listeria monocytogenes*. (1) Inhibition of pathogen adhesion; (2) competitive exclusion approach toward pathogenic bacteria; (3) production of antimicrobial substances: (3a) bacteriocins interact with lipid- II cell wall component (serves as a recognition "anchor") and working as cytoplasmic membrane perturbators to promote the dissociation of lipid II molecules. These actions inhibit the normal cell cycle and cell wall synthesis of targeted cell and also cause the formation of pores in bacterial cytoplasmic membrane, leading to the cell death via dissipation of proton motive force of the bacterial system; (3b) bacteriocins inhibit the DNA replication and cause DNA damage, ultimately resulting in cell death by inhibition of DNA gyrase, RNA polymerase, and aspartyl–tRNA synthetase; (3c) organic acids provide an unfavorable acidic condition that can inhibit the pathogen growth, proceeding to the cell death via dissipation of proton motive force of the bacterial system; (3d) hydrogen peroxide (H_2O_2) interferes with targeted bacterial DNA, RNA, and protein metabolism, which results in cell death via DNA damages. *LAB*, lactic acid bacteria. *Reprinted with permission from Elsevier, LWT Food Science and Technology, Inhibitory effects of probiotic potential lactic acid bacteria isolated from kimchi against* Listeria monocytogenes *biofilm on lettuce, stainless-steel surfaces, and MBEC™ biofilm device, Hossain et al. (2020).*

Lactobacillus, *Enterococcus*, and *Pediococcus*, and these have been frequently isolated from not only fermented foods, such as cheese and milk (e.g., kefir), but also meat and vegetables. Their action not only is limited to the inhibition of planktonic growth of the target bacteria but has been expanded to include antipathogenic traits, such as inhibition of biofilm formation and expression of other virulence factors as well (Graham et al., 2017). It is still known that the expression of many of these traits by pathogenic bacteria is adjusted by quorum sensing (QS) mechanisms, through which bacteria coordinate their behavior by sensing not only their own population density but also those of their surroundings (Giaouris et al., 2015). Interestingly, LAB species have been shown to secrete metabolites with anti-QS activity (Kiymaci et al., 2018).

Undoubtedly, the application of LAB and/or their purified (or semipurified) metabolites against foodborne pathogenic biofilms is considered an environmentally friendly approach (limiting the use of synthetic chemicals), while the use of such metabolites at subinhibitory concentrations for planktonic growth of the target bacteria (mainly through interference with cell-to-cell interactions) is believed to exert less selective pressure to the latter and therefore limit the likelihood for resistance development. This last is quite important considering the great resistance numerous pathogens are currently displaying against some of the most common antibiotics and/or other sanitizers (Hutchings and Truman, 2019).

The most recent studies exploring LAB and/or their purified metabolites against biofilms formed by foodborne bacterial pathogens are reviewed in this chapter. To facilitate reading, these studies have been grouped to those involving (1) the in situ presence of LAB, (2) the action of CFSs of the LAB species, and those involving their purified, (3) bacteriocins, (4) EPS, and (5) biosurfactants. Finally, (6) some studies reporting the antibiofilm action of LAB through inhibiting the QS systems of the target bacteria are briefly discussed.

9.2 Antibiofilm activities of lactic acid bacteria and their metabolites against foodborne bacterial pathogens

9.2.1 Antibiofilm activity of the in situ lactic acid bacteria presence

Several studies have explored the effects of the in situ LAB presence against biofilm formation by common foodborne bacterial pathogens, including *Salmonella enterica*, *Escherichia coli*, *L. monocytogenes*, *Bacillus cereus*, and *Staphylococcus aureus*. In such a study, Abedi et al. (2013) investigated the in vitro antiadherence properties of a *Lactobacillus delbrueckii* subsp. *bulgaricus* strain against *E. coli*, using Caco-2 cells (to mimic gut infection) and testing three competition conditions, including competitive inhibition, displacement, and exclusion. When the two bacterial species were simultaneously added, the degree of inhibition of *E. coli* binding by *L. delbrueckii* was 77% (competitive inhibition). In displacement assay, where the one bacterial species was first allowed to attach to the human colorectal tumor cells before the addition of the second species, the reduction of *E. coli* attachment by the subsequent binding of *L. delbrueckii* was 81.3%. In exclusion assay, preadhered *L. delbrueckii* was able to eliminate the attachment of *E. coli* by around 43.5%. The CFS of *L. delbrueckii* was also shown (by the disk diffusion method) to present a good inhibitory activity against *E. coli* mainly due to the produced lactic acid, since this activity was diminished when its pH was adjusted to 6.5.

In another much interesting study, Didienne et al. (2012) analyzed microbial biofilms spontaneously formed on wooden vats used for making a protected designation of origin (PDO) French cheese and evaluated their putative role in milk inoculation and in preventing colonization by *L. monocytogenes* and *S. aureus*. The farm vats were found to be colonized by *Lactobacillus* (4.5 − 6.0 log CFU/cm^2), gram-negative bacteria (1.5 − 4.6 log CFU/cm^2), yeasts (2.9 − 5.6 log CFU/cm^2), and molds (1.7 − 4.5 log CFU/cm^2), while no pathogenic microorganisms were recovered. Interestingly, upon using experimental vats colonized by biofilm hosting a great diversity of LAB and on purpose contaminating milk with pathogenic bacteria, these last were found neither on the vat surfaces nor in the cheeses. This positive biofilm was still capable to quickly inoculate pasteurized milk with beneficial starter microflora at levels superior to those encountered in raw milk.

Gómez et al. (2016) evaluated the potential of four bacteriocinogenic (two *Lactococcus lactis*, one *Lactobacillus sakei*, and one *Lactobacillus curvatus*) and four nonbacteriocinogenic (one *L. lactis*, one *Lactobacillus helveticus*, one *Lactobacillus casei*, and one *Weissella viridescens*) LAB strains, isolated from Brazilian foods, to develop protective biofilms against *L. monocytogenes*, *E. coli* O157:H7, and *Salmonella* Typhimurium. To do this, they left pathogens to form biofilms (at 30°C for up to 72 h) on wells of polystyrene microtiter plates (PMPs) precolonized (at 30°C for 48 h) with LAB. The nonbacteriocinogenic strain of *L. lactis*, the bacteriocinogenic *L. sakei*, and *L. curvatus* strains resulted in more than six log reductions in the pathogenic counts when compared with the controls (pure cultures). This inhibition was still time dependent and fluctuated depending on the strains employed. In another similar study, Guerrieri et al. (2009) left *L. monocytogenes* to develop biofilms (at refrigeration for up to 10 days) on the wells of PMP precolonized (at 30°C for 7 days) by each one of four LAB strains of which two were bacteriocin producers (*Lactobacillus plantarum* and *Enterococcus casseliflavus*) and the other two (*L. plantarum* and *Enterococcus faecalis*) were not. Both *L. plantarum* strains reduced the pathogenic sessile populations by ca. 4 logs, with this effect to be mainly associated with the pH reduction. The *Enterococcus casseliflavus* (bac+) similarly reduced *L. monocytogenes* sessile cells by ca. 4 logs, however, due to the production of the bacteriocin, since at this case the pH value was not significantly reduced. Indeed, the nonbacteriocinogenic *E. faecalis* strain did not significantly influence the adherent *L. monocytogenes* population. Interestingly, when the pathogen was cocultured with a *Pseudomonas putida* strain, an important reduction of the antilisterial activity of *Lactobacillus* biofilms was observed due to the pH increase (from 5.1 to 6.3). In another study employing the same pathogen and *Enterococcus* species, da Silva Fernandes et al. (2015) evaluated the ability of *E. faecalis*, *Enterococcus faecium*, and *L. monocytogenes* strains (4 strains/species) to form multispecies biofilms on stainless steel (SS) surfaces at 25 and 39°C over an 8 day period. Alarmingly, at 25°C, the biofilm growth of *L. monocytogenes* was favored (compared with monospecies conditions), reaching counts above 6 log CFU/cm^2, whereas at 39°C, only *Enterococcus* spp. formed biofilm (with counts of 8 log CFU/cm^2), with *L. monocytogenes* biofilm counts staying below the detection limit.

Hossain et al. (2020) screened 34 LAB strains isolated from kimchi, a traditional Korean fermented food, for antilisterial activity (using the spot-on-lawn method) and identified six isolates (three *L. plantarum*, one *L. curvatus*, one *L. sakei*, and one *Leuconostoc mesenteroides*), which were also subsequently tested for inhibition of pathogenic biofilm on SS coupons, lettuce, and minimum biofilm eradication concentration (MBEC) biofilm (Calgary) device. Following 24 h coculture with LAB, the biofilm population of *L. monocytogenes* was reduced by up to 2.2 log CFU/cm^2, 1.6 log CFU/cm^2, and 1.1 log CFU/peg on SS, lettuce, and MBEC, respectively. Jalilsood et al. (2015) identified an *L. plantarum* isolate (from a tropical plant commonly used in Southeast Asian cuisines) that was able to form a strong biofilm in pure culture and tested it for its ability to inhibit biofilm formation by the foodborne pathogenic and spoilage bacteria *S. enterica*, *B. cereus*, *Pseudomonas fluorescens*, and *Aeromonas hydrophila*. To do this, they left target bacteria to develop biofilms (for up to 6 days) on the wells of a microtiter plate precolonized with a mature biofilm of the LAB strain (incubated at 30°C for 7 days). The sessile populations of *P. fluorescens*, *A. hydrophila*, and *B. cereus* were reduced on the second day of incubation by 1.7, 2.2, and 5.9 log, respectively, and these were not detected at the sixth day. However, *S. enterica* exhibited a greater resistance in the presence of the LAB biofilm presenting a 2.4 log reduction at the end of the experiment. Authors also showed that this LAB strain could be used as host for genetic engineering (e.g., to produce recombinant proteins) since it was capable to constitutively express green fluorescence protein (GFP) when transformed with an expression vector harboring the *gfp* gene (as a reporter).

Kumar et al. (2017) investigated the antibiofilm activity of an *L. casei* and an *L. plantarum* strain (isolated from a traditional Indian home-made curd and kefir and selected based on their antimicrobial action) against methicillin-resistant *S. aureus* (MRSA). To do this, the MRSA strain was left to develop biofilm (at 35°C for up to 6 days) on the wells of PMP precolonized by the tested LAB (through anaerobic incubation at 35 °C for 7 days). Results showed a time-dependent inhibition, with *L. casei* exhibiting the highest reduction of MRSA biofilms, by 3.5 log at 48 h, whereas *L. plantarum* provoked the highest reduction of 2.6 log at 36 h. However, nonsignificant differences in the pathogenic counts (at both planktonic and sessile phases) were recorded at the end of the incubation period, between pure MRSA culture and coculture with the LAB. Leriche et al. (1999) investigated the survival of *L. monocytogenes* in a dual-species biofilm with a bacteriocinogenic *L. lactis* strain by either depositing the pathogen on a 24 h-old biofilm of *L. lactis* or leaving both bacterial species to simultaneously adhere. At both cases, biofilms were formed on SS surfaces incubated at 25°C for up to 48 h. Results showed that the antilisterial activity of the *L. lactis* (always inoculated at 10^8 CFU/mL) was dependent on the initial inoculum size of the pathogenic bacteria. Thus, in the case of small inoculum (10^{6-7} CFU/mL) and at both simulating conditions (deferred/simultaneous adhesion), the pathogen was not detected following 24 h of incubation (neither in the adherent or planktonic population). However, in the case of the high inoculum (10^8 CFU/mL), the reduction of the pathogenic counts was relatively slow, leading to the establishment of a stable sessile population (10^{5-6} CFU/cm^2).

The effects of three lactobacilli isolates from kefir grains (two *Lactobacillus kefiri* and one *L. plantarum*) and their surface proteins on biofilm development by a *Salmonella* Enteritidis strain were investigated by Merino et al. (2019). The preincubation of the tested surfaces (24-well PMPs) with 2×10^8 CFU/mL of LAB for 2 h or the coincubation of the pathogenic bacteria with the three lactobacilli strains (both at 2×10^8 CFU/mL) was found to significantly reduce (about 1 log) *Salmonella* biofilm population on PMP (incubation at 28°C for 48 h), and this was also confirmed by confocal microscopy (by forming biofilms on glass and staining them with the LIVE/DEAD^R bacterial viability kit) (Fig. 9.2). Similarly, when the surface proteins of the LAB (0.5 μg/mL) were preincubated or coincubated with *S.* Enteritidis, biofilm formation by the pathogen was significantly decreased. Authors also showed that the pathogenic cells could aggregate with the LAB, in a strain-specific manner, by testing 15 *Salmonella* strains representing three different serotypes commonly found in poultry (i.e., Enteritidis, Typhimurium, and Gallinarum).

Ndahetuye et al. (2012) investigated the ability of an LAB cocktail (10^8 CFU/mL) of three strains (i.e., *Pediococcus acidilactici*, *Lactobacillus amylovorus*, and *Lactobacillus animalis*, mixed in a ratio of 1:1:2), isolated from

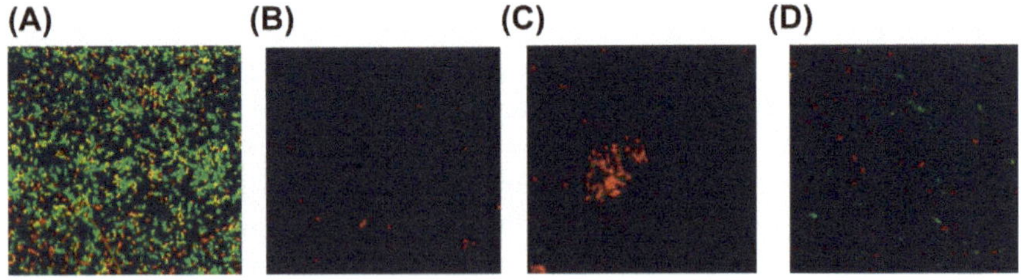

FIGURE 9.2 CLSM images (at 63X magnification) of biofilm formation by *Salmonella* Enteritidis on glass surfaces. (A) Only *S.* Enteritidis cells (control), (B) coincubation with *Lactobacillus kefiri* (first strain), (C) coincubation with *L. kefiri* (second strain), (D) coincubation with *Lactobacillus plantarum*. At all cases, sessile cells were stained with propidium iodide and SYTO 9, to differentiate between live (green) and dead (red) bacteria. *CLSM*, confocal laser scanning microscopy. *Reprinted with permission from Elsevier, Food Research International, Lactobacillus strains inhibit biofilm formation of Salmonella sp. isolates from poultry, Merino, L., Trejo, F.M., De Antoni, G., Golowczyc, M.A., 2019.* Lactobacillus *strains inhibit biofilm formation of* Salmonella *sp. isolates from poultry. Food Research International 123, 258–265.*

ready-to-eat meat and exhibiting antilisterial action, to attach to deli slicer SS coupons and inhibit the attachment of *L. monocytogenes* (incubation in TSB-YE at 23°C for up to 72 h). To do this, the coupons were (1) pretreated with the LAB cocktail for 24 h prior to the addition of the pathogenic bacteria (10^3 CFU/mL), (2) simultaneously treated with all the bacteria, or (3) pretreated for 24 h with *L. monocytogenes* prior to the addition of the LAB (displacement assay). Under all these conditions, the LAB cocktail was found to significantly decrease the attachment of the pathogen, depending on the sequence of the bacterial attachment onto the coupons. A colorimetric method also revealed that the attached LAB produced significantly more extracellular carbohydrates than their planktonic counterparts. Tan et al. (2017) examined the ability of a *P. acidilactici* isolate from fermented kimchi, able to produce the antimicrobial peptide pediocin, to inhibit biofilm formation by *E. coli*, *S.* Typhimurium, *S. aureus*, and *L. monocytogenes* (and their mixture) on abiotic surfaces (SS, polyvinyl chloride [PVC] and glass coupons) in exclusion, displacement, and competition assays. To do this, coupons with attached bacteria (following their incubation in saline cellular suspensions of 10^8 CFU/mL at 4°C for 24 h) were vertically placed in 50 mL conical centrifuge tubes containing TSB and incubated at 25°C for 6 days (except for the displacement assay where LAB were added to already established pathogenic biofilms and incubated at 25°C for 24 h). Under all simulating conditions, the sessile pathogenic populations, on all tested surfaces, were efficaciously reduced by ca. 4 log CFU/coupon.

The effect of a *Lactobacillus paraplantarum* cheese isolate on biofilm formation by *L. monocytogenes* on SS coupons, by using culture and fluorescence in situ hybridization (FISH) methods, was evaluated by Winkelströter et al. (2015). To do this, the coupons were placed in brain heart infusion (BHI) broth inoculated with both bacterial species at 10^6 CFU/mL and incubated at 25°C for up to 72 h (without shaking). The adhered pathogenic cells on the coupons were significantly reduced up to 2.4 log CFU/mL at 24 and 48 h of incubation and 1.9 log CFU/mL at 72 h, whereas the presence of the pathogenic bacteria did not influence the sessile LAB population (in comparison with the pure culture). In addition, sodium dodecyl sulfate polyacrylamide gel electrophoresis (SDS-PAGE) revealed the presence (in CFS of the LAB strain) of a bacteriocin with a molecular mass around 3900 Da, whereas PCR-DNA sequencing showed that bacteria harbored the plantaricin NC8 structural gene. Woo and Ahn (2013) evaluated the inhibitory effects of probiotic *Lactobacillus acidophilus, L. casei, Lactobacillus Paracasei*, and *Lactobacillus rhamnosus* strains on biofilm formation by *S.* Typhimurium and *L. monocytogenes* (on 12-well PMP incubated at 37°C for 24 h), in competition, exclusion, and displacement assays. Under competitive coculture (both species at 10^7 CFU/mL), no significant reductions were observed in the numbers of sessile *S.* Typhimurium cells, whereas *L. monocytogenes* biofilm cells were reduced by more than 2 logs when cocultured with the probiotic strains. These last effectively reduced biofilm formation by both foodborne pathogens, when the latter were left to adhere on surfaces precolonized with the LAB (exclusion assays). Concerning the displacement activity of probiotic bacteria, this was not effective against *S.* Typhimurium, whereas the number of sessile *L. monocytogenes* cells was efficiently reduced (1.5–3.9 logs) by all the LAB species.

Zhao et al. (2004) identified nine LAB isolates (six *Enterococcus durans*, two *L. lactis* subsp. *lactis*, and one *L. plantarum*) upon screening 413 microbial isolates (obtained from floor drains in food processing facilities with a history of no *L. monocytogenes* presence) for microorganisms producing antilisterial compounds. The antibiofilm efficiency of these isolates was evaluated by leaving a five-strain mixture of the pathogenic cells (10^{3-4} CFU/mL) together with the candidate LAB (10^{6-8} CFU/mL) to form mixed-culture biofilms on SS coupons at 37°C (for 24 h), 15 and 8°C (for up to 28 days), and 4°C (for up to 35 days) with TSB as the growth medium (renewed at appropriate time intervals following an initial 6 h attachment step). Results revealed that two of these isolates (*E. durans* and *L. lactis*) were able to drastically decrease (>5 log CFU/cm^2) the sessile *L. monocytogenes* populations under all the tested conditions. It should also be noted that the *L. lactis* strain was found to contain the genes for nisin A and B, and this was still able to control *L. monocytogenes* biofilm growth at 4°C for a long time period (35 days) despite the fact that this was unable to grow at such a low temperature. In a later study by the same authors, the ability of these two competitive-exclusion LAB isolates to in situ reduce *Listeria* spp. contamination in floor drains of a poultry processing plant was evaluated (Zhao et al., 2006). To do this, they selected six floor drains, which presented different colonization patterns with the pathogenic bacteria (with respect to occurrence, persistence, and cellular population), and treated them (for 20 min following the routine sanitation procedures) with in an enzyme foam–based cleaner containing a mixture of the two LAB species (10^7 CFU/mL) (four times in the first week and twice a week for the next 3 weeks). Results revealed that the sessile *Listeria* spp. counts at all the sampling sites (3–26°C) were efficiently reduced following the LAB application, whereas the routine sanitation alone was unable to affect these populations. However, at the same time, the aerobic plate counts (APCs) remained approximately the same without being affected by this treatment, indicating a specific interaction of the LAB solely with the *Listeria* spp. These two LAB isolates were also efficient in reducing preformed (at 37°C for 24 h) *L. monocytogenes* biofilms on coupons composed of different materials (stainless steel, plastic, rubber, glass, and silicone), as this was determined at 4 and 8°C following a 3 week LAB exposure (Zhao et al., 2013).

9.2.2 Antibiofilm activity of lactic acid bacteria cell-free culture supernatants

Numerous studies have explored the antibiofilm action of LAB CFSs, with either their original acidic pH (3.5–4.0; mainly due to the lactic acid production) or their neutralization (to 6.5–7.0 with NaOH) to exclude the organic acid effects, against common foodborne pathogens. In such a study, Camargo et al. (2016) tested CFSs from nine LAB strains (including three *L. curvatus*, one *Lactobacillus fermentum*, one *L. delbrueckii*, one *P. acidilactici*, and three *E. faecium* strains) for their efficiency to inhibit biofilm formation (on 96-well PMP) by six *L. monocytogenes* strains. To do this, they collected CFSs following the growth of each LAB strain in MRS broth at 37°C for 24 h, centrifugation, pH adjustment to 6.0, and heat treatment at 80°C for 10 min and subsequently left pathogenic cells to form biofilms in TSB containing 11.1% of each CFS (bacteriocins at 12,800 AU/mL) at 37°C for 24 h. In addition, they tested the effect of CFS (150 min exposure) on already-formed biofilms. Under the coinoculation conditions, a significant inhibition in biofilm production was observed. On the contrary, the CFSs were unable to cause the elimination of preformed biofilms. In another experiment by the same authors, CFS (6.25–6400 AU/mL) from the *L. curvatus* strain was also investigated for its ability to eliminate biofilms of three *L. monocytogenes* strains, in combination with EDTA (0.39–25 mM). In general, low concentrations of CFS combined with EDTA caused a significant reduction in already-formed biofilms, but this combination was not still able to fully eliminate them (Camargo et al., 2016).

In another study, Ait Ouali et al. (2014) recovered 130 LAB isolates from milk tanks and milking machine surfaces (at two distinct Algerian farms). Among these LAB isolates, three *Lactobacillus brevis* and two *Lactobacillus pentosus* strains were found to present high biofilm-forming abilities (on polystyrene tissue culture plates) and in addition were capable of producing bacteriocins-like inhibitory substances against *E. coli*, *Salmonella*, *S. aureus*, and *Pseudomonas aeruginosa* species (also isolated from the same farms), when grown in MRS or TSB-YE media. Further experiments with the neutralized CFSs of these five LAB isolates showed that these strongly hampered the adhesion and subsequently the biofilm formation of *S. aureus* on two abiotic surfaces: polystyrene and SS. For the last surface, neutralized CFSs were deposited on SS slides and maintained there for 2 h at 37°C to create conditioning films, before the inoculation with the pathogenic bacteria.

Bacteria of the genus *Enterococcus* are ubiquitous cocci that can be found everywhere, while these are frequently isolated in large numbers from foods, such as dairy products (Ben Braïek and Smaoui, 2019). The antibacterial and antiadherence activities of an *E. durans* strain, originally isolated from a milking machine surface, were investigated against *E. coli*, *S. aureus*, *P. aeruginosa*, and *Listeria innocua* (Amel et al., 2015). All these bacteria had been previously shown as biofilm producers on both plastic and SS surfaces. Both the cells of *E. durans* and their CFS presented a significant inhibitory activity against all the targeted bacterial species found either on solid media or in broth cocultures. In addition, authors showed that the antiadherence activity of the CFS was completely lost following its treatment with pepsin, something that reveals the proteinaceous nature of the antibacterial substance (being most probably a bacteriocin).

Ben Slama et al. (2013) tested the extracts of three *L. plantarum* strains isolated from traditional Tunisian fermented food for their ability to inhibit biofilm formation by four *L. monocytogenes* strains (including ATCC 19115 and three other strains isolated from meat), by using the 2,3-*bis* (2-methoxy-4-nitro-5-sulfophenyl)-5-[(phenylamino) carbonyl]-2*H*-tetrazolium hydroxide (XTT) reduction method (dependent on the microbial respiratory activity). All the extracts were prepared following the cultivation of the bacteria in MRS broth for 24 h at 37°C, centrifugation, and sterilization of the supernatant (through filtration). These authors found biofilm inhibitory concentrations $(BIC)_{50}$ ranging from 5% to 30%, depending on the LAB and the targeted strains employed. However, the eradication of preformed biofilms was quite more difficult to accomplish, with biofilm eradication concentrations $(BEC)_{90}$ ranging from 60%–90%. Isolation and identification of the constituents that exhibited the antibiofilm properties was not done by these authors.

Bolocan et al. (2017) screened mushroom substrates to isolate strains presenting antilisterial activity and identified 11 as potential bacteriocin producers. Among these, four were identified as *L. lactis* and were all able to produce nisin Z, as confirmed by mass spectrometry, PCR, and sequencing. When three *L. monocytogenes* strains (selected from a total of 42 isolates of different serotypes based on their biofilm-forming ability) were left to produce biofilm on PMP at 15°C for 72 h, it was shown that their biofilm-forming ability was significantly reduced when grown in the presence of neutralized CFS of one selected *L. lactis* strain (producing nisin Z). However, the same CFS (applied for 24 h at 15°C) could not reduce the preformed biofilms (at 15°C for 72 h) by the same pathogenic strains. Similar biofilm-forming experiments on SS coupons by one *L. monocytogenes* strain (with the strongest biofilm-forming ability) showed that this *L. lactis* isolate could reduce the total number of sessile pathogenic cells by 3.6 log CFU/cm^2 under coculture (competitive exclusion) biofilm-forming conditions (in brain–heart infusion broth supplemented with 0.6% yeast extract; BHI-YE). Interestingly, the neutralized CFS from this bacteriocin-producing strain could also reduce *L. monocytogenes* preformed biofilm (after 72 h at 15°C in BHI-YE) on SS coupons by 1.53 log CFU/cm^2. In another similar screening study of mushroom growth substrates to isolate bacteria able to inhibit or eliminate *L. monocytogenes* biofilms, Dygico et al. (2019) identified eight antilisterial strains, all belonging to *L. lactis* subsp. *lactis*. All these strains produced nisin Z (as again shown by colony mass spectrometry), were strong biofilm formers (as shown by the crystal violet [CV] assay), and able to cause a 4-log reduction of *L. monocytogenes* cell counts upon coculture biofilm-forming conditions on SS surfaces at 25°C for 120 h (through competitive exclusion). The CFS of the tested strains also provoked significant reductions on most of the 72 h biofilms formed on microtiter plates by the three *L. monocytogenes* strains, whereas this did not present any effect on pathogenic biofilm counts on SS.

The effects of a bacteriocin-containing crude extract produced by an *L. lactis* subsp. *lactis* biovar. *diacetylactis* strain against biofilm formation by 8 coagulase negative staphylococci (CoNS, including four *Staphylococcus epidermidis*, two *Staphylococcus hominis*, one *Staphylococcus lugdunensis*, and one *Staphylococcus haemolyticus* strain) and 12 *L. monocytogenes* isolates were evaluated by Cirkovic et al. (2016). To do this, they tested five subinhibitory concentrations of the crude extract (1/2 to 1/32 × MIC, with minimum inhibitory concentrations [MICs] values ranging from 400 to 3200 AU/mL) and left biofilms to be formed on 96-well PMP for 24 h at 35°C. Significant activity of the crude extract against biofilm formation by *L. monocytogenes* was obtained upon its application at 1/2 × and 1/4 × MIC. However, all the tested concentrations had no significant inhibitory effects on biofilm formation by the CoNS strains. Nevertheless, the application of the crude extract (for 24 h at 35°C) at either 100 or 200 AU/mL significantly reduced the amount of 24 h-old (already formed) biofilms by all the tested strains.

Cui et al. (2018) identified 12 LAB strains (by screening 321 LAB strains isolated from traditional artisanal milk cheeses from Northeast China) whose neutralized CFSs inhibited the growth of various enteropathogenic bacteria (including *L. monocytogenes*, *S.* Typhimurium, *Shigella flexneri*, *Yersinia enterocolitica*, and *Clostridium perfringens*) and exhibited antibiofilm action against *S. aureus* and/or *E. coli*. These 12 strains were also resistant to gastric lumen conditions (low pH, trypsin, and bile salts) and were assigned (through 16S rRNA gene sequencing) to *L. plantarum* (7), *L. helveticus* (3), *P. acidilactici* (1), and *E. faecium* (1) species. Three of these strains (two *L. plantarum* and one *L. helveticus*) were susceptible to most of the tested antibiotics (15 in total), presented high hydrophobicity (similar to or higher than the reference strain *L. rhamnosus* GG), and were confirmed safe through nonhemolytic activities and bacterial translocation (in mice model).

Das et al. (2013) evaluated the efficacy of an *L. plantarum* strain (isolated from a traditional fermented milk product of India), in suppressing the in vitro growth and pathogenicity of *S.* Enteritidis. CFS (11%) of this strain was found to inhibit the growth of *S.* Enteritidis without affecting the growth of other *Lactobacillus* strains. The isolated strain also provoked a significant reduction in the biofilm-forming ability of the pathogen (by 2 log CFU/mL; 96-well PMP, 37°C for 24 h), when the last one was added 1 h after the addition of the LAB strain (1:1 ratio). Moreover, an 80% reduced invasion of HCT-116 colon epithelial cell line by the pathogen was observed under similar delayed addition conditions, together with a downregulation of the expression of the *hilA* gene (regulator of other genes implicated in the invasion of intestinal epithelial cells) of *Salmonella* Pathogenicity Island 1 (SPI1) in the presence of CFS of the *L. plantarum* strain.

The antibiofilm activities of four *L. plantarum* and two *L. fermentum* strains that were isolated from fermented clover silage and that were found to present strong antagonistic properties against various pathogenic bacteria were characterized by Gavrilova et al. (2019). The MRS broth suspensions of all of them were able to efficiently eradicate 48 h preformed biofilms of *S. aureus* and *E. coli* (24 h exposure), while these were also able to prevent the sessile growth of the pathogenic bacteria in cocultivation experiments. Their antimicrobial properties were linked to the acidification of the growth media, since the reduction of glucose from 2% to 0.2% (and as thus pH increase due to the lower acid production) drastically diminished their antagonistic properties. The effects of fermented broths of five LAB strains (two *L. lactis*, one *L. casei*, one *L. paracasei*, and one *L. rhamnosus*) on the adhesion (on 96-well PMP for 12 h) of seven potentially pathogenic and spoilage microorganisms (*E. coli*, *B. cereus*, *L. innocua*, *P. fluorescens*, *P. putida*, *S. aureus*, and *S. epidermidis*) was evaluated by Gutiérrez et al. (2016). They observed species-specific effects on the adhesion with some broths reducing the adhesion of the target bacteria, whereas some others were found to increase that.

Kaur et al. (2018) tested seven *Lactobacillus* spp. strains (selected following screening of 55 lactobacilli isolates from fecal samples of healthy children for antimicrobial action against gram-negative gut pathogens) for their efficiency to prevent biofilm formation by *Vibrio cholerae* and *Vibrio parahaemolyticus* and disperse their preformed biofilms. The nonneutralized CFSs of all the isolates (pH 3.5) inhibited biofilm formation (on 96-well PMP; at 37°C for 48 h; CV assay) by *V. cholerae* by more than 90%, while their neutralization (pH adjusted to 6.5) was found to reduce the biofilm inhibitory potential of only two of the isolates. In addition, all CFSs provoked the dispersion of 24-h-old preformed *V. cholerae* biofilms (by 62%−85%); however, their neutralization significantly reduced the biofilm dispersing action of four isolates. Concerning *V. parahaemolyticus*, the nonneutralized CFS of five isolates inhibited its biofilm formation by 62%−82%, whereas none of these CFSs could disperse its preformed biofilms. The CFSs of the tested isolates were also found to inhibit the adherence of *V. cholerae* to the intestinal epithelial cell line HCT-15. Based on these results, authors concluded that the biofilm inhibitive and dispersive actions of CFSs of lactobacilli depend both on the strains employed (LAB and target) and on the pH.

Koohestani et al. (2018) investigated the biofilm removal potential of CFSs (40%−100%) of an *L. acidophilus* and an *L. casei* strain (following their growth in MRS broth at 37°C for 48 h) against 24 h-old biofilms of *S. aureus* developed on polystyrene and glass surfaces and found that the 30 min exposure to both CFSs significantly removed the biofilms on both surfaces in a concentration-dependent manner. The antibiofilm action of CFSs produced by *L. fermentum* and *L. plantarum* probiotic isolates (from fermented fine cocoa seeds), against *S. aureus*, was investigated by Melo et al. (2016). Both CFSs (obtained following bacterial growth in MRS broth at 37°C for 48 h, centrifugation, lyophilization of the supernatant, and rehydration in deionized water) inhibited *S. aureus* growth, with MIC values equal to 20 and 2.5 mg/mL for *L. fermentum* and *L. plantarum*, respectively. Their application at 90% of their

MICs significantly reduced biofilm formation of the pathogen (on 96-well PMP, incubation at 37°C for 24 h in TSB containing 1% glucose), with the action of *L. fermentum* CFS to be, however, significantly higher to that of *L. plantarum*. The former CFS was also able to remarkably reduce biofilm biomass accumulated on glass coverlips even when this was applied at ½ of its MIC, as observed by scanning electron microscopy (SEM) (Fig. 9.3). This was also confirmed by confocal laser scanning microscopy (CLSM) (with a reduction of biofilm thickness from 14 to 5.2 mm), whereas qPCR revealed that at this CFS concentration (10 mg/mL), *icaA* was downregulated, and at the same time, *icaR*, coding for a transcriptional repressor of the *icaADBC* operon (involved in the synthesis of the intercellular adhesion polysaccharide PIA of the biofilm matrix), was significantly upregulated.

Singh et al. (2020) characterized the antibiofilm efficiency of CFSs of 14 *Lactobacillus* spp. strains (of goat milk origin) against *Cronobacter sakazakii* and *L. monocytogenes*. The neutralized (pH 6.5) and heat-treated (at 100°C for 15 min) CFSs were capable to both prevent biofilm formation and disrupt preformed biofilms of the tested pathogens (on 96-well PMP, incubation at 37°C for 24 h; CV and MTT viability assays) in a strain-specific manner. This antibiofilm action was also confirmed by light and fluorescent microscopy observations, following the staining of preformed biofilms (on glass coverslips

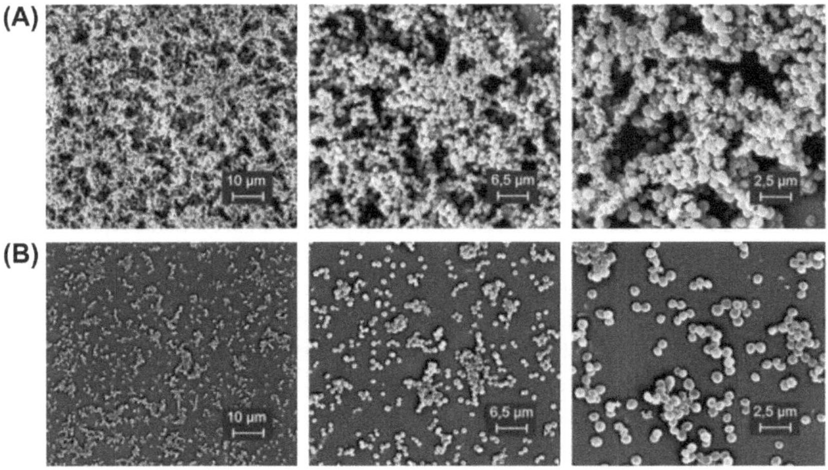

FIGURE 9.3 Scanning electron microscopy of *Staphylococcus aureus* biofilms on glass coverslips (at various magnifications) treated or not with *Lactobacillus fermentum* cell-free culture supernatant: (A) untreated control; (B) treated with ½ MIC. *MIC*, minimum inhibitory concentration. *Reprinted from article distributed under the terms of the Creative Commons Attribution 4.0 International License (http://creativecommons.org/licenses/by/4.0/), Springer Nature, BMC Microbiology, Inhibition of* Staphylococcus aureus *biofilm by Lactobacillus isolated from fine cocoa, Melo, T.A., Dos Santos, T.F., de Almeida, M.E., Junior, L.A., Andrade, E.F., Rezende, R.P., Marques, L.M., Romano, C.C., 2016. Inhibition of Staphylococcus aureus biofilm by* Lactobacillus *isolated from fine cocoa. BMC Microbiology 16(1):250. doi:10.1186/s12866-016-0871-8.*

at 37°C for 48 h in the presence of CFSs) by CV and LIVE/DEAD BacLight Bacterial Viability Kit, respectively. Todorov et al. (2018) tested the individual and combined effects of bacteriocin-containing CFS (2.25–12,800 AU/mL) produced by an *L. plantarum* strain, vancomycin (glycopeptide antibiotic; 0.015–1.0 μg/mL), propolis (resinous mixture produced by honeybees; non-diluted to 1/128 diluted), and EDTA (chelating agent; 0.15–10 mM) for controlling biofilm formation by four *L. monocytogenes* strains (on 96-well PMP, incubation in TSB at 37°C for 24 h; CV assay). The obtained results revealed a concentration-dependent pathogenic biofilm reduction by all the tested substances (except EDTA), and in parallel a synergistic effect of CFS and each of the other agents (vancomycin, propolis, EDTA), with the combined application of CFS and vancomycin to be the most effective approach.

Winkelströter et al. (2011) investigated the ability of *L. monocytogenes* to adhere to SS coupons (incubated in TSB-YE at 25°C for up to 48 h, under constant stirring at 15 rpm) in combination with a bacteriocin-producing *L. sakei* foodborne strain or its neutralized CFS (pH 7.0) containing sakacin 1 and compared the results with those obtained using a nonbacteriocinogenic *L. sakei* strain and its bacteriocin-free CFS. The coculture of the pathogenic bacteria with the bacteriocinogenic *L. sakei* (both initially inoculated at 10^8 CFU/mL) reduced their sessile populations to counts below 3.0 log CFU/cm^2 during all the sampling period (Fig. 9.4). Similar results were also obtained with the bacteriocin-containing CFS (50%) for up to 24 h of incubation; however, the number of adhered *L. monocytogenes* cells was increased at 48 h of incubation, probably due to the lack of nutrients competition or even the emergence of bacteriocin-tolerant bacteria. On the contrary, neither the non-bacteriocinogenic *L. sakei* strain nor its CFS could reduce the number of adhered *L. monocytogenes* cells (with counts around 5 log CFU/cm^2 by the 12 h of incubation and afterward) (Fig. 9.4).

The antibiofilm effects of CFSs of *L. plantarum*, *L. casei*, and *L. acidophilus* strains against eight resistant to ciprofloxacin *E. coli* isolates from urinary tract–infected patients were shown by Yeganeh et al. (2017) who left the target uropathogenic strains to adhere on 96-well PMP in pasteurized milk containing CFS (50%) at 37°C for 24 h and quantified their biomasses through CV staining. Yong et al. (2015) showed the antibiofilm effect of CFS of an *L. plantarum* foodborne isolate against biofilm formation by *S. aureus* on porcine skin (at 37°C for 24 h) by observing sessile biomasses through CLSM. To do this, pathogenic bacteria (10^6 CFU/mL) were either coincubated with the CFS (prevention group), or the latter was added to surfaces following their colonization with the pathogen (treatment group). In another study, the CFS from a *Lactobacillus gasseri* human gut isolate was found to inhibit *E. coli* biofilm formation (on 96-well PMP; incubation at 30°C for 36 h; CV assay) when this was applied at subinhibitory concentrations (0.75%–2.5% v/v) for planktonic growth (Zanfardino et al., 2017). In addition, these authors synthesized a bioactive octa-peptide, which still presented the antibiofilm action, based on the

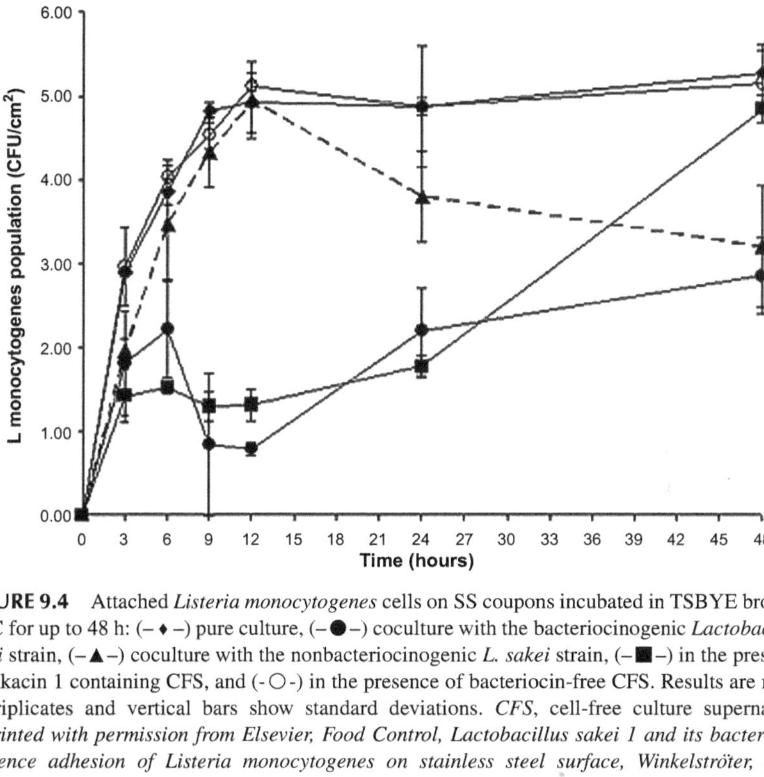

FIGURE 9.4 Attached *Listeria monocytogenes* cells on SS coupons incubated in TSBYE broth at 25°C for up to 48 h: (– ♦ –) pure culture, (–●–) coculture with the bacteriocinogenic *Lactobacillus sakei* strain, (–▲–) coculture with the nonbacteriocinogenic *L. sakei* strain, (–■–) in the presence of sakacin 1 containing CFS, and (-○-) in the presence of bacteriocin-free CFS. Results are mean of triplicates and vertical bars show standard deviations. *CFS*, cell-free culture supernatant. *Reprinted with permission from Elsevier, Food Control, Lactobacillus sakei 1 and its bacteriocin influence adhesion of Listeria monocytogenes on stainless steel surface, Winkelströter, L.K., Gomes, B.C., Thomaz, M.R.S., Souza, V.M., De Martinis, E.C.P., 2011.* Lactobacillus sakei *1 and its bacteriocin influence adhesion of* Listeria monocytogenes *on stainless steel surface. Food Control 22(8), 1404–1407.*

identification of a 1.3-kDa secreted peptide in this CFS, indicating that this should be the mail molecule responsible for the antibiofilm action.

9.2.3 Antibiofilm activity of purified lactic acid bacteriocins

The most studied antimicrobial metabolites produced by LAB are bacteriocins. These are ribosomally synthesized antimicrobial peptides, posttranslationally modified or not, that are released extracellularly by various bacteria (both gram positive and negative), and present action mostly toward other gram-positive bacteria, such as *L. monocytogenes* and *S. aureus*. Their production is considered strategically important as this enables producing bacteria to compete other bacteria found in the same environmental niche. Bacteriocins produced by LAB are grouped into three distinct classes based on their size and modification status: small posttranslationally modified peptides less than 10 kDa (class I), unmodified bacteriocins less than 10 kDa (class II), and unmodified (thermolabile) bacteriocins larger than 10 kDa (class III)

(Alvarez-Sieiro et al., 2016). The most well-known bacteriocin, which is also approved as a food preservative in over 50 countries, is nisin (E234), produced by *L. lactis* strains. This is known to inhibit cell wall synthesis by blocking the lipid II cycle. Other well-characterized bacteriocins are those produced by lactobacilli (e.g., sakacin, curvacin, glycosin), enterococci (e.g., enterocin), streptococci (e.g., streptolysin, thermophilin), and pediococci (e.g., pediocin, also commercialized as food additive).

E. faecalis B3A-B3B collected from infant feces has been reported to produce the bacteriocin B3A-B3B with activity against *L. monocytogenes*, *S. aureus*, *C. perfringens*, and *S.* Newport. This bacteriocin has similar amino acid sequence to the class II MR10A-MR10B enterocin. Interestingly, the combined antimicrobial treatment of SS slides with the bacteriocin nisin (0.256 mg/mL) and enterocin B3A-B3B (0.008 mg/mL) was found to decrease by at least 2 logs (after 0, 3, 6, and 24 h of incubation at 37°C) the sessile counts of an *L. monocytogenes* strain, previously isolated from a smoked salmon production factory (Al-Seraih et al., 2017). In addition, these two bacteriocins (applied again in combination at 4 and 0.008 mg/mL, respectively) were also found to be effective against already established biofilms of a nisin-resistant derivative of this strain (162R) on glass slides, significantly reducing the number of biofilm cells and in parallel modifying the bacterial structure, as confirmed by SEM (Fig. 9.5).

García-Almendárez et al. (2008) prepared a spray-dried crude bacteriocin fermentate (CBF) of a bacteriocinogenic *L. lactis* strain isolated from a Mexican-style cheese, able to produce nisin A, by culturing this strain on supplemented whey medium. A dose of 1.4 g CBF/mL caused a 2.7 log reduction of *L. monocytogenes* Scott A biofilm counts on SS coupons (incubation at 37°C for 96 h). In addition, competitive exclusion tests, performed by coinoculating both bacteria on the SS surface, showed that *L. lactis* reduced more than 5 log cycles the pathogenic biofilm counts per coupon (after 6 h of antagonism at 37°C). In addition, FISH using labeled specific probes for each species showed that *L. lactis* significantly outnumbered *Listeria* sessile cells in dual-species biofilms formed on a glass slide (for 3 days).

Minei et al. (2008) investigated the effect of nisin and coculture with *E. faecium* (either able or unable to produce bacteriocin) on *L. monocytogenes* biofilm formation on SS coupons (vertically incubated in BHI broth at 37°C for up to 48 h under constant agitation at 15 rpm). When the pathogen was cocultured with the nonbacteriocinogenic *E. faecium* strain, the number of its adhered cells was initially reduced by ca. 2 log, and stabilized to ca. 4 log CFU/cm^2 following 48 h of incubation. On the contrary, when the pathogen was cocultured with the bacteriocinogenic *E. faecium* strain, no adherent cells could be detected from 6 to 24 h of incubation, while around 2 log CFU/cm^2 were recovered from surfaces at the end of incubation. In the presence of nisin (10^3 IU/mL), sessile *L. monocytogenes* growth was decreased up to ca. 5 logs at 12 h of incubation, but this difference was greatly limited to ca. 2 logs after

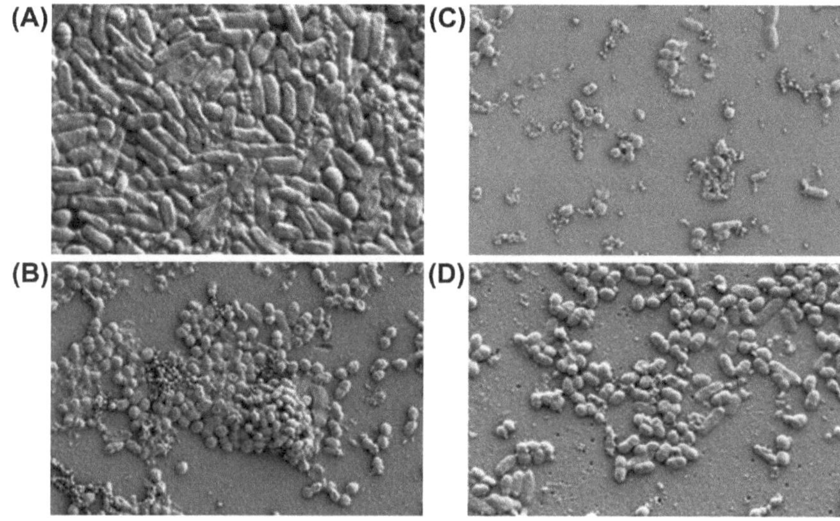

FIGURE 9.5 Scanning electron microscopy images of biofilm formation (A) by *Listeria monocytogenes* 162 R, (B) following treatment with nisin (16 mg/mL), (C) following treatment with enterocin B3A-B3B (0.064 mg/mL), and (D) following combined treatment with nisin/enterocin B3A-B3B (4/0.008 mg/mL, respectively). *Reprinted with permission from Springer International Publishing, Antonie van Leeuwenhoek, Enterocin B3A-B3B produced by LAB collected from infant feces: potential utilization in the food industry for* Listeria monocytogenes *biofilm management, Al-Seraih, A., Belguesmia, Y., Baah, J., Szunerits, S., Boukherroub, R., Drider, D., 2017. Enterocin B3A-B3B produced by LAB collected from infant faeces: potential utilization in the food industry for* Listeria monocytogenes *biofilm management. Antonie van Leeuwenhoek 110(2), 205–219.*

48 h. Pérez-Ibarreche et al. (2016) tested the efficiency of a meat-borne bacteriocinogenic *L. sakei* strain and its semipurified bacteriocin solution (curvacin A; 266.7 AU/mL) to inhibit *L. monocytogenes* biofilm formation on SS and polytetrafluoroethylene (PTFE) surfaces at 10°C, by using displacement, exclusion, and competition assays (Fig. 9.6). Under all simulating conditions, *L. sakei* efficiently inhibited pathogenic sessile growth on both surfaces, with the simultaneous coculture of both bacteria (competition for 6 days) being found to display the highest efficiency (reduction of pathogenic counts by 5.5 and 4.5 logs on PTFE and SS, respectively).

The inhibitory effects of partially purified bacteriocins produced by three *E. faecium* and one *E. faecalis* strain on either developing or preformed biofilms of *L. monocytogenes*, *Listeria ivanovii*, and *L. innocua* were investigated by Rocha et al. (2019). All the bacteriocins tested were able to decrease the biomasses and viabilities of both developing and preformed biofilms of all *Listeria* species (on 96-well PMP, incubation in BHI broth at either 30 or 37°C for 24 h; CV combined with XTT-reduction viability assays). This was also confirmed by confocal microscopy of the *Listeria* biofilms formed on glass coverslips (stained with DAPI), while fluorescence microscopy verified that

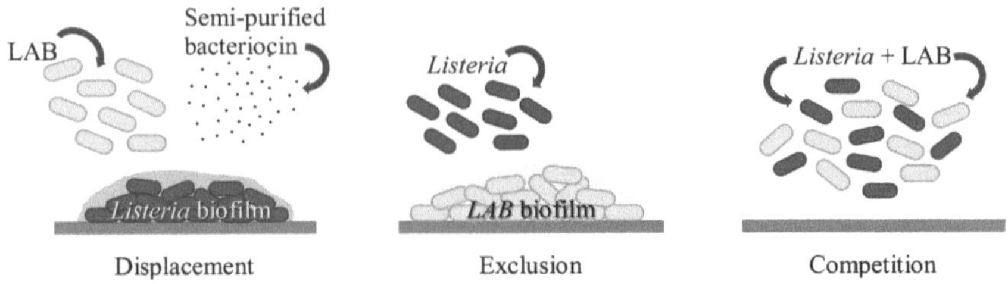

FIGURE 9.6 Scheme of the inhibition strategies of *Listeria* by *Lactobacillus sakei* (CRL1862) and/or its bacteriocin. Light gray rods represent *L. sakei* cells, whereas gray rods correspond to *Listeria* cells and small points represent semipurified bacteriocin solution. "Displacement": *L. sakei* cells and/or semipurified bacteriocin solution were added to previously formed *Listeria* biofilm. "Exclusion": *Listeria* cells were added to previously formed *L. sakei* biofilm. "Competition": planktonic *L. sakei* and *Listeria* cells were cocultured on the coupon surfaces. *Reprinted with permission from Oxford University Press, FEMS Microbiology Letters, Control of* Listeria monocytogenes *biofilms on industrial surfaces by the bacteriocin-producing Lactobacillus sakei CRL1862, Pérez-Ibarreche, M., Castellano, P., Leclercq, A., Vignolo, G., 2016. Control of* Listeria monocytogenes *biofilms on industrial surfaces by the bacteriocin-producing* Lactobacillus sakei *CRL1862. FEMS Microbiology Letters 363(12).*

Listeria biofilm cells lose their viability following the bacteriocin treatment (propidium iodide staining). SEM also showed that this treatment provoked protuberances on the surface of these cells, indicative of cellular leakages.

9.2.4 Antibiofilm activity of lactic acid bacteria exopolysaccharides

EPS are produced by many LAB species varying widely in their chemical composition and structural characteristics. Besides their involvement in the sensorial and textural properties of the fermented foods, the stress tolerance they confer to the producing bacteria (e.g., to gastrointestinal tract conditions, food-associated stresses), these may also bring important health benefits to human (e.g., immunomodulatory potential, pathogen antagonism, antioxidant activity, cholesterol-lowering action, prebiotics), making promising their exploitation in the production of added-value functional food products (Caggianiello et al., 2016; Lynch et al., 2018). Toward these probiotic actions, several studies have also shown the antiadhesive and antibiofilm potential of LAB EPS against microorganisms of clinical significance, mostly probably due to their physicochemical interactions with surfaces (both cellular and abiotic ones) and their resulting negative effect on the aggregation of microbial cells. In such a study, Kanmani et al. (2011) evaluated the optimum culture conditions for the synthesis of a novel EPS from an LAB marine strain of *Streptococcus phocae* and investigated the potential antibiofilm activity of the purified EPS against *L. monocytogenes*, *S.* Typhi, *P. aeroginosa*, *B. cereus*, and *S. aureus*. Biofilm formation by all the tested pathogens (on PMP) was inhibited in the presence of the purified EPS in a dose-dependent manner, with the sessile counts of *L. monocytogenes* and *S. aureus* showing the highest reduction (67% and 51%, respectively). In another study, EPS from an *L. helveticus* strain inhibited biofilm formation by *P. aeruginosa*, *S. aureus*, and *E. coli* at rates 43.6%, 65.8%, and 33.4%, respectively (Li et al., 2014).

Liu et al. (2017) extracted and purified EPS from a potential probiotic *L. plantarum* strain, isolated from human breast milk, and tested its anti-adhesion/biofilm activities. EPS at 1.0 mg/mL significantly inhibited the adhesion of *E. coli* O157:H7 to human colon cancer HT-29 cells in competition (i.e., simultaneous addition of EPS and *E. coli*), replacement (i.e., addition of EPS to HT-29 cells preincubated with *E. coli*), and inhibition assays (i.e., addition of *E. coli* to HT-29 cells preincubated with EPS), with inhibition rates of 20.2%, 29.7%, and 30.6%, respectively. In addition, the EPS strongly inhibited biofilm formation (on 96-well PMP, incubation at 37°C for 24 h; CV assay) by *P. aeruginosa*, *E. coli* O157:H7, *S.* Typhimurium, and *S. aureus*, in a concentration-dependent manner (0.2–5 mg/mL). Riaz Rajoka et al. (2018) showed that EPS produced by six *L. rhamnosus* isolates, again from human breast milk, exhibited strong inhibitory activity against *E. coli* and *S.* Typhimurium biofilm formation (on 96-well PMP; incubation at 37°C for 24 h), again in a concentration-dependent manner (EPS applied at 0.2–5 mg/mL). In another study, the EPS produced by an *L. plantarum* isolate from kefir grains inhibited in a concentration-dependent manner (0.2–5.0 mg/mL) biofilm formation by *S. flexneri*, *S. aureus*, and *S.* Typhimurium (on 96-well PMP, incubation in Luria–Bertani [LB] at 37°C for 24 h; CV assay) (Wang et al., 2015).

The EPS secreted by another *L. plantarum* strain isolated from a traditional Tunisian food (ricotta cheese) was found to reduce *E. coli* biofilm formation (on 96-well PMP, incubation at 37°C for 24 h; CV assay), at a subinhibitory concentration of 512 μg/mL, by more than 50% (Mahdhi et al., 2018). Moreover, at this concentration, the tested EPS inhibited the efflux of ethidium bromide (EtBr), the production of indole (a signal molecule implicated in QS), and reduced the hydrophobicity of the tested bacteria. The same subinhibitory EPS concentration was also found to significantly (>50%) reduce the biomasses of 24-h-old preformed biofilms of *S. aureus*, *L. monocytogenes*, *S.* Typhimurium, and *P. aeruginosa* (upon their incubation with EPS for 24 h), while it also reduced the surface hydrophobicity of all these bacteria (Mahdhi et al., 2017). In another study, the released EPS (1.0 mg/mL) from *L. acidophilus* strain decreased the biofilms formed by an enterohemorrhagic *E. coli* (EHEC) O157:H7 strain on polystyrene and PVC surfaces by 87% and 94%, respectively, by repressing the expression of genes related to curli production (*crl*, *csgA*, and *csgB*) and chemotaxis (*cheY*) and without affecting neither the growth nor the autoinducer-2 (AI-2)-like activity of the target bacteria (Kim et al., 2009). Similar effect was still recorded when the released EPS was applied to a continuous-flow chamber biofilm model.

9.2.5 Antibiofilm activity of lactic acid bacteria biosurfactants

Biosurfactants are amphiphilic molecules (e.g., neutral lipids, fatty acids, phospholipids, polysaccharide–protein complexes, lipopeptides) displaying surface and emulsifying properties, which are produced by various microorganisms, either staying on their surface or are extracellularly secreted. These are also known for their antimicrobial and antibiofilm actions against various microorganisms (Nitschke and Silva, 2018). In such a study, Sharma and Saharan (2016) characterized the antiadhesive and antibiofilm properties of a biosurfactant produced by an *L. helveticus* isolate from an ethnic fermented food (cheese) against various pathogenic microorganisms, including *E. coli*, *Salmonella*, *S. aureus*, *L. monocytogenes*, and *B. cereus*. The antiadhesive properties were checked on 96-well PMP pretreated with the biosurfactant (at 4°C for 18 h) before the addition of the target bacteria for attachment (at 4°C for 4 h), while biofilm inhibition was checked on (medical grade) silicon tubes incubated overnight at 37°C in inoculated LB medium containing the biosurfactant. At both cases, CV staining was used to quantify the sessile biomasses. The application of biosurfactant (at 25 mg/mL) was found to significantly inhibit the attachment of all the tested bacteria, whereas biofilm formation was almost absolutely diminished. In addition, this low concentration of the biosurfactant strongly inhibited the planktonic bacterial growth, something that should likewise account for its antibiofilm action. In another study, cell-bound biosurfactants extracted from *P. acidilactici* and *L. plantarum* strains (isolated from traditional artisanal cheese from Northeast China) were found to inhibit adhesion (at 4°C for 4 h) and biofilm formation (at 37°C for 24 h) of *S. aureus* (on 96-well PMP; CV assays) in a dose-dependent manner (12.5–50 mg/mL) (Yan et al., 2019). This was also confirmed by SEM, while the biosurfactants were found to affect the expression of biofilm-related genes (*cidA*, *icaA*, *dltB*, *agrA*, *sortaseA*, and *sarA*) (RT-qPCR assay) and interfere with the release of AI-2 signaling molecules (bioluminescence assay) by the pathogenic cells.

9.2.6 Biofilm inhibition trough quorum sensing interference by lactic acid bacteria

Many bacterial pathogens are known to use QS to induce pathogenicity (i.e., production of virulence factors including biofilm formation), by synthesizing, detecting, and responding to small signals called autoinducers (AIs) (Machado et al., 2020). For instance, the QS system employing AI-2 is quite ubiquitous, being encountered in both gram-positive and gram-negative bacteria, and is believed to be involved in both intra- and interspecies communication. Indeed, QS disruption is nowadays considered as a much promising approach to combat pathogenic bacteria, with lower likelihood for resistance development. Interestingly, LAB species have been shown to secrete metabolites with anti-QS activity and in this way inhibit pathogenic biofilm formation, among other antipathogenic traits (Kiymaci et al., 2018). In such a study, Park et al. (2014) investigated the AI-2 inhibition activity of a putative probiotic strain of *L. sakei* isolated from kimchi, and its effect on the virulence of EHEC O157:H7, aiming to develop a quorum quenching probiotic strain. The coculturing of the two strains was found to significantly reduce AI-2 production by the EHEC strain (bioluminescence assay using *Vibrio harveyi* reporter strain), while the CFS of the LAB strain (applied at 10%) also decreased motility, biofilm formation (on polystyrene and PVC microplates, incubation in 1/10 LB broth at 30°C for 48 h; CV assay), attachment to human epithelial type 2 (Hep-2) cells, and the expression of some critical virulence genes, by AI-2 signaling inhibition and without affecting the viability of the pathogenic bacteria. In another study, Pelyuntha et al. (2019) isolated two LAB strains from fermented grapes, namely *W. viridescens* and *Weissella confusa* (by screening 146 LAB strains from various fermented foods for antimicrobial activity), whose CFSs were found to reduce (> 90%) biofilm formation of *S.* Typhi (on PMP, incubation in LB broth at 37°C for 48 h; CV assay), both presenting a BIC value of 20%. These also significantly reduced biofilm formation of *S.* Typhimurium, with BIC values 15% and 20% for *W. viridescens* and *W. confusa*, respectively, while their application (at 5%) also decreased (by ca. 99%) the AI-2 activity of both pathogenic serovars (bioluminescence assay using *V. harveyi* reporter strain). It is known that AI-2 is excreted into the environment during the exponential growth of *Salmonella*, which uses this QS system to control the expression of many of its virulence genes (Walters and Sperandio, 2006).

9.3 Conclusions

Numerous studies have explored the antibiofilm action of LAB and/or their purified (or semipurified) metabolites (such as bacteriocins, EPS, and biosurfactants) against foodborne bacterial pathogens with quite promising results. These have been reviewed in this chapter, whereas several others have still been occupied with the antibiofilm activities of LAB against other clinically significant microorganisms, such as oral streptococci (Ciandrini et al., 2017) and respiratory tract pathogens (Shokri et al., 2018; van den Broek et al., 2018). To show antibiofilm action, researchers have (1) coincubated LAB and the target pathogenic microorganisms simultaneously forming mixed-culture biofilms (competition/inhibition assays), (2) left pathogens to form biofilms on surfaces precolonized with LAB (exclusion assays), and (3) inoculated LAB on surfaces precolonized with the pathogenic cells (displacement/eradication assays). Besides the application of LAB, some of these studies used just their produced metabolites (without the bacteria producing them), either in the form of crude extracts (CFSs) or following their purification. When the CFSs were used, before their application, their original acidic pH was frequently neutralized (to 6.5–7.0 to exclude any organic acid effect), and in parallel sometimes treated with catalase (to exclude the action of H_2O_2), so that to be able to finally target any antibiofilm action observed only to that conferred by other CFS-containing substances, being most probably bacteriocins. However, it should be noted that in some of these competition studies, the dilution of the biofilm supporting growth media with the CFSs could have also significantly influenced the planktonic growth of the cells (due to the lower nutrients availability). Thus, the biofilm inhibition subsequently observed could be just a result of the lower growth rate of the cells (both before and following their attachment to surfaces). Nevertheless, several other studies have managed to show the antibiofilm action of LAB metabolites or crude CFSs upon their use at sub-MIC values, something that undoubtedly reveals that this should be a result of biofilm-specific inhibition mechanisms (e.g., inhibition of cell-to-cell interactions and autoaggregation, QS interference, influence of biofilm-associated genes expression) and not due to the reduced cellular growth rate. These last studies are quite important since, as already stated, this targeted antibiofilm action is believed to provoke less selective pressure to the target pathogens and thus limit the possibilities for them to develop resistance, as that already achieved against several antibiotics and/or sanitizers (Koluman and Dikici, 2013). The engineering of bacteriocins (e.g., by modifying their amino acids) could also be a valuable strategy to enhance their antibacterial and antibiofilm activities (Field et al., 2015). However, one other important point that should be also considered is the quite complex nature of biofilms that may be encountered in real food processing environments, mainly with respect to the number of different types of microorganisms being present (Giaouris et al., 2015). This last may influence the resistance of the enclosed microorganisms, and thus the antibiofilm action of LAB and/or their metabolites which might result even in its elimination. For instance, a study examining the background microbiota established on surfaces in some small-scale cheese production facilities (including LAB among several other microorganisms) did not succeed to show any inhibitive action of it on the growth and surface attachment of *L. monocytogenes*, demonstrating that this resident environmental flora cannot on its own contribute to control the growth and survival of this important foodborne pathogen (Schirmer et al., 2013). Nevertheless, in another study previously reviewed in this chapter, some selected LAB isolates have been shown enough capable to in situ reduce *Listeria* spp. contamination in floor drains of a poultry processing plant, something which could not be achieved by the solely application of the routine sanitation (Zhao et al., 2006). Surely, more in situ studies employing the active microflora present in each food processing environment should be in the future performed to better evaluate the efficiently of LAB and/or their metabolites to control unwanted biofilm formation and strengthen the in vitro observed results. Finally, even in that prosperous case of pathogenic eradication, the safety of such an application should be firstly thoroughly verified with the specific strains each time employed. This is because even though several of the LAB species have the generally recognized as safe (GRAS) status (e.g., lactococci, lactobacilli), the uncontrolled release of some of others (e.g., enterococci) may pose public health risks (e.g., horizontal transfer of genes responsible for drug resistance and/or infectivity, excessive immune system stimulation, cytotoxicity against human cells, other gastrointestinal disturbances) (Zawistowska-Rojek and Tyski, 2018). Hopefully, when all the above have been successfully resolved, the safe and successful application of LAB biocontrol in food production, either alone or better in combination with the routine sanitization procedures (employing chemicals and/or other agents), could reduce the risk of cross-contamination with pathogenic bacteria, and as thus limit foodborne outbreaks, improving public health in a sustainable and environmentally friendly way.

References

Abedi, D., Feizizadeh, S., Akbari, V., Jafarian-Dehkordi, A., 2013. *In vitro* anti-bacterial and anti-adherence effects of *Lactobacillus delbrueckii* subsp. *bulgaricus* on *Escherichia coli*. Research in Pharmaceutical Sciences 8 (4), 260–268.

Ait Ouali, F., Al Kassaa, I., Cudennec, B., Abdallah, M., Bendali, F., Sadoun, D., Chihib, N.E., Drider, D., 2014. Identification of lactobacilli with inhibitory effect on biofilm formation by pathogenic bacteria on stainless steel surfaces. International Journal of Food Microbiology 191, 116–124.

Al-Seraih, A., Belguesmia, Y., Baah, J., Szunerits, S., Boukherroub, R., Drider, D., 2017. Enterocin B3A-B3B produced by LAB collected from infant faeces: potential utilization in the food industry for *Listeria monocytogenes* biofilm management. Antonie van Leeuwenhoek 110 (2), 205–219.

Alvarez-Sieiro, P., Montalbán-López, M., Mu, D., Kuipers, O.P., 2016. Bacteriocins of lactic acid bacteria: extending the family. Applied Microbiology and Biotechnology 100 (7), 2939–2951.

Amel, A.M., Farida, B., Djamila, S., 2015. Anti-adherence potential of *Enterococcus durans* cells and its cell-free supernatant on plastic and stainless steel against foodborne pathogens. Folia Microbiologica (Praha) 60 (4), 357–363.

Ben Braïek, O., Smaoui, S., 2019. Enterococci: between emerging pathogens and potential probiotics. BioMed Research International. https://doi.org/10.1155/2019/5938210.

Ben Slama, R., Kouidhi, B., Zmantar, T., Chaieb, K., Bakhrouf, A., 2013. Anti-listerial and anti-biofilm activities of potential probiotic *Lactobacillus* strains isolated from Tunisian traditional fermented food. Journal of Food Safety 33 (1), 8–16.

Bolocan, A.S., Pennone, V., O'Connor, P.M., Coffey, A., Nicolau, A.I., McAuliffe, O., Jordan, K., 2017. Inhibition of *Listeria monocytogenes* biofilms by bacteriocin-producing bacteria isolated from mushroom substrate. Journal of Applied Microbiology 122 (1), 279–293.

Bridier, A., Sanchez-Vizuete, P., Guilbaud, M., Piard, J.C., Naïtali, M., Briandet, R., 2015. Biofilm-associated persistence of food-borne pathogens. Food Microbiology 45 (Pt B), 167–178.

Caggianiello, G., Kleerebezem, M., Spano, G., 2016. Exopolysaccharides produced by lactic acid bacteria: from health-promoting benefits to stress tolerance mechanisms. Applied Microbiology and Biotechnology 100 (9), 3877–3886.

Camargo, A.C., de Paula, O.A., Todorov, S.D., Nero, L.A., 2016. *In Vitro* evaluation of bacteriocins activity against *Listeria monocytogenes* biofilm formation. Applied Biochemistry and Biotechnology 178 (6), 1239–1251.

Camargo, A.C., Todorov, S.D., Chihib, N.E., Drider, D., Nero, L.A., 2018. Lactic acid bacteria (LAB) and their bacteriocins as alternative biotechnological tools to control *Listeria monocytogenes* biofilms in food processing facilities. Molecular Biotechnology 60 (9), 712–726.

Castellano, P., Pérez Ibarreche, M., Blanco Massani, M., Fontana, C., Vignolo, G.M., 2017. Strategies for pathogen biocontrol using lactic acid bacteria and their metabolites: a focus on meat ecosystems and industrial environments. Microorganisms 5, E38. https://doi.org/10.3390/microorganisms5030038.

Ciandrini, E., Campana, R., Baffone, W., 2017. Live and heat-killed *Lactobacillus* spp. interfere with *Streptococcus mutans* and *Streptococcus oralis* during biofilm development on titanium surface. Archives of Oral Biology 78, 48–57.

Cirkovic, I., Bozic, D.D., Draganic, V., Lozo, J., Beric, T., Kojic, M., Arsic, B., Garalejic, E., Djukic, S., Stankovic, S., 2016. Licheniocin 50.2 and bacteriocins from *Lactococcus lactis* subsp. *lactis* biovar. diacetylactis BGBU1-4 inhibit biofilms of coagulase negative staphylococci and *Listeria monocytogenes* clinical isolates. PLoS One 11 (12), e0167995. https://doi.org/10.1371/journal.pone.0167995.

Cui, X., Shi, Y., Gu, S., Yan, X., Chen, H., Ge, J., 2018. Antibacterial and antibiofilm activity of lactic acid bacteria isolated from traditional artisanal milk cheese from Northeast China against enteropathogenic bacteria. Probiotics and Antimicrobial Proteins 10 (4), 601–610.

da Silva Fernandes, M., Kabuki, D.Y., Kuaye, A.Y., 2015. Behavior of *Listeria monocytogenes* in a multi-species biofilm with *Enterococcus faecalis* and *Enterococcus faecium* and control through sanitation procedures. International Journal of Food Microbiology 200, 5–12.

Das, J.K., Mishra, D., Ray, P., Tripathy, P., Beuria, T.K., Singh, N., Suar, M., 2013. *In vitro* evaluation of anti-infective activity of a *Lactobacillus plantarum* strain against *Salmonella enterica* serovar Enteritidis. Gut Pathogens 5 (1), 11. https://doi.org/10.1186/1757-4749-5-11.

Didienne, R., Defargues, C., Callon, C., Meylheuc, T., Hulin, S., Montel, M.C., 2012. Characteristics of microbial biofilm on wooden vats ('gerles') in PDO Salers cheese. International Journal of Food Microbiology 156 (2), 91–101.

Dygico, L.K., O'Connor, P.M., Hayes, M., Gahan, C.G.M., Grogan, H., Burgess, C.M., 2019. *Lactococcus lactis* subsp. *lactis* as a natural anti-listerial agent in the mushroom industry. Food Microbiology 82, 30–35.

Field, D., Gaudin, N., Lyons, F., O'Connor, P.M., Cotter, P.D., Hill, C., Ross, R.P., 2015. A bioengineered nisin derivative to control biofilms of *Staphylococcus pseudintermedius*. PLoS One 10 (3), e0119684. https://doi.org/10.1371/journal.pone.0119684.

García-Almendárez, B.E., Cann, I.K.O., Martin, S.E., Guerrero-Legarreta, I., Regalado, C., 2008. Effect of *Lactococcus lactis* UQ2 and its bacteriocin on *Listeria monocytogenes* biofilms. Food Control 19 (7), 670–680.

Gavrilova, E., Anisimova, E., Gabdelkhadieva, A., Nikitina, E., Vafina, A., Yarullina, D., Bogachev, M., Kayumov, A., 2019. Newly isolated lactic acid bacteria from silage targeting biofilms of foodborne pathogens during milk fermentation. BMC Microbiology 19 (1), 248. https://doi.org/10.1186/s12866-019-1618-0.

Giaouris, E., Heir, E., Desvaux, M., Hébraud, M., Møretrø, T., Langsrud, S., Doulgeraki, A., Nychas, G.J., Kačániová, M., Czaczyk, K., Ölmez, H., Simões, M., 2015. Intra- and interspecies interactions within biofilms of important foodborne bacterial pathogens. Frontiers in Microbiology 6, 841. https://doi.org/10.3389/fmicb.2015.00841.

Gómez, N.C., Ramiro, J.M., Quecan, B.X., de Melo Franco, B.D., 2016. Use of potential probiotic lactic acid bacteria (LAB) biofilms for the control of *Listeria monocytogenes, Salmonella* Typhimurium, and *Escherichia coli* O157:H7 biofilms formation. Frontiers in Microbiology 7, 863. https://doi.org/10.3389/fmicb.2016.00863.

Graham, C.E., Cruz, M.R., Garsin, D.A., Lorenz, M.C., 2017. *Enterococcus faecalis* bacteriocin EntV inhibits hyphal morphogenesis, biofilm formation, and virulence of *Candida albicans*. Proceedings of the National Academy of Sciences of the United States of America 114 (17), 4507–4512.

Guerrieri, E., de Niederhäusern, S., Messi, P., Sabia, C., Iseppi, R., Anacarso, I., Bondi, M., 2009. Use of lactic acid bacteria (LAB) biofilms for the control of *Listeria monocytogenes* in a small-scale model. Food Control 20 (9), 861–865.

Gutiérrez, S., Martínez-Blanco, H., Rodríguez-Aparicio, L.B., Ferrero, M.A., 2016. Effect of fermented broth from lactic acid bacteria on pathogenic bacteria proliferation. Journal of Dairy Science 99 (4), 2654–2665.

Hossain, M.I., Mizan, M.F.R., Ashrafudoulla, M., Nahar, S., Joo, H.-J., Jahid, I.K., Park, S.H., Kim, K.-S., Ha, S.-D., 2020. Inhibitory effects of probiotic potential lactic acid bacteria isolated from kimchi against *Listeria monocytogenes* biofilm on lettuce, stainless-steel surfaces, and MBEC™ biofilm device. LWT Food Science and Technology 118. https://doi.org/10.1016/j.lwt.2019.108864 (in press).

Hutchings, M.I., Truman, A.W., 2019. Antibiotics: past, present and future. Current Opinion in Microbiology 51, 72–80.

Jalilsood, T., Baradaran, A., Song, A.A., Foo, H.L., Mustafa, S., Saad, W.Z., Yusoff, K., Rahim, R.A., 2015. Inhibition of pathogenic and spoilage bacteria by a novel biofilm-forming *Lactobacillus* isolate: a potential host for the expression of heterologous proteins. Microbial Cell Factories 14, 96. https://doi.org/10.1186/s12934-015-0283-8.

Kanmani, P., Satish kumar, R., Yuvaraj, N., Paari, K.A., Pattukumar, V., Arul, V., 2011. Production and purification of a novel exopolysaccharide from lactic acid bacterium *Streptococcus phocae* PI80 and its functional characteristics activity *in vitro*. Bioresource Technology 102 (7), 4827–4833.

Kaur, S., Sharma, P., Kalia, N., Singh, J., Kaur, S., 2018. Anti-biofilm properties of the fecal probiotic lactobacilli against *Vibrio* spp. Frontiers in Cellular and Infection Microbiology 8, 120. https://doi.org/10.3389/fcimb.2018.00120.

Kim, Y., Oh, S., Kim, S.H., 2009. Released exopolysaccharide (r-EPS) produced from probiotic bacteria reduce biofilm formation of enterohemorrhagic *Escherichia coli* O157:H7. Biochemical and Biophysical Reseach Communications 379 (2), 324–329.

Kiymaci, M.E., Altanlar, N., Gumustas, M., Ozkan, S.A., Akin, A., 2018. Quorum sensing signals and related virulence inhibition of *Pseudomonas aeruginosa* by a potential probiotic strain's organic acid. Microbial Pathogenesis 121, 190–197.

Koluman, A., Dikici, A., 2013. Antimicrobial resistance of emerging foodborne pathogens: status quo and global trends. Critical Reviews in Microbiology 39 (1), 57–69.

Koohestani, M., Moradi, M., Tajik, H., Badali, A., 2018. Effects of cell-free supernatant of *Lactobacillus acidophilus* LA5 and *Lactobacillus casei* 431 against planktonic form and biofilm of *Staphylococcus aureus*. Veterinary Research Forum 9 (4), 301–306.

Kumar, L.M., Zuhainis Saad, W., Mohamad, R., Abdul Rahim, R., 2017. Influence of biofilm-forming lactic acid bacteria against methicillin-resistant *Staphylococcus aureus* (MRSA S547). Asian Pacific Journal of Tropical Biomedicine 7 (12), 1107–1115.

Leriche, V., Chassaing, D., Carpentier, B., 1999. Behaviour of *L. monocytogenes* in an artificially made biofilm of a nisin-producing strain of *Lactococcus lactis*. International Journal of Food Microbiology 51 (2–3), 169–182.

Li, W., Ji, J., Rui, X., Yu, J., Tang, W., Chen, X., Jiang, M., Dong, M., 2014. Production of exopolysaccharides by *Lactobacillus helveticus* MB2–1 and its functional characteristics *in vitro*. LWT - Food Science and Technology 59 (2), 732–739.

Liu, Z., Zhang, Z., Qiu, L., Zhang, F., Xu, X., Wei, H., Tao, X., 2017. Characterization and bioactivities of the exopolysaccharide from a probiotic strain of *Lactobacillus plantarum* WLPL04. Journal of Dairy Science 100 (9), 6895–6905.

Lynch, K.M., Zannini, E., Coffey, A., Arendt, E.K., 2018. Lactic acid bacteria exopolysaccharides in foods and beverages: isolation, properties, characterization, and health benefits. Annual Review of Food Science and Technology 9, 155–176.

Machado, I., Silva, L.R., Giaouris, E.D., Melo, L.F., Simões, M., 2020. Quorum sensing in food spoilage and natural-based strategies for its inhibition. Food Research International 127, 108754. https://doi.org/10.1016/j.foodres.2019.108754.

Mahdhi, A., Leban, N., Chakroun, I., Bayar, S., Mahdouani, K., Majdoub, H., Kouidhi, B., 2018. Use of extracellular polysaccharides, secreted by *Lactobacillus plantarum* and *Bacillus* spp., as reducing indole production agents to control biofilm formation and efflux pumps inhibitor in *Escherichia coli*. Microbial Pathogenesis 125, 448–453.

Mahdhi, A., Leban, N., Chakroun, I., Chaouch, M.A., Hafsa, J., Fdhila, K., Mahdouani, K., Majdoub, H., 2017. Extracellular polysaccharide derived from potential probiotic strain with antioxidant and antibacterial activities as a prebiotic agent to control pathogenic bacterial biofilm formation. Microbial Pathogenesis 109, 214–220.

Melo, T.A., Dos Santos, T.F., de Almeida, M.E., Junior, L.A., Andrade, E.F., Rezende, R.P., Marques, L.M., Romano, C.C., 2016. Inhibition of *Staphylococcus aureus* biofilm by *Lactobacillus* isolated from fine cocoa. BMC Microbiology 16 (1), 250. https://doi.org/10.1186/s12866-016-0871-8.

Merino, L., Trejo, F.M., De Antoni, G., Golowczyc, M.A., 2019. *Lactobacillus* strains inhibit biofilm formation of *Salmonella* sp. isolates from poultry. Food Research International 123, 258–265.

Minei, C.C., Gomes, B.C., Ratti, R.P., D'Angelis, C.E., De Martinis, E.C., 2008. Influence of peroxyacetic acid and nisin and coculture with *Enterococcus faecium* on *Listeria monocytogenes* biofilm formation. Journal of Food Protection 71 (3), 634–638.

Ndahetuye, J.B., Koo, O.K., O'Bryan, C.A., Ricke, S.C., Crandall, P.G., 2012. Role of lactic acid bacteria as a biosanitizer to prevent attachment of *Listeria monocytogenes* F6900 on deli slicer contact surfaces. Journal of Food Protection 75 (8), 1429–1436.

Nitschke, M., Silva, S.S.E., 2018. Recent food applications of microbial surfactants. Critical Reviews in Food Science and Nutrition 58 (4), 631–638.

Park, H., Yeo, S., Ji, Y., Lee, J., Yang, J., Park, S., Shin, H., Holzapfel, W., 2014. Autoinducer-2 associated inhibition by *Lactobacillus sakei* NR28 reduces virulence of enterohaemorrhagic *Escherichia coli* O157: H7. Food Control 45, 62–69.

Pelyuntha, W., Chaiyasut, C., Kantachote, D., Sirilun, S., 2019. Cell-free supernatants from cultures of lactic acid bacteria isolated from fermented grape as biocontrol against *Salmonella* Typhi and *Salmonella* Typhimurium virulence via autoinducer-2 and biofilm interference. PeerJ 7, e7555. https://doi.org/10.7717/peerj.7555.

Pérez-Ibarreche, M., Castellano, P., Leclercq, A., Vignolo, G., 2016. Control of *Listeria monocytogenes* biofilms on industrial surfaces by the bacteriocin-producing *Lactobacillus sakei* CRL1862. FEMS Microbiology Letters 363 (12). https://doi.org/10.1093/femsle/fnw118.

Riaz Rajoka, M.S., Jin, M., Haobin, Z., Li, Q., Shao, D., Jiang, C., Huang, Q., Yang, H., Shi, J., Hussain, N., 2018. Functional characterization and biotechnological potential of exopolysaccharide produced by *Lactobacillus rhamnosus* strains isolated from human breast milk. LWT Food Science and Technology 89, 638–647.

Rocha, K.R., Perini, H.F., de Souza, C.M., Schueler, J., Tosoni, N.F., Furlaneto, M.C., Furlaneto-Maia, L., 2019. Inhibitory effect of bacteriocins from enterococci on developing and preformed biofilms of *Listeria monocytogenes, Listeria ivanovii* and *Listeria innocua*. World Journal of Microbiology and Biotechnology 35 (7), 96. https://doi.org/10.1007/s11274-019-2675-0.

Schirmer, B.C., Heir, E., Møretrø, T., Skaar, I., Langsrud, S., 2013. Microbial background flora in small-scale cheese production facilities does not inhibit growth and surface attachment of *Listeria monocytogenes*. Journal of Dairy Science 96 (10), 6161−6171.

Sharma, D., Saharan, B.S., 2016. Functional characterization of biomedical potential of biosurfactant produced by *Lactobacillus helveticus*. Biotechnology Reports 11, 27−35.

Shokri, D., Khorasgani, M.R., Mohkam, M., Fatemi, S.M., Ghasemi, Y., Taheri-Kafrani, A., 2018. The inhibition effect of lactobacilli against growth and biofilm formation of *Pseudomonas aeruginosa*. Probiotics and Antimicrobial Proteins 10 (1), 34−42.

Singh, N., Kaur, R., Singh, B.P., Rokana, N., Goel, G., Puniya, A.K., Panwar, H., 2020. Impairment of *Cronobacter sakazakii* and *Listeria monocytogenes* biofilms by cell-free preparations of lactobacilli of goat milk origin. Folia Microbiologica (Praha) 65, 185−196.

Tan, X., Han, Y., Xiao, H., Zhou, Z., 2017. Pediococcus acidilactici inhibit biofilm formation of food-borne pathogens on abiotic surfaces. Transactions of Tianjin University 23 (1), 70−77.

Todorov, S.D., de Paula, O.A.L., Camargo, A.C., Lopes, D.A., Nero, L.A., 2018. Combined effect of bacteriocin produced by *Lactobacillus plantarum* ST8SH and vancomycin, propolis or EDTA for controlling biofilm development by *Listeria monocytogenes*. Revista Argentina de Microbiología 50 (1), 48−55.

van den Broek, M.F.L., De Boeck, I., Claes, I.J.J., Nizet, V., Lebeer, S., 2018. Multifactorial inhibition of lactobacilli against the respiratory tract pathogen *Moraxella catarrhalis*. Beneficial Microbes 9 (3), 429−439.

Walters, M., Sperandio, V., 2006. Quorum sensing in *Escherichia coli* and *Salmonella*. International Journal of Medical Microbiology 296 (2−3), 125−131.

Wang, J., Zhao, X., Yang, Y., Zhao, A., Yang, Z., 2015. Characterization and bioactivities of an exopolysaccharide produced by *Lactobacillus plantarum* YW32. International Journal of Biological Macromolecules 74, 119−126.

Winkelströter, L.K., Gomes, B.C., Thomaz, M.R.S., Souza, V.M., De Martinis, E.C.P., 2011. *Lactobacillus sakei* 1 and its bacteriocin influence adhesion of *Listeria monocytogenes* on stainless steel surface. Food Control 22 (8), 1404−1407.

Winkelströter, L.K., Tulini, F.L., De Martinis, E.C.P., 2015. Identification of the bacteriocin produced by cheese isolate *Lactobacillus paraplantarum* FT259 and its potential influence on *Listeria monocytogenes* biofilm formation. LWT − Food Science and Technology 64 (2), 586−592.

Woo, J., Ahn, J., 2013. Probiotic-mediated competition, exclusion and displacement in biofilm formation by food-borne pathogens. Letters in Applied Microbiology 56 (4), 307−313.

Yan, X., Gu, S., Cui, X., Shi, Y., Wen, S., Chen, H., Ge, J., 2019. Antimicrobial, anti-adhesive and anti-biofilm potential of biosurfactants isolated from *Pediococcus acidilactici* and *Lactobacillus plantarum* against *Staphylococcus aureus* CMCC26003. Microbial Pathogenesis 127, 12−20.

Yeganeh, M., Hosseini, H., Mehrabian, S., Torbati, E.S., Zamir, S.M., 2017. Antibiofilm effects of lactobacilli against ciprofloxacin-resistant uropathogenic *Escherichia coli* strains in pasteurized milk. Applied Food Biotechnology 4 (4), 241−250.

Yong, C.-C., Khoo, B.-Y., Sasidharan, S., Piyawattanametha, W., Kim, S.-H., Khemthongcharoen, N., Chuah, L.-O., Ang, M.-Y., Liong, M.-T., 2015. Activity of crude and fractionated extracts by lactic acid bacteria (LAB) isolated from local dairy, meat, and fermented products against *Staphylococcus aureus*. Annals of Microbiology 65 (2), 1037−1047.

Zanfardino, A., Criscuolo, G., Di Luccia, B., Pizzo, E., Ciavatta, M.L., Notomista, E., Carpentieri, A., Pezzella, A., Varcamonti, M., 2017. Identification of a new small bioactive peptide from *Lactobacillus gasseri* supernatant. Beneficial Microbes 8 (1), 133–141.

Zawistowska-Rojek, A., Tyski, S., 2018. Are probiotic really safe for humans? Polish Journal of Microbiology 67 (3), 251–258.

Zhao, T., Doyle, M.P., Zhao, P., 2004. Control of *Listeria monocytogenes* in a biofilm by competitive-exclusion microorganisms. Applied and Environmental Microbiology 70 (7), 3996–4003.

Zhao, T., Podtburg, T.C., Zhao, P., Chen, D., Baker, D.A., Cords, B., Doyle, M.P., 2013. Reduction by competitive bacteria of *Listeria monocytogenes* in biofilms and *Listeria* bacteria in floor drains in a ready-to-eat poultry processing plant. Journal of Food Protection 76 (4), 601–607.

Zhao, T., Podtburg, T.C., Zhao, P., Schmidt, B.E., Baker, D.A., Cords, B., Doyle, M.P., 2006. Control of *Listeria* spp. by competitive-exclusion bacteria in floor drains of a poultry processing plant. Applied and Environmental Microbiology 72 (5), 3314–3320.

Chapter 10

Role of equipment design in biofilm prevention

Thierry Bénézech, Christine Faille
Univ. Lille, CNRS, INRAE, ENSCL, UMET, Villeneuve d'Ascq, France

10.1 Introduction

Food contact surfaces in processing equipment are considered to be major factors in the risk of food contamination for the food and beverage industries. Over recent years, apart from numerous studies of surface disinfections, more and more interest has been shown in understanding both the mechanisms of equipment surface contamination and cleaning. This has led to a better definition of equipment geometry and construction materials for biofilm prevention. Hygienic design of food processing equipment is indeed now considered to be mandatory in the reduction of the risk of microbial food contamination. It is now urgent to find combined strategies combining hygienic design with, for instance, new cleansing processes to remove any remaining contaminants, to provide cost-effective results without causing any adverse effect on human health or on the environment.

This chapter mainly concentrates on closed (processing lines) or semiclosed equipment (tanks), looking at the design consequences (geometries and surfaces) on hygiene for (1) fully immersed surfaces and (2) surfaces concerned by air–liquid–material (ALM) interfaces. It also highlights, in the cases of both open and closed equipment, the consequences on hygiene of surfaces prone to drying after contamination (splash areas), which are also considered to be of hygienic design concern.

10.2 Simple equipment geometries

Hydrodynamics refer to both processing and cleaning operations for closed systems or possibly semiclosed ones, such as storage tanks and washing tanks. The interaction between the equipment design and the nature of the fluid flow and its consequences at the wall, in terms of adhesion/removal, is the main concern. The vast majorities of studies focus on cleaning, and the mechanisms described are simple geometries, such as pipes with or without dead ends,

Recent Trends in Biofilm Science and Technology. https://doi.org/10.1016/B978-0-12-819497-3.00010-6

protruding sensors, connections, and bends, which are still required, even in brand new processing lines. Indeed, pipework constitutes a significant part of processing lines as a whole.

Despite the number of studies dealing with turbulence generation and transport near the wall, few authors have looked into the adhesion and detachment mechanisms of bacterial cells in the confined area near wall zones. However, detachment was observed under a turbulent flow regime of lycopodium spores with an average diameter 10 times larger than the bacteria (Braaten et al., 1988). Generally, the size and the density of particles are considered as major parameters, which dictate the kind of interaction with coherent structures such as equipment walls. Larger particles reaching the external region of the boundary layer are subjected to the effect of the large-scale motions in the external region. This contrasts to smaller particles, probably located deeper in the viscous sublayer, where the instantaneous velocity distribution is linear, as in the case of bacteria spores or initial biofilm stages composed of a few cells directly attached to the surface or in small, flat clusters (Ziskind et al., 2000). In this zone, the effect of the large-scale motion on particle resuspension is excluded. Nevertheless, it was shown (Corino and Brodkey, 1969) that ejections and sweeps of coherent structures contribute to increases in the wall shear stress. Horseshoe or hairpin vortices are the most important flow patterns inducing the detachment of small particles from the wall due to their contribution to the shear stress forces generated (Ziskind et al., 2000). Thus, whatever the size of the particles, eddies generated near the wall induce fluctuating shear forces. When these hydrodynamic forces are greater than the adhesion ones, resuspension occurs.

A detailed comparison of the local wall shear stress and the cleanability experiments was proposed using different pieces of equipment in various circuit arrangements representative of production lines, including straight pipes of different diameters and sudden or gradual contraction or expansion pipes (Lelièvre et al., 2002). The effect of the mean wall shear stress on the removal efficiency had been confirmed in older studies, such as that by Grasshoff (Institution of Chemical Engineers (Great Britain). Food and Drink Subject Group. and EFCE Food Working Party. 1992). Hence, high values of the wall shear stress can be achieved using larger flow velocities and are used in some commercial cleaning in place (CIP) systems to reduce cleaning time (Celnik et al., 2006). However, more recently, Lelièvre et al. (2003) demonstrated a clear positive effect of an increase of the fluctuation rate on the spore detachment. Low mean wall shear stress zones as in sudden expansion at around 0.15 Pa (in processing lines the mean wall shear stress ranges from 1 up to 20 Pa) could be cleanable because of the high fluctuation rate (around 21%). These results underline the importance of the design and the flow configuration. Local wall shear stress compounds (mean value and fluctuations) and velocity profile measurements were used to determine flow repartition through loops containing bends and gradual expansions and contractions separated by straight pipes, as found in many pipework systems (Blel et al., 2007). The velocity profiles obtained showed a variation in local wall shear stress values between the lower, the upper, and the right sides of the systems, owing to the flow asymmetry inside the equipment tested. Boundary layer detachment inducing this asymmetry explains the nonuniform shear stress profiles observed essentially in the straight pipes. The effect of the recirculation zone occurring after bends was transmitted through the loops by means of other characteristics of the rig, thereby preventing flow establishment. Flow disturbance was also seen to be induced by sudden or gradual contractions or expansions, which affected the level of residual contamination of pipes placed downstream. These results were confirmed by means of velocity profiles and turbulence intensity repartitions through computational fluid dynamics (CFD) calculations. The wall shear stress and velocity profile measurements explained the levels of adhering spores before and after cleaning of a section of a dairy production line consisting of geometrical changes like gradual asymmetric expansion or contraction pipes and bends (Blel et al., 2008). For example, the confined zone observed near the gradual expansion is characterized by low-mean and fluctuating shear stress values that explain their poor cleanability. Soiling experiments emphasized the role of recirculation zones in the potential adhesion of spores due to low-shear stress components. Transport of spores from the solution to the pipe wall is a major parameter in surface contamination depending on the flow rate. Experiments also showed that a homogeneous contamination or cleanability level could be obtained depending on the applied turbulence intensity at the straight section far from geometrical changes. Cleanability experiments emphasized that the gradual expansion presents low-cleanability levels due to fluid separation, which leads to flow disturbance. This result depends essentially on equipment design, in particular the opening angle of the gradual expansion and the bend coil curvature. These parameters must be selected to avoid disturbing the flow and to induce a streamline separation. Electrochemical measurements showed that the limited flow disturbance with no boundary layer detachment from the wall observed at gasket and welding zones induced an increase in the mean wall shear stress. However, cleanability analyses at different positions in the horizontal and the vertical lines showed that these zones constitute deposit spots for spores due to the high surface roughness. Despite the high shear stress values generated by the applied flow, low-cleanability level is observed in these zones. In this case, cleanability is likely to be governed by the surface roughness parameters, especially when the average roughness exceeds the standard value of 0.8 μm. As a result, it is necessary to promote slight disturbances in the boundary layer through longer straight pipes. The use of progressive geometrical changes with small opening angles and high-surface quality could be recommended. The local wall shear stress repartition through a test loop was studied comparing numerical and experimental approaches (Blel et al., 2009). Misalignment can affect the wall shear stress and jeopardize any attempt through CFD

FIGURE 10.1 CFD calculation showing the flow organization after a bend in an enlargement of the pipe diameter (Laurent Bouvier, INRA PIHM). *CFD*, computational fluid dynamics.

approaches to improve piping cleanability in a food processing line. An example of CFD results describing the wall shear stress distribution is presented Fig. 10.1. Bends were also shown to cause nonhomogenous cleanability in the following pipe.

10.3 Complex equipment design

10.3.1 Immersed surfaces

Hydrodynamic conditions are known to impact biofilm dynamics at any stage, from the adhesion step to the appearance of fully developed biofilms. Much literature is available on this topic, mainly in environmental sciences. In bioreactors, the effect of the hydrodynamic conditions under which biofilms are formed has very recently been shown (Araújo et al., 2016; Lemos et al., 2015). The authors demonstrated that higher flow velocities can give rise to more complex and denser biofilms. All of these studies were done at the laboratory scale. Indeed, few works have looked into what happens in complex food equipment or processing lines. When working on the soiling of an industrial two-way valve by a suspension of *Bacillus cereus* spores in water, it was observed that in turbulent flow conditions, the transport process of particles is mainly governed by convection and diffusion. However, when viscous forces near the wall are strong enough, the transport mechanism is mainly carried out by diffusion throughout the boundary layer (Blel et al., 2010). The adhesion process occurs due to attraction forces (van der Waals and electrostatic forces) between particles and the wall. Moreover, for bacterial spores of approximately 1 μm width and 2 μm long, adhesion forces are complex due to their surface energies and the potential presence of specific 3D surface structures. When soiling the valve, interactions between suspended spores and coherent turbulent structures of different sizes allowed the authors to explain the eventual readhesion of spores in confined zones of the loop. These interactions also helped to explain the role of the bursting phenomenon in the increase of the wall shear stress, which induces spore detachment.

Concerning semiclosed equipment, biofilm growth dynamics have been studied under turbulent flow conditions on internal surfaces of mock-ups mimicking the design of washing tanks widely used in fresh-cut food industries (Cunault et al., 2015). These vats were filled with water containing *Pseudomonas fluorescens* cells and agitated by a Rushton impeller inducing a turbulent flow regime. The flow pattern inside the vats was significantly modified by the design of the vats (corners, folds, or flat surfaces). The authors demonstrated a strong link between the flow pattern induced by the design and the biofilm development dynamics. Using CFD, the different zones were ranked according to the contamination load. The differences observed were compared with the flow arrangement inside the vats and with the CFD evaluation of the wall shear stress. Recirculation areas' impact on the contamination load was emphasized by low wall shear stress conditions, such as in corners inducing high levels of attached bacteria over 107 CFU cm^{-2} after 72 h. Conversely, for flat and vertical walls affected by the highest mean wall shear stress values, which were 10 times more, the observed surface contamination load was reduced to less than 10^2 CFU cm^{-2} after 72 h. As previously observed (Blel et al., 2010), the wall shear stress parameter did not appear to explain this difference sufficiently. PIV (particle image velocimetry) measurements have shown that fluctuations in the local flow velocity occurred directly due to the impeller blade rotation. Such fluctuations over time, which have been shown to be a key factor in understanding the surface contamination removal (Blel et al., 2007), would probably deeply affect the initial contamination and the further biofilm growth dynamics, as demonstrated in the recent study (Araújo et al., 2016). In the aforementioned vats, biofilm structures depended on the location in the tanks, which could range from a discontinuous monolayer to large thick clusters (Cunault et al., 2019). Biofilms grown in zones subjected to shear were significantly more resistant to mechanical stress. Consequently, poor design features such as horizontal surfaces, right angles, welds, and corners at the vat bottom appeared critical in terms of hygiene, being more resistant to enzymatic cleaning.

Hygienic design and biofilm installation issues are of concern to the food industry and to related equipment suppliers. Close cooperation was found to be mandatory in the 1990s. The European hygienic engineering and design group (www.ehedg.org) has over the past three decades proposed 50 guides regarding hygienic design principles from the basics (Document 8) to those relating to specific machineries and the most recent being devoted to CIP installations (Document 50). The latter relies on research (such as that) by Le Gentil et al. (2010). These authors demonstrated a deposition phenomenon of bacteria contaminants during CIP procedures following experiments focused on the surface contamination during the CIP procedure of pipes and of a two-way valve (cleaned and disinfected before the test) coming from soiled pipes (around 5×10^6 CFU cm^{-2}) inserted in a clean loop. They further observed that readhesion was governed by the flow pattern, the valve being significantly contaminated after CIP.

10.3.2 Air–liquid–material interfaces

In contrast to fully immersed surfaces, biofilms at ALM interfaces (ALM biofilms) have received relatively little attention in the literature. Furthermore, the investigation of bacterial contamination of ALM interfaces has almost exclusively been dedicated to biofilms produced in laboratory conditions, for example in microtiter plates or in tubes (Srinandan et al., 2015). It is now widely accepted that a ringlike structure often forms at the air–liquid interface on the surface of wells or tubes and that these biofilms might sometimes (Scher et al., 2005), but not always (Römling et al., 2000) provide an attachment site for the subsequent pellicle formation at the liquid surface. This interfacial area would indeed be favorable to bacterial adhesion and biofilm formation, probably because bacteria thrive better in environments where oxygen is available than on submerged surfaces. Other environmental parameters could affect the formation of rings (vs. totally immersed biofilms), such as the source of carbon or energy (Sutrina et al., 2019) or the growth medium (Mosharaf et al., 2018). Some bacterial properties play also a major role, such as motility (Majed et al., 2016; Wijman et al., 2007) or multicellular behavior resulting from the presence of thin aggregative fimbriae (Römling et al., 2000) or curli (Weiss-Muszkat et al., 2010). Under laboratory conditions, bacterial species able to produce ALM biofilms include foodborne pathogens such as *Salmonella* Typhimurium (Sutrina et al., 2019), *Salmonella* Enteritidis (Giaouris and Nychas, 2006), or *Klebsiella pneumoniae* (Mosharaf et al., 2018), as well as nonpathogenic bacteria often isolated from food environments such as *Pseudomonas* sp. (Mosharaf et al., 2018; Robertson et al., 2013). However, marked differences in the ability to form ALM biofilms (as with most bacterial properties) have been also reported between strains belonging to a single species, e.g., *Bacillus cereus* (Wijman et al., 2007). Some authors evaluated the benefit from the interface air–liquid in terms of amount of biofilm produced on surfaces. For example, a *S.* Enteritidis strain was shown to produce 102 to 103 times more biofilm on half-immersed stainless steel coupons, than on fully submerged coupons (Giaouris and Nychas, 2006). Similar observations were made on *B. cereus* strains (Wijman et al., 2007), which clearly form large amounts of ALM biofilms with local bacterial concentration at the air–liquid interface up to 108 cells cm^{-2}.

This type of air–liquid interfaces are omnipresent, particularly in food workshops, in partly filled devices, such as washing or storage tanks and piping systems that are only partly filled during a production cycle, or where some residual fluid has remained after operation (Wijman et al., 2007). As regards *B. cereus* in dairy factories, for example, biofilms are found mainly at the air–liquid interface with a typical ring attached to the wall, from which the bacterial biofilm matrix protrudes onto the liquid surface (Fagerlund et al., 2014). However, phenomena occurring at interfaces in conditions close to real environments have so far received only little attention despite the challenge in terms of surface hygiene. To our knowledge, with the exception of a few experiments on biofilm formation on partially submerged stainless steel coupons (Wijman et al., 2007), the only data available in the literature are from our laboratory. We investigated the formation, structure, and resistance to cleaning of ALM biofilm in conditions close to those encountered in the fresh-cut food environment and more precisely in the industrial washing tanks. Our experiments were performed on a pilot rig composed of a series of test vats filled by cascade flow providing 3D flow conditions, where fluid was agitated by a Rushton impeller (Cunault et al., 2015). In these conditions, a *P. fluorescens* strain was shown to produce large amounts of ALM biofilms (corresponding in this study to intermittently submerged areas) and the structure of the biofilms greatly differed from that produced in fully submerged conditions. Indeed, thick patches were easily observed under microscope in the ALM biofilms, but not in submerged biofilms (Cunault et al., 2018). Even more seriously, these ALM biofilms were highly resistant to rinsing and cleaning procedures (Cunault et al., 2019). Works in progress in our laboratory also clearly demonstrate the main role of material surface properties on the amount and organization at the interface (Fig. 10.2).

Therefore, in addition to those areas identified above as poorly hygienic (welds, corners, horizontal surfaces), interface zones should also be considered as places conducive to the installation of resistant bacterial contamination.

10.3.3 Splash areas

In many environments, contaminated surfaces are subjected to drying, and many bacteria are able to withstand such periods of desiccation (Faille et al., 2018). In food processing facilities, these surfaces include open processing surfaces such as conveyor belts or cutting boards, and areas splashed with food or water during the process or even with detergents or other products used in the course of hygiene procedures.

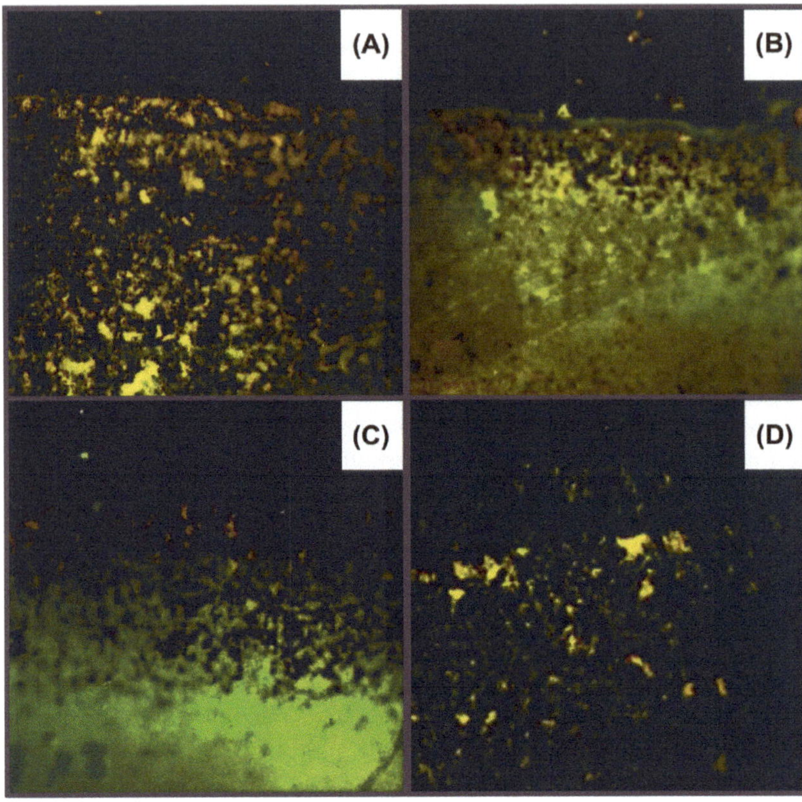

FIGURE 10.2 Images by epifluorescence microscopy of biofilms of *Bacillus cereus* CUETM 98/4 formed at the interface air–liquid on stainless steel 2B (A), on stainless steel 2B-Scotch Brite (B), on stainless steel 2R (C), and on polypropylene (D).

Unfortunately, many bacteria can easily survive periods of desiccation. This is the case, for example, for pathogenic bacteria subjected to desiccation in food environments (Ronan et al., 2013; Stone et al., 2016) or in the medical sector (Xie et al., 2006). This ability to withstand desiccation is of course strain/species dependent, but other parameters such as relative humidity and material hygroscopicity could play a crucial role both in the persistent metabolic rate and in the higher survival rate (Stone et al., 2016). Drying of contaminated surfaces could result in major consequences in terms of hygiene. Indeed, a recent study performed in our laboratory on bacterial spores evidenced that the interaction strength between material and spores increased with drying (time and temperature), which in turn resulted in a reduced ease of cleaning (Faille et al., 2016). The shape and organization of the bacterial deposit after drying is obviously affected by the material surface properties, topography versus physicochemistry. An example of droplets containing Polybeads microspheres (Polysciences Inc.) is given in Fig. 10.3. 1-μL droplets were placed at the surface of glass

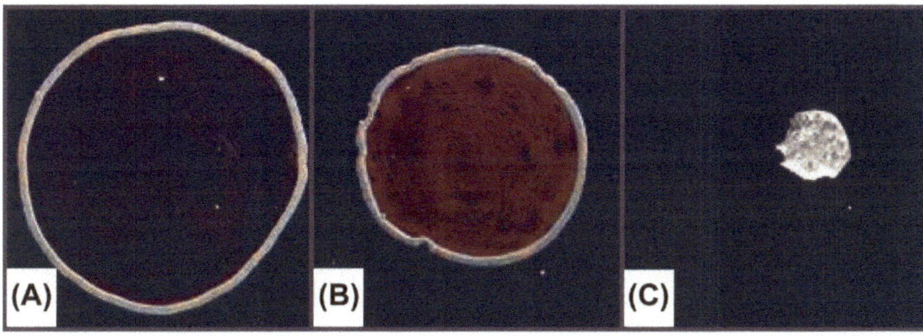

FIGURE 10.3 Images by stereoscopic microscopy of the residue patterns formed from evaporating drops on glass (A), stainless steel 2R (B), and polypropylene (C).

(hydrophilic), stainless steel (intermediate), and Teflon (hydrophobic). After complete drying at 30°C, a ringlike structure was clearly observed on glass and stainless steel, and the ring diameter is significantly greater on glass. Within the external ring, the inner surface was also contaminated, but no clear structure could be observed. Conversely, the deposition pattern is smaller on polypropylene, almost all the microspheres being assembled on a narrow area more or less at the center of the initial interface liquid–solid of the droplet. It therefore seems highly probable that the width of the residual deposit would be affected by the material properties, which may result in a higher protection against inactivation of the bacteria hidden in the depth of the deposit and probably in a higher resistance to detachment. Other works reported in the literature evidenced the main role of material properties on the deposition pattern of bacteria. The hydrophilic character of the material (between 5 degrees and 20 degrees) was demonstrated to affect the size of the deposit of *P. aeruginosa* as well as the inner pattern, from multirings to honeycomb or cellular patterns (Baughman et al., 2010). What has been more unexpected is the role played by the bacterial motility on the ring formation, first highlighted on various strains of *E. coli* and *Lactobacillus salivarius*, both motile and nonmotile (Nellimoottil et al., 2007), and confirmed more recently on swimming/nonswimming *E. coli* (Kasyap et al., 2014). It is thus possible that the physiological state of the bacteria, for example, before or after a disinfection procedure might affect the deposition pattern and perhaps the further resistance to cleaning and disinfection procedures.

That being said, surface contamination of splash areas in food environments is assumed to be unable to return to the product (ISO 14159:2008) or to food-contact surfaces. However, cross-contaminations could occur through draining, condensation, or splash-back. These areas where a drying process occurs must therefore be of concern in the hygiene management in the food industries. Indeed, institutions and stakeholders involved in food hygiene take the potential risk very seriously that these newly contaminated surfaces may be responsible for food poisoning. Advice can be found in various guidance documents and regulatory information. One good example is a guide proposed by the Food and Drug Administration in 1998, which indicates that spray wash treatments may spread pathogens by splashing or by aerosol, or on food contact surfaces. That was reinforced by experimental assays on the spread of contamination by aerosols. For example, the spread of *Listeria monocytogenes* was shown to occur from contaminated drains to food surfaces or food through high-pressure cleaning, which generates aerosols (Saini et al., 2012).

Another concern in surface hygiene is that these dried deposits may be difficult to detach by the tools (swabs, sponges, Rodac plates, and so on) used routinely to monitor the efficiency of surface cleaning. Therefore, some bacteria contamination may go undetected, thereby allowing them to act as reservoirs for further cross-contamination.

Accordingly, it is essential to avoid splashing equipment surfaces in particular with aerosols from floor and wall cleaning.

10.4 Material properties

The ability of microorganisms to contaminate all kinds of surfaces in food industry plants varies greatly according to the substrata. Along with their chemical composition, surface properties of materials used in the food industry also vary greatly. Some of them, e.g., physicochemistry and topography, have been assumed to affect interactions between microorganisms and materials, in terms of surface contamination and/or further ease of cleaning.

10.4.1 Physicochemistry

Concerning physicochemical interactions, e.g., acid–base, electrostatic, or hydrophobic interactions, two major theories have been proposed for predicting the level of bacterial attachment to inert surfaces: The Derjaguin–Landau–Verwey–Overbeek (DLVO) theory describes the total Gibbs energy of interaction between a microorganism and a surface as a function of the separation distance (Hermansson, 1999) and the thermodynamic theory based on the comparison of the interfacial surface free energies for bacteria in suspension and bacteria in an adhered state (van Oss, 1997). According to van Oss, hydrophobic interactions would even be the strongest long-range noncovalent interactions in biological systems. Indeed, hydrophobic interactions play a major role, due to the unusually strong attraction of hydrophobic surfaces in water (Meyer et al., 2006). However, even if much has been written on the role of substratum hydrophobicity in the attachment of bacteria and in the formation of biofilms, there are a number of inherent contradictions. Such discrepancies could be partly explained by bacterial properties, e.g., the presence of fimbriae (Otto et al., 1999) or of a mucous layer (Faille et al., 2014). Another hypothesis explaining these discrepancies between models and real phenomena is that they probably reflect the nonhomogeneous character of surface free energies providing localized sites favoring bacterial attachment or retention, e.g., at stainless steel grain boundaries (Geesey et al., 1996). Micro- or nanoscale heterogeneities in substratum physicochemistry have been proven to affect bacterial adhesion (Bos, 2000; Galopin et al., 2010; Kalasin et al., 2010). These aspects have as yet been poorly investigated, despite their possible consequences in terms of bacterial contamination of heterogeneous substrata.

To complexify this topic, substratum is continually subjected to surface conditioning resulting from the rapid adsorption of materials from the liquids in contact (proteins, fat, minerals, detergents, surfactants, disinfectants, etc.) affecting the physicochemistry of the materials' surfaces. To condition stainless steel surfaces, some authors (Jullien et al., 2008) subjected stainless steel surfaces to up to 25 successive soiling (with milk or meat) and cleaning (with detergent) runs in a pilot rig. They observed a significant decrease in the hygienic status of the stainless steel surfaces for both conditioning procedures (milk or meat). Hence, such modifications resulting from the conditioning film were shown to be bacteria dependent (Palmer et al., 2007).

To conclude, surface physicochemistry and consequently surface hygiene in food environments may differ radically, depending on the food chain and the related processes.

10.4.2 Topography

In addition to the physicochemical properties, substratum topography has been recognized to largely influence bacterial contamination. Despite numerous data provided by the literature, the mechanisms by which surface topography modulates bacterial contamination remain largely unclear. One of the difficulties in describing these mechanisms could be that for most reported works, the surface topography was solely described by the parameter Ra (average roughness, arithmetical mean deviation of the absolute ordinate values within a sampling length). Some results suggested that bacterial adhesion/colonization of stainless steel surfaces (Ra ranging from 0.06 to 0.89 μm) decreases with Ra, until a threshold average roughness of 0.16 μm, below which no further reduction could be expected (Medilanski et al., 2002), while others (Hilbert et al., 2003) failed to observe any effect on the adhesion and removal by a cleaning procedure (Ra ranging from 0.01 to 0.9 μm). If Ra is also reported as a relevant parameter when labeling a surface as hygienic (0.8 μm) according to the EHEDG principles, this parameter is unable to account for the shape of the surface irregularities, or even the presence of some rare irregularities, even though these may play a significant role in surface hygiene (Faille et al., 2000). Furthermore, adherent bacteria have often been associated with surface features similar in size to, or slightly larger than the cells or spores involved, as observed in the bottom of crevices (Díaz et al., 2007), or in the depth of the grain boundaries and crevices (personal communication, Fig. 10.4). The relevance of other roughness parameters than Ra, such as those derived from the Abbott-Firestone curve, e.g., reduced valley depth (Rvk) or reduced peak

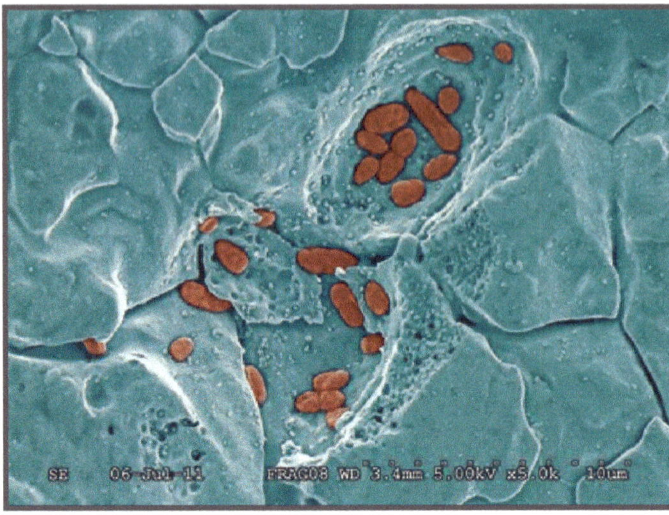

FIGURE 10.4 Observation by scanning electron microscopy of residual Pseudomonas fragi cells trapped in a surface default of stainless steel with a 2B finish.

height (Rpk) described in the standard ISO 13565-2:1996 have proven to be better linked to the hygienic status of the materials (Faille et al., 2000; Jullien et al., 2003).

To gain the best possible understanding of the role of specific features in bacterial adhesion, surfaces with regular, well-defined topographical patterns were produced and investigated for their hygienic status. Using microbial retention assays with a panel of differently sized microorganisms, it was possible to highlight the links between the sizes of both surface defects and bacterial cells explaining the resistance to removal of adherent bacteria (Whitehead et al., 2005). Spatially organized microtopographic surface patterns reduced bacterial adhesion (30%–45%) more than the smooth control samples, whatever the bacterial shape or surface energy (Perera-Costa et al., 2014). Further study is still needed to validate the effectiveness of such new patterned materials when potentially subjected to soiling by various food residues or polymeric substances from biofilms, which are sometimes resistant to cleaning.

10.5 Conclusion

Hygienic design of food processing equipment is nowadays considered to be mandatory in the reduction of the risk of microbial food contamination. Equipment design indeed plays a major role in contamination trapping or in the ease of removal via the flow arrangements, notably during cleaning and rinsing operations. More detailed knowledge of the flow mechanical actions combined with the unavoidable chemical action of the detergents would allow equipment manufacturers to go further in enhancing certain "good" phenomena by means of changing geometries, e.g., fluctuating wall shear stress induced by actively creating flow recirculation phenomena. The presentation of potential roles of (1) interfaces between ALM, (2) materials properties on the remaining bacterial soil, after soiling and cleaning, will also provide new insights when envisaging any hygienic improvements by means of geometries and material in use. Last but not least are those possibilities offered by a hygienically designed processing line, in the quest for environmental impact mitigation of cleaning operations, i.e., drinking water and energy reductions, or the use of new "green" mild detergents.

Acknowledgments

This work was funded by the European Research Project SUSCLEAN (contract number FP7-KBBE-2011-5, project number: 28514) and the project Veg-I-Tec (programme Interreg V France-Wallonia-Flanders GoToS3NuTeX).

References

Araújo, P.A., Malheiro, M., Machado, I., Mergulhão, F., Melo, L., Simões, M., 2016. Influence of flow velocity on the characteristics of *Pseudomonas fluorescens* biofilms. Journal of Environmental Engineering 142 (7), 04016031.

Baughman, K.F., Maier, R.M., Norris, T.A., Beam, B.M., Mudalige, A., Pemberton, J.E., Curry, J.E., 2010. Evaporative deposition patterns of bacteria from a sessile drop: effect of changes in surface wettability due to exposure to a laboratory atmosphere. Langmuir 26 (10), 7293–7298.

Blel, W., Bénézech, T., Legentilhomme, P., Legrand, J., Le Gentil-Lelièvre, C., 2007. Effect of flow arrangement on the removal of *Bacillus* spores from stainless steel equipment surfaces during a cleaning in place procedure. Chemical Engineering Science 62 (14), 3798–3808.

Blel, W., Legentilhomme, P., Legrand, J., Bénézech, T., Le Gentil-Lelièvre, C., 2008. Hygienic design: effect of hydrodynamics on the cleanability of a food processing line. AIChE Journal 54 (10), 2553–2566.

Blel, W., Pierrat, D., Le Gentil, C., Legentilhomme, P., Legrand, J., Hermon, C., Faille, C., Bénézech, T., 2009. Numerical and experimental investigations of the flow structures through a gradual expansion pipe. Trends in Food Science and Technology 20 (Suppl. 1).

Blel, W., Legentilhomme, P., Le Gentil-Lelièvre, C., Faille, C., Legrand, J., Bénézech, T., 2010. Cleanability study of complex geometries: interaction between *Bacillus cereus* spores and the different flow eddies scales. Biochemical Engineering Journal 49 (1), 40–51.

Bos, R., 2000. Retention of bacteria on a substratum surface with micro-patterned hydrophobicity. FEMS Microbiology Letters 189 (2), 311–315.

Braaten, D.A., Paw, U.K.T., Shaw, R.H., 1988. Coherent turbulent structures and particle detachment in boundary layer flows. Journal of Aerosol Science 19 (7), 1183–1186.

Celnik, M.S., Patel, M.J., Pore, M., Scott, D.M., Wilson, D.I., 2006. Modelling laminar pulsed flow for the enhancement of cleaning. Chemical Engineering Science 61 (6), 2079–2084.

Corino, E.R., Brodkey, R.S., 1969. A visual investigation of the wall region in turbulent flow. Journal of Fluid Mechanics 37 (1), 1–30.

Cunault, C., Faille, C., Bouvier, L., Föste, H., Augustin, W., Scholl, S., Debreyne, P., Benezech, T., 2015. A novel set-up and a CFD approach to study the biofilm dynamics as a function of local flow conditions encountered in fresh-cut food processing equipments. Food and Bioproducts Processing 93, 217–223.

Cunault, C., Faille, C., Briandet, R., Postollec, F., Desriac, N., Benezech, T., 2018. *Pseudomonas* sp. biofilm development on fresh-cut food equipment surfaces – a growth curve – fitting approach to building a comprehensive tool for studying surface contamination dynamics. Food and Bioproducts Processing 107, 70–87.

Cunault, C., Faille, C., Calabozo-Delgado, A., Benezech, T., 2019. Structure and resistance to mechanical stress and enzymatic cleaning of *Pseudomonas* fluorescens biofilms formed in fresh-cut ready to eat washing tanks. Journal of Food Engineering 262, 154–161.

Díaz, C., Schilardi, P.L., Salvarezza, R.C., Lorenzo de Mele, M.F., 2007. Nano/microscale order affects the early stages of biofilm formation on metal surfaces. Langmuir 23 (22), 11206–11210.

Fagerlund, A., Dubois, T., Økstad, O.A., Verplaetse, E., Gilois, N., Gominet, M., Kolstø, A.B., 2014. SinR controls enterotoxin expression in *Bacillus thuringiensis* biofilms. PLoS One 9 (1), e96707.

Faille, C., Membre, J.M., Tissier, J.P., Bellon-Fontaine, M.N., Carpentier, B., Laroche, M.A., Benezech, T., 2000. Influence of physicochemical properties on the hygienic status of stainless steel with various finishes. Biofouling 15 (4), 261–274.

Faille, C., Ronse, A., Dewailly, E., Slomianny, C., Maes, E., Krzewinski, F., Guerardel, Y., 2014. Presence and function of a thick mucous layer rich in polysaccharides around *Bacillus subtilis* spores. Biofouling 30 (7), 845–858.

Faille, C., Bihi, I., Ronse, A., Ronse, G., Baudoin, M., Zoueshtiagh, F., 2016. Increased resistance to detachment of adherent microspheres and *Bacillus* spores subjected to a drying step. Colloids and Surfaces B: Biointerfaces 143, 293–300.

Faille, C., Cunault, C., Dubois, T., Bénézech, T., 2018. Hygienic design of food processing lines to mitigate the risk of bacterial food contamination with respect to environmental concerns. Innovative Food Science and Emerging Technologies 46, 65–73.

Galopin, E., Piret, G., Szunerits, S., Lequette, Y., Faille, C., Boukherroub, R., 2010. Selective adhesion of *Bacillus cereus* spores on heterogeneously wetted silicon nanowires. Langmuir 26 (5), 3479–3484.

Geesey, G.G., Gillis, R.J., Avci, R., Daly, D., Hamilton, M., Shope, P., Harkin, G., 1996. The influence of surface features on bacterial colonization and subsequent substratum chemical changes of 316L stainless steel. Corrosion Science 38 (1), 73–95.

Giaouris, E.D., Nychas, G.J.E., 2006. The adherence of *Salmonella* enteritidis PT4 to stainless steel: the importance of the air-liquid interface and nutrient availability. Food Microbiology 23 (8), 747–752.

Hermansson, M., 1999. The DLVO theory in microbial adhesion. Colloids and Surfaces B: Biointerfaces 14 (1–4), 105–119.

Hilbert, L.R., Bagge-Ravn, D., Kold, J., Gram, L., 2003. Influence of surface roughness of stainless steel on microbial adhesion and corrosion resistance. International Biodeterioration & Biodegradation 52 (3), 175–185.

Institution of Chemical Engineers (Great Britain) Food and Drink Subject Group and EFCE Food Working Party, 1992. Food Engineering in a Computer Climate. Institution of Chemical Engineers.

Jullien, C., Bénézech, T., Carpentier, B., Lebret, V., Faille, C., 2003. Identification of surface characteristics relevant to the hygienic status of stainless steel for the food industry. Journal of Food Engineering 56 (1), 77–87.

Jullien, C., Benezech, T., Gentil, C.L., Boulange-Petermann, L., Dubois, P.E., Tissier, J.P., Traisnel, M., Faille, C., 2008. Physico-chemical and hygienic property modifications of stainless steel surfaces induced by conditioning with food and detergent. Biofouling 24 (3), 163–172.

Kalasin, S., Dabkowski, J., Nusslein, K., Santore, M.M., 2010. The role of nano-scale heterogeneous electrostatic interactions in initial bacterial adhesion from flow: a case study with *Staphylococcus aureus*. Colloids and Surfaces Part B-Biointerfaces 76 (2), 489–495.

Kasyap, T.V., Koch, D.L., Wu, M., Kasyap, T.V., Koch, D.L., Wu, M., 2014. Bacterial collective motion near the contact line of an evaporating sessile drop bacterial collective motion near the contact line of an evaporating sessile drop. Physics of Fluids 26 (11), 1–7.

Le Gentil, C., Sylla, Y., Faille, C., 2010. Bacterial re-contamination of surfaces of food processing lines during cleaning in place procedures. Journal of Food Engineering 96 (1), 37–42.

Lelièvre, C., Antonini, G., Faille, C., Bénézech, T., 2002. Cleaning-in-place: modelling of cleaning kinetics of pipes soiled by *Bacillus* spores assuming a process combining removal and deposition. Food and Bioproducts Processing: Transactions of the Institution of Chemical Engineers, Part C 80 (4), 305–311.

Lelièvre, C., Legentilhomme, P., Legrand, J., Faille, C., Bénézech, T., 2003. Hygienic design: influence of local wall shear stress variations on the cleanability of a three-way valve. Chemical Engineering Research and Design 81 (9), 1071–1076.

Lemos, M., Mergulhão, F., Melo, L., Simões, M., 2015. The effect of shear stress on the formation and removal of *Bacillus cereus* biofilms. Food and Bioproducts Processing 93, 242–248.

Majed, R., Faille, C., Kallassy, M., Gohar, M., 2016. *Bacillus cereus* biofilms-same, only different. Frontiers in Microbiology 7, 1–16.

Medilanski, E., Kaufmann, K., Wick, L.Y., Wanner, O., Harms, H., 2002. Influence of the surface topography of stainless steel on bacterial adhesion. Biofouling 18 (3), 193–203.

Meyer, E.E., Rosenberg, K.J., Israelachvili, J., 2006. Recent progress in understanding hydrophobic interactions. Proceedings of the National Academy of Sciences 103 (43), 15739–15746.

Mosharaf, M.K., Tanvir, M.Z.H., Haque, M.M., Haque, M.A., Khan, M.A.A., Molla, A.H., Alam, M.Z., Islam, M.S., Talukder, M.R., 2018. Metal-adapted bacteria isolated from wastewaters produce biofilms by expressing proteinaceous curli fimbriae and cellulose nanofibers. Frontiers in Microbiology 9, 1–17.

Nellimoottil, T.T., Rao, P.N., Ghosh, S.S., Chattopadhyay, A., 2007. Evaporation-induced patterns from droplets containing motile and nonmotile bacteria. Langmuir 23 (17), 8655–8658.

Otto, K., Elwing, H., Hermansson, M., 1999. The role of type 1 fimbriae in adhesion of *Escherichia coli* to hydrophilic and hydrophobic surfaces. Colloids and Surfaces B: Biointerfaces 15 (1), 99–111.

Palmer, J., Flint, S., Brooks, J., 2007. Bacterial cell attachment, the beginning of a biofilm. Journal of Industrial Microbiology & Biotechnology 34 (9), 577–588.

Perera-Costa, D., Bruque, J.M., González-Martín, M.L., Gómez-García, A.C., Vadillo-Rodríguez, V., 2014. Studying the influence of surface topography on bacterial adhesion using spatially organized microtopographic surface patterns. Langmuir 30 (16), 4633–4641.

Robertson, M., Hapca, S.M., Moshynets, O., Spiers, A.J., 2013. Air-liquid interface biofilm formation by psychrotrophic Pseudomonads recovered from spoilt meat. Antonie van Leeuwenhoek 103 (1), 251–259.

Römling, U., Rohde, M., Olsen, A., Normark, S., Reinköster, J., 2000. AgfD, the checkpoint of multicellular and aggregative behaviour in *Salmonella typhimurium* regulates at least two independent pathways. Molecular Microbiology 36 (1), 10–23.

Ronan, E., Yeung, C.W., Hausner, M., Wolfaardt, G.M., 2013. Interspecies interaction extends bacterial survival at solid-air interfaces. Biofouling 29 (9), 1087–1096.

Saini, J.K., Marsden, J.L., Fung, D.,Y.C., Crozier-dodson, B.A., 2012. Evaluation of potential for translocation of *Listeria monocytogenes* from floor drains to food contact surfaces in the surrounding environment using *Listeria innocua* as a surrogate. Advances in Microbiology 2 (4), 565–570.

Scher, K., Romling, U., Yaron, S., 2005. Effect of heat, acidification, and chlorination on *Salmonella enterica* serovar *typhimurium* cells in a biofilm formed at the air-liquid interface. Applied and Environmental Microbiology 71 (3), 1163–1168.

Srinandan, C.S., Elango, M., Gnanadhas, D.P., Chakravortty, D., 2015. Infiltration of matrix-non-producers weakens the *Salmonella* biofilm and impairs its antimicrobial tolerance and pathogenicity. Frontiers in Microbiology 6, 1–12.

Stone, W., Kroukamp, O., Korber, D.R., McKelvie, J., Wolfaardt, G.M., 2016. Microbes at surface-air interfaces: the metabolic harnessing of relative humidity, surface hygroscopicity, and oligotrophy for resilience. Frontiers in Microbiology 7, 1–15.

Sutrina, S.L., Callender, S., Grazette, T., Scantlebury, P., Neal, S.O., Thomas, K., Harris, D.C., Mota-Meira, M., 2019. The quantity and distribution of biofilm growth of *Escherichia coli* strain ATCC 9723 depends on the carbon/energy source. Microbiology 165 (1), 47–64.

van Oss, C.J., 1997. Kinetics and energetics of specific intermolecular interactions. Journal of Molecular Recognition 10 (5), 203–216.

Weiss-Muszkat, M., Shakh, D., Zhou, Y., Pinto, R., Belausov, E., Chapman, M.R., Sela, S., 2010. Biofilm formation by and multicellular behavior of *Escherichia coli* O55:H7, an atypical enteropathogenic strain. Applied and Environmental Microbiology 76 (5), 1545–1554.

Whitehead, K.A., Colligon, J., Verran, J., 2005. Retention of microbial cells in substratum surface features of micrometer and sub-micrometer dimensions. Colloids and Surfaces B: Biointerfaces 41 (2–3), 129–138.

Wijman, J.G., de Leeuw, P.P., Moezelaar, R., Zwietering, M.H., Abee, T., 2007. Air-Liquid interface biofilms of *Bacillus cereus*: formation, sporulation, and dispersion. Applied and Environmental Microbiology 73 (5), 1481–1488.

Xie, X., Li, Y., Zhang, T., Fang, H.H., 2006. Bacterial survival in evaporating deposited droplets on a teflon-coated surface. Applied Microbiology and Biotechnology 73, 703–712.

Ziskind, G., Fichman, M., Gutfinger, C., 2000. Particle behavior on surfaces subjected to external excitations. Journal of Aerosol Science 31 (6), 703–719.

Chapter 11

Biofilm control with enzymes

Anabela Borges, Ana Meireles, Filipe Mergulhão, Luís Melo, Manuel Simões
LEPABE — Laboratory for Process Engineering Environment, Biotechnology and Energy, Department of Chemical Engineering, Faculty of Engineering, University of Porto, Porto, Portugal

11.1 Biofilms and problems associated with their control

The resistance of bacteria to current antibiotics is a serious public health problem that has been increasingly documented. This situation is aggravated when bacteria become resistant to several antimicrobials simultaneously and when they form a biofilm (Davies and Davies, 2010; Mah and O'Toole, 2001). Indeed, in most of the cases, the limited efficacy of antimicrobial treatments is related to the presence of biofilms. Biofilms are defined as a structured community of microorganisms that are irreversibly attached to an inanimate or living surface and encapsulated within a self-produced polymeric matrix (Stoodley et al., 2002; Simões, 2011; Costerton et al., 1995; Cos et al., 2010). These sessile communities represent the prevalent mode of microbial life, occurring in both natural and man-made environments on a wide variety of surfaces, including living tissues, indwelling medical devices, industrial/potable water system piping, and natural aquatic systems (Donlan, 2002; Percival et al., 2011).

The development of a mature biofilm is a dynamic and multicellular process that is affected largely by the characteristics of the surface where the attachment occurs (e.g., texture, roughness, chemical composition, physicochemical properties), the bacterial cells involved, the environmental conditions (e.g., oxygen level, shear force, nutrients, pH), and genetic factors (expression of biofilm essential genes) (Donlan and Costerton, 2002; Percival et al., 2011). Biofilm formation includes a series of physical, biological, and chemical processes, occurring in various sequential steps that involve development of a conditioning film, movement of microorganisms to surfaces followed by initial microbial attachment, cell−cell communication, formation of microcolonies, production of extracellular polymeric substances (EPS), biofilm maturation, and detachment (Simões, 2011; Simões et al., 2009). Therefore, biofilm formation is a developmental process mediated by a combination of adhesion mechanisms, EPS synthesis, bacterial motility, and quorum sensing (QS) phenomena, which are intrinsically related (Landini et al., 2010).

Recent Trends in Biofilm Science and Technology. https://doi.org/10.1016/B978-0-12-819497-3.00011-8

Biofilms can be formed by a single or multiple bacterial species of both commensal and pathogenic bacteria. However, in natural environments, biofilm communities are normally inhabited by different species in close proximity. Hence, spatial and metabolic interactions between species may contribute to the organization of multispecies biofilms and to the production of dynamic local environment (Nadell et al., 2009). This lifestyle is very different from the planktonic mode of growth in terms of behavior, structure, and physiology (Stoodley et al., 2002). Indeed, bacteria within biofilms experiment many alterations in gene regulation that lead biofilm cells to become phenotypically and metabolically different from their planktonic counterparts (Stoodley et al., 2002; Patel, 2005). In biofilms, microbial cells perform several functions (e.g., increased access to nutrients/water; mutualistic or synergistic associations—community benefits; increased expression of beneficial genes) that would not occur if the cells were alone or outside of this sessile community. Moreover, biofilm-growing microorganisms are more protected from environmental stresses, predators, and host defenses, thereby creating the ideal conditions for their establishment and success (Simões et al., 2009; Davies, 2003; Davies and Davies, 2010).

A particular characteristic of the cells in biofilms is their increased resistance to antimicrobial products (disinfectants, biocides, and antibiotics) compared with planktonic cells (Mah and O'Toole, 2001; Donlan and Costerton, 2002; Simões, 2011). Consequently, the concentrations required to eradicate bacterial biofilms often exceed the highest deliverable dose, thus impeding efficient control based on conventional antibacterials (Drenkard, 2003). In this context, biofilms can cause serious problems in medical fields, industrial processes, and water systems (Hall-Stoodley et al., 2004). In a clinical point of view, biofilm formation and persistence has enormous implications for the human health. The National Institutes of Health (NIH) has estimated that over 80% of human bacterial infections involve biofilms. Common diseases associated with the formation of biofilms include lung infections in cystic fibrosis patients, wound infections, endocarditis, periodontitis, ear infections, urinary tract infections, and infections related to surgical implants/medical devices (Richards and Melander, 2009; Drenkard, 2003; Hancock et al., 2010; Khatoon et al., 2018). Due to the recalcitrant properties of bacteria in biofilm, many biofilm infections are particularly difficult to treat becoming frequently recurrent or chronic. This leads to an increase of morbidity and mortality of affected individuals (Bryers, 2008; Anderson and O'Toole, 2008).

In industry, equipment damages, product contamination, blockages in membrane systems, and increase of energy costs are examples of detrimental effects of biofilms that imply substantial economic losses (Cloete et al., 1998; Beech, 2004; Shi and Zhu, 2009; Gilbert et al., 2003; Yuan et al., 2019; Alexandrovich et al., 2018).

Biofilms in drinking water distribution systems (DWDS) affect the microbiological quality of water and may lead to a number of undesirable effects on their organoleptic properties (turbidity, taste, odor, and color). The problems associated with biofilm formation are not only in DWDS but also in other water supply systems (e.g., dental unit waterlines, dialysis units, reverse osmosis systems) that can be colonized by biofilms (Chaves Simoes and Simoes, 2013). Thus, due to their protective effect against stress conditions, biofilms can constitute a reservoir of pathogenic microorganisms, which are responsible for various waterborne diseases (Huq et al., 2008). Besides, these microbial consortiums are responsible for an enormous increase in deterioration and obstruction of metallic pipes, which leads to decrease in water velocity and hinders the efficient disinfection/cleaning of water distribution systems (Chaves Simoes and Simoes, 2013).

The concept that biofilm formation can explain the increased microbial resistance has captivated the interest of researchers in exploring the main mechanisms involved in biofilm formation, maintenance, and dispersion (Hall-Stoodley et al., 2004; Mah and O'Toole, 2001; Smith, 2005; Stewart, 1996; Xu et al., 2000). This can help in the discovery of new inhibitors and/or more promising strategies to prevent and eliminate these structures.

11.2 Biofilm structure and mechanisms of bacterial resistance

The resistance to antimicrobial compounds is one of the most important features of the biofilm mode of growth. However, the reasons for this higher resistance are not entirely understood (Mah and O'Toole, 2001). It is known that the mechanisms involved in resistance of planktonic bacteria (e.g., efflux pumps, modifying enzymes, and target mutation) (Walsh, 2000; Stewart and Costerton, 2001) cannot fully explain the protection of biofilm cells (Stewart and Costerton, 2001; Patel, 2005). Solid evidence suggests that biofilm resistance is usually multifactorial, and therefore, a combination of multiple mechanisms is the possible explanation for their increased resistance (Gilbert et al., 2003; Aslam, 2008). These include (1) poor penetration or inactivation of antimicrobials in the extracellular polymeric matrix; (2) nutrient/oxygen limitation that leads to an altered bacterial metabolic state; (3) the presence of persister cells; (4) the expression of specific biofilm resistance genes; and (5) biofilm-specific phenotype induced by the antimicrobial itself following the use of sublethal concentrations and the upregulation of efflux pumps (Gilbert et al., 2003; Anderson and O'Toole, 2008; Stewart, 2002).

The EPS matrix is one of the main constituents of biofilms, comprising approximately 50%–90% of the total organic matter. Beyond the polysaccharides, biofilms also contain proteins, nucleic acids, lipids, phospholipids, and humic substances (Flemming et al., 2007; Allison, 2003). This extracellular matrix constitutes a protective barrier that limits or retards the access of antimicrobial agents to the cells within biofilms. In addition to their action as a physical barrier, their components can interact with antimicrobials, making them inactive. Such inactivation can be the result of the combination of ionic interactions and molecular sieving events (Allison et al., 2000; Costerton et al., 1999; Stewart, 2002). The reaction–diffusion limitation capacity of the EPS matrix can be improved through the production of extracellular enzymes, which get accumulated within this matrix (Davies, 2003; Gilbert et al., 2002; Simões, 2011; Cloete, 2003).

EPS are also regarded as a key factor in the biofilm formation process. They provide mechanical stability to biofilms that enables the assembly of structured and complex communities (Allison, 2003; Flemming and Wingender, 2010; Donlan, 2002). In addition, the biosynthesis of EPS is involved in other aspects, such as initial attachment of cells to solid surfaces, formation of microcolonies, and development of mature biofilms with a three-dimensional structure. In some cases, the EPS matrix also allows the bacteria to capture nutrients and contributes to water retention maintaining the environment around biofilm cells highly hydrated (Vu et al., 2009; Dunne, 2002; Flemming and Wingender, 2010). Since biofilm cells are not fully immobilized and are maintained in close proximity to each other, the exchange of genetic information (horizontal gene transfer) is facilitated, increasing their resistance (Savage et al., 2013). Moreover, these intense interactions can favor cell–cell communication (Flemming et al., 2007; Flemming and Wingender, 2010). In this sense, the several functions attributed to EPS have clearly demonstrated a wide range of advantages for the biofilm mode of life (Flemming and Wingender, 2010).

Although EPS have an important role in biofilm-associated antimicrobial resistance, it is not known if EPS are the predominant responsible mechanism for this resistance. Another important mechanism is associated with nutrient and oxygen limitation that can lead to reduction in growth rate. Dormant or slow-growing cells are less susceptible to almost all antimicrobials that have a growth-dependent activity (Mah and O'Toole, 2001; Gilbert et al., 1990; Smith et al., 2003). In the periphery of a biofilm, cells have growth rates close to planktonic forms. However, in the deeper zones of biofilm, cells enter into a nongrowing state. Consequently, distinct gradients are generated along the biofilm community (Allison et al., 2000; Fux et al., 2005). This physiological heterogeneity enables the cells in different regions of the biofilm to exhibit altered patterns of gene expression (Costerton et al., 1999). This process is regulated by production of signaling molecules in QS phenomena (Davies et al., 1998).

As a consequence of the two mechanisms referred earlier, the action of antimicrobials is attenuated; the concentration that reaches biofilm cells, particularly in the deeper zones, is lower than the desired; and more protected/tolerant biofilm phenotypes are selected. In this context, cells in deeper layers can be exposed to subinhibitory concentrations of the antimicrobial agent for a long period of time, contributing to the appearance of resistant phenotype within the biofilm population (Gilbert and McBain, 2003; Gilbert et al., 2002; Anderson and O'Toole, 2008). Furthermore, the upregulation of efflux pumps and the expression of specific biofilm resistance genes also contributes to development of resistant phenotypes (Anderson and O'Toole, 2008; Cloete, 2003; Patel, 2005; Costerton et al., 1999).

The occurrence of persister cells is the more recent explanation for increased biofilm resistance to antimicrobials. These cells survive to antimicrobial treatments without undergoing mutations that confer resistance. They are phenotypic variants of the wild type that can exist in both planktonic and biofilm populations (Davies, 2003; Lewis, 2007, 2010).

Each mechanism individually is insufficient to fully explain the observed resistance. It is therefore more likely that together they complement each another to create decreased susceptibility and an adequate environment for the emergence of tolerant phenotypes and genotypes within biofilms.

As referred above, due to their "special" physiology and morphology, biofilm cells are consistently described as being more resistant and consequently more difficult to eliminate. Hence, conventional methods for biofilm prevention and control present limitations and are often ineffective, contributing to the dissemination of resistance.

To overcome biofilm resistance, new sources of antimicrobials and new strategies are urgently needed to effectively control biofilms.

11.3 Emergent strategies of biofilm control and eradication

It is known that prevention of biofilm growth and specifically biofilm control is very difficult to attain. In fact, the control of an already established biofilm is more difficult to achieve than preventing its formation (Sandasi et al., 2010; Borges et al., 2012).

Several antibiofilm approaches have been used for many years due to the necessity to eradicate detrimental biofilms. Their focus has mostly been concentrated on the inhibition of bacterial biofilm contamination. However, a better understanding of the cellular processes involved in biofilm formation, such as the initial adhesion, bacterial motility, production of EPS, and QS phenomena, can lead to the discovery of novel and more effective strategies to inhibit, reduce, and/or remove biofilms (Landini et al., 2010). Indeed, the success of any antibiofilm strategy lies in its efficacy to decrease initial attachment and establishment and promote disintegration of the biofilm cells (Singh and Singh, 2013). Hence, these approaches should include the application of antimicrobials capable of suppressing bacterial resistance mechanisms in this mode of growth and with different mode of action/targets. Another important strategy involves the use of techniques with the capacity to increase the susceptibility of the microorganisms to previously established methods and with ability to interfere with the structure of the biofilm or disturb cell-to-cell communication (Abreu et al., 2012; Saleem et al., 2010; Landini et al., 2010; Beitelshees et al., 2018). In this context, several researchers have identified improved strategies for biofilm control (Simões, 2011; Simões et al., 2009, 2010; Landini et al., 2010; Kostakioti et al., 2013; Taraszkiewicz et al., 2013; Sharahi et al., 2019; Borges et al., 2016).

Currently, a wide range of antibiofilm approaches have been proposed, and these include the use of physical, chemical, and biological methods. These can include the use of natural compounds from plants (phytochemicals), QS inhibitors, bacteriophages, antimicrobial surfaces, photodynamic therapy (PDT), and enzymes. Each one of these methods comprises several techniques and has its own advantages and disadvantages that provide diverse effects in biofilm control (Simões et al., 2010; Simões, 2011; Taraszkiewicz et al., 2013; Singh and Singh, 2013; Chen et al., 2013; Kostakioti et al., 2013; Sharahi et al., 2019; Borges et al., 2016). Due to the complexity of biofilms, a single strategy is often insufficient. So, it is crucial to apply a combination of various strategies to reach different targets on the biofilm (Taraszkiewicz et al., 2013).

Chemical products with antimicrobial properties (e.g., antibiotic, biocides, and disinfectants) are the most commonly used methods to control detrimental biofilms. Nevertheless, there are no compounds with complete efficacy. Moreover, the eradication of biofilms with single target antimicrobials is difficult to perform, and frequently, the combination of different bacterial targets is needed (Simões, 2011). Due to the enormous tolerance of sessile microorganisms to antimicrobial agents, high concentrations are often required, leading to the emergence of resistant strains. Natural compounds from plants are a green and sustainable source of new antimicrobials that have shown to be effective inhibitors of biofilms (Borges et al., 2012, 2014). Besides, these products present several advantages over synthetic molecules, including different mechanisms of action and multiple targets in bacterial cell, which might help to overcome the resistance problem (Dixon, 2001; Samy and Gopalakrishnakone, 2010; Saleem et al., 2010; Borges et al., 2016). In recent years, some reports demonstrated the capacity of phytochemicals to inhibit QS (Gao et al., 2003; Vattem et al., 2007; Adonizio et al., 2006; Zhu and Sun, 2008; Khan et al., 2009; Packiavathy et al., 2019; Bodede et al., 2018; Aswathanarayan and Vittal, 2018; Prateeksha et al., 2019; Zhou et al., 2019). As QS is an important regulatory mechanism in biofilm formation and differentiation, various studies focused on the inhibition of this system. In addition to its role in biofilms, QS regulates the expression of genes involved in processes related to cell survival, virulence, and pathogenicity. Therefore, the disruption of QS signaling pathways can affect biofilm development and make bacteria more susceptible to antimicrobials (Borges et al., 2013, 2016; Subhadra et al., 2018; Quecán et al., 2018).

The use of bacteriophages for biofilm control is a very promising strategy that can be applied in alternative to antibiotics or in combination.

Bacteriophages have the capability to infect bacteria within the biofilm, because they can penetrate the intact biofilm matrix or express EPS degrading enzymes. Bacteriophages have other advantage over chemical agents; they are cheap, specific to host or host range, easy to isolate, and eco-friendly and have minimum side effects (Simões et al., 2010; Singh and Singh, 2013).

As initial adhesion is the initial stage of biofilm formation, the modification of surfaces is one of the principal strategies used for biofilm prevention. The changes can be done through incorporation of a substance in the surface polymer itself or by coating the surface (Singh and Singh, 2013; Awad et al., 2018). Some approaches entail the production of coatings that can prevent biofilm formation but also capable to kill the bacterial cells (Cheng et al., 2008; Vishwakarma et al., 2009; Cao et al., 2018). Biosurfactants are surface active compounds of microbial origin that can be used for modification of surfaces. The preconditioning of surfaces with biosurfactants can alter their physicochemical properties avoiding bacterial adhesion and consequent biofilm formation (Banat et al., 2010; Satpute et al., 2019). Rhamnolipid is a glycolipid produced by *Pseudomonas aeruginosa* that has been extensively studied. This biosurfactant has shown antimicrobial and antiadhesive properties against some microorganisms. Other well-known biosurfactant with good efficacy on inhibition of adhesion and biofilm disruption is the surfactin synthetized by *Bacillus subtilis* (Nitschke et al., 2010; Zezzi do Valle Gomes and Nitschke, 2012). Biosurfactants obtained from *Lactobacillus* (e.g., *L. acidophilus*) also due to their antibiofilm and antiadhesive properties has been demonstrated importance in the biomedical and therapeutic field. This is of relevance since they be part of the human microbiota (Satpute et al., 2018, 2019). Due to their characteristics, low toxicity, and high biodegradability, biosurfactants have advantages over synthetic surfactants attracting the interest of different sectors. However, the high cost of production limits their application (Banat et al., 2010).

Nanoparticles have shown strong antimicrobial and antibiofilm properties. Silver, magnesium, chitosan, lipid, and magnetic nanoparticles have been used in several studies for biofilm control. Inhibition of biofilm formation by disruption of EPS matrix and prevention of microbial adhesion is often observed (Ferreira et al., 2011; de Alteriis et al., 2018; Viswanathan et al., 2019). The encapsulation of antimicrobials into nanoparticles enables their release in a sustained and controlled manner. Besides, this strategy has other advantages such as the high reactivity, unique interactions with biological systems, the small size of the particles, and the large surface of contact (Ferreira et al., 2011; de Alteriis et al., 2018; Viswanathan et al., 2019). Nanoparticles can be used in combination with other techniques, such as modification of surfaces with its incorporation into polymers or on surface coating, and also in PDT (Allaker, 2010; Khalid et al., 2019).

Antimicrobial photodynamic therapy (APDT) is a relatively new approach that has been extensively studied as a potential alternative to conventional antibiotics. Indeed, it has been demonstrated that APTD is successful in eradication of antibiotic resistant bacteria and biofilms (Biel et al., 2011a; Silva et al., 2018b). Its principle is based on the excitation of a light-sensitive agent known as photosensitizer. The APDT effect is target specific because it only acts in microorganisms that have absorbed the photosensitizer and are exposed to a specific wavelength of light. Moreover, it has to be emphasized that its mechanism of action is completely different from common antimicrobials, thus avoiding the development of resistance (Silva et al., 2018b). Due to the lack of toxicity and well-known photoreactive behavior, xanthene dyes (e.g., rose bengal and erythrosine), methylene blue, and other phenothiazines have been frequently used as photosensitizer agents in numerous investigations (Silva et al., 2018a, 2019; Tokubo et al., 2018; Yang et al., 2019). APDT is reported as a noninvasive, safe and effective technique, which is effective against a wide variety of microorganisms. Furthermore, there is no evidence of bacterial photoresistance after repeated APDT treatment cycles (Biel et al., 2011a, 2011b; Soukos and Goodson, 2011).

Another interesting way to prevent biofilm formation or destroy already formed biofilms is the use of enzymes that target different features of biofilms (Thallinger et al., 2013; Kovach et al., 2019; Chen et al., 2018). Given its relevance, this strategy will be discussed in more detail further.

11.4 Antibiofilm enzymes

Enzymes are proteins that can be associated with nonproteins (cofactors) and act as biological catalysts increasing the reaction velocity. The cell metabolism depends on a complex network of chemical reactions involving specific enzymes, and any modification on the enzyme behavior may have consequences for any living organism (Dixon and Webb, 1964; Cabral et al., 2003). The enzyme specificity varies according to the substrate, the presence or absence of cofactors, activators, or inhibitors, the pH, and temperature. Temperature is usually one of the most important affecting enzymatic activities since elevated temperatures can promote enzyme denaturation (Cabral et al., 2003; Copeland, 2004; Aurand et al., 1987).

Enzymes can be classified into six classes: oxidoreductases, transferases, hydrolases, lyases, isomerases, and ligases. Oxidoreductases are enzymes that transfer oxygen or hydrogen atoms through redox reaction, whereas transferases allow the transfer of an atom or a group of atoms. Hydrolases catalyze hydrolytic reactions, and lyases can remove a group of atoms. Isomerization can be catalyzed by isomerases, and ligases, or synthetases, join two molecules together with a covalent bond (Cabral et al., 2003; Shen and Chou, 2007).

Enzymes are well known and used in many industrial applications (food, drugs, detergents, baking, animal feed, textiles, and beverages) (Srey et al., 2013). More recently, the application of enzymes to remove biofilms from industrial surfaces is increasing and in development (Xavier et al., 2005). This new direction is commanded by a conscious society that demands for "greener" alternatives with minimal environmental impacts. These alternatives imply the use of biodegradable and less harmful biocides for the environment (Cortés et al., 2011; Srey et al., 2013).

The enzymes usually target the EPS, since the EPS forms the biofilms and influences their shape, structure, and resistance to shear stress forces (Lequette et al., 2010; Asker et al., 2018). Enzymes can attack the biofilms directly or interfere with their formation, catalyze the formation of antimicrobials, affect the QS activity, or even destroy the biofilm. An enzymatic treatment to remove biofilms from the surfaces of tanks and pipes has already proved its efficacy by degrading the key components of the biofilm matrix (Simões et al., 2010; Thallinger et al., 2013). These strategies are related to the biofilm intrinsic properties and composition (Augustin et al., 2004; Donlan, 2002). The green biocides can be used as aids to the common biocides, reducing their concentrations without compromising the inactivation of the harmful microorganism. As enzymes have the EPS as target, to remove and/or inactivate the biofilm, they also facilitate the access to the common disinfectant (Cortés et al., 2011; Srey et al., 2013). Additionally, several commercial biocides used in antifouling coatings have been recently banned. Consequently the search for alternative eco-friendly biocides is urgent (Lequette et al., 2010; Camps et al., 2011).

Nevertheless, enzymes can also be used to inhibit biofilm formation and not only to remove it. Enzymes can act on the adhesive properties of the cells avoiding their attachment to surfaces (Torres et al., 2012; Asker et al., 2018). Torres et al. (2012). prevented biofilm formation in the paper industry with the use of multiple enzymes. Moreover, Ghanwate et al. (2012). studied the prevention of biofilm formation in urinary catheters with the combination of enzymes, gentamycin, and EDTA (ethylenediaminetetraacetic acid).

11.4.1 Biofilm disruption by enzymes

Different biofilms usually require different types of enzymes (Kumar and Anand, 1998; Thallinger et al., 2013). Normally, the enzymatic formulations have more than one enzyme or other antimicrobial agents associated. This is due to the heterogeneity of the biofilm, as it is formed by a diversity of components. According to the different biofilm compositions, a cocktail of enzymes may be applied, and the concentration and enzyme type must be defined (Augustin et al., 2004; Thallinger et al., 2013).

Enzymes act by decreasing the physical integrity of the biofilm by weakening the different structures that form the EPS. To obtain an efficient biofilm removal, it is important that the EPS structural components are identified prior to application of the enzymes (Molobela et al., 2010). The EPS comprises various multi structural components, which are mainly derived from polysaccharides, carbohydrates, lipids, proteins, glycoproteins, nucleic acids, glycolipids, and phospholipids, among others (Liu et al., 2004). Consequently, the enzyme application is targeted to the EPS molecules converting them into smaller units. These smaller units can be transported through the cell membranes and be metabolized by the cell (Molobela et al., 2010).

There are four types of enzymes that are applied in biofilm removal: proteolytic enzymes, polysaccharide-degrading enzymes (Johansen et al., 1997; Thallinger et al., 2013), oxidative enzymes, and anti-QS enzymes (Thallinger et al., 2013; Bzdrenga et al., 2017). Proteases are proteolytic enzymes that hydrolyze proteins. Proteases are used in pipelines and in contact lenses for protein removal (Augustin et al., 2004). Amylases, lysozymes, hydrolases, and lyases are examples of polysaccharide-degrading enzymes. These enzymes are commercialized for food-industry application such as fruit juice extraction (Lequette et al., 2010; Thallinger et al., 2013). These types of enzymes can be combined being more efficient in biofilm control (Johansen et al., 1997; Yamasaki et al., 2005; Orgaz et al., 2006). Glucose oxidase and peroxidases are oxidative enzymes applied in health and environmental industries to mimic the human defense system, as these enzymes produce reactive oxygen species that destroy pathogens. The anti-QS enzymes are lactonases that hydrolyze the bond of the homoserine ring, preventing acyl homoserine lactones (AHLs) from binding to transcriptional regulators (Thallinger et al., 2013).

One of the disadvantages is the fact that the EPS is different and heterogeneous from biofilm to biofilm, and some studies report that carbohydrates are the main constituents of the EPS (Zhang et al., 2005), whereas other studies highlight the abundance of proteins (Orgaz et al., 2006). The EPS components of the biofilms differ in quantity, structure, and nature depending on the microorganisms forming the biofilm (Liu et al., 2004). Therefore, a single enzyme may not eliminate a specific biofilm, but it may be used in combination with other enzymes or with other treatments (Augustin et al., 2004; Lequette et al., 2010). Another drawback is the high cost of the enzymes (Thallinger et al., 2013), which is aggravated by the fact that a multiple enzymatic treatment is always advised (Augustin et al., 2004; Simões et al., 2010; Thallinger et al., 2013). In the medical areas, the enzymatic treatments have additional costs, since they must be used in high-purity formulations. Furthermore, their application in certain environments may be ineffective due to their inactivation at high pH extremes or at high temperatures or even in hostile environments (Thallinger et al., 2013). Nevertheless, the use of enzymatic treatments to break down EPS in biofilms is a possible alternative when standard cleaning agents do not yield satisfactory results in removing biofilms (Augustin et al., 2004). Stiefel et al. (2016) studied the combined effect of commercial cleaners with enzymes and found improved performance in terms of biofilm and blood removal against *Staphylococcus aureus* and *P. aeruginosa*. It should be noted that a mixture of enzymes (proteases, polysaccharidases, lipases, and DNases) was required to be effective against *P. aeruginosa*, whereas for *S. aureus*, an active protease was sufficient.

As enzymes are costly, they can be immobilized to enable their reuse and make the process profitable. The enzymes are immobilized on a matrix that serves as a support and have to be different from the substrates and products used in the process. Usually, the materials used for the matrix should be affordable, inert, and stable and, if possible, increase enzymatic activity (Datta et al., 2013). Additionally, the use of bioinformatics and genetic engineering to develop new and improved enzymes is a potential tool for successful enzymatic treatments. To achieve maximum biofilm removal, genetically modified enzymes are seen as the future in enzyme-based biofilm control (Thallinger et al., 2013).

11.4.2 Example applications

Enzymes have been well studied, and their application as antibiofilms has increased over the years (Thallinger et al., 2013; Taraszkiewicz et al., 2013; Loiselle and Anderson, 2003). Biofilms are a problem in several areas, including the paper industry (Torres et al., 2012). Marcato-Romain et al. tested four enzymes with different concentrations (trypsin [0.06, 0.12, 0.25, 4.44, 8.88, 17.76, 23.80, 35.52 UP/mL], pepsin [2.14, 4.29, 8.57, 17.15, 34.30, 68.60, 91.46, 137.19 UP/mL], Alcalase [0.56, 2.80, 14.00 UP/mL], Savinase [0.003, 0.015, 0.031, 1.545, 3.090, 6.180, 30.900, 154.500 UP mL]), on a 24-h-aged multispecies biofilm. Since the biofilm is mainly comprised of proteins, proteases were tested. Savinase was the most effective (Marcato-Romain et al., 2012). However, the use of enzymes in the paper industry is limited, since it is not widely studied and applied. Monocomponent enzymes can be used, but the biofilm heterogeneity limits the efficacy of the treatments (Torres et al., 2012).

The use of both amylase and protease has been explored and compared. Molobela et al. stated that amylase enzymes were less effective than proteases, in *Pseudomonas fluorescens* biofilm degradation. This is related to the dominance of proteins in the EPS and high biofilm heterogeneity. Therefore, the biofilm structural composition is important (Molobela et al., 2010). The same conclusion was drawn by Huang et al. These authors observed that proteases were better in removing an aging biofilm. They also observed that the proteases tested had different mechanisms in removing the biofilm, such as enzymatic hydrolysis of biopolymers and adsorption of enzyme macromolecules onto biofilm (Huang et al., 2014).

Some studies have proven that the same type of enzyme applied had different results in different scenarios. In the study of Leroy et al. (2008), it was concluded that the application of two hydrolases had diverse outcomes, since Savinase (0.5–10 g/L) was effective in preventing *Pseudoalteromonas* sp. adhesion, whereas other hydrolase increased adhesion by this bacterium. Likewise, enzymes can be efficient against one microorganism and ineffective for another. Craigen et al. (2011) tested α-amylase (100 mg/mL) on *S. aureus* and *Staphylococcus epidermidis* and concluded that it was only capable of reducing the *S. aureus* biofilm by 79% after a 5 min exposure assay. The role of *P. aeruginosa* exopolysaccharides Pel and Psl in the biofilm formation and maintenance is well known. Considering this aspect, Baker et al. (2016) studied the effect of specific glycoside hydrolases, PelAh and PslGh, on *P. aeruginosa* exopolysaccharide components and biofilms. The application of these enzymes resulted in exopolysaccharides degradation, inhibition of biofilm formation, and disruption of preformed biofilms. Moreover, PelAh and PslGh enzymes were able to potentiate the activity of the antibiotic colistin. More recently, Asker et al. (2018) tested the influence of PslGh, covalently bounded to different chemical surfaces applying amine functionalization and glutaraldehyde linking, against *P. aeruginosa* biofilms. They found that any adsorption of the Psl on the PslGh-bound surfaces did not occur. Besides, a significant reduction of the *P. aeruginosa* attachment to this conditioned surfaces and biofilm formation was achieved. The observed effect can be related with the enzymatic degradation of the Psl exopolysaccharide.

Lactonase is an anti-QS enzyme that has been applied to reduce *P. aeruginosa* biofilm. The biofilm formed by this microorganism is increasingly reduced (69%−77%) in the presence of lactonase (0.01 −10 μM). Moreover, this enzyme decreased *P. aeruginosa* virulence factors (Kiran et al., 2011). There are also studies with enzymes excreted from microorganisms. Kaplan et al. found that *Actinobacillus actinomycetemcomitans* released an enzyme that removed *S. epidermidis* biofilms from plastic surfaces. The enzyme causes *S. epidermidis* to produce an extracellular slime that disables biofilm formation on plastic surfaces (Kaplan et al., 2004). In the case of *S. epidermidis*, the enzymatic treatment for biofilm removal is related to the biofilm constituents, but it is also dependent on the clinical isolate (Chaignon et al., 2007).

Despite the fact that there are not many applications of immobilized enzymes in the biofilm removal, Cordeiro et al. have this strategy to prevent cell adhesion. The authors developed an immobilized method to prevent the attachment of *P. aeruginosa* and *S. epidermidis*. The enzymes used were subtilisin A and cellulase, and they were immobilized in a copolymer, poly(ethylene-alt-maleic) anhydride. They were able to reduce the attachment in 67% and 44% of *S. epidermidis* with immobilized cellulase and *P. aeruginosa* with immobilized subtilisin A, respectively. Cellulase had no effect on *P. aeruginosa* attachment, and subtilisin A had no effect on *S. epidermidis* attachment (Cordeiro et al., 2011).

The combination of enzymes was found to be more effective by Johansen et al. (1997). When using oxidoreductases with polysaccharide-hydrolyzing enzymes, the biofilm was removed and the bacterial activity reduced, whereas in separate oxidoreductases, bactericidal and polysaccharide-hydrolyzing enzymes were efficient in removing the biofilm (Johansen et al., 1997). Based on Wiatr (1991), a blend of enzymes consisting of protease, α-amylase, and β-glucanase was found effective in cleaning a simulated industrial biofilm formed during paper pulp manufacture. In another work, silicone urinary catheters were coated with a mixture of acylase and α-amylase enzymes and with each one in separate. The results showed that both enzymes when used alone inhibited the biofilm formation by *P. aeruginosa* and *S. aureus*. However, when a combination of both enzymes was used, an improved activity was obtained (Ivanova et al., 2015).

Recently, the use of hydrolytic enzymes was proposed, to act on EPS components as an environmentally friendly strategy to prevent mainly marine biofouling. A number of several enzymes such as proteases and carbohydrases have been studied for the prevention of adhesion of marine microorganisms to solid surfaces, whereas proteases such as subtilisin have shown to inhibit biofilm formation by cultures of *P. fluorescens* (Zanaroli et al., 2011).

Similarly, the combination with other treatments has also proven to be efficient. Oulahal-Lagsir et al. demonstrated the synergism between ultrasonic waves and enzymes in *Escherichia coli* biofilm removal from stainless steel surfaces. The sonication alone (40 kHz) caused a 30% of biofilm removal, but when combined with enzymes, it could reach up to 96% removal and a minimum of 61%, depending on the conditions (enzyme [protease or crystalline trypsin], enzyme concentration [36 U/mL for protease or 7600 U/mL for crystalline trypsin], time of exposure) (Oulahal-Lagsir et al., 2003). Combining mechanical/physical treatments with enzymes was also performed by Pechaud et al., (2012). These authors found that increasing the shear stress by 2.5 Pa combined with an enzymatic treatment (Savinase at 6 UP/mL) leads to 90% biofilm removal, whereas a treatment with sodium hypochlorite (individually 150 mg/L) only reached 35% biofilm removal (Pechaud et al., 2012). The combination of antibiotics with enzymes was also studied for *P. aeruginosa* biofilms that were completely removed when alginate lyase (20 U/mL) and gentamycin (64 μg/mL) were applied in combination for 96 h (Alkawash et al., 2006). The effectiveness of α-mannosidase, β-mannosidase, and trypsin enzymes was evaluated on the ability to degrade biofilms of *P. aeruginosa* bacterial strains, isolated from burn wound infections. The increase of the susceptibility to ceftazidime antibiotic was assessed. All tested enzymes revealed good results in terms of biofilm degradation and improvement of ceftazidime activity. In general, trypsin showed worse results than α-mannosidase but lesser cytotoxic effects to A-431 human epidermoid carcinoma cell lines. In another study, the combination of enzymatic agents (dispersin B, lysostaphin, alpha amylase, V8 protease, and serrapeptase) with conventional antibiotics (vancomycin and rifampicin) demonstrated higher efficacy than enzymatic or antibiotic treatment alone, regarding the disruption and biofilm control of *S. aureus* strains resistant and susceptible to methicillin (Hogan et al., 2017).

There are also some studies with enzymes in the food industry. Biofilms can rapidly become resistant to the chemical treatments applied in the processing lines in the food industry. Lequette et al. concluded that with the diversity of components in EPS, a combination of enzymes, surfactants, and chelating agents would be appropriate and as effective as the chemical cleaning agents. This conclusion was drawn since different enzymes were specific for different microorganisms, and in the processing lines, a wide variety of microorganisms can be found. It was determined that enzymatic treatment is efficient in decreasing the biofilm cohesion by destroying the physical integrity of the matrix while having no identified negative impact on the environment (Lequette et al., 2010).

Oulahal et al. have also shown that a combination of treatments is necessary to have a significant biofilm removal. Those researchers combined ultrasounds (40 kHz) with a chelating agent and enzymes (protease at 36 U/mL and crystalline trypsin at 7600 U/mL, amyloglucosidase at 50 U/mL, papain at 3 U/mL, and lysozyme at 6000 U/mL), obtaining a 75% removal of *E. coli* and *S. aureus* biofilms. The efficacy of the treatment was dependent on the surfaces, the food matrix, and the microorganisms involved (Oulahal et al., 2007). There are two main companies that sell products with enzymes commonly applied for biofilm control purposes: Realco (food and beverage industry) (Realco, 2016) and Novozymes (food and beverage industry, pulp and paper industry, agriculture, and wastewater) (Novozymes, 2019).

11.5 Conclusions

The complexity of sessile communities provides to biofilm-embedded cells a privileged environment where they are more protected from various stresses, making them more difficult to eradicate. The fast establishment of resistance to antimicrobial agents prompted the interest of researchers in the search for alternative strategies with improved efficacy. However, finding a completely effective method is a hard task, since biofilm resistance is caused by a combination of different mechanisms.

Conventional methods are often ineffective in biofilm control because they are incapable to surpass its physical barrier. Moreover, they can interfere with cellular processes essential to microbial growth, imposing a selective pressure that leads to the development of resistance. Overall, new approaches should promote the penetration of antimicrobial compounds through the EPS matrix and interfere with the different structural components of biofilms, promoting their disaggregation. Besides, they must also be capable to interfere with the different cellular processes involved in biofilm formation and maturation. Biofilm-disrupting enzymes are an emergent approach with high antimicrobial activity and specificity that satisfies these requisites. They can be employed to target different components of the biofilm matrix and QS molecules, impairing biofilm formation or promoting their inactivation/destruction. The use of green strategies has also increasingly assumed an important role in society, in which enzyme application is promising. More advances in genetically modified enzymes are required to improve their application in biofilm removal to overcome enzymatic limitations, particularly specificity and cost.

Acknowledgments

This work was financially supported by project UID/EQU/00511/2019—Laboratory for Process Engineering, Environment, Biotechnology and Energy (LEPABE) funded by national funds through FCT/MCTES (PIDDAC); project "LEPABE-2-ECO-INNOVATION"—NORTE-01-0145-FEDER-000005, funded by Norte Portugal Regional Operational Programme (NORTE 2020), under PORTUGAL 2020 Partnership Agreement, through the European Regional Development Fund (ERDF); projects POCI-01-0145-FEDER-030219, POCI-01-0145-FEDER-028397, POCI-01-0145-FEDER-006939, funded by FEDER funds through COMPETE2020—Programa Operacional Competitividade e Internacionalização (POCI) and by national funds (PIDDAC) through FCT/MCTES. This work was also the result of the grant awarded to Manuel Simões (SFRH/BSAB/150379/2019) from national funds through FCT—Fundação para a Ciência e Tecnologia.

References

Abreu, A.C., McBain, A.J., Simões, M., 2012. Plants as sources of new antimicrobials and resistance-modifying agents. Natural Product Reports 29 (9), 1007–1021.

Adonizio, A.L., Downum, K., Bennett, B.C., Mathee, K., 2006. Anti-quorum sensing activity of medicinal plants in southern Florida. Journal of Ethnopharmacology 105 (3), 427–435.

Alexandrovich, S.O., Yurievna, V.S., Olegovich, P.I., Yurievna, K.N., Ilyinichna, V.L., Pavlovna, Y.T., 2018. Provision of microbiological safety in the food industry based on special technological supporting solutions. International Journal of Pharmaceutical Research and Allied Sciences 7 (1).

Alkawash, M.A., Soothill, J.S., Schiller, N.L., 2006. Alginate lyase enhances antibiotic killing of mucoid *Pseudomonas aeruginosa* in biofilms. APMIS 114 (2), 131–138.

Allaker, R.P., 2010. The use of nanoparticles to control oral biofilm formation. Journal of Dental Research 89 (11), 1175–1186.

Allison, D.G., 2003. The biofilm matrix. Biofouling 19 (2), 139–150.

Allison, D.G., Maira-Litran, T., Gilbert, P., 2000. Antimicrobial resistance of biofilms. In: Evans, L.V. (Ed.), Biofilms: Recent Advances in Their Study and Control. Harwood Academic Publishers, Amsterdam.

Anderson, G.G., O'Toole, G.A., 2008. Innate and induced resistance mechanisms of bacterial biofilms. Current Topics in Microbiology and Immunology 322, 85–105.

Asker, D., Awad, T.S., Baker, P., Howell, P.L., Hatton, B.D., 2018. Non-eluting, surface-bound enzymes disrupt surface attachment of bacteria by continuous biofilm polysaccharide degradation. Biomaterials 167, 168–176.

Aslam, S., 2008. Effect of antibacterials on biofilms. American Journal of Infection Control 36 (10). S175.e9–S175.e11.

Aswathanarayan, J.B., Vittal, R.R., 2018. Inhibition of biofilm formation and quorum sensing mediated phenotypes by berberine in *Pseudomonas aeruginosa* and *Salmonella typhimurium*. RSC Advances 8 (63), 36133–36141.

Augustin, M., Ali-Vehmas, T., Atroshi, F., 2004. Assessment of enzymatic cleaning agents and disinfectants against bacterial biofilms. Journal of Pharmacy & Pharmaceutical Sciences 7 (1), 55–64.

Aurand, L., Woods, A.E., Wells, M., 1987. Enzymes. In: Food Composition and Analysis. Springer, Netherlands.

Awad, T.S., Asker, D., Hatton, B.D., 2018. Food-safe modification of stainless steel food-processing surfaces to reduce bacterial biofilms. ACS Applied Materials & Interfaces 10 (27), 22902–22912.

Baker, P., Hill, P.J., Snarr, B.D., Alnabelseya, N., Pestrak, M.J., Lee, M.J., Jennings, L.K., Tam, J., Melnyk, R.A., Parsek, M.R., 2016. Exopolysaccharide biosynthetic glycoside hydrolases can be utilized to disrupt and prevent *Pseudomonas aeruginosa* biofilms. Science Advances 2 (5), e1501632.

Banat, I., Franzetti, A., Gandolfi, I., Bestetti, G., Martinotti, M., Fracchia, L., Smyth, T., Marchant, R., 2010. Microbial biosurfactants production, applications and future potential. Applied Microbiology and Biotechnology 87 (2), 427–444.

Beech, I.B., 2004. Corrosion of technical materials in the presence of biofilms—current understanding and state-of-the art methods of study. International Biodeterioration & Biodegradation 53 (3), 177–183.

Beitelshees, M., Hill, A., Jones, C., Pfeifer, B., 2018. Phenotypic variation during biofilm formation: implications for anti-biofilm therapeutic design. Materials 11 (7), 1086.

Biel, M.A., Sievert, C., Usacheva, M., Teichert, M., Balcom, J., 2011a. Antimicrobial photodynamic therapy treatment of chronic recurrent sinusitis biofilms. International Forum of Allergy and Rhinology 1 (5), 329–334.

Biel, M.A., Sievert, C., Usacheva, M., Teichert, M., Wedell, E., Loebel, N., Rose, A., Zimmermann, R., 2011b. Reduction of endotracheal tube biofilms using antimicrobial photodynamic therapy. Lasers in Surgery and Medicine 43 (7), 586–590.

Bodede, O., Shaik, S., Chenia, H., Singh, P., Moodley, R., 2018. Quorum sensing inhibitory potential and in silico molecular docking of flavonoids and novel terpenoids from *Senegalia nigrescens*. Journal of Ethnopharmacology 216, 134–146.

Borges, A., Abreu, A., Dias, C., Saavedra, M., Borges, F., Simões, M., 2016. New perspectives on the use of phytochemicals as an emergent strategy to control bacterial infections including biofilms. Molecules 21 (7), 877.

Borges, A., Saavedra, M.J., Simões, M., 2012. The activity of ferulic and gallic acids in biofilm prevention and control of pathogenic bacteria. Biofouling 28 (7), 755–767.

Borges, A., Serra, S., Cristina Abreu, A., Saavedra, M.J., Salgado, A., Simões, M., 2013. Evaluation of the effects of selected phytochemicals on quorum sensing inhibition and in vitro cytotoxicity. Biofouling 30 (2), 183–195.

Borges, A., Simões, L.C., Saavedra, M.J., Simões, M., 2014. The action of selected isothiocyanates on bacterial biofilm prevention and control. International Biodeterioration & Biodegradation 86 (Part A(0)), 25–33.

Bryers, J.D., 2008. Medical biofilms. Biotechnology and Bioengineering 100 (1), 1–18.

Bzdrenga, J., Daudé, D., Rémy, B., Jacquet, P., Plener, L., Elias, M., Chabrière, E., 2017. Biotechnological applications of quorum quenching enzymes. Chemico-Biological Interactions 267, 104–115.

Cabral, J., Gama, M., Aires-Barros, M., 2003. Introdução. In: Cabral, J., Aires-Barros, M., Gama, M. (Eds.), Engenharia Enzimática. Lidel.

Camps, M., Briand, J.-F., Guentas-Dombrowsky, L., Culioli, G., Bazire, A., Blache, Y., 2011. Antifouling activity of commercial biocides vs. natural and natural-derived products assessed by marine bacteria adhesion bioassay. Marine Pollution Bulletin 62 (5), 1032–1040.

Cao, Y., Su, B., Chinnaraj, S., Jana, S., Bowen, L., Charlton, S., Duan, P., Jakubovics, N.S., Chen, J., 2018. Nanostructured titanium surfaces exhibit recalcitrance towards *Staphylococcus epidermidis* biofilm formation. Scientific Reports 8 (1), 1071.

Chaignon, P., Sadovskaya, I., Ragunah, C., Ramasubbu, N., Kaplan, J.B., Jabbouri, S., 2007. Susceptibility of staphylococcal biofilms to enzymatic treatments depends on their chemical composition. Applied Microbiology and Biotechnology 75 (1), 125–132.

Chaves Simoes, L., Simoes, M., 2013. Biofilms in drinking water: problems and solutions. RSC Advances 3 (8), 2520–2533.

Chen, M., Yu, Q., Sun, H., 2013. Novel strategies for the prevention and treatment of biofilm related infections. International Journal of Molecular Sciences 14 (9), 18488–18501.

Chen, Z., Wang, Z., Ren, J., Qu, X., 2018. Enzyme mimicry for combating bacteria and biofilms. Accounts of Chemical Research 51 (3), 789–799.

Cheng, G., Xue, H., Zhang, Z., Chen, S., Jiang, S., 2008. A switchable biocompatible polymer surface with self-sterilizing and nonfouling capabilities. Angewandte Chemie International Edition 47 (46), 8831–8834.

Cloete, T.E., 2003. Resistance mechanisms of bacteria to antimicrobial compounds. International Biodeterioration & Biodegradation 51 (4), 277–282.

Cloete, T.E., Jacobs, L., Brözel, V.S., 1998. The chemical control of biofouling in industrial water systems. Biodegradation 9 (1), 23–37.

Copeland, R.A., 2004. Enzymes: A Practical Introduction to Structure, Mechanism, and Data Analysis, second ed. John Wiley & Sons.

Cordeiro, A.L., Hippius, C., Werner, C., 2011. Immobilized enzymes affect biofilm formation. Biotechnology Letters 33 (9), 1897–1904.

Cortés, E., Consuegra, J., Sinisterra, R., Mendez-Vilas, A., 2011. Biofilm formation, control and novel strategies for eradication. Science Against Microbial Pathogens: Communicating Current Research and Technological Advances 2, 896–905.

Cos, P., Toté, K., Horemans, T., Maes, L., 2010. Biofilms: an extra hurdle for effective antimicrobial therapy. Current Pharmaceutical Design 16 (20), 2279–2295.

Costerton, J.W., Lewandowski, Z., Caldwell, D.E., Korber, D.R., Lappin-Scott, H.M., 1995. Microbial biofilms. Annual Review of Microbiology 49, 711–745.

Costerton, J.W., Stewart, P.S., Greenberg, E.P., 1999. Bacterial biofilms: a common cause of persistent infections. Science 284 (5418), 1318–1322.

Craigen, B., Dashiff, A., Kadouri, D.E., 2011. The use of commercially available alpha-amylase compounds to inhibit and remove *Staphylococcus aureus* biofilms. The Open Microbiology Journal 5, 21.

Datta, S., Christena, L.R., Rajaram, Y.R.S., 2013. Enzyme immobilization: an overview on techniques and support materials. 3 Biotech 3 (1), 1–9.

Davies, D., 2003. Understanding biofilm resistance to antibacterial agents. Nature Reviews Drug Discovery 2 (2), 114–122.

Davies, D.G., Parsek, M.R., Pearson, J.P., Iglewski, B.H., Costerton, J.W., Greenberg, E.P., 1998. The involvement of cell-to-cell signals in the development of a bacterial biofilm. Science 280 (5361), 295–298.

Davies, J., Davies, D., 2010. Origins and evolution of antibiotic resistance. Microbiology and Molecular Biology Reviews 74 (3), 417–433.

de Alteriis, E., Maselli, V., Falanga, A., Galdiero, S., Di Lella, F.M., Gesuele, R., Guida, M., Galdiero, E., 2018. Efficiency of gold nanoparticles coated with the antimicrobial peptide indolicidin against biofilm formation and development of *Candida* spp. clinical isolates. Infection and Drug Resistance 11, 915.

Dixon, M., Webb, E.C., 1964. Enzymes, second ed. Academic Press, New York.

Dixon, R.A., 2001. Natural products and plant disease resistance. Nature 411 (6839), 843–847.

Donlan, R.M., 2002. Biofilms: microbial life on surfaces. Emerging Infectious Diseases 8 (9), 881–890.

Donlan, R.M., Costerton, J.W., 2002. Biofilms: survival mechanisms of clinically relevant microorganisms. Clinical Microbiology Reviews 15 (2), 167–193.

Drenkard, E., 2003. Antimicrobial resistance of *Pseudomonas aeruginosa* biofilms. Microbes and Infection 5 (13), 1213–1219.

Dunne, W.M., 2002. Bacterial adhesion: seen any good biofilms lately? Clinical Microbiology Reviews 15 (2), 155–166.

Ferreira, C., Pereira, A.M., Pereira, M.C., Melo, L.F., Simões, M., 2011. Physiological changes induced by the quaternary ammonium compound benzyldimethyldodecylammonium chloride on *Pseudomonas fluorescens*. Journal of Antimicrobial Chemotherapy 66 (5), 1036–1043.

Flemming, H.-C., Neu, T.R., Wozniak, D.J., 2007. The EPS matrix: the "house of biofilm cells". Journal of Bacteriology 189 (22), 7945–7947.

Flemming, H.C., Wingender, J., 2010. The biofilm matrix. Nature Reviews Microbiology 8, 623–633.

Fux, C.A., Costerton, J.W., Stewart, P.S., Stoodley, P., 2005. Survival strategies of infectious biofilms. Trends in Microbiology 13 (1), 34–40.

Gao, M., Teplitski, M., Robinson, J.B., Bauer, W.D., 2003. Production of substances by *Medicago truncatula* that affect bacterial quorum sensing. Molecular Plant-Microbe Interactions 16 (9), 827–834.

Ghanwate, N.A., Thakare, P., Bhise, P., Dhanke, A., Apotikar, S., 2012. Prevention of biofilm formation in urinary catheter by coating enzymes/gentamycin/EDTA. In: Proceedings of World Academy of Science, Engineering and Technology. World Academy of Science, Engineering and Technology.

Gilbert, P., Allison, D.G., McBain, A.J., 2002. Biofilms in vitro and in vivo: do singular mechanisms imply cross-resistance? Journal of Applied Microbiology 92 (1), 98S–110S.

Gilbert, P., Collier, P.J., Brown, M.R., 1990. Influence of growth rate on susceptibility to antimicrobial agents: biofilms, cell cycle, dormancy, and stringent response. Antimicrobial Agents and Chemotherapy 34 (10), 1865–1868.

Gilbert, P., McBain, A.J., 2003. Potential impact of increased use of biocides in consumer products on prevalence of antibiotic resistance. Clinical Microbiology Reviews 16 (2), 189–208.

Gilbert, P., McBain, A.J., Rickard, A.H., 2003. Formation of microbial biofilm in hygienic situations: a problem of control. International Biodeterioration & Biodegradation 51 (4), 245–248.

Hall-Stoodley, L., Costerton, J.W., Stoodley, P., 2004. Bacterial biofilms: from the natural environment to infectious diseases. Nature Reviews Microbiology 2 (2), 95–108.

Hancock, V., Dahl, M., Klemm, P., 2010. Abolition of biofilm formation in urinary tract *Escherichia coli* and *klebsiella* isolates by metal interference through competition for fur. Applied and Environmental Microbiology 76 (12), 3836–3841.

Hogan, S., Zapotoczna, M., Stevens, N.T., Humphreys, H., O'Gara, J.P., O'Neill, E., 2017. Potential use of targeted enzymatic agents in the treatment of *Staphylococcus aureus* biofilm-related infections. Journal of Hospital Infection 96 (2), 177–182.

Huang, H., Ren, H., Ding, L., Geng, J., Xu, K., Zhang, Y., 2014. Aging biofilm from a full-scale moving bed biofilm reactor: characterization and enzymatic treatment study. Bioresource Technology 154, 122–130.

Huq, A., Whitehouse, C.A., Grim, C.J., Alam, M., Colwell, R.R., 2008. Biofilms in water, its role and impact in human disease transmission. Current Opinion in Biotechnology 19 (3), 244–247.

Ivanova, K., Fernandes, M.M., Francesko, A., Mendoza, E., Guezguez, J., Burnet, M., Tzanov, T., 2015. Quorum-quenching and matrix-degrading enzymes in multilayer coatings synergistically prevent bacterial biofilm formation on urinary catheters. ACS Applied Materials & Interfaces 7 (49), 27066–27077.

Johansen, C., Falholt, P., Gram, L., 1997. Enzymatic removal and disinfection of bacterial biofilms. Applied and Environmental Microbiology 63 (9), 3724–3728.

Kaplan, J.B., Ragunath, C., Velliyagounder, K., Fine, D.H., Ramasubbu, N., 2004. Enzymatic detachment of *Staphylococcus epidermidis* biofilms. Antimicrobial Agents and Chemotherapy 48 (7), 2633–2636.

Khalid, H.F., Tehseen, B., Sarwar, Y., Hussain, S.Z., Khan, W.S., Raza, Z.A., Bajwa, S.Z., Kanaras, A.G., Hussain, I., Rehman, A., 2019. Biosurfactant coated silver and iron oxide nanoparticles with enhanced anti-biofilm and anti-adhesive properties. Journal of Hazardous Materials 364, 441–448.

Khan, M.S.A., Zahin, M., Hasan, S., Husain, F.M., Ahmad, I., 2009. Inhibition of quorum sensing regulated bacterial functions by plant essential oils with special reference to clove oil. Letters in Applied Microbiology 49 (3), 354–360.

Khatoon, Z., McTiernan, C.D., Suuronen, E.J., Mah, T.-F., Alarcon, E.I., 2018. Bacterial biofilm formation on implantable devices and approaches to its treatment and prevention. Heliyon 4 (12), e01067.

Kiran, S., Sharma, P., Harjai, K., Capalash, N., 2011. Enzymatic quorum quenching increases antibiotic susceptibility of multidrug resistant *Pseudomonas aeruginosa*. Iranian Journal of Microbiology 3 (1), 1.

Kostakioti, M., Hadjifrangiskou, M., Hultgren, S.J., 2013. Bacterial biofilms: development, dispersal, and therapeutic strategies in the dawn of the postantibiotic era. Cold Spring Harbor Perspectives in Medicine 3 (4).

Kovach, K.N., Fleming, D., Rumbaugh, K.P., Gordon, V.D., 2019. Specific disruption of established *P. aeruginosa* biofilms using polymer-attacking enzymes. Langmuir 36 (6), 1585–1595.

Kumar, C.G., Anand, S.K., 1998. Significance of microbial biofilms in food industry: a review. International Journal of Food Microbiology 42 (1–2), 9–27.

Landini, P., Antoniani, D., Burgess, J.G., Nijland, R., 2010. Molecular mechanisms of compounds affecting bacterial biofilm formation and dispersal. Applied Microbiology and Biotechnology 86 (3), 813–823.

Lequette, Y., Boels, G., Clarisse, M., Faille, C., 2010. Using enzymes to remove biofilms of bacterial isolates sampled in the food-industry. Biofouling 26 (4), 421–431.

Leroy, C., Delbarre, C., Ghillebaert, F., Compere, C., Combes, D., 2008. Effects of commercial enzymes on the adhesion of a marine biofilm-forming bacterium. Biofouling 24 (1), 11–22.

Lewis, K., 2007. Persister cells, dormancy and infectious disease. Natural Reviews Microbiology 5, 48–56.

Lewis, K., 2010. Persister cells. Annual Reviews in Microbiology 64, 357–372.

Liu, Y., Yang, S.F., Li, Y., Xu, H., Qin, L., Tay, J.H., 2004. The influence of cell and substratum surface hydrophobicities on microbial attachment. Journal of Biotechnology 110 (3), 251–256.

Loiselle, M., Anderson, K.W., 2003. The use of cellulase in inhibiting biofilm formation from organisms commonly found on medical implants. Biofouling 19 (2), 77–85.

Mah, T.F.C., O'Toole, G.A., 2001. Mechanisms of biofilm resistance to antimicrobial agents. Trends in Microbiology 9 (1), 34–39.

Marcato-Romain, C.E., Pechaud, Y., Paul, E., Girbal-Neuhauser, E., Dossat-Letisse, V., 2012. Removal of microbial multi-species biofilms from the paper industry by enzymatic treatments. Biofouling 28 (3), 305–314.

Molobela, I.P., Cloete, T.E., Beukes, M., 2010. Protease and amylase enzymes for biofilm removal and degradation of extracellular polymeric substances (EPS) produced by *Pseudomonas fluorescens* bacteria. African Journal of Microbiology Research 4 (14), 1515–1524.

Nadell, C.D., Xavier, J.B., Foster, K.R., 2009. The sociobiology of biofilms. FEMS Microbiology Reviews 33 (1), 206–224.

Nitschke, M., Costa, S.V.A.O., Contiero, J., 2010. Structure and applications of a rhamnolipid surfactant produced in soybean oil waste. Applied Biochemistry and Biotechnology 160 (7), 2066–2074.

Novozymes, 2019. Biosolutions – Giving You Industrial Efficiency and Product Improvements with Cost Savings. Denmark. Available from: www.novozymes.com/en/solutions/Pages/default.aspx. (Accessed 15 January 2019).

Orgaz, B., Kives, J., Pedregosa, A.M., Monistrol, I.F., Laborda, F., SanJosé, C., 2006. Bacterial biofilm removal using fungal enzymes. Enzyme and Microbial Technology 40 (1), 51–56.

Oulahal-Lagsir, N., Martial-Gros, A., Bonneau, M., Blum, L.J., 2003. "*Escherichia coli*-milk" biofilm removal from stainless steel surfaces: synergism between ultrasonic waves and enzymes. Biofouling 19 (3), 159–168.

Oulahal, N., Martial-Gros, A., Bonneau, M., Blum, L.J., 2007. Removal of meat biofilms from surfaces by ultrasounds combined with enzymes and/or a chelating agent. Innovative Food Science & Emerging Technologies 8 (2), 192–196.

Packiavathy, I.A.S., Maruthamuthu, S., Gnanaselvan, G., Manoharan, S., Paul, J.B.J., Annapoorani, A., Kannappan, A., Ravi, A.V., 2019. The control of microbially induced corrosion by methyl eugenol – a dietary phytochemical with quorum sensing inhibitory potential. Bioelectrochemistry 128, 186–192.

Patel, R., 2005. Biofilms and antimicrobial resistance. Clinical Orthopaedics and Related Research 437, 41–47.

Pechaud, Y., Marcato-Romain, C.E., Girbal-Neuhauser, E., Queinnec, I., Bessiere, Y., Paul, E., 2012. Combining hydrodynamic and enzymatic treatments to improve multi-species thick biofilm removal. Chemical Engineering Science 80 (0), 109–118.

Percival, S.L., Malic, S., Cruz, H., Williams, D.W., 2011. Introduction to biofilms. In: Percival, S.L.K., Derek, C., Cochrane, C.A. (Eds.), Biofilms and Veterinary Medicine. Springer-Verlag Berlin Heidelberg.

Prateeksha, Rao, C.V., Das, A.K., Barik, S.K., Singh, B.N., 2019. ZnO/Curcumin nanocomposites for enhanced inhibition of *Pseudomonas aeruginosa* virulence via LasR-RhlR quorum sensing systems. Molecular Pharmaceutics 16 (8), 3399–3413.

Quecán, B.X.V., Rivera, M.L.C., Pinto, U.M., 2018. Bioactive phytochemicals targeting microbial activities mediated by quorum sensing. In: Kalia, V.C. (Ed.), Biotechnological Applications of Quorum Sensing Inhibitors. Springer Singapore, Singapore.

Realco, 2016. Our Markets. Belgium. Available from: www.realco.be/en/our-markets. (Accessed 15 January 2016).

Richards, J.J., Melander, C., 2009. Controlling bacterial biofilms. ChemBioChem: European Journal of Chemical Biology 10 (14), 2287–2294.

Saleem, M., Nazir, M., Ali, M.S., Hussain, H., Lee, Y.S., Riaz, N., Jabbar, A., 2010. Antimicrobial natural products: an update on future antibiotic drug candidates. Natural Product Reports 27 (2), 238–254.

Samy, R.P., Gopalakrishnakone, P., 2010. Therapeutic potential of plants as anti-microbials for drug discovery. Evidence-based Complementary and Alternative Medicine 7 (3), 283–294.

Sandasi, M., Leonard, C.M., Viljoen, A.M., 2010. The in vitro antibiofilm activity of selected culinary herbs and medicinal plants against *Listeria monocytogenes*. Letters in Applied Microbiology 50 (1), 30–35.

Satpute, S.K., Mone, N.S., Das, P., Banat, I.M., Banpurkar, A.G., 2019. Inhibition of pathogenic bacterial biofilms on PDMS based implants by *L. acidophilus* derived biosurfactant. BMC Microbiology 19 (1), 39.

Satpute, S.K., Mone, N.S., Das, P., Banpurkar, A.G., Banat, I.M., 2018. *Lactobacillus acidophilus* derived biosurfactant as a biofilm inhibitor: a promising investigation using microfluidic approach. Applied Sciences 8 (9), 1555.

Savage, V.J., Chopra, I., O'Neill, A.J., 2013. *Staphylococcus aureus* biofilms promote horizontal transfer of antibiotic resistance. Antimicrobial Agents and Chemotherapy 57 (4), 1968–1970.

Sharahi, J.Y., Azimi, T., Shariati, A., Safari, H., Tehrani, M.K., Hashemi, A., 2019. Advanced strategies for combating bacterial biofilms. Journal of Cellular Physiology 234 (9), 14689–14708.

Shen, H.-B., Chou, K.-C., 2007. EzyPred: a top–down approach for predicting enzyme functional classes and subclasses. Biochemical and Biophysical Research Communications 364 (1), 53–59.

Shi, X., Zhu, X., 2009. Biofilm formation and food safety in food industries. Trends in Food Science & Technology 20 (9), 407–413.

Silva, A., Borges, A., Freitas, C., Hioka, N., Mikcha, J., Simões, M., 2018a. Antimicrobial photodynamic inactivation mediated by rose bengal and erythrosine is effective in the control of food-related bacteria in planktonic and biofilm states. Molecules 23 (9), 2288.

Silva, A.F., Borges, A., Giaouris, E., Graton Mikcha, J.M., Simões, M., 2018b. Photodynamic inactivation as an emergent strategy against foodborne pathogenic bacteria in planktonic and sessile states. Critical Reviews in Microbiology 44 (6), 667–684.

Silva, A.F., dos Santos, A.R., Trevisan, D.A.C., Bonin, E., Freitas, C.F., Batista, A.F.P., Hioka, N., Simões, M., Graton Mikcha, J.M., 2019. Xanthene dyes and green LED for the Inactivation of foodborne pathogens in planktonic and biofilm states. Photochemistry and Photobiology.

Simões, M., 2011. Antimicrobial strategies effective against infectious bacterial biofilms. Current Medicinal Chemistry 18 (14), 2129–2145.

Simões, M., Bennett, R.N., Rosa, E.A., 2009. Understanding antimicrobial activities of phytochemicals against multidrug resistant bacteria and biofilms. Natural Product Reports 26 (6), 746–757.

Simões, M., Simões, L.C., Vieira, M.J., 2010. A review of current and emergent biofilm control strategies. LWT - Food Science and Technology 43 (4), 573–583.

Singh, A.P., Singh, J., 2013. Antibiofilm strategies. In: Simões, M., Mergulhão, F. (Eds.), Biofilms in Bioengineering. Nova Science Pub Inc, New York.

Smith, A.W., 2005. Biofilms and antibiotic therapy: is there a role for combating bacterial resistance by the use of novel drug delivery systems? Advanced Drug Delivery Reviews 57 (10), 1539–1550.

Smith, A.W., Brown, M.R.W., Smith, A.W., Brown, M.R.W., 2003. Biofilms, dormancy and resistance. In: Dormancy and Low-Growth States in Microbial Disease. Cambridge University Press.

Soukos, N.S., Goodson, J.M., 2011. Photodynamic therapy in the control of oral biofilms. Periodontology 2000 55 (1), 143–166.

Srey, S., Jahid, I.K., Ha, S.-D., 2013. Biofilm formation in food industries: a food safety concern. Food Control 31 (2), 572–585.

Stewart, P.S., 1996. Theoretical aspects of antibiotic diffusion into microbial biofilms. Antimicrobial Agents and Chemotherapy 40 (11), 2517–2522.

Stewart, P.S., 2002. Mechanisms of antibiotic resistance in bacterial biofilms. International Journal of Medical Microbiology 292 (2), 107–113.

Stewart, P.S., Costerton, J.W., 2001. Antibiotic resistance of bacteria in biofilms. Lancet 358 (9276), 135–138.

Stiefel, P., Mauerhofer, S., Schneider, J., Maniura-Weber, K., Rosenberg, U., Ren, Q., 2016. Enzymes enhance biofilm removal efficiency of cleaners. Antimicrobial Agents and Chemotherapy 60 (6), 3647–3652.

Stoodley, P., Sauer, K., Davies, D.G., Costerton, J.W., 2002. Biofilms as complex differentiated communities. Annual Review of Microbiology 56, 187–209.

Subhadra, B., Kim, H.D., Woo, K., Surendran, S., Choi, H.C., 2018. Control of biofilm formation in healthcare: recent advances exploiting quorum-sensing interference strategies and multidrug efflux pump inhibitors. Materials 11 (9).

Taraszkiewicz, A., Fila, G., Grinholc, M., Nakonieczna, J., 2013. Innovative strategies to overcome biofilm resistance. BioMed Research International 2013, 13.

Thallinger, B., Prasetyo, E.N., Nyanhongo, G.S., Guebitz, G.M., 2013. Antimicrobial enzymes: an emerging strategy to fight microbes and microbial biofilms. Biotechnology Journal 8 (1), 97–109.

Tokubo, L.M., Rosalen, P.L., de Cássia Orlandi Sardi, J., Freires, I.A., Fujimaki, M., Umeda, J.E., Barbosa, P.M., Tecchio, G.O., Hioka, N., de Freitas, C.F., Suga Terada, R.S., 2018. Antimicrobial effect of photodynamic therapy using erythrosine/methylene blue combination on *Streptococcus mutans* biofilm. Photodiagnosis and Photodynamic Therapy 23, 94–98.

Torres, C.E., Negro, C., Fuente, E., Blanco, A., 2012. Enzymatic approaches in paper industry for pulp refining and biofilm control. Applied Microbiology and Biotechnology 96 (2), 327–344.

Vattem, D.A., Mihalik, K., Crixell, S.H., McLean, R.J.C., 2007. Dietary phytochemicals as quorum sensing inhibitors. Fitoterapia 78 (4), 302–310.

Vishwakarma, V., Josephine, J., George, R.P., Krishnan, R., Dash, S., Kamruddin, M., Kalavathi, S., Manoharan, N., Tyagi, A.K., Dayal, R.K., 2009. Antibacterial copper–nickel bilayers and multilayer coatings by pulsed laser deposition on titanium. Biofouling 25 (8), 705–710.

Viswanathan, A., Rangasamy, J., Biswas, R., 2019. Functionalized antibacterial nanoparticles for controlling biofilm and intracellular infections. In: Pathak, Y.V. (Ed.), Surface Modification of Nanoparticles for Targeted Drug Delivery. Springer International Publishing, Cham.

Vu, B., Chen, M., Crawford, R., Ivanova, E., 2009. Bacterial extracellular polysaccharides involved in biofilm formation. Molecules 14 (7), 2535–2554.

Walsh, C., 2000. Molecular mechanisms that confer antibacterial drug resistance. Nature 406 (6797), 775–781.

Wiatr, C.L., 1991. Enzyme Blend Containing Cellulase to Control Industrial Slime.

Xavier, J.B., Picioreanu, C., Rani, S.A., van Loosdrecht, M.C.M., Stewart, P.S., 2005. Biofilm-control strategies based on enzymic disruption of the extracellular polymeric substance matrix – a modelling study. Microbiology 151 (12), 3817–3832.

Xu, K.D., McFeters, G.A., Stewart, P.S., 2000. Biofilm resistance to antimicrobial agents. Microbiology 146 (3), 547–549.

Yamasaki, M., Ogura, K., Hashimoto, W., Mikami, B., Murata, K., 2005. A structural basis for depolymerization of alginate by polysaccharide lyase family-7. Journal of Molecular Biology 352 (1), 11–21.

Yang, S.M., Lee, D.W., Park, H.J., Kwak, M.H., Park, J.M., Choi, M.-G., 2019. Hydrogen peroxide enhances the antibacterial effect of methylene blue-based photodynamic therapy on biofilm-forming bacteria. Photochemistry and Photobiology 95 (3), 833–838.

Yuan, L., Hansen, M.F., Røder, H.L., Wang, N., Burmølle, M., He, G., 2019. Mixed-species biofilms in the food industry: current knowledge and novel control strategies. Critical Reviews in Food Science and Nutrition 1–17.

Zanaroli, G., Negroni, A., Calisti, C., Ruzzi, M., Fava, F., 2011. Selection of commercial hydrolytic enzymes with potential antifouling activity in marine environments. Enzyme and Microbial Technology 49 (6–7), 574–579.

Zezzi do Valle Gomes, M., Nitschke, M., 2012. Evaluation of rhamnolipid and surfactin to reduce the adhesion and remove biofilms of individual and mixed cultures of food pathogenic bacteria. Food Control 25 (2), 441–447.

Zhang, T., Ke, S.Z., Liu, Y., Fang, H.P., 2005. Microbial characteristics of a methanogenic phenol-degrading sludge. Water Science and Technology 52 (1–2), 73–78.

Zhou, J.-W., Ruan, L.-Y., Chen, H.-J., Luo, H.-Z., Jiang, H., Wang, J.-S., Jia, A.-Q., 2019. Inhibition of quorum sensing and virulence in *Serratia marcescens* by hordenine. Journal of Agricultural and Food Chemistry 67 (3), 784–795.

Zhu, H., Sun, S.J., 2008. Inhibition of bacterial quorum sensing-regulated behaviors by *Tremella fuciformis* extract. Current Microbiology 57 (5), 418–422.

Chapter 12

The potential of phytochemical products in biofilm control

Mafalda Andrade[1,a], Joana Malheiro[1,2,3,a], Fernanda Borges[3], Maria José Saavedra[4], Manuel Simões[1]

[1]*LEPABE — Laboratory for Process Engineering Environment, Biotechnology and Energy, Department of Chemical Engineering, Faculty of Engineering, University of Porto, Porto, Portugal;* [2]*CIQUP/Department of Chemistry and Biochemistry, Faculty of Sciences, University of Porto, Porto, Portugal;* [3]*Cardiff School of Pharmacy and Pharmaceutical Sciences, Cardiff University, Cardiff, Wales, United Kingdom;* [4]*CITAB-Centre for Research and Technology of Agro-Environmental and Biological Sciences, Veterinary Science Department, University of Trás-os-Montes e Alto Douro, Vila Real, Portugal*

12.1 Antimicrobial properties of phytochemicals

The increased bacterial resistance observed with classical antimicrobial treatments makes biofilm elimination a major challenge with serious economic and health repercussions. Biofilms have been implicated in medicine, as the cause of many chronic and biomaterial-associated infections, and in the industrial and environmental settings, as the cause for biofouling, biocorrosion, and biodeterioration, especially in food processing and water distribution systems (Budzynska et al., 2011; Gilbert et al., 2002; Madigan et al., 2009; Oniciuc et al., 2019). However, the successful use of any therapeutic agent is compromised by the potential development of drug tolerance or resistance, a problem that is often associated to the treatment of microbial infections and of chronic diseases. The rise of antibiotic resistance in hospital environments has also been related with the formation of microbial biofilms being a serious contributor for the therapeutic shortcomings.

The current concerns over bacterial multiresistance, as well as biofilm resistance, not only to the conventional treatments but also to the new generation drugs, and toxicity to some of the current employed antimicrobials, have stimulated the screening and development of new active products and new improved alternative strategies for biofilm control (Borges et al., 2013; Song et al., 2017; Trentin Dda et al., 2011). These types of studies are

[a] Both authors contributed equally.

Recent Trends in Biofilm Science and Technology. https://doi.org/10.1016/B978-0-12-819497-3.00012-X

particularly focused in the discovery of new types of antimicrobials based on different scaffolds or operating by different mechanisms of action that can be less susceptible to bacterial resistance mechanisms (Bellik et al., 2012; Limsuwan et al., 2009).

Plants, especially medicinal plants, are an interesting source of bioactive compounds; conventionally, they have been extensively used for basic healthcare needs, such as treatment of skin infection. In addition, diverse phytochemicals have also been described to behave as coadjutants of antibiotics, enhancing their activity as resistance-modifying agents (Abreu et al., 2012; Hemaiswarya et al., 2008). Obtaining these phytochemicals includes a process of extraction from the plant. The extraction is the most crucial step, since it is necessary to obtain the desired phytochemical components from the plant to be possible their separation and characterization. The quality of the plant extract will always depend on the plant material, choice of solvent, and extraction methods (Brusotti et al., 2014; Das et al., 2010). Considering the plant material, a homogeneous sample should be obtained by prewashing and drying the material. Next, the solvents choice will depend on the bioactive nature of the target chemical since the solvents allow the diffusion into the solid plant tissues and the solubilization of compounds with similar polarities. Several other techniques can also be used, such as sonication, heating under reflux, as well as more modern methods such as solid-phase extraction, microextraction, supercritical–fluid extraction, pressurized–liquid extraction, microwave-assisted extraction, and surfactant-mediated techniques. Modern techniques possess several advantages such as decrease in solvent consumption and sample degradation (Ignat et al., 2011; Sasidharan et al., 2011). Plant extracts are a combination of various types of phytochemicals; therefore, their identification and characterization is a big challenge. Considering the isolation and characterization of the phytochemicals, several techniques can be used such as thin-layer, column, flash, and Sephadex chromatography as well as high-performance liquid chromatography. In addition, immunoassays, monoclonal antibodies, and Fourier transform infrared spectroscopy can also be used to isolate and characterize the compounds (Brusotti et al., 2014; Ignat et al., 2011; Sasidharan et al., 2011).

12.1.1 Phytochemical classes

Phytochemicals classification is not universal as it can be grounded on different features, such as botanical or biosynthetic in origin and based on their chemical structure or biological properties. A simple phytochemical classification has been proposed by Scalbert et al. (2011) (Das et al., 2010; Scalbert et al., 2011) based on the presence of characteristic chemical structural motifs or by its chemical functionality that determines which class they belong to, such as 2-phenyl-1,4-benzopyrone in the case of flavonoids and phytosterols that have a steroid backbone that is hydroxylated in the 3-position of the A-ring or alkaloids that contain nitrogen atoms in complex and highly diverse structures. However, the same phytochemicals can be included in more than one class, once it can present motifs that fit several classes such as oleuropein that can be classified as a terpene or a phenolic compound since it contains substructures for the two classes. The opposite scenario also exists with respect to classification, where structurally different phytochemicals can be incorporated into the same class when bioactivity is considered as a rule (Cowan, 1999; Scalbert et al., 2011). A classic phytochemical classification is shown in Fig. 12.1 (Bellik et al., 2012; Liu, 2004). Despite the variety of classes and subclasses of phytochemicals, some of the most common and studied are polyphenols (e.g., flavonoids and terpenes) and alkaloids, but others must also be considered, such as carotenoids and organosulfur-based compounds (Abreu et al., 2012; Dillard and German, 2000; Saxena et al., 2013; Scalbert et al., 2011).

Polyphenols are compounds that possess aromatic rings in their chemical structure with one or more hydroxyl groups. This class of secondary plant metabolites can contribute to the color of several parts of plants and participate in essential reproduction and plant growth functions, as well as acting upon the plant's defense mechanisms (Liu, 2004; Pandey and Rizvi, 2009). It is estimated that two-thirds of the phenolic content in our diet are flavonoids and the remaining one-third are phenolic acids (Liu, 2004). Flavonoids are abundant in photosynthesizing cells, and they can be found in fruit, vegetables, seeds, nuts, stems, and flowers (Cushnie and Lamb, 2005a; Liu, 2004). They have several functions depending on the organ in which they are inserted, such as promoting plant physiological survival, photosensitization, energy transfer, action of growth hormones and regulators, control of respiration, photosynthesis, morphogenesis, and sex determination (Cushnie and Lamb, 2005a).

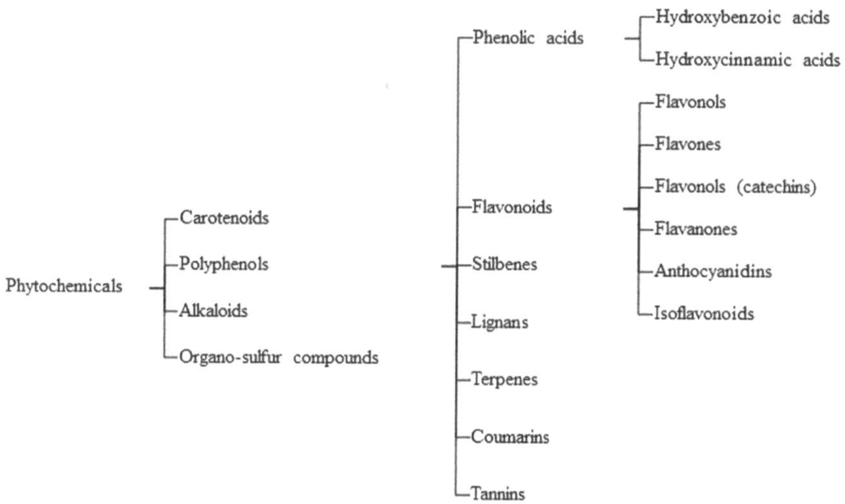

FIGURE 12.1 Phytochemicals chemical classification (Bellik et al., 2012; Liu, 2004).

The basic structure of a flavonoid is a flavane nucleus, which consists of two benzene rings linked through a heterocyclic pyrane ring. The chemical decoration of the core leads to a diversity of flavonoid-based structures (Fig. 12.1).

Phenolic acids can be found in plants in both their free and esterified forms and may be classified into two subclasses: hydroxybenzoic acids and hydroxycinnamic acids. Hydroxybenzoic acids, such as gallic or protocatechuic acids, are generally found in very low quantities in plants and are components of complex structures such as hydrolysable tannins. On the other hand, hydroxycinnamic acids, such as cinnamic, *p*-coumaric, ferulic, caffeic, and synaptic acids, are more common than hydroxybenzoic acids (Manach et al., 2004). Coumarins are phenolic compounds that occur by fusion of benzene and α-pyrone ring (Cowan, 1999). Moreover, the phenolic acids are commonly present in plants in the bound form (esterified with sugar or organic acids) and are typically part of complex structures such as lignins and hydrolysable tannins (Liu, 2004). Lignans are diphenolic compounds, with a 2,3-dibenzylbutane structure, formed by dimerization of two cinnamic acid residues, and tannins are polymeric phenolics with a molecular weight of 500–3000 Da (Pandey and Rizvi, 2009). Furthermore, stilbenes have two phenyl moieties linked by carbon methylene bridges. Their occurrence in plants is low as they are only synthetized in response to infection or injury, acting as antifungal phytoalexins (Pandey and Rizvi, 2009).

Alkaloids are a huge class of phytochemicals containing heterocyclic nitrogen compounds that form salts when exposed to acids. They comprise a wide range of different structures, which makes their classification difficult, therefore being grouped into several subclasses depending on the type of the heterocyclic ring system (Saxena et al., 2013). Another important class of phytochemicals are the terpenoids that are derived from five-carbon isoprene units and have basic multicyclic structures (Saxena et al., 2013).

Carotenoids consist of eight isoprenoid units joined. They are fat-soluble natural pigments that are known for their antioxidant properties and physiological functions such as immune stimulation. In humans, they have vitamin A activity and can act as antioxidants (Campos-Vega and Oomah, 2013). Organosulfur compounds are known for their antioxidant, antiinflammatory properties, inhibition of platelet aggregation, reduction of systolic blood pressure, and cholesterol. The most studied organosulfur compound is isothiocyanate that is formed by replacing the sulfur for oxygen in the isocyanate group (Vazquez-Prieto and Miatello, 2010).

12.1.2 Modes of antimicrobial action

Phytochemical compounds can display single or several modes of action, being distinct from the ones displayed by current antibiotics. In general, phytochemicals are known to inhibit microbial growth, interfere with its metabolic processes, and modulate gene expression and signal transduction pathways. Therefore, their usage in synergism with antibiotics has been widely studied. An example is the case of epigallocatechin gallate, a flavonoid derivative, which can inhibit penicillinase produced by *S. aureus*, thus allowing penicillin to be effective (Hemaiswarya et al., 2008).

12.1.2.1 Interference with DNA synthesis or expression

There is a wide list of phytochemicals that can act as inhibitors of DNA synthesis or expression, consequently affecting bacterial metabolism. Flavonoids, such as robinetin, myricetin, and (−)-epigallocatechin, are capable of inhibiting DNA synthesis of *Proteus vulgaris* and RNA synthesis of *S. aureus*, by intercalation of their B ring and/or by establishing hydrogen bonds with the stacking of nucleic acid bases (Cushnie and Lamb, 2005a; Mori et al., 1987). In addition, DNA gyrase was shown to be inhibited by several flavonoids, as observed by Ohemeng and coworkers and Plaper et al. (Cushnie and Lamb, 2005a; Ohemeng et al., 1993; Plaper et al., 2003). Furthermore, other phytochemicals have been shown to target other biomolecules: rutin inhibits topoisomerase IV, myricetin inhibits DnaB helicase, allicin inhibits RNA synthesis, and several other compounds from *Polygonum cuspidatum* can inhibit DNA primase (Bernard et al., 1997; Feldberg et al., 1988; Griep et al., 2007; Hegde et al., 2004).

12.1.2.2 Destabilization of cytoplasmic membrane function

Plant alkaloids such as berberine and piperine can inhibit efflux pumps after they intercalate with DNA in *S. aureus* (Khan et al., 2006; Simões et al., 2009). In addition, Ikigai et al. (1993) hypothesized for the first time the action of "catechins," especially (−)-epigallocatechin gallate, causing damage of bacterial membranes by direct penetration and destabilization or even by membrane fusion, which results in leakage of intracellular material and leads to aggregation (Cushnie and Lamb, 2005a; Ikigai et al., 1993). Furthermore, galangin also promotes membrane damage and potassium leakage in *S. aureus* (Cushnie and Lamb, 2005b), as well as sophoraflavanone G and naringenin that demonstrated antibacterial activity as a consequence of the reduction in membrane fluidity of bacteria (Tsuchiya and Iinuma, 2000). Another example of membrane permeabilizers are prenylated (iso)flavonoids and essential oils, such as thymol and carvacrol that lead to increase in membrane permeability, disturbance of membrane proteins, and alteration of ion transport pathways (Araya-Cloutier et al., 2018; Helander et al., 1998; Trombetta et al., 2005). Glycoside saponins may induce porelike structures, and consequently, membrane permeability is affected (Melzig et al., 2001; Simões et al., 2009).

In addition to the physical barrier provided by membranes, another defense mechanism used by bacteria is the development of multidrug resistance pumps, which prevents the accumulation of antimicrobials inside the cells (Hemaiswarya et al., 2008). Certain plant-derived natural compounds can control efflux pumps, such as carnosic acid and penta-substituted pyridine that are able to inhibit NorA efflux pump and reserpine that inhibits an ABC efflux systems (Hemaiswarya et al., 2008; Marquez et al., 2005; Oluwatuyi et al., 2004).

12.1.2.3 Destabilization of metabolic pathways

Another target of action of phytochemicals on bacterial cells is by regulating the enzymes involved in metabolic pathways. Several phenols and phenolic acids can disturb energy production by inhibiting enzyme's activity, such as dehydrogenase on the plasma membrane (Lin et al., 2005). Glycoside saponins also have the capacity to interfere with membrane metabolism as a consequence of the interaction with catabolic enzymes and the electron transport chain (Mandal et al., 2005).Haraguchi et al. (1998) also hypothesized that licochalcones may interfere with energy metabolism, especially coenzyme Q, since energy is necessary for active uptake of several metabolites and for the biosynthesis of macromolecules (Haraguchi et al., 1998).

12.1.2.4 Medicinal properties

Phytochemicals or phytoextracts, despite their proven efficacy as antibacterial chemicals, are recognized for other medicinal purposes, such as anticancer agents, as already reviewed in detail (HemaIswarya and Doble, 2006; Hemaiswarya and Doble, 2010). One example are flavonoids that, despite their antimicrobial activity, have other properties such as antifungal, antiviral, antiinflammatory, estrogenic, antiallergic, antioxidant, vascular and cytotoxic antitumor activities, and even enzyme inhibition capacity (Cushnie and Lamb, 2005a; Gautam and Jachak, 2009). In addition, antiinflammatory properties were also observed for other phytochemicals, such as phenolics, alkaloids, and terpenoids; however, their mode of action is not yet fully understood (Bellik et al., 2012). Phenolic compounds are also associated with health benefits such as the reduction of chronic diseases. The phenolic groups can accept electrons to form relatively stable phenoxyl radicals, therefore disrupting chain oxidation reactions in cellular components (Pandey and Rizvi, 2009; Scalbert et al., 2005). There are several evidences that, as antioxidants, phenolic phytochemicals may protect cell constituents against oxidative damage and consequently diminish the risk of various degenerative diseases associated with oxidative stress (Pandey and Rizvi, 2009). Furthermore, phenolic compounds are also associated with reduction and protection against the progression of several chronic pathological diseases such as cancer, diabetes, cardiovascular complications, and aging (Harman, 2006; Rizvi and Zaid, 2005; Scalbert et al., 2005; Yang et al., 2001).

Although plant products can exhibit promising outcomes for medicinal purposes, they also possess cytotoxic effects, such as triggering allergic reactions, cardiovascular diseases (by increasing blood pressure), and homocysteinemia (Bellik et al., 2012). Phytochemical studies on medicinal properties proved their potential as anticancers (Kawasaki et al., 2008), antioxidants (Son et al., 2008), antimicrobials (Cowan, 1999; Khoury et al., 2019; Song et al., 2017), and fungicidals (Boulogne et al., 2012; Khoury et al., 2019).

12.2 Phytochemicals as biofilm-controlling agents

Presence of biofilms possesses a huge economic burden in several environments (Kuehn et al., 2010; Van Houdt and Michiels, 2010). Therefore, studies to identify new and improved strategies for biofilm control are of extreme importance. Phytochemicals have been tested for their potential antimicrobial activity and mode of action against several kinds of bacteria in the planktonic state, as well as for their potential ability to act as inhibitors of biofilm formation or even biofilm controllers.

Currently, there is no known technique or chemical that completely prevents or controls biofilm development. However, considering the stages of biofilm formation, antibiofilm strategies can be based on inhibition of microbial adhesion to the surface, interference with signal molecules involved in biofilm development, and disaggregation of the biofilm matrix (Francolini and Donelli, 2010). In fact, the control of an already established biofilm is far more difficult to achieve than to prevent or inhibit the initial stages of biofilm formation, namely bacterial adhesion (Borges et al., 2014a, 2014b). This way, cells are kept in a planktonic state, switching off expression of virulence, which is typical of biofilms and making the microorganisms more susceptible to the action of other antimicrobials (Sharahi et al., 2019; Trentin Dda et al., 2011).

Bacterial adhesion is a complex process that is affected by many factors such as the biological properties of the bacteria (presence of fimbriae or flagella, production of expandable polystyrene, etc.), the physicochemical characteristics of the bacteria (hydrophobicity, surface charge, etc.), the adhesion material surface properties (chemical composition, surface charge, hydrophobicity, roughness, or texture), and environmental factors (temperature, pH, time of exposure, bacterial concentration, presence of chemical or antimicrobial treatment, and flow conditions) (Simões et al., 2010b). Considerable resources have been directed toward technologies designed to inhibit microbial attachment. Prospects have included surface material coatings that prevent adhesion and responsive surfaces that phase change upon command or controlled orientation of surface-tethered adhesion molecules (Bryers, 2008; Francolini and Donelli, 2010; Simões et al., 2010a), but, nowadays, the most popular approaches to prevent bacterial adhesion are through quorum sensing (QS) and motility inhibition, which are both important steps of biofilm formation and development (Simões et al., 2010a; Venkadesaperumal, 2015; Wojnicz et al., 2012).

Bacterial motility and, in particular, swimming and swarming are dependent on flagella and contribute to cell adhesion to biotic and abiotic surfaces. Swarming has long been recognized to be important for both the initial interactions with surfaces and for the movement along these and, therefore, for the early steps of biofilm formation (Lee et al., 2014b; Wojnicz et al., 2012). Biofilm formation is invariably preceded by attachment mediated by the abovementioned flagellar motilities, while later stages are due to twitching motility (implicated in cell recruitment from adjacent monolayers and cell aggregate formation), which is related to type IV pili (Borges et al., 2012; Kumar et al., 2013). All considered, the importance of flagella in biofilm formation makes it an attractive target for the development of alternative biofilm control chemicals (Vikram et al., 2013). In fact, several phytochemical compounds have been shown to inhibit biofilm formation by affecting the microorganisms' motility, such as galactoside and glucoside derivatives in *Escherichia coli* (Kawasaki et al., 2008; Yang et al., 2001) and coumarin derivatives, gallic acid, and catechins in both *E. coli* and *P. aeruginosa* (Boulogne et al., 2012; Wojnicz et al., 2012).

Another important event that affects bacterial biofilm growth and development is related with cell–cell interactions (signaling pathways). This is due to the fact that bacterial communities have developed sophisticated ways to interact and associate with the environment in which they are inserted, such as allowing bacteria to monitor their population density and accordingly control a variety of physiological processes by releasing and receiving small signaling molecules, named autoinducers (Nazzaro et al., 2013; Zhang et al., 2014). The coordination is established by bacterium–bacterium interactions and associations with higher organisms through intercellular communication systems that involve the expression of new genes known as QS systems (Nazzaro et al., 2013). This cell-to-cell communication mechanism influences both initiation and maturation of bacterial biofilms and has been showed to be an important regulatory mechanism in biofilm formation/differentiation, rendering the interference with QS systems as a highly attractive and promising target to tackle biofilm control (Borges et al., 2014a; Nazzaro et al., 2013; Zhang et al., 2014). The inhibition of QS may occur through several mechanisms including inhibition of acylated homoserine lactone (AHL) synthesis, inhibition of AHL transport and/or secretion, sequestrations of AHL, and antagonistic action or inhibition of targets downstream of AHL receptor binding (Chevrot et al., 2006; Nazzaro et al., 2013; Niu et al., 2006; Song et al., 2017; Vandeputte et al., 2011). On the other hand, QS can regulate several bacterial activities, such as bioluminescence, virulence factor expression, swarming motility, sporulation, and biofilm production (Nazzaro et al., 2013). Several quorum-quenching compounds have been identified from plants and were summarized by Nazzaro et al. (2013). It is possible to observe that the majority of the compounds presented are categorized as phenolic compounds, such as flavonoids, cinnamaldehyde, and curcumin. Among the quorum-quenching phytochemicals, compounds such as cinnamaldehyde derivatives and catechin derivatives have been shown to be successful in inhibiting biofilm formation in *P. aeruginosa*, while others such as furocoumarins, salicylic acid, rosmarinic acid, and ursolic acid have also been proven to effectively inhibit biofilm formation in *E. coli* (Son et al., 2008). Song et al. (2017) also summarized several plant extracts that presented anti-QS activity such as *Polyporus squamosus* and *Carex dimorpholepis* methanolic extracts.

Another attractive approach is the use of chemicals capable of disintegrating the biofilm matrix once, as a consequence of the loss of the highly protective barrier (exopolysaccharide matrix), sessile microbial cells are directly exposed to the biocide (Francolini and Donelli, 2010).

Regarding all the possible strategies to prevent and control biofilm development, phytochemicals have been employed and studied for a variety of purposes, including inhibition of bacterial adhesion through interference of these compounds in QS systems, bacterial motility, or bacterial surface properties (i.e., membrane destabilization) Table 12.1.

Overall, the results presented in the literature regarding in vitro studies on the use of phytochemicals as antimicrobials and biofilm inhibitors are very enthusiastic, specially taking into account the green and natural character of these compounds (many of which are found in human dietary products, thus rendering a safe cytotoxic status) and their diverse and broad spectrum modes of action, which would not further inflate the bacterial multiresistance as a problematic factor (Borges et al., 2013; Manach et al., 2004). The therapeutic potential of phytochemical products as alternatives or even potentiators of common antibiotics has also been recognized (Abreu et al., 2012).

12.3 Conclusions

Phytochemical products, naturally produced by plants, are structurally diverse and present a wide range of modes of action as antimicrobial products against several microorganisms in the planktonic state. The increasing interest in phytochemicals as biofilm-controlling agents derives from the pressing need to find new alternative and effective antimicrobials that overcome the increased bacterial resistance against the classical antimicrobial treatments and, thus, help to solve a major challenge in several industrial and hospital settings: biofilm formation. Antibiofilm strategies based upon inhibition of the first stages of the biofilm formation process are becoming increasingly important and present better results than the control of fully matured biofilms. In fact,

TABLE 12.1 Examples of phytochemicals that prevent and/or control biofilm development.

	Phytochemical	Bacterium	Mode of action	Ref.
Prevention	Paeonidin-3-O-galactoside, paeonidin-3-O-arabinoside, cyanidin-3-O-galactoside, and cyanidin-3-O-glucoside	*E coli*	The influence in bacteria hydrophobicity and motility decreased cell capacity for biofilm development	Wojnicz et al. (2012)
	β-Sitosterol glucoside	*E. coli*	Biofilm formation and bacteria motility is inhibited by repression of flhDC via a mechanism involving hns and rssAB	Vikram et al. (2013)
	Coumarin derivatives	*S. aureus* *E. coli*	Repression of curli genes and motility genes, which reduces fimbriae production and swarming motility	Lee et al. (2014a)
	Gallic acid, catechin, epicatechin, epigallocatechin gallate, benzoic acid, quercetin, tannins, and kaempferol	*E. coli* *P. aeruginosa* *C. violaceum*	The influence in bacteria motility and QS decreased cell capacity for biofilm development	Zhang et al. (2014)
	Resveratrol	*Vibrio cholerae*	Prevents biofilm formation by interaction with AphB protein	Augustine et al. (2014)
	Sesquiterpenes	*S. aureus*	Inhibition of biofilm formation	Elmasri et al. (2014)

	Caffeic acid and rosmarinic acid	*P. aeruginosa*	Inhibition of biofilm formation	Scognamiglio et al. (2016)
	Petunidin	*Klebsiella pneumoniae*	Inhibition of biofilm formation by inhibiting LasR receptor pathway	Gopu et al. (2016)
	Asiatic and ursolic acid	*Enterococcus faecalis*	Reduction of growth and enzyme synthesis and inhibition of biofilm formation	Wojnicz et al. (2017)
	6,8-Trihydroxy-7-methoxy-2-(4-methoxyphenyl)-4H-chromen-4-one	*S. mutans*	Inhibition of biofilm formation	Ngabaza et al. (2018)
	Anacardic acid	*S. aureus*	Impregnated silicon catheters efficiently inhibited biofilm formation on its surface	Sajeevan et al. (2018)
	Rutin	*P. aeruginosa*	Biofilm reduction by reduction of EPS and production of reactive oxygen species	Sathiya Deepika et al. (2018)
Prevention and control	Salvipisone	*S. aureus* *Staphylococcus epidermidis*	Decrease in viable biofilm cells or limiting activity by surfactant properties affecting surface hydrophobicity or communication systems	Kuźma et al. (2007)
	Gallic acid and ferulic acid	*E. coli* *P. aeruginosa* *S. aureus* *Listeria monocytogenes*	Influence in bacteria hydrophobicity and motility and consequently decrease in cell capacity to form biofilm	Simões et al. (2010a)

Continued

TABLE 12.1 Examples of phytochemicals that prevent and/or control biofilm development.—cont'd

	Phytochemical	Bacterium	Mode of action	Ref.
	4-Hydroxybenzoic acid, vanillin, gallic acid; ferulic acid, synaptic acid, cinnamic acid, epicatechin, and chlorogenic acid	*P. aeruginosa*	Inhibition of biofilm formation	Plyuta et al. (2013)
	Zingerone	*P. aeruginosa*	The reduction in motility may lead to ineffective attachment and consequent delay or inhibition of biofilm development	Kumar et al. (2013)
	Isothiocyanates	*E. coli* *P. aeruginosa* *S. aureus* *L. monocytogenes*	Biofilm prevention and control related with antimicrobial action, interference with motility, and cell surface properties modification	(Abreu et al., 2013; Borges et al., 2014b)
	Plumbagin (5-hydroxy-2-methyl-1, 4-naphthoquinone)	*S. aureus*	Biofilm prevention and dispersion related with antimicrobial activity	Nair et al. (2016)
	Morin, pyrrolidine, and quercetin	*S. aureus*	Biofilm prevention and control related with antimicrobial action	Abreu et al. (2016)
	Catechol, veratrol, guaiacol, 2-ethoxphenol, 4-methylcatechol, 4-tert-butylcatechol, pyrogallol, 3-methoxycatechol and o-phenylene-phosphochloridite	*E. coli*	Biofilm prevention and control related with metabolic inactivation, sessile cells' membrane disruption and death, inhibition of swimming motility, and reduction of cell surface hydrophobicity	Baptista et al. (2019)

Control	Polyprenylated benzophenone 7-epiclusianone	*S. mutans*	Reduction in biofilm viability	Almeida et al. (2008)
	Linalool, linalyl acetate, α-terpineol, and terpinen-4-ol	*S. aureus* *E. coli*	Reduction in biofilm cell metabolic activity	Budzynska et al. (2011)
	7-(2-Oxohexyl)-taxodione	*S. aureus*	Reduction of biofilm viability in dose-dependent manner	Kuźma et al. (2012)
	Icariin, resveratrol, and salidroside	*Propionibacterium acnes*	Reduction in biofilm viability	Coenye et al. (2012)
	Geraniol and citral	*S. mutans*	Reduction in biofilm viability	Tofiño-Rivera et al. (2016)
	Isoquercetin and cassinopin	Methicillin-resistant *S. aureus*	The nanoparticles caused altered membrane permeability and reduced surface hydrophobicity	Lotha et al. (2019)
Quorum quenching	Sesquiterpene lactones	*P. aeruginosa*	Reduction AHLs production	Amaya et al. (2012)
	Allyl-isothiocyanate, benzyl-isothiocyanate, and 2-phenylethyl-isothiocyanate	*C. violaceum*	QS inhibitors by modulation of AHLs activity and synthesis	Borges et al. (2014a)
	Chlorogenic acid, isoorientin, orientin, isovitexin, vitexin, and rutin	*C. violaceum* *E. coli*	QS inhibition without affecting bacteria growth	Brango-Vanegas et al. (2014)
	2′,5-Di-O-galloyl-D-hamamelose	*S. aureus*	Affects biofilm susceptibility through the TraP receptor by affecting cell wall biosynthesis and extracellular DNA release	Brackman et al. (2016)

Continued

TABLE 12.1 Examples of phytochemicals that prevent and/or control biofilm development.—cont'd

	Phytochemical	Bacterium	Mode of action	Ref.
	Carvacrol and eugenol	*Pectobacterium*	Biofilm inhibition by affecting QS due to direct interaction with homoserine lactone synthase (ExpI) and the regulatory protein (ExpR)	Joshi et al. (2016)
	Carvacrol and thymol	*Pseudomonas fluorescens KM121*	QS inhibition by reduction of AHLs production, suppression of bacterial motility, and reduction of mRNA levels of flagella (*flgA*) gene	Myszka et al. (2016)
	Methyl trans-cinnamate	*C. violaceum*	QS inhibition	Malheiro et al. (2019)
	3,5,7-trihydroxyflavone	*P. aeruginosa*	Reduction of biofilm formation through the inhibition of QS-dependent phenotypes such as pyocyanin production, proteolytic, swimming motility and EPS activities, and modulation of QS gene regulation	Abinaya and Gayathri (2019)

AHLs, acylated homoserine lactones; *EPS*, expandable polystyrene; *QS*, quorum sensing.

several classes of phytochemicals have been described as presenting effective antibiofilm activities, by acting upon one or more of the microbial properties that enable biofilm formation, namely through inhibition of microbial adhesion to a surface, motility, and QS mechanisms. The broad-spectrum antimicrobial action of phytochemicals and their green status render them as interesting alternatives and/or potentiators to common antibiotics in the biofilm prevention and control scenarios. Nonetheless, additional studies on their exact mode of action, cytotoxicity, and, in some cases, improvement of cost and isolation yields must be optimized to allow phytochemicals to be used as antibiofilm agents or antibiotic coadjutants.

Acknowledgments

This work was financially supported by project UID/EQU/00511/2019—Laboratory for Process Engineering, Environment, Biotechnology and Energy (LEPABE) funded by national funds through FCT/MCTES (PIDDAC); project POCI-01-0145-FEDER-030219; POCI-01-145-FEDER-006939; POCI-01-0247-FEDER-035234, funded by FEDER funds through COMPETE2020—Programa Operacional Competitividade e Internacionalização (POCI) and by national funds (PIDDAC) through FCT/MCTES; project "LEPABE-2-ECO-INNOVATION"—NORTE-01-0145-FEDER-000005, funded by Norte Portugal Regional Operational Programme (NORTE 2020), under PORTUGAL 2020 Partnership Agreement, through the European Regional Development Fund (ERDF); National Funds by FCT—Portuguese Foundation for Science and Technology, under the projects UID/AGR/04033/2019 and UID/QUI/50006/2019. This work was the result of the grants awarded to Joana Malheiro (SFRH/BD/103843/2014) and Manuel Simões (SFRH/BSAB/150379/2019) from national funds through FCT—Fundação para a Ciência e Tecnologia.

References

Abinaya, M., Gayathri, M., 2019. Inhibition of biofilm formation, quorum sensing activity and molecular docking study of isolated 3, 5, 7-trihydroxyflavone from *Alstonia scholaris* leaf against *P.aeruginosa*. Bioorganic Chemistry 87, 291–301.

Abreu, A.C., Borges, A., Simões, L.C., Saavedra, M.J., Simões, M., 2013. Antibacterial activity of phenyl isothiocyanate on *Escherichia coli* and *Staphylococcus aureus*. Medicinal Chemistry 9 (5), 756–761.

Abreu, A.C., McBain, A.J., Simões, M., 2012. Plants as sources of new antimicrobials and resistance-modifying agents. Natural Product Reports 29 (9), 1007–1021.

Abreu, A.C., Saavedra, M.J., Simões, L.C., Simões, M., 2016. Combinatorial approaches with selected phytochemicals to increase antibiotic efficacy against *Staphylococcus aureus* biofilms. Biofouling 32 (9), 1103–1114.

Almeida, L.S.B., Murata, R.M., Yatsuda, R., Dos Santos, M.H., Nagem, T.J., Alencar, S.M., Koo, H., Rosalen, P.L., 2008. Antimicrobial activity of *Rheedia brasiliensis* and 7-epiclusianone against *Streptococcus mutans*. Phytomedicine 15 (10), 886–891.

Amaya, S., Pereira, J.A., Borkosky, S.A., Valdez, J.C., Bardón, A., Arena, M.E., 2012. Inhibition of quorum sensing in *Pseudomonas aeruginosa* by sesquiterpene lactones. Phytomedicine 19 (13), 1173–1177.

Araya-Cloutier, C., Vincken, J.-P., van de Schans, M.G.M., Hageman, J., Schaftenaar, G., den Besten, H.M.W., Gruppen, H., 2018. QSAR-based molecular signatures of prenylated (iso) flavonoids underlying antimicrobial potency against and membrane-disruption in Gram positive and Gram negative bacteria. Scientific Reports 8 (1), 1–14.

Augustine, N., Goel, A.K., Sivakumar, K.C., Kumar, R.A., Thomas, S., 2014. Resveratrol - a potential inhibitor of biofilm formation in *Vibrio cholerae*. Phytomedicine 21 (3), 286–289.

Baptista, J., Simões, M., Borges, A., 2019. Effect of plant-based catecholic molecules on the prevention and eradication of *Escherichia coli* biofilms: a structure activity relationship study. International Biodeterioration and Biodegradation 141, 101–113.

Bellik, Y., Boukraâ, L., Alzahrani, H.A., Bakhotmah, B.A., Abdellah, F., Hammoudi, S.M., Iguer-Ouada, M., 2012. Molecular mechanism underlying anti-inflammatory and anti-allergic activities of phytochemicals: an update. Molecules 18 (1), 322–353.

Bernard, F.X., Sablé, S., Cameron, B., Provost, J., Desnottes, J.F., Crouzet, J., Blanche, F., 1997. Glycosylated flavones as selective inhibitors of topoisomerase IV. Antimicrobial Agents and Chemotherapy 41 (5), 992–998.

Borges, A., Ferreira, C., Saavedra, M.J., Simões, M., 2013. Antibacterial activity and mode of action of ferulic and gallic acids against pathogenic bacteria. Microbial Drug Resistance 19 (4), 256–265.

Borges, A., Saavedra, M.J., Simões, M., 2012. The activity of ferulic and gallic acids in biofilm prevention and control of pathogenic bacteria. Biofouling 28 (7), 755–767.

Borges, A., Serra, S., Cristina Abreu, A., Saavedra, M.J., Salgado, A., Simões, M., 2014a. Evaluation of the effects of selected phytochemicals on quorum sensing inhibition and *in vitro* cytotoxicity. Biofouling 30 (2), 183–195.

Borges, A., Simões, L.C., Saavedra, M.J., Simões, M., 2014b. The action of selected isothiocyanates on bacterial biofilm prevention and control. International Biodeterioration and Biodegradation 86 (Part A), 25–33.

Boulogne, I., Petit, P., Ozier-Lafontaine, H., Desfontaines, L., Loranger-Merciris, G., 2012. Insecticidal and antifungal chemicals produced by plants: a review. Environmental Chemistry Letters 10 (4), 325–347.

Brackman, G., Breyne, K., De Rycke, R., Vermote, A., Van Nieuwerburgh, F., Meyer, E., Van Calenbergh, S., Coenye, T., 2016. The quorum sensing inhibitor hamamelitannin increases antibiotic susceptibility of *Staphylococcus aureus* biofilms by affecting peptidoglycan biosynthesis and eDNA release. Scientific Reports 6, 20321.

Brango-Vanegas, J., Costa, G.M., Ortmann, C.F., Schenkel, E.P., Reginatto, F.H., Ramos, F.A., Arévalo-Ferro, C., Castellanos, L., 2014. Glycosylflavonoids from *Cecropia pachystachya* Trécul are quorum sensing inhibitors. Phytomedicine 21 (5), 670–675.

Brusotti, G., Cesari, I., Dentamaro, A., Caccialanza, G., Massolini, G., 2014. Isolation and characterization of bioactive compounds from plant resources: the role of analysis in the ethnopharmacological approach. Journal of Pharmaceutical and Biomedical Analysis 87, 218–228.

Bryers, J.D., 2008. Medical biofilms. Biotechnology and Bioengineering 100 (1), 1–18.

Budzynska, A., Wieckowska-Szakiel, M., Sadowska, B., Kalemba, D., Rózalska, B., 2011. Antibiofilm activity of selected plant essential oils and their major components. Polish Journal of Microbiology 60 (1), 35–41.

Campos-Vega, R., Oomah, B.D., 2013. Chemistry and classification of phytochemicals. In: Handbook of Plant Food Phytochemicals. John Wiley & Sons Ltd.

Chevrot, R., Rosen, R., Haudecoeur, E., Cirou, A., Shelp, B.J., Ron, E., Faure, D., 2006. GABA controls the level of quorum-sensing signal in *Agrobacterium tumefaciens*. Proceedings of the National Academy of Sciences of the United States of America 103 (19), 7460–7464.

Coenye, T., Brackman, G., Rigole, P., De Witte, E., Honraet, K., Rossel, B., Nelis, H.J., 2012. Eradication of *Propionibacterium acnes* biofilms by plant extracts and putative identification of icariin, resveratrol and salidroside as active compounds. Phytomedicine 19 (5), 409–412.

Cowan, M.M., 1999. Plant products as antimicrobial agents. Clinical Microbiology Reviews 12 (4), 564–582.

Cushnie, T.P.T., Lamb, A.J., 2005a. Antimicrobial activity of flavonoids. International Journal of Antimicrobial Agents 26 (5), 343–356.

Cushnie, T.P.T., Lamb, A.J., 2005b. Detection of galangin-induced cytoplasmic membrane damage in *Staphylococcus aureus* by measuring potassium loss. Journal of Ethnopharmacology 101 (1–3), 243–248.

Das, K., Tiwari, R.K.S., Shrivastava, D.K., 2010. Techniques for evaluation of medicinal plant products as antimicrobial agents: current methods and future trends. Journal of Medicinal Plants Research 4 (2), 104–111.

Dillard, C.J., German, J.B., 2000. Phytochemicals: nutraceuticals and human health. Journal of the Science of Food and Agriculture 80 (12), 1744–1756.

Elmasri, W.A., Hegazy, M.-E.F., Aziz, M., Koksal, E., Amor, W., Mechref, Y., Hamood, A.N., Cordes, D.B., Paré, P.W., 2014. Biofilm blocking sesquiterpenes from *Teucrium polium*. Phytochemistry 103 (0), 107–113.

Feldberg, R.S., Chang, S.C., Kotik, A.N., Nadler, M., Neuwirth, Z., Sundstrom, D.C., Thompson, N.H., 1988. *In vitro* mechanism of inhibition of bacterial cell growth by allicin. Antimicrobial Agents and Chemotherapy 32 (12), 1763–1768.

Francolini, I., Donelli, G., 2010. Prevention and control of biofilm-based medical-device-related infections. FEMS Immunology and Medical Microbiology 59 (3), 227–238.

Gautam, R., Jachak, S.M., 2009. Recent developments in anti-inflammatory natural products. Medicinal Research Reviews 29 (5), 767–820.

Gilbert, P., Allison, D.G., McBain, A.J., 2002. Biofilms *in vitro* and *in vivo*: do singular mechanisms imply cross-resistance? Journal of Applied Microbiology 92 (s1), 98S–110S.

Gopu, V., Meena, C.K., Murali, A., Shetty, P.H., 2016. Petunidin as a competitive inhibitor of acylated homoserine lactones in *Klebsiella pneumoniae*. RSC Advances 6 (4), 2592–2601.

Griep, M.A., Blood, S., Larson, M.A., Koepsell, S.A., Hinrichs, S.H., 2007. Myricetin inhibits *Escherichia coli* DnaB helicase but not primase. Bioorganic and Medicinal Chemistry 15 (22), 7203–7208.

Haraguchi, H., Tanimoto, K., Tamura, Y., Mizutani, K., Kinoshita, T., 1998. Mode of antibacterial action of retrochalcones from *Glycyrrhiza inflata*. Phytochemistry 48 (1), 125–129.

Harman, D., 2006. Free radical theory of aging: an update: increasing the functional life span. Annals of the New York Academy of Sciences 1067, 10–21.

Hegde, V.R., Pu, H., Patel, M., Black, T., Soriano, A., Zhao, W., Gullo, V.P., Chan, T.-M., 2004. Two new bacterial DNA primase inhibitors from the plant *Polygonum cuspidatum*. Bioorganic & Medicinal Chemistry Letters 14 (9), 2275–2277.

Helander, I.M., Alakomi, H.-L., Latva-Kala, K., Mattila-Sandholm, T., Pol, I., Smid, E.J., Gorris, L.G.M., von Wright, A., 1998. Characterization of the action of selected essential oil components on gram-negative bacteria. Journal of Agricultural and Food Chemistry 46 (9), 3590–3595.

Hemalswarya, S., Doble, M., 2006. Potential synergism of natural products in the treatment of cancer. Phytotherapy Research 20 (4), 239–249.

Hemaiswarya, S., Doble, M., 2010. Synergistic interaction of phenylpropanoids with antibiotics against bacteria. Journal of Medical Microbiology 59 (12), 1469–1476.

Hemaiswarya, S., Kruthiventi, A.K., Doble, M., 2008. Synergism between natural products and antibiotics against infectious diseases. Phytomedicine 15 (8), 639–652.

Ignat, I., Volf, I., Popa, V.I., 2011. A critical review of methods for characterisation of polyphenolic compounds in fruits and vegetables. Food Chemistry 126 (4), 1821–1835.

Ikigai, H., Nakae, T., Hara, Y., Shimamura, T., 1993. Bactericidal catechins damage the lipid bilayer. Biochimica et Biophysica Acta (BBA) – Biomembranes 1147 (1), 132–136.

Joshi, J., Khazanov, N., Senderowitz, H., Burdman, S., Lipsky, A., Yedidia, I., 2016. Plant phenolic volatiles inhibit quorum sensing in pectobacteria and reduce their virulence by potential binding to ExpI and ExpR proteins. Scientific Reports 6, 38126.

Kawasaki, B.T., Hurt, E.M., Mistree, T., Farrar, W.L., 2008. Targeting cancer stem cells with phytochemicals. Molecular Interventions 8 (4), 174–184.

Khan, I.A., Mirza, Z.M., Kumar, A., Verma, V., Qazi, G.N., 2006. Piperine, a phytochemical potentiator of ciprofloxacin against *Staphylococcus aureus*. Antimicrobial Agents and Chemotherapy 50 (2), 810–812.

Khoury, M., El Beyrouthy, M., Ouaini, N., Eparvier, V., Stien, D., 2019. *Hirtellina lobelii* DC. essential oil, its constituents, its combination with antimicrobial drugs and its mode of action. Fitoterapia 133, 130–136.

Kuehn, C., Graf, K., Heuer, W., Hilfiker, A., Chaberny, I.F., Stiesch, M., Haverich, A., 2010. Economic implications of infections of implantable cardiac devices in a single institution. European Journal of Cardio-Thoracic Surgery 37 (4), 875–879.

Kumar, L., Chhibber, S., Harjai, K., 2013. Zingerone inhibit biofilm formation and improve antibiofilm efficacy of ciprofloxacin against *Pseudomonas aeruginosa* PAO1. Fitoterapia 90, 73–78.

Kuźma, Ł., Różalski, M., Walencka, E., Różalska, B., Wysokińska, H., 2007. Antimicrobial activity of diterpenoids from hairy roots of *Salvia sclarea* L.: salvipisone as a potential anti-biofilm agent active against antibiotic resistant *Staphylococci*. Phytomedicine 14 (1), 31–35.

Kuźma, Ł., Wysokińska, H., Różalski, M., Budzyńska, A., Wieckowska-Szakiel, M., Sadowska, B., Paszkiewicz, M., Kisiel, W., Różalska, B., 2012. Antimicrobial and anti-biofilm properties of new taxodione derivative from hairy roots of *Salvia austriaca*. Phytomedicine 19 (14), 1285–1287.

Lee, J.-H., Kim, Y.-G., Cho, H.S., Ryu, S.Y., Cho, M.H., Lee, J., 2014a. Coumarins reduce biofilm formation and the virulence of *Escherichia coli* O157:H7. Phytomedicine 21 (8), 1037–1042.

Lee, J.-H., Kim, Y.-G., Ryu, S.Y., Cho, M.H., Lee, J., 2014b. Ginkgolic acids and *Ginkgo biloba* extract inhibit *Escherichia coli* O157:H7 and *Staphylococcus aureus* biofilm formation. International Journal of Food Microbiology 174 (17), 47–55.

Limsuwan, S., Trip, E.N., Kouwen, T.R.H.M., Piersma, S., Hiranrat, A., Mahabusarakam, W., Voravuthikunchai, S.P., van Dijl, J.M., Kayser, O., 2009. Rhodomyrtone: a new candidate as natural antibacterial drug from *Rhodomyrtus tomentosa*. Phytomedicine 16 (6), 645–651.

Lin, Y.T., Kwon, Y.I., Labbe, R.G., Shetty, K., 2005. Inhibition of *Helicobacter pylori* and associated urease by oregano and cranberry phytochemical synergies. Applied and Environmental Microbiology 71 (12), 8558–8564.

Liu, R.H., 2004. Potential synergy of phytochemicals in cancer prevention: mechanism of action. Journal of Nutrition 134 (s12), 3479S–3485S.

Lotha, R., Shamprasad, B.R., Sundaramoorthy, N.S., Nagarajan, S., Sivasubramanian, A., 2019. Biogenic phytochemicals (cassinopin and isoquercetin) capped copper nanoparticles (ISQ/CAS@CuNPs) inhibits MRSA biofilms. Microbial Pathogenesis 132, 178–187.

Madigan, M.T., Martinko, J.M., Dunlap, P.V., Clark, D.P., 2009. Brock Biology of Microorganisms. Benjamin Cummings, San Francisco, CA.

Malheiro, J.F., Maillard, J.Y., Borges, F., Simões, M., 2019. Evaluation of cinnamaldehyde and cinnamic acid derivatives in microbial growth control. International Biodeterioration and Biodegradation 141, 71–78.

Manach, C., Scalbert, A., Morand, C., Rémésy, C., Jiménez, L., 2004. Polyphenols: food sources and bioavailability. American Journal of Clinical Nutrition 79 (5), 727–747.

Mandal, P., Sinha Babu, S.P., Mandal, N.C., 2005. Antimicrobial activity of saponins from *Acacia auriculiformis*. Fitoterapia 76 (5), 462–465.

Marquez, B., Neuville, L., Moreau, N.J., Genet, J.-P., dos Santos, A.F., Caño de Andrade, M.C., Goulart Sant'Ana, A.E., 2005. Multidrug resistance reversal agent from *Jatropha elliptica*. Phytochemistry 66 (15), 1804–1811.

Melzig, M.F., Bader, G., Loose, R., 2001. Investigations of the mechanism of membrane activity of selected triterpenoid saponins. Planta Medica 67 (1), 43–48.

Mori, A., Nishino, C., Enoki, N., Tawata, S., 1987. Antibacterial activity and mode of action of plant flavonoids against *Proteus vulgaris* and *Staphylococcus aureus*. Phytochemistry 26 (8), 2231–2234.

Myszka, K., Schmidt, M.T., Majcher, M., Juzwa, W., Olkowicz, M., Czaczyk, K., 2016. Inhibition of quorum sensing-related biofilm of *Pseudomonas fluorescens* KM121 by *Thymus vulgare* essential oil and its major bioactive compounds. International Biodeterioration and Biodegradation 114, 252–259.

Nair, S.V., Baranwal, G., Chatterjee, M., Sachu, A., Vasudevan, A.K., Bose, C., Banerji, A., Biswas, R., 2016. Antimicrobial activity of plumbagin, a naturally occurring naphthoquinone from *Plumbago rosea*, against *Staphylococcus aureus* and *Candida albicans*. International Journal of Medical Microbiology 306 (4), 237–248.

Nazzaro, F., Fratianni, F., Coppola, R., 2013. Quorum sensing and phytochemicals. International Journal of Molecular Sciences 14 (6), 12607–12619.

Ngabaza, T., Moeno, S., Patel, M., 2018. Anti-acidogenic and anti-biofilm activity of 5,6,8-trihydroxy-7-methoxy-2-(4-methoxyphenyl)-4*H*-chromen-4-one. Microbial Pathogenesis 123, 149–152.

Niu, C., Afre, S., Gilbert, E.S., 2006. Subinhibitory concentrations of cinnamaldehyde interfere with quorum sensing. Letters in Applied Microbiology 43 (5), 489–494.

Ohemeng, K.A., Schwender, C.F., Fu, K.P., Barrett, J.F., 1993. DNA gyrase inhibitory and antibacterial activity of some flavones(1). Bioorganic and Medicinal Chemistry Letters 3 (2), 225–230.

Oluwatuyi, M., Kaatz, G.W., Gibbons, S., 2004. Antibacterial and resistance modifying activity of *Rosmarinus officinalis*. Phytochemistry 65 (24), 3249–3254.

Oniciuc, E.-A., Likotrafiti, E., Alvarez-Molina, A., Prieto, M., López, M., Alvarez-Ordóñez, A., 2019. Food processing as a risk factor for antimicrobial resistance spread along the food chain. Current Opinion in Food Science 30, 21–26.

Pandey, K.B., Rizvi, S.I., 2009. Plant polyphenols as dietary antioxidants in human health and disease. Oxidative Medicine and Cellular Longevity 2 (5), 270–278.

Plaper, A., Golob, M., Hafner, I., Oblak, M., Šolmajer, T., Jerala, R., 2003. Characterization of quercetin binding site on DNA gyrase. Biochemical and Biophysical Research Communications 306 (2), 530–536.

Plyuta, V., Zaitseva, J., Lobakova, E., Zagoskina, N., Kuznetsov, A., Khmel, I., 2013. Effect of plant phenolic compounds on biofilm formation by *Pseudomonas aeruginosa*. Acta Pathologica, Microbiologica et Immunologica Scandinavica 121 (11), 1073–1081.

Rizvi, S.I., Zaid, M.A., 2005. Impairment of sodium pump and Na/H exchanger in erythrocytes from non-insulin dependent diabetes mellitus patients: effect of tea catechins. Clinica Chimica Acta 354 (1–2), 59–67.

Sajeevan, S.E., Chatterjee, M., Paul, V., Baranwal, G., Kumar, V.A., Bose, C., Banerji, A., Nair, B.G., Prasanth, B.P., Biswas, R., 2018. Impregnation of catheters with anacardic acid from cashew nut shell prevents *Staphylococcus aureus* biofilm development. Journal of Applied Microbiology 125 (5), 1286–1295.

Sasidharan, S., Chen, Y., Saravanan, D., Sundram, K.M., Yoga Latha, L., 2011. Extraction, isolation and characterization of bioactive compounds from plants' extracts. African Journal of Traditional, Complementary and Alternative Medicines 8 (1), 1–10.

Sathiya Deepika, M., Thangam, R., Sakthidhasan, P., Arun, S., Sivasubramanian, S., Thirumurugan, R., 2018. Combined effect of a natural flavonoid rutin from *Citrus sinensis* and conventional antibiotic gentamicin on *Pseudomonas aeruginosa* biofilm formation. Food Control 90, 282–294.

Saxena, M., Saxena, J., Nema, R., Singh, D., Gupta, A., 2013. Phytochemistry of medicinal plants. Journal of Pharmacognosy and Phytochemistry 1 (6), 168–182.

Scalbert, A., Andres-Lacueva, C., Arita, M., Kroon, P., Manach, C., Urpi-Sarda, M., Wishart, D., 2011. Databases on food phytochemicals and their health-promoting effects. Journal of Agricultural and Food Chemistry 59 (9), 4331–4348.

Scalbert, A., Manach, C., Morand, C., Rémésy, C., Jiménez, L., 2005. Dietary polyphenols and the prevention of diseases. Critical Reviews in Food Science and Nutrition 45 (4), 287–306.

Scognamiglio, M., Buommino, E., Coretti, L., Graziani, V., Russo, R., Caputo, P., Donnarumma, G., D'Abrosca, B., Fiorentino, A., 2016. Phytochemical investigation and antimicrobial assessment of *Bellis sylvestris* leaves. Phytochemistry Letters 17, 6–13.

Sharahi, J.Y., Azimi, T., Shariati, A., Safari, H., Tehrani, M.K., Hashemi, A., 2019. Advanced strategies for combating bacterial biofilms. Journal of Cellular Physiology 522–554.

Simões, L.C., Simões, M., Vieira, M.J., 2010a. Adhesion and biofilm formation on polystyrene by drinking water-isolated bacteria. Antonie van Leeuwenhoek 98 (3), 317–329.

Simões, M., Bennett, R.N., Rosa, E.A., 2009. Understanding antimicrobial activities of phytochemicals against multidrug resistant bacteria and biofilms. Natural Product Reports 26 (6), 746–757.

Simões, M., Simões, L.C., Vieira, M.J., 2010b. A review of current and emergent biofilm control strategies. LWT – Food Science and Technology 43 (4), 573–583.

Son, T.G., Camandola, S., Mattson, M.P., 2008. Hormetic dietary phytochemicals. Neuromolecular Medicine 10 (4), 236–246.

Song, X., Xia, Y.-X., Dan He, Z., Zhang, H., 2017. A review of natural products with anti-biofilm activity. Current Organic Chemistry 22, 788–816.

Tofiño-Rivera, A., Ortega-Cuadros, M., Galvis-Pareja, D., Jiménez-Rios, H., Merini, L.J., Martínez-Pabón, M.C., 2016. Effect of *Lippia alba* and *Cymbopogon citratus* essential oils on biofilms of *Streptococcus mutans* and cytotoxicity in CHO cells. Journal of Ethnopharmacology 194, 749–754.

Trentin Dda, S., Giordani, R.B., Zimmer, K.R., da Silva, A.G., da Silva, M.V., Correia, M.T., Baumvol, I.J., Macedo, A.J., 2011. Potential of medicinal plants from the Brazilian semi-arid region (Caatinga) against *Staphylococcus epidermidis* planktonic and biofilm lifestyles. Journal of Ethnopharmacology 137 (1), 327–335.

Trombetta, D., Castelli, F., Sarpietro, M.G., Venuti, V., Cristani, M., Daniele, C., Saija, A., Mazzanti, G., Bisignano, G., 2005. Mechanisms of antibacterial action of three monoterpenes. Antimicrobial Agents and Chemotherapy 49 (6), 2474–2478.

Tsuchiya, H., Iinuma, M., 2000. Reduction of membrane fluidity by antibacterial sophoraflavanone G isolated from *Sophora exigua*. Phytomedicine 7 (2), 161–165.

Van Houdt, R., Michiels, C.W., 2010. Biofilm formation and the food industry, a focus on the bacterial outer surface. Journal of Applied Microbiology 109 (4), 1117–1131.

Vandeputte, O.M., Kiendrebeogo, M., Rasamiravaka, T., Stévigny, C., Duez, P., Rajaonson, S., Diallo, B., Mol, A., Baucher, M., El Jaziri, M., 2011. The flavanone naringenin reduces the production of quorum sensing-controlled virulence factors in *Pseudomonas aeruginosa* PAO1. Microbiology 157 (Pt 7), 2120–2132.

Vazquez-Prieto, M.A., Miatello, R.M., 2010. Organosulfur compounds and cardiovascular disease. Molecular Aspects of Medicine 31 (6), 540–545.

Venkadesaperumal, G., 2015. Cyanidin inhibits quorum signalling pathway of food borne opportunistic pathogen. Journal of Food Science & Technology 53 (2), 968–976.

Vikram, A., Jayaprakasha, G.K., Uckoo, R.M., Patil, B.S., 2013. Inhibition of *Escherichia coli* O157:H7 motility and biofilm by *β*-sitosterol glucoside. Biochimica et Biophysica Acta (BBA) – General Subjects 1830 (11), 5219–5228.

Wojnicz, D., Sycz, Z., Walkowski, S., Gabrielska, J., Aleksandra, W., Alicja, K., Anna, S.-Ł., Hendrich, A.B., 2012. Study on the influence of cranberry extract Żuravit S·O·S® on the properties of uropathogenic *Escherichia coli* strains, their ability to form biofilm and its antioxidant properties. Phytomedicine 19 (6), 506–514.

Wojnicz, D., Tichaczek-Goska, D., Korzekwa, K., Kicia, M., Hendrich, A., 2017. Anti-enterococcal activities of pentacyclic triterpenes. Advances in Clinical and Experimental Medicine 26 (3), 483–490.

Yang, C.S., Landau, J.M., Huang, M.T., Newmark, H.L., 2001. Inhibition of carcinogenesis by dietary polyphenolic compounds. Annual Review of Nutrition 21, 381–406.

Zhang, J., Rui, X., Wang, L., Guan, Y., Sun, X., Dong, M., 2014. Polyphenolic extract from *Rosa rugosa* tea inhibits bacterial quorum sensing and biofilm formation. Food Control 42, 125–131.

Chapter 13

Photoinactivation of biofilms

Alex Fiori Silva[1], João Vitor de Oliveira Silva[2], Paula Aline Zanetti Campanerut-Sá[2], Jane Martha Graton Mikcha[2]

[1]*Federal Institute of Paraná, Paranavaí, Paraná, Brazil;* [2]*Department of Clinical Analysis and Biomedicine, State University of Maringá, Maringá, Paraná, Brazil*

13.1 Photodynamic therapy

13.1.1 Historical remarks

The original report of photodynamic therapy (PDT) was first published over a hundred years ago. A medical student, Oscar Raab, showed that *Paramecium caudatum* could be inactivated when exposed to the acridine dyes or eosin and illuminated by sunlight, but not when they were kept in the dark. PDT was applied to tumor cells by Raab's supervisors in 1905. Later on, it was discovered that the combined action of dye, light, and oxygen produced the light-mediated effect, which was termed photodynamic effect or photosensitization (Wainwright et al., 2017; Ghorbani et al., 2018). In the 1970s, Dougherty and coworkers introduced a mixture of porphyrins that was later purified and named as Photofrin, the first photosensitizer (PS) used in PDT (Abrahamse and Hamblin, 2016). The PDT potential to cure microbial origin diseases was exploited from the beginning of this century with the emergence of bacterial resistance. Treatments based on photodynamic effect are being developed as an important therapeutic option (Hu et al., 2018). PDT has been successfully applied in medical, dental, industrial, and environmental areas nowadays (Alves et al., 2015; Wainwright et al., 2017; Hu et al., 2018). Lately, attempts to use this technology for food safety have been on the rise (Ghate et al., 2019; Silva et al., 2019). The photoinactivation technique evaluated for antimicrobial purposes is also known as photodynamic inactivation (PDI) or antimicrobial photodynamic therapy (aPDT).

13.1.2 Basic principles and mechanism of photosensitization

The basic principle of photosensitization is the synergistic interaction of three agents: molecular oxygen, light, and a photoactive molecule (called PS). This combination leads to the generation of reactive oxygen species (ROS), such as

Recent Trends in Biofilm Science and Technology. https://doi.org/10.1016/B978-0-12-819497-3.00013-1

singlet oxygen ($^1O_2$), hydrogen peroxide (H_2O_2), hydroxyl radicals ($^{\cdot}OH$), and superoxide anions $O_2^{\cdot -}$, which rapidly inactivate nearby cells oxidatively (Alves et al., 2015). When exposed to some light under an appropriate wavelength, PS absorbs energy forming the excited singlet state. This state can return to the singlet ground by fluorescence emission or undergo an intersystem crossing to a long-lived triplet state. The PS excited to a triplet state can emit phosphorescence decaying to the ground state or react with oxygen following two pathways: type I involves electron transfer from excited PS, generating ROS; type II in which PS directly transfers energy to molecular oxygen, leading to the formation of singlet oxygen (Fig. 13.1) (Wainwright et al., 2017). It is reported that these two mechanisms can occur simultaneously and the ratio between these processes depends on the PS characteristics, oxygen concentration, and the PS affinity to the substrate (Wainwright and Crossley, 2004). The generated reactive molecules present a broad spectrum of activity and can destroy numerous molecular microbial targets (e.g., proteins, lipids, and nucleic acids) driving to cell death (Hamblin and Hasan, 2004). Once the antimicrobial effect is provided by simultaneous interactions with multiple targets, the development of antimicrobial resistance is not supposed to occur, and microorganisms are equally killed regardless of their intrinsic or acquired antibiotic resistance (Almeida et al., 2015).

13.1.3 Photosensitizers and light sources

Photosensitizers are essential to the aPDT effectiveness. For a molecule to be considered a good PS, it should have some characteristics, such as high

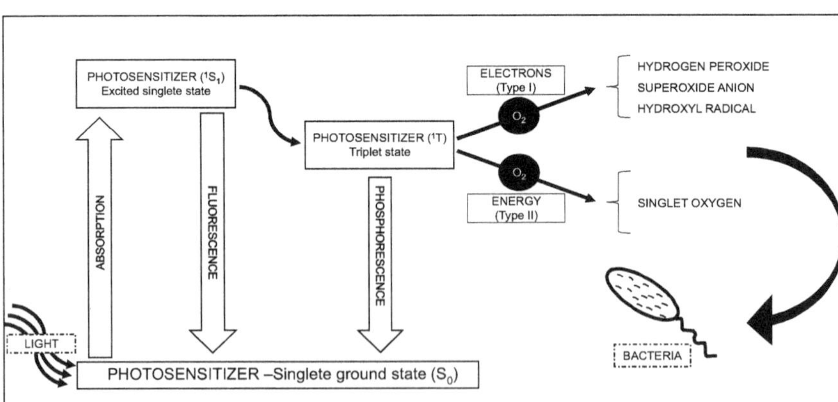

FIGURE 13.1 Jablonski scheme for conventional photon excitation of a photosensitizer molecule. *From: A. F. Silva et al. (2018b). Photodynamic inactivation as an emergent strategy against foodborne pathogenic bacteria in planktonic and sessile states, Critical Reviews in Microbiology 44(6), 667–684.*

absorption coefficient, low aggregation tendency, high quantum yield of triplet state, capacity of absorbing and using energy to excite oxygen to its singlet state, a long-lasting triplet state, high solubility, and stability (Luksiene and Brovko, 2013).

Many photoactivatable molecules have been found to be useful candidates for photodynamic applications (Roguin et al., 2019). The main classes of PS are synthetic dyes, tetrapyrrole structures, and natural photosensitizers (Table 13.1).

The wavelength of light necessary for the photoinactivation of microorganisms depends on the structure and electron absorption spectrum of the PS. Therefore, it is important to ensure that the light delivery system provides radiation in the region of the maximum absorption wavelength of the PS to be effective (Cieplik et al., 2014).

Photosensitization has been performed with conventional gas discharge and incandescent lamps equipped with filters, metal halide lamps, the short-arc

TABLE 13.1 The main classes of photosensitizers and their spectra of absorption.

Class	Example	Spectra absorption
Synthetic dyes	Methylene blue	632 nm
	Toluidine blue	410 nm
	Rose bengal	532 nm
	Dimethyl methylene blue	635–652 nm
	New methylene blue	635–652 nm
Tetrapyrrole structures	Porphyrin	446 nm
	Phthalocyanine	670 nm
	Zinc Pc derivatives	690 nm
	Chlorine—neutral	660 nm
	Chlorine–cationic	532 nm
	Bacteriochlorin	732 nm
Natural PS	Curcumin	547 nm
	Hypericin	593 nm
	Flavin derivatives	450 nm

Based on: Abrahamse, H., Hamblin, M.R., 2016. New photosensitizers for photodynamic therapy. Biochemical Journal 473(4), 347–364; Ghorbani, J., Rahban, D., Aghamiri, S., Teymouri, A., Bahador, A., 2018. Photosensitizers in antibacterial photodynamic therapy: an overview. Laser Therapy 27(4), 293–302.

xenon lamp, and UV lamp for the activation of PS (Calin and Parasca, 2009). More recently, light-emitting diodes (LEDs) and LASERs (light amplification by stimulated emission of radiation) have also been applied to kill microorganisms (Luksiene and Brovko, 2013). LEDs are between LASERs and conventional lamps in terms of spectral and radiation features, and they have been widely applied in photodynamic studies (Souza et al., 2015; Bonin et al., 2018; Silva et al., 2019).

LEDs have significant advantages over some other alternatives: the low driving voltage, shock and vibration resistance, the nonexistence of hazardous agents (heavy metals), narrow-band emission, robustness, and absence of undesirable spectral components (Luksiene and Brovko, 2013). Little to no thermal effects are observed, which makes them extremely suitable for several areas (Brancaleon and Moseley, 2002; Souza et al., 2015). This light system provides high-energy doses, and the apparatus is cheaper than the ones from alternative sources, like LASERs (Brancaleon and Moseley, 2002).

13.2 Photoinactivation of biofilms

The prevalence of biofilms is a significant problem in several important fields, such as industry, medicine, and environment (Alvarez-Ordóñez et al., 2019; Hu et al., 2018; Simões et al., 2010; Borges et al., 2016; Sadekuzzaman et al., 2015). Biofilm formation can lead to substantial economic and health problems. Microbial biofilms are responsible for infections that are difficult to be eradicate (Hu et al., 2018), and they are also associated with contamination and spoilage in food processing, biofouling, and corrosion in the environment (Silva et al., 2019; Borges et al., 2016). It is well known that biofilm bacteria are highly resistant to antimicrobials, sanitizers, and other stresses compared with their planktonic counterparts (Silva et al., 2019; Borges et al., 2016).

In the food industry, biofilms are recognized to contribute to persistence of spoilage and pathogenic microorganisms and can serve as a source of cross-contamination in foods, compromising food quality and safety (Alvarez-Ordóñez et al., 2019; Silva et al., 2019; Sadekuzzaman et al., 2015). Besides food contamination, the presence of microbial biofilms in food processing environments creates other problems, including damage of metal surface pipelines (e.g., corrosion), reduction of the heat transfer in heat exchange equipment, reduction of the filtration systems permeability, and even mechanical blockage (Alvarez-Ordóñez et al., 2019; Simões et al., 2010) In addition, the development of effective strategies to control biofilms in food processing environments is challenging.

Despite having several methods, physical removal and/or the use of antimicrobials, sanitizers, or disinfectants to eliminate biofilm organisms, their control is a challenge due to the higher resistance (Ghate et al., 2019; Borges et al., 2016). Thus, the search for more efficient antimicrobial techniques has been a subject of permanent research. The photodynamic approach has emerged as an innovative alternative for eradicating biofilms in food science area, and their application has shown superior results compared with the conventional approaches (Ghate et al., 2019). A general scheme representing the aPDT in biofilm is demonstrated in Fig. 13.2.

For instance, Bonifácio et al. (2018) investigated the efficacy of commercial curcumin and *Curcuma longa* extract (rich in curcumin) for the inactivation of *Listeria innocua* biofilms on polypropylene surface. Irradiation was performed with a 400–500 nm light apparatus (150 mW/cm^2). After 30 min of light exposure, the commercial PS at 0.6 and 3.7 mg/L decreased 3 and 5 log CFU/mL, respectively. On the other hand, lower reductions were found when the natural extract was employed (2-3 log reduction).

The photoinactivation of *L. monocytogenes* biofilms formed on packing yellow trays (polyolefine) using Na-chlorophyllin (E140) as PS was investigated by Luksiene and Paskeviciute (2011). The 2-day-old biofilms were incubated with E140 and irradiated with an LED light source for 5 min. A slight reduction was observed when sessile cells were treated with Na-chlorophyllin at 1.5×10^{-5} M. However, the microbial population was completely eliminated when the PS concentration was increased to 1.5×10^{-4} M. In another study, the same research group found out that both freshly attached and mature biofilms formed by *L. monocytogenes* on polyolefin surface were reduced (3.0 log CFU/cm^2) according to the 5-aminolevulinic acid (ALA) concentration employed: 7.5 and 10 mM (Buchovec et al., 2010).

Recently, Castro et al. (2017) compared the effect of meso-tetraarylporphyrins (0.2 mM) to phenyl or pentafluorophenyl groups at the meso positions with or without acid groups and immobilized in chitosan films

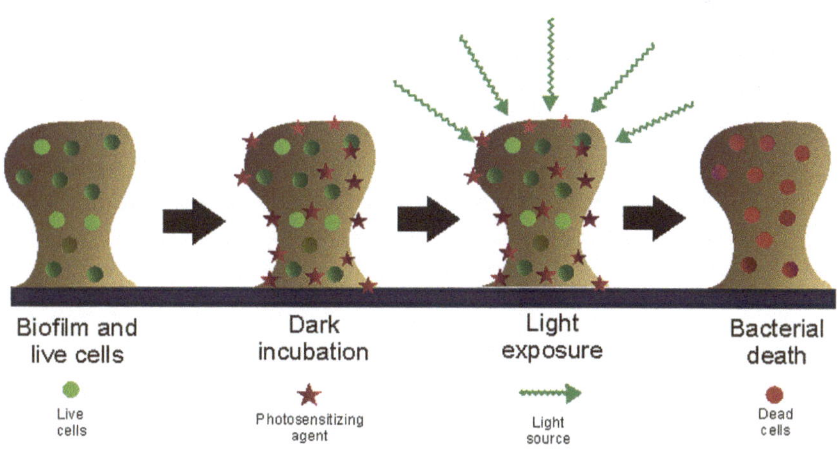

FIGURE 13.2 General scheme representing the aPDT against biofilm cells. *aPDT*, antimicrobial photodynamic therapy. *From: Silva, A. et al. (2018a) Antimicrobial photodynamic inactivation mediated by rose Bengal and erythrosine is effective in the control of food-related bacteria in planktonic and biofilm states,* Molecules, *23(9), p. 2288.*

(CFs) to prevent *L. innocua* attachment and future biofilm formation. It was observed that all tested porphyrinic CFs reduced bacterial attachment during 24 h of exposure to white light. The subsequent biofilm formation was almost impaired in two of the PS-CFs using the same treatment and incubation in the dark for another 48 h. The aPDT effectiveness was found to be related to the chemical properties of each porphyrin used.

Antimicrobial photodynamic therapy with a novel-designed synthetic porphyrin was evaluated against *Staphylococcus aureus* biofilms formed onto polystyrene microtiter plates in a study of Mamone et al. (2016). The incubation with 20 μM porphyrin during 10 min followed by irradiation (180 J/cm^2) with a tungsten halogen light source decreased about 3 log CFU of bacterial counts. Biofilms stained with BacLight Live/Dead observed on confocal laser scanning microscopy (CLSM) presented dead cells throughout the biofilm matrix, revealing that no PS hindrance occurred. The unimpaired penetration of the PS molecules was attributed to the cationic character of the porphyrin. Ke et al. (2012) also observed that biofilms of *S. aureus* formed on stainless steel surfaces, which is a common surface used in food processing environments, were completely eliminated after photoinactivation mediated by the food-approved additive erythrosine.

Considering that plastics are also used for food handling, Miñán et al. (2015) proposed the use of aPDT for preventing *S. aureus* surface adherence. They demonstrated that planktonic cells treated with subinhibitory concentrations of 6-carboxypterin (0.1 μM) and exposed to UV-A irradiation were unable to adhere and form biofilms onto polystyrene surfaces. Recently, Rout et al. (2017) also demonstrated that aPDT was able to prevent *S. aureus* biofilm formation on polystyrene surface using encapsulated toluidine blue (TBO) as PS. It was hypothesized that high supply of ROS in the early stages of biofilm formation can interfere within the process. They also observed that TBO encapsulated in the microemulsion induced greater damage in biofilm than the nonencapsulated. The improvement in antibiofilm activity was related to the nano-sized delivery vehicles.

The application of nano-sized vehicles has the potential of overcoming several limitations associated with the conventional methods (Wang et al., 2017). The reduced penetration of the light-activated PS into the deep regions of the biofilm matrix caused by EPS is one of the main causes of the low susceptibility of biofilm to aPDT. Nanoparticles with ability to disrupting the three-dimensional organization of biofilm can be a promising choice for this goal (Shrestha et al., 2010; Chávez de Paz et al., 2011).

Parasuraman et al. (2019) encapsulated TBO in mesoporous silica nanoparticles (TB-MSNs) to evaluate *P. aeruginosa* and *S. aureus* biofilm inhibition after irradiation with a 125-mW diode laser (λ 670 nm). It was observed that after 3 h of dark incubation and 5 min of light exposure, TB MSN-treated *S. aureus* showed ~76.22% biofilm inhibition, whereas for *P. aeruginosa*, 66.39% inhibition in biofilm formation was noticed. Rout et al. (2017)

proposed that the inhibition of biofilm formation by TBO was related to quorum sensing interruption. The authors observed that, after photosensitization, the concentration of lactones produced by *P. aeruginosa* was reduced. Such effect was attributed to lactonolysis induced by hydroxyl radicals during photosensitization.

Considering that chitosan nanoparticles (CSNPs) are known to possess broad-spectrum antibacterial activity (Perinelli et al., 2018), Darabpour et al. (2016) proposed its use to improve methylene blue (MB)−mediated aPDT against preformed biofilms of *S. aureus* and *P. aeruginosa*. Less than 1 log CFU reduction was found when biofilms were incubated with MB in the dark for 7 min and irradiated with a diode laser during 10 min (22.93 J/cm^2). However, when CSNPs were coincubated with the PS, about 3 log CFU reduction was observed in both species. In another study, Shrestha et al. (2012) conjugated RB with CSNPs, observing a complete elimination of 7-day-old biofilms formed by *P. aeruginosa*. They used laser scanning confocal microscopy to observe that photoinactivation decreased both biofilm thicknesses and cells viability. Chitosan was able to enhance the uptake of RB into the biofilms, which possibly occurred due to its positively charged characteristic as CSNPs. Indeed, this finding is exciting since the effective penetration of photosensitizer into the biofilm structure is a critical parameter for the aPDT performance.

Recently, Silva et al. (2018a,b) also demonstrated a complete eradication of *Escherichia coli* sessile cells by rose bengal (RB) and erythrosine (ERY) when using a light-emitting diode (LED) source for illumination. The study showed that RB and ERY excited by green LED light also eradicated biofilms formed by *Enterococcus hirae*, *L. innocua*, and *S. aureus*. Live/Dead staining was used to observe that biofilm cells subjected to aPDI presented membrane permeabilization. It was also noticed that higher concentration of the PS agents was needed to inactive gram-negative sessile cells compared with the concentrations employed for the gram-positive ones.

The study mentioned earlier also demonstrated that biofilms formed by the gram-negative foodborne pathogen *Salmonella* Typhimurium were more tolerant to photoinactivation than sessile cells of *S. aureus*. While RB at 100 μM completely eliminated the biofilm formed by *S. aureus* after 30 min of LED light irradiation, 10-fold higher concentration of RB decreased about 3 log CFU/cm^2 of *S.* Typhimurium counts. Scanning electron microscopy images (Fig. 13.3) showed that photodynamic treatments also disintegrated the biofilm structure (Silva et al., 2019).

Nowadays, antimicrobial resistance is one of the most relevant threats faced by humanity (Vikesland et al., 2019). Alarmingly, a remarkable increase in the number of resistant foodborne pathogens isolated from foodstuffs or food processing environments has been observed(Olaimat et al., 2018; Obaidat and Stringer, 2019; Zeng et al., 2019). Considering that biofilm formation and antimicrobial resistance in gram-negative bacteria may be connected

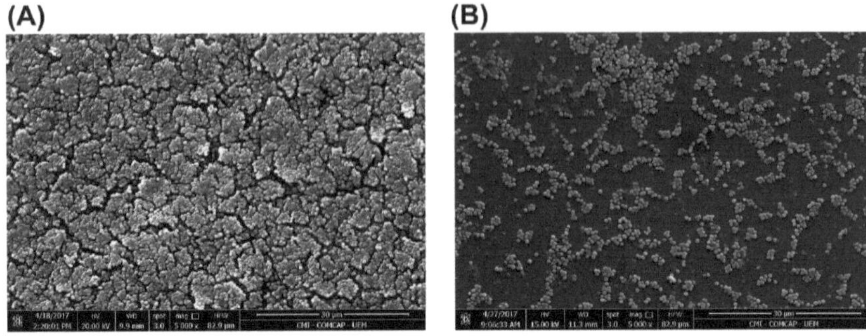

FIGURE 13.3 Scanning electron microscopy photomicrograph of a *Staphylococcus aureus* biofilm formed on plastic surface, where (A) is the control group and (B) represents the sessile cells subjected to aPDT with erythrosine and green LED source. aPDT, antimicrobial photodynamic therapy; LED, light-emitting diode. *Adapted from: Silva, A. F. et al. (2019) Xanthene Dyes and Green LED for the Inactivation of Foodborne Pathogens in Planktonic and Biofilm States,* Photochemistry and Photobiology, *95(5), pp. 1230–1238.*

(Cepas et al., 2019) and that aPDT is not supposed to induce antimicrobial resistance (Al-Mutairi et al., 2018), photosensitization could be useful for controlling biofilms formed by antimicrobial-resistant bacteria. In this regard, Cossu et al. (2016) investigated the effect of gallic acid (GA) and UV-A light (λ365 nm) on rifampicin-resistant *E. coli* O157:H7 biofilms. Biofilms formed on polystyrene surfaces were exposed to 10-mM GA and irradiated for 60 min. The application of resazurin fluorescence assay revealed that GA-UV-A treatment was able to decrease 70% of biofilm metabolic activity (Cossu et al., 2016).

De Sordi et al. (2015) (Cossu et al., 2016) investigated the photoinactivation capability of four light-activated molecules: methylene blue, phthalocyanines S4, chlorin e6, and talaporfin against 6-day-old biofilms of *Clostridium difficile* formed on microplate surface. They found that irradiation of 1 mM PS during 10 s (0.24 J/cm^2) with a laser diode light induced a 23 log10 reduction in bacterial numbers. It was suggested that PDT-resistant spores eventually presented within the biofilm could germinate after treatment. Thus, a complete inactivation of biofilm cells was not observed.

Another study by Luksiene and Paskeviciute (2011) evaluated the photoinactivation capability of Na-chlorophyllin against *Bacillus cereus* spores attached on polyolefin surface. They observed a reduction from 6 to 1 log in spore counts when 7.5×10^{-5} M were irradiated with an LED-based light source ($\lambda = 405$ nm, 12 mW/cm^2) during 5 min.

Despite the remarkable antibacterial activity of aPDT, it has been shown that higher PS concentrations and light doses are required to control biofilms in comparison with their planktonic counterparts (Silva et al., 2019). It is mainly attributed to the decrease of ROS generation due to the lower penetration of light and PS molecule throughout biofilm structure; therefore, aPDT has been proved to be an effective method for controlling biofilm cells.

13.3 Concluding remarks

Foodborne illness remains a worldwide problem in which bacterial biofilms on food processing surfaces can represent critical sources of contamination. In addition, biofilm formation has serious implication in industrial process in terms of obstruction, corrosion, and decrease of heat transfer. Thus, aPDT has been extensively studied for biofilm control and prevention in this area of science. As mentioned, several studies noticed that biofilms formed by foodborne pathogens can be efficiently controlled by using aPDT technology. Several types of photosensitizers can be employed against both sessile gram-positive and gram-negative for EPS disruption, quorum sensing molecules destruction, and cell membrane destabilization. Despite the latest reports showing that aPDT is safe and effective, this technology has not yet been established as a standard antimicrobial procedure against foodborne microorganisms in food industry.

References

Abrahamse, H., Hamblin, M.R., 2016. New photosensitizers for photodynamic therapy. Biochemical Journal 473 (4), 347–364.

Al-Mutairi, R., Tovmasyan, A., Batinic-Haberle, I., Benov, L., 2018. Sublethal photodynamic treatment does not lead to development of resistance. Frontiers in Microbiology 9 (7), 1–9.

Almeida, A., Faustino, M.A., Tomé, J.P., 2015. Photodynamic inactivation of bacteria: finding the effective targets. Future Medicinal Chemistry 7 (10), 1221–1224.

Alvarez-Ordóñez, A., Coughlan, L.M., Briandet, R., Cotter, P.D., 2019. Biofilms in food processing environments: challenges and opportunities. Annual Review of Food Science and Technology 10 (1), 173–195.

Alves, E., Faustino, M.A.F., Neves, M.G.P.M.S., Cunha, Â., Nadais, H., Almeida, A., 2015. Potential applications of porphyrins in photodynamic inactivation beyond the medical scope. Journal of Photochemistry and Photobiology C: Photochemistry Reviews 22, 34–57.

Bonifácio, D., Martins, C., David, B., Lemos, C., Neves, M.G.P.M., Almeida, A., Pinto, D.C.G., Faustino, M.A., Cunha, Â., 2018. Photodynamic inactivation of *Listeria innocua* biofilms with food-grade photosensitizers: a curcumin-rich extract of Curcuma longa vs commercial curcumin. Journal of Applied Microbiology 125 (1), 282–294.

Bonin, E., dos Santos, A.R., Fiori da Silva, A., Ribeiro, L.H., Favero, M.E., Campanerut-Sá, P.A.Z., de Freitas, C.F., Caetano, W., Hioka, N., Mikcha, J.M.G., 2018. Photodynamic inactivation of foodborne bacteria by eosin Y. Journal of Applied Microbiology 124 (6), 1617–1628.

Borges, A., Abreu, A., Dias, C., Saavedra, M., Borges, F., Simões, M., 2016. New perspectives on the use of phytochemicals as an emergent strategy to control bacterial infections including biofilms. Molecules 21 (7), 877.

Brancaleon, L., Moseley, H., 2002. Laser and non-laser light sources for photodynamic therapy. Lasers in Medical Science 17 (3), 173–186.

Buchovec, I., Paskeviciute, E., Luksiene, Z., 2010. Photodynamic inactivation of food pathogen *Listeria monocytogenes*. *Food Technology and Biotechnology* 48 (2), 207–213.

Calin, M.A., Parasca, S.V., 2009. Light sources for photodynamic inactivation of bacteria. Lasers in Medical Science 24 (3), 453–460.

Castro, K.A.D.F., Moura, N.M.M., Fernandes, A., Faustino, M.A.F., Simões, M.M.Q., Cavaleiro, J.A.S., Nakagaki, S., Almeida, A., Cunha, Â., Silvestre, A.J.D., Freire, C.S.R., Pinto, R.J.B., Neves, M. da G.P.M.S., 2017. Control of *Listeria innocua* biofilms by biocompatible photodynamic antifouling chitosan based materials. Dyes and Pigments 137, 265–276.

Cepas, V., López, Y., Muñoz, E., Rolo, D., Ardanuy, C., Martí, S., Xercavins, M., Horcajada, J.P., Bosch, J., Soto, S.M., 2019. Relationship between biofilm formation and antimicrobial resistance in gram-negative bacteria. Microbial Drug Resistance 25 (1), 72–79.

Chávez de Paz, L.E., Resin, A., Howard, K.A., Sutherland, D.S., Wejse, P.L., 2011. Antimicrobial effect of chitosan nanoparticles on *Streptococcus mutans* biofilms. Applied and Environmental Microbiology 77 (11), 3892–3895.

Cieplik, F., Tabenski, L., Buchalla, W., Maisch, T., 2014. Antimicrobial photodynamic therapy for inactivation of biofilms formed by oral key pathogens. Frontiers in Microbiology 5 (8), 1–17.

Cossu, A., Ercan, D., Wang, Q., Peer, W.A., Nitin, N., Tikekar, R.V., 2016. Antimicrobial effect of synergistic interaction between UV-A light and gallic acid against *Escherichia coli* O157:H7 in fresh produce wash water and biofilm. Innovative Food Science & Emerging Technologies 37, 44–52.

Darabpour, E., Kashef, N., Mashayekhan, S., 2016. Chitosan nanoparticles enhance the efficiency of methylene blue-mediated antimicrobial photodynamic inactivation of bacterial biofilms: an in vitro study. Photodiagnosis and Photodynamic Therapy 14 (14), 211–217.

Ghate, V.S., Zhou, W., Yuk, H.-G., 2019. Perspectives and Trends in the application of photodynamic inactivation for microbiological food safety. Comprehensive Reviews in Food Science and Food Safety 18 (2), 402–424.

Ghorbani, J., Rahban, D., Aghamiri, S., Teymouri, A., Bahador, A., 2018. Photosensitizers in antibacterial photodynamic therapy: an overview. Laser Therapy 27 (4), 293–302.

Hamblin, M.R., Hasan, T., 2004. Photodynamic therapy: a new antimicrobial approach to infectious disease? Photochemical and Photobiological Sciences 3 (5), 436.

Hu, X., Huang, Y., Wang, Y., Wang, X., Hamblin, M.R., 2018. Antimicrobial photodynamic therapy to control clinically relevant biofilm infections. Frontiers in Microbiology 9 (6), 1–24.

Ke, E.S., Nazzal, S., Tseng, Y.H., Chen, C.P., Tsai, T., 2012. Erythrosine-mediated photodynamic inactivation of bacteria and yeast using green light-emitting diode light. Journal of Food and Drug Analysis 20 (4), 951–956.

Luksiene, Z., Brovko, L., 2013. Antibacterial photosensitization-based treatment for food safety. Food Engineering Reviews 5 (4), 185–199.

Luksiene, Z., Paskeviciute, E., 2011. Microbial control of food-related surfaces: Na-Chlorophyllin-based photosensitization. Journal of Photochemistry and Photobiology B: Biology 105 (1), 69–74.

Mamone, L., Ferreyra, D.D., Gándara, L., Di Venosa, G., Vallecorsa, P., Sáenz, D., Calvo, G., Batlle, A., Buzzola, F., Durantini, E.N., Casas, A., 2016. Photodynamic inactivation of planktonic and biofilm growing bacteria mediated by a meso-substituted porphyrin bearing four basic amino groups. Journal of Photochemistry and Photobiology B: Biology 161 (161), 222–229.

Miñán, A., Lorente, C., Ipiña, A., Thomas, A.H., Fernández Lorenzo de Mele, M., Schilardi, P.L., 2015. Photodynamic inactivation induced by carboxypterin: a novel non-toxic bactericidal strategy against planktonic cells and biofilms of *Staphylococcus aureus*. Biofouling 31 (5), 459–468.

Obaidat, M.M., Stringer, A.P., 2019. Prevalence, molecular characterization, and antimicrobial resistance profiles of *Listeria monocytogenes*, *Salmonella enterica*, and *Escherichia coli* O157:H7 on dairy cattle farms in Jordan. Journal of Dairy Science 102 (10), 8710–8720.

Olaimat, A.N., Al-Holy, M.A., Shahbaz, H.M., Al-Nabulsi, A.A., Abu Ghoush, M.H., Osaili, T.M., Ayyash, M.M., Holley, R.A., 2018. Emergence of antibiotic resistance in *Listeria monocytogenes* isolated from food products: a comprehensive review. Comprehensive Reviews in Food Science and Food Safety 17 (5), 1277−1292.

Parasuraman, P., Antony, A.P., B, S.L.S., Sharan, A., Siddhardha, B., Kasinathan, K., Bahkali, N.A., Dawoud, T.M.S., Syed, A., 2019. Antimicrobial photodynamic activity of toluidine blue encapsulated in mesoporous silica nanoparticles against *Pseudomonas aeruginosa* and *Staphylococcus aureus*. Biofouling 35 (1), 89−103.

Perinelli, D.R., Fagioli, L., Campana, R., Lam, J.K.W., Baffone, W., Palmieri, G.F., Casettari, L., Bonacucina, G., 2018. Chitosan-based nanosystems and their exploited antimicrobial activity. European Journal of Pharmaceutical Sciences 117, 8−20.

Roguin, L.P., Chiarante, N., García Vior, M.C., Marino, J., 2019. Zinc(II) phthalocyanines as photosensitizers for antitumor photodynamic therapy. The International Journal of Biochemistry & Cell Biology 114 (5), 105575.

Rout, B., Liu, C.-H., Wu, W.-C., 2017. Photosensitizer in lipid nanoparticle: a nano-scaled approach to antibacterial function. Scientific Reports 7 (1), 7892.

Sadekuzzaman, M., Yang, S., Mizan, M.F.R., Ha, S.D., 2015. Current and recent advanced strategies for combating biofilms. Comprehensive Reviews in Food Science and Food Safety 14 (4), 491−509.

Shrestha, A., Hamblin, M.R., Kishen, A., 2012. Characterization of a conjugate between rose bengal and chitosan for targeted antibiofilm and tissue stabilization effects as a potential treatment of infected dentin. Antimicrobial Agents and Chemotherapy 56 (9), 4876−4884.

Shrestha, A., Zhilong, S., Gee, N.K., Kishen, A., 2010. Nanoparticulates for antibiofilm treatment and effect of aging on its antibacterial activity. Journal of Endodontics 36 (6), 1030−1035.

Silva, A., Borges, A., Freitas, C., Hioka, N., Mikcha, J., Simões, M., 2018a. Antimicrobial photodynamic inactivation mediated by rose bengal and erythrosine is effective in the control of food-related bacteria in planktonic and biofilm states. Molecules 23 (9), 2288.

Silva, A.F., Borges, A., Giaouris, E., Graton Mikcha, J.M., Simões, M., 2018b. Photodynamic inactivation as an emergent strategy against foodborne pathogenic bacteria in planktonic and sessile states. Critical Reviews in Microbiology 44 (6), 667−684.

Silva, A.F., dos Santos, A.R., Trevisan, D.A.C., Bonin, E., Freitas, C.F., Batista, A.F.P., Hioka, N., Simões, M., Graton Mikcha, J.M., 2019. Xanthene dyes and green LED for the inactivation of foodborne pathogens in planktonic and biofilm states. Photochemistry and Photobiology 95 (5), 1230−1238.

Simões, M., Simões, L.C., Vieira, M.J., 2010. A review of current and emergent biofilm control strategies. LWT - Food Science and Technology 43 (4), 573−583.

De Sordi, L., Butt, M.A., Pye, H., Kohoutova, D., Mosse, C.A., Yahioglu, G., Stamati, I., Deonarain, M., Battah, S., Ready, D., Allan, E., Mullany, P., Lovat, L.B., 2015. Development of photodynamic antimicrobial chemotherapy (PACT) for *Clostridium difficile*. PLoS One 10 (8) e0135039.

Souza, C.D., Yuk, H., Khoo, G.H., Zhou, W., 2015. Application of light-emitting diodes in food production , postharvest preservation , and microbiological food safety. Comprehensive Reviews in Food Science and Food Safety 14 (6), 719−740.

Vikesland, P., Garner, E., Gupta, S., Kang, S., Maile-Moskowitz, A., Zhu, N., 2019. Differential drivers of antimicrobial resistance across the world. Accounts of Chemical Research 52 (4), 916−924.

Wainwright, M., Crossley, K.B., 2004. Photosensitising agents—circumventing resistance and breaking down biofilms: a review. International Biodeterioration & Biodegradation 53 (2), 119−126.

Wainwright, M., Maisch, T., Nonell, S., Plaetzer, K., Almeida, A., Tegos, G.P., Hamblin, M.R., 2017. Photoantimicrobials—are we afraid of the light? The Lancet Infectious Diseases 17 (2), e49–e55.

Wang, L., Hu, C., Shao, L., 2017. The antimicrobial activity of nanoparticles: present situation and prospects for the future. International Journal of Nanomedicine 12, 1227–1249.

Zeng, Y.-B., Xiong, L.-G., Tan, M.-F., Li, H.-Q., Yan, H., Zhang, L., Yin, D.-F., Kang, Z.-F., Wei, Q.-P., Luo, L.-G., 2019. Prevalence and antimicrobial resistance of *Salmonella* in pork, chicken, and duck from retail markets of China. Foodborne Pathogens and Disease 16 (5), 339–345.

Chapter 14

The potential of drug repurposing to face bacterial and fungal biofilm infections

Isabel M. Oliveira, Anabela Borges, Manuel Simões
LEPABE – Laboratory for Process Engineering Environment, Biotechnology and Energy, Department of Chemical Engineering, Faculty of Engineering, University of Porto, Porto, Portugal

14.1 Introduction

The spectrum of bacterial and fungal biofilm infections is wide, and they are difficult to treat due to the structure and physiological attributes of biofilms that confer resistance to host immune responses and action of antimicrobial drugs (Costerton et al., 1999; Desai et al., 2014; Lebeaux et al., 2014; Romling et al., 2014; Cavalheiro and Teixeira, 2018). For this reason, long-term treatment with high doses of antimicrobial agents and often using a combination of drugs with different mechanisms of action is applied (Wu et al., 2014). In addition, the emergence of multidrug-resistant (MDR) pathogens is rapidly increasing, forcing the World Health Organization (WHO) to warn that antimicrobial resistance is one of the most important problems for human health. However, on the one hand, there is a lack of new antibiotics being developed, and on the other hand, the treatment options for invasive fungal infections are limited (Brooks and Brooks, 2014; Fisher et al., 2018). The decreased interest of pharmaceutical industries to search and develop new drugs to treat infectious diseases is mainly due to the time-consuming process and increased costs of putting a new drug on the market. Moreover, there is insufficient investment in this market due to the poor economic returns for the pharmaceutical sector and lack of incentives by the medical and scientific communities. This can be in part related to the excessive use of antimicrobial agents—considered one of the main reasons for the emergence of resistance (Projan, 2003; Piddock, 2012).

Drug repurposing arises an attractive approach to face the absence of effective antimicrobials. It can be defined as the process of finding new uses outside the scope of the original medical indication for existing drugs, and it is also known as redirecting, repositioning, and reprofiling (Ashburn and Thor, 2004).

Recent Trends in Biofilm Science and Technology. https://doi.org/10.1016/B978-0-12-819497-3.00014-3

The main advantage of using approved or investigational drugs is the lower risk of failure because the safety and pharmacokinetic profiles are known. Eventually, if early-stage trials have been completed, it is poorly probable that the repurposed drug will fail in subsequent efficacy trials, at least from a safety point of view (Chong and Sullivan, 2007; Oprea and Mestres, 2012; Pushpakom et al., 2018). This strategy has also a shorter development timeline compared with traditional de novo drug discovery and development. It is generally recognized that de novo drug discovery and development process is a 10- to 17-year pathway to the market, whereas drug repurposing offers the possibility to reduce this process to 3−12 years (Ashburn and Thor, 2004). Additionally, depending on the stage and process of development, the overall investment can potentially be reduced, principally in preclinical and phase I and II costs (Chong and Sullivan, 2007; Pushpakom et al., 2018). Altogether, drug repurposing can reduce risk, time, and costs associated with drug development pipeline.

Historically, drug repurposing has been largely serendipitous, with unintentional fortunate observations, but there are different systematic approaches for the identification of new repurposing opportunities (Cavalla, 2013). The systematic approaches include experimental and computational strategies. The former can be subdivided into binding assays and phenotypic screening of compounds using in vitro or in vivo models or even clinical observations. The latter involves the analysis of data such as gene expression, chemical structure, genotype, and proteomic data or electronic health records (Pushpakom et al., 2018; Parvathaneni et al., 2019).

Potential antimicrobial and antibiofilm activities have been observed in several drugs used in the treatment of noninfectious human diseases, as is the case of antineoplastic, antipsychotic, antidepressant, and antiinflammatory compounds (Cederlund and Mardh, 1993; Rangel-Vega et al., 2015; Das et al., 2016; Richter et al., 2017; Soo et al., 2017; Yssel et al., 2017; Zimmermann and Curtis, 2017; Miró-Canturri et al., 2019; Moraes and Ferreira-Pereira, 2019). These drugs can have direct antimicrobial activity and/or increase the efficacy of antibiotics and antifungal agents or even help to treat the infections by interacting with host targets to activate the host defense mechanisms. This chapter summarizes the antimicrobial effects of different drug classes against bacteria and fungi, with a special focus on antibiofilm activity.

14.2 Antimicrobial activity among drugs used for noninfectious human diseases

Some compounds that were originally developed as antineoplastic drugs are effective at attenuating the growth of microbial pathogens, and some of them have shown promising results through in vitro and in vivo models (Soo et al., 2017). Gallium compounds, 5-fluorouracil, and floxuridine are some examples of such drugs.

There are studies that have explored the possibility to use iron mimetics as antimicrobial agents, as is the case of gallium compounds. The solution and coordination chemistries of Ga^{3+} are very similar to those of Fe^{3+}; however, the Ga^{3+} is irreducible under physiological conditions (Bernstein, 1998). For this reason, gallium is able to repress the activity of iron-containing enzymes involved in DNA synthesis, metabolism, and oxidative stress response by substitution of Fe^{3+} in their active site (Bonchi et al., 2014). Therefore, the use of gallium as an antimicrobial agent has been tested against different important pathogenic bacteria. Gallium compounds inhibited the growth of *Mycobacterium tuberculosis* in broth and intracellularly human macrophages and even showed efficacy in murine tuberculosis models (Olakanmi et al., 2000, 2013). The antibacterial activity of gallium compounds has also been tested against the recalcitrant *Pseudomonas aeruginosa* and the MDR *Acinetobacter baumannii* pathogen both through in vitro and in vivo models (Kaneko et al., 2007; Antunes et al., 2012; de Léséleuc et al., 2012). Additionally, Antunes et al. (2012) demonstrated that gallium nitrate, besides being active alone, also synergizes with colistin against both colistin-sensitive and colistin-resistant *A. baumannii* isolates. Colistin has been a limited option and used as a "last-line" therapy for infections caused by MDR gram-negative bacteria, in particular, *P. aeruginosa*, *A. baumannii*, and *Klebsiella pneumoniae*, when essentially no other options are available (Nation and Li, 2009). Therefore, the combination of gallium nitrate–colistin can represent a promising therapeutic option against pan-resistant *A. baumannii*. In fact, it would provide the benefit of reducing the colistin dosages required to treat the infections and reducing the probability of the emergence of resistant strains to this antibiotic.

The uracil analog 5-fluorouracil is commonly used as a systemic or topical treatment for a variety of cancers, but it has also shown potent antimicrobial effects against several bacterial pathogens. Gieringer et al. (1986) tested the effect of 4 antineoplastic agents against 56 strains of 7 bacterial species and observed that 5-fluorouracil had a half minimal inhibitory concentration (MIC_{50}) lower than 0.8 mg/L for all the strains of *Staphylococcus aureus* and *Staphylococcus epidermidis*. Additionally, the combination of 5-fluorouracil with β-lactam antibiotics promoted a synergetic inhibition of gram-negative strains (Ueda et al., 1983; Gieringer et al., 1986). Nyhlén et al. (2002) also demonstrated the synergy of 5-fluorouracil with tobramycin against *S. aureus*.

Floxuridine, an analog of 5-fluorouracil, showed bactericidal activity with MIC_{90} of 0.0039 mg/L, against MDR *Staphylococcus* spp. isolates including methicillin-resistant *S. aureus* (MRSA), vancomycin-resistant *S. aureus* (VRSA), and vancomycin-intermediate *S. aureus* (VISA). Additionally, it enhanced mouse survival in a lethal model of septicemic MRSA infection (Younis et al., 2015). Floxuridine had also fungicidal activity against *Exserohilum rostratum* with a significant reduction of hyphae-derived colony-forming units at twice its half maximal inhibitory concentration value ($IC_{50} = 846$ nM) (Sun et al., 2013).

The antimicrobial activity of antipsychotic drugs, especially phenothiazines, and antidepressant compounds has been the subject of some studies. Thioridazine and triflupromazine presented consistent antimicrobial activity against methicillin-susceptible *S. aureus* (MSSA) and MRSA with MIC between 16 and 32 mg/L (Hendricks et al., 2003; Hendricks, 2006). In addition, an in vitro cell study conducted by Ordway et al. (2002) demonstrated that thioridazine at a concentration of 0.1 mg/L inhibited intracellular growth of *S. aureus* in human macrophages. Phenothiazines and tricyclic antidepressants have also shown synergy between conventional antibiotics. Promethazine has limited inhibitory activity against *Escherichia coli* and *S. epidermidis*; however, its combination with either tetracycline or erythromycin produced significant synergistic activity (Gunics et al., 2000). The tricyclic antidepressant clomipramine had also synergistic interaction with the same two antibiotics against resistant *S. epidermidis* (Gunics et al., 2000). A combination of prochlorperazine with oxacillin reduced the resistance of MRSA in a range of 16 to 128-fold, and a 128-fold reduction of MRSA resistance with the combination of oxacillin with thioridazine was observed (Hendricks, 2006). Additionally, chlorpromazine, trans-chlorprothixene, and amitriptyline were able to reduce or reverse resistance of gram-positive and gram-negative bacterial strains to penicillin, methicillin, tobramycin, and cefuroxim (Kristiansen et al., 2010).

Regarding the effects of antipsychotic and antidepressant drugs against fungi, sertraline was reported to be fungicidal against *Candida albicans*, *Candida glabrata, Candida tropicalis*, and *Candida parapsilosis* with minimal fungicidal concentration between 3 and 29 mg/L after 48 h of incubation (Lass-Flörl et al., 2001). Silva et al. (2017) also demonstrated that sertraline and two other antidepressant drugs, fluoxetine and paroxetine, had the ability to inhibit the growth of fluconazole-resistant *Candida* spp., inducing cellular death by apoptosis. Additionally, the synergistic combination of fluoxetine with fluconazole was demonstrated against *C. tropicalis, C. albicans, C. parapsilosis,* and *Candida krusei,* with fractional inhibitory index values between 0.15 and 0.31 (Oliveira et al., 2014). Besides the synergy with fluconazole, fluoxetine was also reported to have in vitro synergistic activity with itraconazole and voriconazole (Gu et al., 2016). Furthermore, a *Galleria mellonella* infection model was used to confirm the efficacy of the combination of fluoxetine with fluconazole against resistant *C. albicans* in vivo (Gu et al., 2016). Holbrook et al. (2017) demonstrated by checkerboard and time-kill assays the synergy of bromperidol and a series of its derivatives with posaconazole and voriconazole against *C. albicans*, *C. glabrata,* and *Aspergillus terreus.*

The primary effects of antipsychotic and antidepressant drugs are on the cytoplasmic membranes of prokaryotes and eukaryotes, especially on efflux pumps and energy-providing enzymes, such as ATPase (Dastidar et al., 2013). Phenothiazines were reported to inhibit the NorA-mediated fluoroquinolone efflux of *S. aureus* and reduce oxacillin resistance by a suggested deactivation of a variant (PBP2a) of penicillin-binding protein PBP2 (Kaatz et al., 2003; Kristiansen et al., 2003, 2007). Eilam (1984) also observed that phenothiazines caused a substantial increase in the membrane potential and strongly inhibited the activity of the plasma membrane ATPase. Through a high-throughput screening using recombinant *Saccharomyces cerevisiae* strains overexpressing *Candida* spp. ATP-binding cassette (ABC), and major facilitator superfamily (MFS) transporters, and by secondary in vitro assays, the antidepressant drug clorgyline was identified as a broad-spectrum inhibitor of the referred two classes of fungal efflux pumps (Holmes et al., 2012). In addition, phenothiazines were also recognized as effective antiplasmid agents, which render the bacterial carrier of the plasmid sensitive to antibiotics (Molnár et al., 2003; Amaral et al., 2004; Wolfart et al., 2006). On the other hand, Gu et al. (2016) observed that the underlying mechanism of the synergistic effects of fluconazole and fluoxetine was a downregulation of secreted aspartyl proteinase (*SAP*) genes with inhibition of extracellular phospholipase activity of resistant *C. albicans*.

The statins, lipid-lowering agents used in the treatment of atherosclerotic cardiovascular disease and hypercholesterolemia, have also been investigated for their antimicrobial effects. Masadeh et al. (2012) revealed that atorvastatin, simvastatin, and rosuvastatin were able to induce variable degrees of antibacterial activity against diverse standard bacterial strains and bacterial clinical isolates. Simvastatin and atorvastatin were the most potent ones, and these statins induced susceptibility to *S. aureus*, *Enterococci*, *A. baumannii*, *S. epidermidis*, *Enterobacter aerogenes*, MRSA, and vancomycin-resistant *Enterococci* (VRE). Simvastatin was also reported to reduce the *M. tuberculosis* infection in a human macrophage in vitro model and in vivo murine model (Parihar et al., 2014; Skerry et al., 2014). Additionally, Lobato et al. (2014) demonstrated that atorvastatin and simvastatin induced a bactericidal effect in *M. tuberculosis* and *Mycobacterium leprae* in a macrophage in vitro infection model and in vivo Shepard's mouse footpad *M. leprae* infection model. Synergistic interaction between atorvastatin and rifampin was also observed against *M. leprae* in vivo (Lobato et al., 2014).

In the case of fungi, simvastatin and atorvastatin inhibited the growth of *C. albicans*, *C. glabrata*, *C. tropicalis*, *C. parapsilosis*, and *Aspergillus fumigatus* (Macreadie et al., 2006). A specificity of statins for the mevalonate synthesis pathway was suggested due to the recovery of *A. fumigatus* and *C. albicans* growth from the inhibition by statins. Moreover, another study reported that simvastatin reduced the *C. glabrata* ergosterol levels by 30% and promoted a total loss of mtDNA (Westermeyer and Macreadie, 2007).

The antipyretic and antiinflammatory drugs comprise other drug classes with antimicrobial activity against bacteria and fungi (Zimmermann and Curtis, 2017). Celecoxib, diclofenac, and ibuprofen are examples of such drugs. The nonsteroidal antiinflammatory drug celecoxib exhibited broad-spectrum antimicrobial activity against gram-positive pathogens, such as MRSA, VRSA, *Streptococcus pneumoniae*, *Bacillus anthracis*, *Bacillus subtilis*, *Listeria monocytogenes*, and *Mycobacterium smegmatis*, with MICs between 16 and 64 mg/L (Thangamani et al., 2015a). Although celecoxib alone was not active against gram-negative bacteria, when it was applied in combination with colistin at subinhibitory concentration, celecoxib presented antimicrobial activity against the gram-negative pathogens *P. aeruginosa*, *E. coli*, *A. baumannii*, *Salmonella enterica* serovar Typhimurium, and *K. pneumoniae* with MIC values ranging from 8 to 32 mg/L (Thangamani et al., 2015a). Kalle and Rizvi (2011) also demonstrated that celecoxib increased the sensitivity of bacteria to the antibiotics ampicillin, kanamycin, chloramphenicol, and ciprofloxacin reversing antibiotic resistance in bacteria. Moreover, its topical application significantly reduced the mean bacterial count in a mouse model of MRSA skin infection (Thangamani et al., 2015a). The broad-spectrum antibacterial activity of celecoxib was presumably due to the inhibition of DNA, RNA, and protein synthesis (Thangamani et al., 2015a).

Diclofenac has been demonstrated to have in vitro antimicrobial activity against gram-positive and gram-negative bacteria; for instance, the MIC values against *S. aureus*, *E. coli*, *Bacillus* spp., *Salmonella* spp., *Shigella* spp., and *Vibrio cholerae* were found to be between 50 and 200 mg/L (Dastidar et al., 2000; Dutta et al., 2007). Additionally, diclofenac was also effective against clinically isolated strains of *E. coli* of urinary tract infections with MIC values ranging from 5 to 50 mg/L (Mazumdar et al., 2006). It was also able to protect mice from the lethality of a virulent *Salmonella* infection and presented synergic activity in combination with streptomycin (Dastidar et al., 2000; Dutta et al., 2007). Furthermore, Dastidar et al. (2000) indicated that the mechanism of antibacterial action of diclofenac was through inhibition of DNA synthesis. Regarding the effects on fungi, diclofenac at 500 mg/L completely inhibited hypha formation of *C. albicans* by repressing the cAMP-EFG1 pathway, specifically by repressing the expression of *EFG1* target genes (Ghalehnoo et al., 2010).

Elvers and Wright (1995) investigated the in vitro antibacterial activity of ibuprofen against *S. aureus*, *Micrococcus luteus*, *B. subtilis*, *S. epidermidis*, *E. coli*, and *P. fluorescens* and verified that ibuprofen inhibited the growth of the gram-positive species, being its activity affected by pH. At pH 7, the MIC value for *S. aureus* exceeded 600 mg/L, whereas at pH 6, the value was 350 mg/L. In addition, ibuprofen has been reported to attenuate chronic *Pseudomonas* endobronchial infection in a rat model and improve the survival of mice infected with *M. tuberculosis* (Konstan et al., 1989; Vilaplana et al., 2013). The in vivo combination of ibuprofen with antibiotics amoxicillin, erythromycin, and ampicillin also resulted in the improvement of the therapy of experimental pneumococcal acute otitis media (Diven et al., 1995; del Prado et al., 2006). On the other hand, Sanyal et al. (1993) demonstrated the antifungal activity of ibuprofen against dermatophyte fungi, *Trichophyton rubrum, Trichophyton metagrophytes, Trichophyton tonsurans, Microsporum fulva, Epidermophyton floccosum,* and *Mucor* sp., with MIC values ranging from 5 to 40 mg/L at pH 5. Furthermore, the antifungal activity of ibuprofen in combination with fluconazole against *Candida* species was investigated by Scott et al. (1995) and Pina-Vaz et al. (2000). They observed that the combination of ibuprofen with fluconazole resulted in synergic activity, including for fluconazole-resistant strains. Pina-Vaz et al. (2005) demonstrated that the synergic effect between ibuprofen and azoles resulted from blockade of efflux pumps.

Along with the antineoplastic, antipsychotic, antidepressant, statins, and antiinflammatory drugs, there are other drug classes that have also antimicrobial activity. Terfenadine is an antihistaminic used for the treatment of allergic conditions, and it has been investigated as a potential antibacterial drug. Perlmutter et al. (2014) reported that terfenadine inhibited the growth of *S. aureus, E. coli, Enterococcus faecium, Enterococcus faecalis,* and *M. tuberculosis,* with MIC values ranging from 8 to 16 mg/L. Besides, the synthesized terfenadine-based analogs displayed improved activity with MIC values between 1 and 4 mg/L. Mechanism of action studies suggested that the antibacterial effects showed are related with the inhibition of bacterial type II topoisomerases by targeting both DNA gyrase and topoisomerase IV.

Another example is ivacaftor, a drug prescribed for the treatment of cystic fibrosis because it stimulates chloride ion influx through cell membrane channels. Given that ivacaftor structurally resembles quinolone antibiotics, Reznikov et al. (2014) tested the hypothesis that ivacaftor possessed antibacterial properties. In fact, ivacaftor was active against *S. aureus* and *S. pneumoniae*. Besides, being active alone, ivacaftor displayed positive interactions with ceftriaxone, vancomycin, trimethoprim-sulfamethoxazole, moxifloxacin, and linezolid antibiotics against *S. aureus* and *S. pneumoniae* (Reznikov et al., 2014).

Auranofin, an oral gold-containing drug used for the therapy of rheumatoid arthritis, has been reported to produce significant bactericidal effects against *Staphylococcus* spp. including MRSA, VISA, and VRSA, with MIC ranging from 0.0625 to 0.5 mg/L (Cassetta et al., 2014; Thangamani et al., 2016a). Inhibition of multiple biosynthetic pathways, cell wall, DNA, and protein synthesis was indicated as its mechanism of action (Thangamani et al., 2016a). Additionally, auranofin has also been demonstrated as a potent agent against multiresistant *S. pneumoniae* and MRSA in mouse peritonitis—sepsis and skin infection models (Aguinagalde et al., 2015; Thangamani et al., 2016a, 2016b).

Ejim et al. (2011) screened a collection of drugs to identify compounds that enhance the activity of the antibiotic minocycline and found synergistic drug combinations that exhibited in vitro and in vivo activity against bacterial pathogens. Disulfiram and benserazide alone had weak antibacterial activity against *S. aureus* and *P. aeruginosa*, respectively, but these drugs showed strong synergistic interactions with minocycline. The loperamide—minocycline pair revealed synergistic growth inhibition against *P. aeruginosa, A. baumannii, E. coli*, and *K. pneumoniae* as well, despite loperamide alone had not detectable antibacterial activity.

Ebselen, an investigational drug for the treatment of bipolar disorder, has shown potent bactericidal activity against MRSA, VRSA, and VISA. The MIC at which 90% of clinical isolates of *S. aureus* were inhibited was 0.25 mg/L (Younis et al., 2015). The primary antibacterial mechanism of action of ebselen in *S. aureus* was the inhibition of protein synthesis, at a concentration equivalent to the MIC. However, additional secondary effects on DNA, RNA, and lipid synthesis are observed at higher concentrations (Thangamani et al., 2015b).

14.3 Drug repurposing—an alternative strategy against biofilm infections

The formation of biofilms can be regarded as a universal microbial strategy for survival (Costerton et al., 1987). Moreover, since these sessile communities have an inherent resistance to antimicrobial agents, biofilms are responsible for many persistent and chronic infections (Romling et al., 2014). Therefore, the development of effective strategies for biofilm control is of utmost importance. Keeping in mind the effects of antineoplastic, antipsychotic, antidepressant, and antiinflammatory compounds against planktonic cells, the antibiofilm properties of these drugs have also been evaluated (Table 14.1).

Regarding the antineoplastic drugs, Kaneko et al. (2007) demonstrated that gallium nitrate at low concentrations (0.5 μM) prevented *P. aeruginosa* biofilm formation, and at high levels (100 μM), it was able to kill established biofilms. In fact, gallium was particularly active against centrally located bacteria, which are usually resistant to conventional antibiotics. Additionally, in a murine-wound model of MDR *K. pneumoniae* infection, gallium citrate treatment prevented biofilm formation on dressings, contributing to wound healing (Thompson et al., 2015).

Ueda et al. (2009) demonstrated that uracil plays an important role in the regulation of quorum sensing pathways of *P. aeruginosa*, for instance, LasR, RhlR, and PQS, influencing biofilm formation and virulence. Therefore, the authors screened uracil analogs for biofilm inhibition and verified that 5-fluorouracil abolished quorum sensing phenotypes in *P. aeruginosa*, reduced its virulence, and prevented biofilm formation. Additionally, 5-fluorouracil inhibited *S. epidermidis* biofilms at levels below its MIC and decreased *E. coli* biofilm formation by around fivefold at 25 μM (Hussain et al., 1992; Attila et al., 2009). Moreover, a human trial was conducted to compare the efficacy of central venous catheters externally coated with 5-fluorouracil or chlorhexidine/silver sulfadiazine, and the results indicated that 5-fluorouracil is a safe and effective alternative to prevent catheter colonization (Walz et al., 2010).

TABLE 14.1 Antibiofilm effects of drugs used in the treatment of noninfectious human diseases.

Drugs	Microorganism(s)	Antibiofilm effects	References
Antineoplastic drugs			
Gallium compounds	*Pseudomonas aeruginosa*	- Gallium nitrate at 0.5 μM prevents biofilm formation - Gallium nitrate at 100 μM was able to kill established biofilms	(Kaneko et al., 2007; Thompson et al., 2015)
	Klebsiella pneumonia	- Gallium citrate prevented biofilm formation on dressings in a murine-wound model	
5-Fluorouracil	*Staphylococcus epidermidis*	- At levels below, MIC greatly diminished biofilm formation	(Hussain et al., 1992; Attila et al., 2009; Ueda et al., 2009; Walz et al., 2010)
	P. aeruginosa	- Prevented biofilm formation by repression of quorum sensing phenotypes	
	Escherichia coli	- At 25 μM, biofilm formation was decreased by around fivefold	
Finasteride	*Candida albicans*	- Prevented biofilm formation with metabolic reduction by more than 60% and had moderate action against established biofilms	Chavez-Dozal et al. (2014)
Antipsychotic and antidepressant drugs			
Fluoxetine	*Candida* spp.	- Active against established biofilms with metabolic activity reduction more than 50% at concentrations ≥40 mg/L - Synergistic effect with fluconazole against established biofilms	(Gu et al., 2016; Silva et al., 2017)
Bromperidol	*C. albicans*	- Combination with posaconazole or voriconazole displayed strong additive effect on prevention of biofilm formation	Holbrook et al. (2017)

Continued

TABLE 14.1 Antibiofilm effects of drugs used in the treatment of noninfectious human diseases.—cont'd

Drugs	Microorganism(s)	Antibiofilm effects	References
Statins			
Simvastatin	*P. aeruginosa*	- At 100 μM, decreased the bacterial swarming motility and prevented bacterial adhesion	(Liu et al., 2009; Hennessy et al., 2013; Graziano et al., 2015)
	Staphylococcus aureus (including clinical isolates of MRSA)	- Inhibited biofilm formation - Reduced cell viability and extrapolysaccharide production in established biofilms	
	C. albicans (including bloodstream isolates)	- Inhibited biofilm formation up to 90%	
Lovastatin	*C. albicans* (including itraconazole-resistant strains)	- Combination with itraconazole presented synergistic activity against biofilm formation	Zhou et al. (2018)
Antiinflammatory drugs			
Ibuprofen	*E. coli*	- Inhibited biofilm formation from 37.2% to 44.8%	(Naves et al., 2010; del Prado et al., 2010; Reśliński et al., 2015; Oliveira et al., 2019)
	Streptococcus pneumoniae	- Inhibited biofilm formation	
	S. aureus (including antibiotic-resistant strains)	- Inhibited biofilm formation - Reduced metabolic activity up to 80% and promoted total loss of culturability of preadhered cells and established biofilms	
Aspirin	*P. aeruginosa*	- Reduced cell viability in established biofilms	(Alem and Douglas, 2004; Al-Bakri et al., 2009; Zhou et al., 2012)
	E. coli	- Reduced cell viability in established biofilms	
	C. albicans and *Candida parapsilosis*	- At 1 mM, inhibited biofilm formation up to 95% - At 75–200 μM, reduced established biofilms from 20% to 80% - Combination with amphotericin B presented strong synergistic activity against established biofilms	

Antiinflammatory drugs (cont.)				
Diclofenac	*S. aureus*	-	Inhibited biofilm formation	(Bink et al., 2012; Reśliński et al., 2015)
	E. coli	-	Inhibited biofilm formation	
	C. albicans	-	Potentiated the in vivo activity of caspofungin against catheter-associated biofilm model in rats	
Other drugs				
Niclosamide	*P. aeruginosa*	-	Suppressed swarming motility and reduced biofilm formation	Imperi et al. (2013)
Auranofin	*S. aureus* and *Staphylococcus epidermidis*	-	Reduced established biofilms up to 60% (more than traditional antibiotics linezolid and vancomycin)	Thangamani et al. (2016b)
Ebselen	*S. aureus* and *S. epidermidis*	-	Reduced established biofilms up to 60% (more than traditional antibiotics linezolid, mupirocin, vancomycin, and rifampicin)	Thangamani et al. (2015b)
	P. aeruginosa	-	Reduced biofilm formation up to 40% by inhibition of diguanylate cyclases	Lieberman et al. (2014)
Terfenadine	*S. aureus*	-	Reduced established biofilms (comparable with the activity of ciprofloxacin)	Jacobs et al. (2013)
Valproic acid	*C. albicans* (including azole-resistant strains)	-	Inhibited biofilm formation	Chaillot et al. (2017)
GABA receptor agonists	*C. albicans*	-	Inhibited biofilm formation	Kathwate et al. (2015)
Glatiramer acetate	*P. aeruginosa*	-	Reduced the average bacterial survival to 37% in established biofilms	Christiansen et al. (2017)

MRSA, Methicillin-resistant Staphylococcus aureus.

Another antineoplastic drug with antibiofilm activity is finasteride, which is commonly used for the treatment of benign prostatic hyperplasia. Although finasteride did not display antifungal activity against planktonic *C. albicans*, Chavez-Dozal et al. (2014) observed the prevention of biofilm formation with metabolic activity reduction by more than 60% at finasteride concentrations $\geq$ 16 mg/L and a moderate action on 24-h-old *C. albicans* biofilms. The combination of finasteride with fluconazole also promoted a synergistic activity in the prevention of *C. albicans* biofilm formation. Furthermore, the authors verified an inhibition of fungi filamentation, suggesting it as a mechanism of finasteride antibiofilm action.

In respect of antidepressant and antipsychotic drugs, fluoxetine and bromperidol are examples of such drugs with reported antibiofilm activities. The antifungal effects of fluoxetine were tested against biofilm-forming isolates of *C. parapsilosis*, *C. albicans*, *C. tropicalis*, and *C. glabrata*, and a metabolic inactivation of mature biofilms more than 50% was observed for fluoxetine at concentrations $\geq$40 mg/L (Silva et al., 2017). Additionally, Gu et al. (2016) demonstrated that the combination of fluoxetine with fluconazole had a synergistic action against *C. albicans* biofilms formed over 4, 8, and 12 h. On the other hand, the dual combination of bromperidol with two representative azoles, posaconazole and voriconazole, caused a strong additive antifungal effect against biofilm formation of *C. albicans* (Holbrook et al., 2017).

Concerning the statins, they have also been reported as antibiofilm agents against bacteria and fungi. The influence of statins on the motility of *P. aeruginosa* was studied by Hennessy et al. (2013). These authors found that simvastatin, lovastatin, and mevastatin at 100 μM decreased swarming motility of bacterial cells. At this concentration, simvastatin also prevented the bacterial adhesion. Moreover, simvastatin had antimicrobial activity against biofilms of laboratory and clinical isolates of *S. aureus* and MRSA, inhibiting their formation and reducing the cell viability and extracellular polysaccharide production of established biofilms (Graziano et al., 2015). In the case of the effects against fungi, simvastatin inhibited biofilm formation of *C. albicans*, including laboratory strain and clinical isolates from blood, up to 90% (Liu et al., 2009). Zhou et al. (2018) also reported a synergistic combination of lovastatin with itraconazole in the prevention of biofilm formation of *C. albicans*, including itraconazole-resistant strains.

Regarding the antipyretic and antiinflammatory drugs, Alem and Douglas (2004) tested nine nonsteroidal antiinflammatory drugs against biofilms of *C. albicans*. The authors reported that aspirin, etodolac, and diclofenac presented strong inhibition of biofilm formation and, in particular, aspirin at 1 mM inhibited biofilm formation up to 95%. Moreover, at concentrations that can be achieved in humans by the use of pharmacological doses, aspirin was able to reduce established biofilms up to 80%. Aspirin has also been reported to reduce cell viability of established biofilms of *P. aeruginosa* and *E. coli* and present synergistic activity with amphotericin B against established biofilms of *C. albicans* and *C. parapsilosis* (Al-Bakri et al., 2009; Zhou et al., 2012).

Ibuprofen and diclofenac are other antiinflammatory drugs with antibiofilm properties. del Prado et al. (2010) and Naves et al. (2010) investigated the effects of ibuprofen against biofilm formation by *S. pneumoniae* and pathogenic *E. coli* strains, respectively. Ibuprofen at a concentration of 128 mg/L inhibited biofilm formation of *S. pneumoniae* in 5 out of 11 strains and displayed *E. coli* biofilm reduction from 37.2% to 44.8%. Additionally, Reśliński et al. (2015) verified that ibuprofen and diclofenac inhibited biofilm formation by *S. aureus* and *E. coli* on the surface of monofilament polypropylene mesh, a biomaterial that is often applied in hernia surgery. Moreover, ibuprofen was also active against preadhered and established biofilms of *S. aureus*, including antibiotic-resistant strains, promoting metabolic inactivation up to 80% and total loss of culturability (Oliveira et al., 2019). An in vivo study using a catheter-associated biofilm model in rats has demonstrated that diclofenac can potentiate the activity of caspofungin against *C. albicans* (Bink et al., 2012).

Other drugs with different clinical indications have been investigated as antibiofilm agents. Using a quorum sensing inhibitor assay, Imperi et al. (2013) identified the anthelmintic drug niclosamide as a strong inhibitor of *P. aeruginosa* quorum sensing response. Subsequently, the authors verified that niclosamide suppressed the swarming motility of *P. aeruginosa* and reduced biofilm formation. However, the significant biofilm prevention was only observed at concentrations remarkably higher than those found active against the quorum sensing response, indicating that the biofilm inhibitory action of niclosamide is through other mechanisms. Another example is glatiramer acetate, an immunomodulatory drug used in the treatment of multiple sclerosis. Christiansen et al. (2017) reported that glatiramer acetate was effective against gram-negative bacteria and in 3 h was able to reduce the bacterial survival to 37% in established *P. aeruginosa* biofilms.

Given the potent antibacterial activity of ebselen, auranofin, and terfenadine against planktonic bacteria, these drugs were also tested against biofilms. Actually, ebselen reduced the biomass of established biofilms formed by *S. aureus* and *S. epidermidis* by 60% and 50%, respectively, with an antibiofilm action remarkably superior to traditional antibiotics (Thangamani et al., 2015b). Similarly, auranofin significantly disrupted established in vitro biofilms of *S. aureus* and *S. epidermidis* (Thangamani et al., 2016b). Moreover, ebselen inhibited the production of the signal molecule cyclic-di-GMP by diguanylate cyclases, which is a key regulator of biofilm formation. Consequently, the biofilm formation by *P. aeruginosa* was reduced by up to 40% (Lieberman et al., 2014). Finally, Jacobs et al. (2013) reported that terfenadine elicited a 1.1-log reduction in cell viability of 48-h-old *S. aureus* biofilms, an antibiofilm action comparable with the activity of ciprofloxacin under the same assay conditions.

Concerning the effects on fungal biofilms, valproic acid and GABA receptor agonists have been tested against biofilm formation. Valproic acid is prescribed to treat epilepsy, bipolar disorder, and uncontrolled seizures, but the inhibitory effect on biofilm formation by susceptible and azole-resistant *C. albicans* was also demonstrated (Chaillot et al., 2017). In the case of γ-aminobutyric acid (GABA) receptor agonists, which are used in antiepileptic disorders, lorazepam, midazolam, and diazepam exhibited inhibitory effects on *C. albicans* biofilm formation (Kathwate et al., 2015). In particular, among the GABA receptor agonists tested, lorazepam promoted complete biofilm prevention.

14.4 Conclusions

There is an urgent need to discover new antimicrobials and antibiofilm agents, to overcome the issues of the current epidemiological scenario of infections, mainly caused by MDR microorganisms and usually characterized by the presence of biofilms. This chapter emphasizes drug repurposing as a promising approach to face this alarming situation, by describing the antimicrobial and antibiofilm activity of several drugs currently used in the treatment of noninfectious human diseases. Nevertheless, further research is still needed to understand the mechanisms of antimicrobial action and the real applicability of these compounds. On the other hand, careful management in the doses applied for the in vivo antimicrobial activity must be taken. Ideally, equal or lower concentrations than those approved for the original medical indication are preferred, so as to avoid new pharmacokinetic and toxicity studies (Oprea and Mestres, 2012).

Acknowledgments

This work was financially supported by project UID/EQU/00511/2019—Laboratory for Process Engineering, Environment, Biotechnology and Energy (LEPABE) funded by national funds through FCT/MCTES (PIDDAC); project "LEPABE-2-ECO-INNOVATION"—NORTE-01-0145-FEDER-000005, funded by Norte Portugal Regional Operational Programme (NORTE 2020), under PORTUGAL 2020 Partnership Agreement, through the European Regional Development Fund (ERDF); projects POCI-01-0145-FEDER-030219, POCI-01-0145-FEDER-006939, funded by FEDER funds through COMPETE2020—Programa Operacional Competitividade e Internacionalização (POCI) and by national funds (PIDDAC) through FCT/MCTES, and Ph.D. grant awarded to Isabel Maria Oliveira (SFRH/BD/138117/2018) and the grant awarded to Manuel Simões (SFRH/BSAB/150379/2019) from national funds through FCT—Fundação para a Ciência e Tecnologia.

References

Aguinagalde, L., Díez-Martínez, R., Yuste, J., Royo, I., Gil, C., Lasa, Í., Martín-Fontecha, M., Marín-Ramos, N.I., Ardanuy, C., Liñares, J., García, P., García, E., Sánchez-Puelles, J.M.,

2015. Auranofin efficacy against MDR *Streptococcus pneumoniae* and *Staphylococcus aureus* infections. Journal of Antimicrobial Chemotherapy 70, 2608–2617. https://doi.org/10.1093/jac/dkv163.

Al-Bakri, A.G., Othman, G., Bustanji, Y., 2009. The assessment of the antibacterial and antifungal activities of aspirin, EDTA and aspirin-EDTA combination and their effectiveness as anti-biofilm agents. Journal of Applied Microbiology 107 (1), 280–286. https://doi.org/10.1111/j.1365-2672.2009.04205.x.

Alem, M.A.S., Douglas, L.J., 2004. Effects of aspirin and other nonsteroidal anti-inflammatory drugs on biofilms and planktonic cells of *Candida albicans*. Antimicrobial Agents and Chemotherapy 48 (1), 41–47. https://doi.org/10.1128/AAC.48.1.41.

Amaral, L., Viveiros, M., Molnar, J., 2004. Antimicrobial activity of phenothiazines. In Vivo 18, 725–732. https://doi.org/10.5829/idosi.gjp.2013.7.2.74179.

Antunes, L.C.S., Imperi, F., Minandri, F., Visca, P., 2012. In vitro and in vivo antimicrobial activities of gallium nitrate against multidrug-resistant *Acinetobacter baumannii*. Antimicrobial Agents and Chemotherapy 56 (11), 5961–5970. https://doi.org/10.1128/AAC.01519-12.

Ashburn, T.T., Thor, K.B., 2004. Drug repositioning: identifying and developing new uses for existing drugs. Nature Reviews Drug Discovery 3 (8), 673–683. https://doi.org/10.1038/nrd1468.

Attila, C., Ueda, A., Wood, T.K., 2009. 5-Fluorouracil reduces biofilm formation in *Escherichia coli* K-12 through global regulator AriR as an antivirulence compound. Applied Microbiology and Biotechnology 82 (3), 525–533. https://doi.org/10.1007/s00253-009-1860-8.

Bernstein, L.R., 1998. Mechanisms of therapeutic activity for gallium. Pharmacological Reviews 50 (4), 665–682.

Bink, A., Kucharíková, S., Neirinck, B., Vleugels, J., Dijck, P.V., Cammue, B.P.A., Thevissen, K., 2012. The nonsteroidal antiinflammatory drug diclofenac potentiates the *in vivo* activity of caspofungin against *Candida albicans* biofilms. The Journal of Infectious Diseases 206, 1790–1797. https://doi.org/10.1093/infdis/jis594.

Bonchi, C., Imperi, F., Minandri, F., Visca, P., Frangipani, E., 2014. Repurposing of gallium-based drugs for antibacterial therapy. Biofactors 40 (3), 303–312. https://doi.org/10.1002/biof.1159.

Brooks, B.D., Brooks, A.E., 2014. Therapeutic strategies to combat antibiotic resistance. Advanced Drug Delivery Reviews 78, 14–27. https://doi.org/10.1016/j.addr.2014.10.027.

Cassetta, M.I., Marzo, T., Fallani, S., Vovelli, A., Messori, L., 2014. Drug repositioning: auranofin as a prospective antimicrobial agent for the treatment of severe staphylococcal infections. Biometals 27 (4), 787–791. https://doi.org/10.1007/s10534-014-9743-6.

Cavalheiro, M., Teixeira, M.C., 2018. *Candida* biofilms: threats, challenges, and promising strategies. Frontiers of Medicine 5 (28), 1–15. https://doi.org/10.3389/fmed.2018.00028.

Cavalla, D., 2013. Predictive methods in drug repurposing: gold mine or just a bigger haystack? Drug Discovery Today 18 (11–12), 523–532. https://doi.org/10.1016/j.drudis.2012.12.009.

Cederlund, H., Mardh, P., 1993. Antibacterial activities of non-antibiotic drugs. Journal of Antimicrobial Chemotherapy 32, 355–365. https://doi.org/10.1093/jac/32.3.355.

Chaillot, J., Tebbji, F., García, C., Wurtele, H., Pelletier, R., Sellam, A., 2017. pH-Dependant antifungal activity of valproic acid against the human fungal pathogen *Candida albicans*. Frontiers in Microbiology 8 (1956), 1–10. https://doi.org/10.3389/fmicb.2017.01956.

Chavez-Dozal, A.A., Lown, L., Jahng, M., Walraven, C.J., Lee, S.A., 2014. *In vitro* analysis of finasteride activity against *Candida albicans* urinary biofilm formation and filamentation. Antimicrobial Agents and Chemotherapy 58 (10), 5855–5862. https://doi.org/10.1128/AAC.03137-14.

Chong, C., Sullivan, D., 2007. New uses for old drugs. Nature 448, 645–646. https://doi.org/10.2165/00128413-199308940-00046.

Christiansen, S.H., Murphy, R.A., Juul-Madsen, K., Fredborg, M., Hvam, M.L., Axelgaard, E., Skovdal, S.M., Meyer, R.L., Sørensen, U.B.S., Möller, A., Nyengaard, J.R., Nørskov-Lauritsen, N., Wang, M., Gadjeva, M., Howard, K.A., Davies, J.C., Petersen, E., Vorup-Jensen, T., 2017. The immunomodulatory drug glatiramer acetate is also an effective antimicrobial agent that kills Gram-negative bacteria. Scientific Reports 7 (15653), 1–16. https://doi.org/10.1038/s41598-017-15969-3.

Costerton, J.W., Cheng, K.J., Geesey, G.G., Ladd, T.I., Nickel, J.C., Dasgupta, M., Marrie, T.J., 1987. Bacterial biofilms in nature and disease. Annual Review of Microbiology 41 (1), 435–464. https://doi.org/10.1146/annurev.mi.41.100187.002251.

Costerton, J.W., Steward, P.S., Greenberg, E.P., 1999. Bacterial biofilms: a common cause of persistent infections. Science 284 (5418), 1318–1322. https://doi.org/10.1126/science.284.5418.1318.

Das, S., Dasgupta, A., Chopra, S., 2016. Drug repurposing: a new front in the war against *Staphylococcus aureus*. Future Microbiology. https://doi.org/10.2217/fmb-2016-0021.

Dastidar, S.G., Kristiansen, J.E., Molnar, J., Amaral, L., 2013. Role of phenothiazines and structurally similar compounds of plant origin in the fight against infections by drug resistant bacteria. Antibiotics 2 (1), 58–72. https://doi.org/10.3390/antibiotics2010058.

Dastidar, S.G., Ganguly, K., Chaudhuri, K., Chakrabarty, A.N., 2000. The anti-bacterial action of diclofenac shown by inhibition of DNA synthesis. International Journal of Antimicrobial Agents 14, 249–251. https://doi.org/10.1016/s0924-8579(99)00159-4.

Desai, J.V., Mitchell, A.P., Andes, D.R., 2014. Fungal biofilms, drug resistance and recurrent infection. Cold Spring Harbor Perspectives in Medicine 4, 1–18. https://doi.org/10.1101/cshperspect.a019729.

Diven, W.F., Evans, R.W., Alper, C., Burckart, G.J., Jaffe, R., Doyle, W.J., 1995. Treatment of experimental acute otitis media with ibuprofen and ampicillin. International Journal of Pediatric Otorhinolaryngology 33, 127–139. https://doi.org/10.1016/0165-5876(95)01196-I.

Dutta, N.K., Annadurai, S., Mazumdar, K., Dastidar, S.G., Kristiansen, J.E., Molnar, J., Martins, M., Amaral, L., 2007. Potential management of resistant microbial infections with a novel non-antibiotic: the anti-inflammatory drug diclofenac sodium. International Journal of Antimicrobial Agents 30, 242–249. https://doi.org/10.1016/j.ijantimicag.2007.04.018.

Eilam, Y., 1984. Effects of phenothiazines on inhibition of plasma membrane ATPase and hyperpolarization of cell membranes in the yeast *Saccharomyces*. Biochimica et Biophysica Acta 769, 601–610. https://doi.org/10.1016/0005-2736(84)90059-2.

Ejim, L., Farha, M.A., Falconer, S.B., Wildenhain, J., Coombes, B.K., Tyers, M., Brown, E.D., Wright, G.D., 2011. Combinations of antibiotics and nonantibiotic drugs enhance antimicrobial efficacy. Nature Chemical Biology 7 (6), 348–350. https://doi.org/10.1038/nchembio.559.

Elvers, K.T., Wright, S.J., 1995. Antibacterial activity of the anti-inflammatory compound ibuprofen. Letters in Applied Microbiology 20 (2), 82–84. https://doi.org/10.1111/j.1472-765X.1995.tb01291.x.

Fisher, M.C., Hawkins, N.J., Sanglard, D., Gurr, S.J., 2018. World emergence of resistance to antifungal drugs challenges human health and food security. Science 360 (6390), 739–742. https://doi.org/10.1126/science.aap7999.

Ghalehnoo, Z.R., Rashki, A., Najimi, M., Dominguez, A., 2010. The role of diclofenac sodium in the dimorphic transition in *Candida albicans*. Microbial Pathogenesis 48, 110–115. https://doi.org/10.1016/j.micpath.2009.12.003.

Gieringer, J.H., Wenz, A.F., Just, H.-M., Daschner, F.D., 1986. Effect of 5-fluorouracil, mitoxantrone, methotrexate, and vincristine on the antibacterial activity of ceftriaxone, ceftazidime, cefotiam, piperacillin, and netilmicin. Chemotherapy 32 (1), 418–424. https://doi.org/10.1159/000238445.

Graziano, T.S., Cuzzullin, M.C., Franco, G.C., Schwartz-Filho, H.O., Andrade, E.D., Groppo, F.C., Cogo-Müller, K., 2015. Statins and antimicrobial effects: simvastatin as a potential drug against *Staphylococcus aureus* biofilm. PLoS One 10 (5). https://doi.org/10.1371/journal.pone.0128098.

Gu, W., Guo, D., Zhang, L., Xu, D., Sun, S., 2016. The synergistic effect of azoles and fluoxetine against resistant *Candida albicans* strains is attributed to attenuating fungal virulence. Antimicrobial Agents and Chemotherapy 60 (10), 6179–6188. https://doi.org/10.1128/AAC.03046-15.Address.

Gunics, G., Motohashi, N., Amaral, L., Farkas, S., Molnár, J., 2000. Interaction between antibiotics and non-conventional antibiotics on bacteria. International Journal of Antimicrobial Agents 14 (3), 239–242. https://doi.org/10.1016/S0924-8579(00)00131-X.

Hendricks, O., 2006. Antimicrobial Effects of Selected Non-antibiotics on Sensitivity and Invasion of Gram-Positive Bacteria (Ph.D. thesis).

Hendricks, O., Butterworth, T.S., Kristiansen, J.E., 2003. The *in vitro* antimicrobial effect of non-antibiotics and putative inhibitors of efflux pumps on *Pseudomonas aeruginosa* and *Staphylococcus aureus*. International Journal of Antimicrobial Agents 22 (3), 262–264. https://doi.org/10.1016/S0924-8579(03)00205-X.

Hennessy, E., Mooij, M.J., Legendre, C., Reen, F.J., O'Callaghan, J., Adams, C., O'Gara, F., 2013. Statins inhibit *in vitro* virulence phenotypes of *Pseudomonas aeruginosa*. Journal of Antibiotics 66 (2), 99–101. https://doi.org/10.1038/ja.2012.95.

Holbrook, S.Y.L., Garzan, A., Dennis, E.K., Shrestha, S.K., Garneau-Tsodikova, S., 2017. Repurposing antipsychotic drugs into antifungal agents: synergistic combinations of azoles and bromperidol derivatives in the treatment of various fungal infections. European Journal of Medicinal Chemistry 139, 12–21. https://doi.org/10.1016/j.ejmech.2017.07.030.

Holmes, A.R., Keniya, M.V., Ivnitski-Steele, I., Monk, B.C., Lamping, E., Sklar, L.A., Cannon, R.D., 2012. The monoamine oxidase A inhibitor clorgyline is a broad-spectrum inhibitor of fungal ABC and MFS transporter efflux pump activities which reverses the azole resistance of *Candida albicans* and *Candida glabrata* clinical isolates. Antimicrobial Agents and Chemotherapy 56 (3), 1508–1515. https://doi.org/10.1128/AAC.05706-11.

Hussain, M., Collins, C., Hastings, J.G.M., White, P.J., 1992. Radiochemical assay to measure the biofilm produced by coagulase-negative staphylococci on solid surfaces and its use to quantitate the effects of various antibacterial compounds on the formation of the biofilm. Journal of Medical Microbiology 37 (1), 62–69. https://doi.org/10.1099/00222615-37-1-62.

Imperi, F., Massai, F., Pillai, C.R., Longo, F., Zennaro, E., Rampioni, G., Visca, P., Leoni, L., 2013. New life for an old drug: the anthelmintic drug niclosamide inhibits *Pseudomonas aeruginosa* quorum sensing. Antimicrobial Agents and Chemotherapy 57 (2), 996–1005. https://doi.org/10.1128/AAC.01952-12.

Jacobs, A.C., Didone, L., Jobson, J., Sofia, M.K., Krysan, D., Dunman, P.M., 2013. Adenylate kinase release as a high-throughput-screening-compatible reporter of bacterial lysis for identification of antibacterial agents. Antimicrobial Agents and Chemotherapy 57 (1), 26–36. https://doi.org/10.1128/AAC.01640-12.

Kaatz, G.W., Moudgal, V.V., Seo, S.M., Kristiansen, J.E., 2003. Phenothiazines and thioxanthenes inhibit multidrug efflux pump activity in *Staphylococcus aureus*. Antimicrobial Agents and Chemotherapy 47 (2), 719–726. https://doi.org/10.1128/AAC.47.2.719.

Kalle, A.M., Rizvi, A., 2011. Inhibition of bacterial multidrug resistance by celecoxib, a cyclooxygenase-2 inhibitor. Antimicrobial Agents and Chemotherapy 55 (1), 439–442. https://doi.org/10.1128/AAC.00735-10.

Kaneko, Y., Thoendel, M., Olakanmi, O., Britigan, B.E., Singh, P.K., 2007. The transition metal gallium disrupts *Pseudomonas aeruginosa* iron metabolism and has antimicrobial and anti-biofilm activity. Journal of Clinical Investigation 117 (4), 877–888. https://doi.org/10.1172/JCI30783.

Kathwate, G.H., Shinde, R.B., Karuppayil, S.M., 2015. Antiepileptic drugs inhibit growth, dimorphism, and biofilm mode of growth in human pathogen *Candida albicans*. Assay and Drug Development Technologies 13 (6), 307–312. https://doi.org/10.1089/adt.2015.29007.ghkdrrr.

Konstan, M.W., Vargo, K.M., Davis, P.B., 1989. Ibuprofen attenuates the inflammatory response to *Pseudomonas aeruginosa* in a rat model of chronic pulmonary infection. Implications for antiinflammatory therapy in cystic fibrosis. American Review of Respiratory Disease 141 (1), 186–192. https://doi.org/10.1164/ajrccm/141.1.186.

Kristiansen, J.E., Hendricks, O., Delvin, T., Butterworth, T.S., Aagaard, L., Christensen, J.B., Flores, V.C., Keyzer, H., 2007. Reversal of resistance in microorganisms by help of non-antibiotics. Journal of Antimicrobial Chemotherapy 59 (6), 1271–1279. https://doi.org/10.1093/jac/dkm071.

Kristiansen, J.E., Thomsen, V.F., Martins, A., Viveiros, M., Amaral, L., 2010. Non-antibiotics reverse resistance of bacteria to antibiotics. In Vivo 24 (5), 751–754.

Kristiansen, M.M., Leandro, C., Ordway, D., Martins, M., Viveiros, M., Pacheco, T., Kristiansen, J.E., Amaral, L., 2003. Phenothiazines alter resistance of methicillin-resistant strains of *Staphylococcus aureus* (MRSA) to oxacillin *in vitro*. International Journal of Antimicrobial Agents 22, 250–253. https://doi.org/10.1016/s0924-8579(03)00200-0.

Lass-Flörl, C., Dierich, M.P., Fuchs, D., Semenitz, E., Ledochowski, M., 2001. Antifungal activity against *Candida* species of the selective serotonin-reuptake inhibitor, sertraline. Clinical Infectious Diseases 33, 135–136. https://doi.org/10.1086/324589.

Lebeaux, D., Ghigo, J.-M., Beloin, C., 2014. Biofilm-related infections: bridging the gap between clinical management and fundamental aspects of recalcitrance towards antibiotics. Microbiology and Molecular Biology Reviews 78 (3), 510–543. https://doi.org/10.1128/MMBR.00013-14.

de Léséleuc, L., Harris, G., KuoLee, R., Chen, W., 2012. *In vitro* and *in vivo* biological activities of iron chelators and gallium nitrate against *Acinetobacter baumannii*. Antimicrobial Agents and Chemotherapy 56 (10), 5397–5400. https://doi.org/10.1128/AAC.00778-12.

Lieberman, O.J., Orr, M.W., Wang, Y., Lee, V.T., 2014. High-throughput screening using the differential radial capillary action of ligand assay identifies ebselen as an inhibitor of diguanylate cyclases. ACS Chemical Biology 9, 183–192. https://doi.org/10.1021/cb400485k.

Liu, G., Vellucci, V.F., Kyc, S., Hostetter, M.K., 2009. Simvastatin inhibits *Candida albicans* biofilm *in vitro*. Pediatric Research 66 (6), 600–604. https://doi.org/10.1203/PDR.0b013e3181bd5bf8.

Lobato, L.S., Rosa, P.S., Ferreira, J.S., Neumann, A.S., Silva, M.G., Nascimento, D.C., Soares, C.T., Pedrini, S.C.B., Oliveira, D.S.L., Monteiro, C.P., Pereira, G.M., Ribeiro-Alves, M., Hacker, M.A., Moares, M.O., Pessolani, M.C.V., Duarte, R.S., Lara, F.A., 2014. Statins increase rifampin mycobactericidal effect. Antimicrobial Agents and Chemotherapy 58 (10), 5766–5774. https://doi.org/10.1128/AAC.01826-13.

Macreadie, I.G., Johnson, G., Schlosser, T., Macreadie, P.I., 2006. Growth inhibition of *Candida* species and *Aspergillus fumigatus* by statins. FEMS Microbiology Letters 262 (9–13), 9–13. https://doi.org/10.1111/j.1574-6968.2006.00370.x.

Masadeh, M., Mhaidat, N., Alzoubi, K., Al-azzam, S., Alnasser, Z., 2012. Antibacterial activity of statins: a comparative study of atorvastatin, simvastatin, and rosuvastatin. Annals of Clinical Microbiology and Antimicrobials 11 (13). https://doi.org/10.1186/1476-0711-11-13.

Mazumdar, K., Dutta, N.K., Dastidar, S.G., Motohashi, N., Shirataki, Y., 2006. Diclofenac in the management of *E.coli* urinary tract infections. In Vivo 20, 613–620.

Miró-Canturri, A., Ayerbe-Algaba, R., Smani, Y., 2019. Drug repurposing for the treatment of bacterial and fungal infections. Frontiers in Microbiology 10 (41), 1–12. https://doi.org/10.3389/fmicb.2019.00041.

Molnár, A., Amaral, L., Molnár, J., 2003. Antiplasmid effect of promethazine in mixed bacterial cultures. International Journal of Antimicrobial Agents 22 (3), 217–222. https://doi.org/10.1016/S0924-8579(03)00206-1.

Moraes, D., Ferreira-Pereira, A., 2019. Insights on the anticandidal activity of non-antifungal drugs. Journal de Mycologie Medicale 29 (3), 253–259. https://doi.org/10.1016/j.mycmed.2019.07.004.

Nation, R.L., Li, J., 2009. Colistin in the 21st century. Current Opinion in Infectious Diseases 22 (6), 535–543. https://doi.org/10.1097/QCO.0b013e328332e672.

Naves, P., del Prado, G., Huelves, L., Rodríguez-Cerrato, V., Ruiz, V., Ponte, M.C., Soriano, F., 2010. Effects of human serum albumin, ibuprofen and N-acetyl-l-cysteine against biofilm formation by pathogenic *Escherichia coli* strains. Journal of Hospital Infection 76 (2), 165–170. https://doi.org/10.1016/j.jhin.2010.05.011.

Nyhlén, A., Ljungberg, B., Nilsson-Ehle, I., Odenholt, I., 2002. Bactericidal effect of combinations of antibiotic and antineoplastic agents against *Staphylococcus aureus* and *Escherichia coli*. Chemotherapy 48, 71–77. https://doi.org/10.1159/000057665.

Olakanmi, O., Britigan, B.E., Schlesinger, L.S., 2000. Gallium disrupts iron metabolism of mycobacteria residing within human macrophages. Infection and Immunity 68 (10), 5619–5627. https://doi.org/10.1128/iai.68.10.5619-5627.2000.

Olakanmi, O., Kesavalu, B., Pasula, R., Abdalla, M.Y., Schlesinger, L.S., Britigan, B.E., 2013. Gallium nitrate is efficacious in murine models of tuberculosis and inhibits key bacterial Fe-dependent enzymes. Antimicrobial Agents and Chemotherapy 57 (12), 6074–6080. https://doi.org/10.1128/AAC.01543-13.

Oliveira, A.S., Gaspar, C.A., Palmeira-de-Oliveira, R., Martinez-de-Oliveira, J., Palmeira-de-Oliveira, A., 2014. Anti-*Candida* activity of fluoxetine alone and combined with fluconazole: a synergistic action against fluconazole-resistant strains. Antimicrobial Agents and Chemotherapy 58 (7), 4224–4226. https://doi.org/10.1128/AAC.02623-13.

Oliveira, I.M., Borges, A., Borges, F., Simões, M., 2019. Repurposing ibuprofen to control *Staphylococcus aureus* biofilms. European Journal of Medicinal Chemistry 166, 197–205. https://doi.org/10.1016/j.ejmech.2019.01.046.

Oprea, T.I., Mestres, J., 2012. Drug repurposing: far beyond new targets for old drugs. The AAPS Journal 14 (4), 759–763. https://doi.org/10.1208/s12248-012-9390-1.

Ordway, D., Viveiros, M., Leandro, C., Arroz, M.J., Amaral, L., 2002. Intracellular activity of clinical concentrations of phenothiazines including thioridiazine against phagocytosed *Staphylococcus aureus*. International Journal of Antimicrobial Agents 20 (1), 34–43. https://doi.org/10.1016/S0924-8579(02)00110-3.

Parihar, S.P., Guler, R., Khutlang, R., Lang, D.M., Hurdayal, R., Mhlanga, M.M., Suzuki, H., Marais, D., Brombacher, F., 2014. Statin therapy reduces the *Mycobacterium tuberculosis* burden in human macrophages and in mice by enhancing autophagy and phagosome maturation. Journal of Infectious Diseases 209, 754–763. https://doi.org/10.1093/infdis/jit550.

Parvathaneni, V., Kulkarni, N.S., Muth, A., Gupta, V., 2019. Drug repurposing: a promising tool to accelerate the drug discovery process. Drug Discovery Today. https://doi.org/10.1016/j.drudis.2019.06.014 pii: S1359-6446(18)30259-9.

Perlmutter, J.I., Forbes, L.T., Krysan, D.J., Ebsworth-Mojica, K., Colquhoun, J.M., Wang, J.L., Dunman, P.M., Flaherthy, D.P., 2014. Repurposing the antihistamine terfenadine for antimicrobial activity against *Staphylococcus aureus*. Journal of Medicinal Chemistry 57 (20), 8540–8562. https://doi.org/10.1021/jm5010682.

Piddock, L.J.V., 2012. The crisis of no new antibiotics — what is the way forward? The Lancet Infectious Diseases 12, 249–253. https://doi.org/10.1016/S1473-3099(11)70316-4.

Pina-Vaz, C., Sansonetty, F., Rodrigues, A.G., Martinez-de-Oliveira, J., Fonseca, A.F., Mårdh, P.-A., 2000. Antifungal activity of ibuprofen alone and in combination with fluconazole against *Candida* species. Journal of Medical Microbiology 49, 831–840. https://doi.org/10.1099/0022-1317-49-9-831.

Pina-Vaz, C., Rodrigues, A.G., Costa-de-Oliveira, S., Ricardo, E., Mårdh, P.-A., 2005. Potent synergic effect between ibuprofen and azoles on *Candida* resulting from blockade of efflux pumps as determined by FUN-1 staining and flow cytometry. Journal of Antimicrobial Chemotherapy 56 (4), 678–685. https://doi.org/10.1093/jac/dki264.

del Prado, G., Martínez-Marín, C., Huelves, L., Gracia, M., Rodrígues-Cerrato, V., Fernández-Roblas, R., Ponte, C., Cenjor, C., Soriano, F., 2006. Impact of ibuprofen therapy in the outcome of experimental pneumococcal acute otitis media treated with amoxicillin or erythromycin. Pediatric Research 60 (5), 555–559. https://doi.org/10.1203/01.PDR.0000242258.52590.b5.

del Prado, G., Ruiz, V., Naves, P., Rodríguez-Cerrato, V., Soriano, F., Ponte, M.C., 2010. Biofilm formation by *Streptococcus pneumoniae* strains and effects of human serum albumin, ibuprofen, N-acetyl-L-cysteine, amoxicillin, erythromycin, and levofloxacin. Diagnostic Microbiology and Infectious Disease 67 (4), 311–318. https://doi.org/10.1016/j.diagmicrobio.2010.03.016.

Projan, S.J., 2003. Why is big pharma getting out of antibacterial drug discovery? Current Opinion in Microbiology 6, 427–430. https://doi.org/10.1016/j.mib.2003.08.003.

Pushpakom, S., Iorio, F., Eyers, P.A., Escott, K.J., Hopper, S., Wells, A., Doig, A., Guilliams, T., Latimer, J., McNamee, C., Norris, A., Sanseau, P., Cavalla, D., Pirmohamed, M., 2018. Drug repurposing: progress, challenges and recommendations. Nature Reviews Drug Discovery 1–18. https://doi.org/10.1038/nrd.2018.168.

Rangel-Vega, A., Bernstein, L.R., Mandujano-Tinoco, E.A., García-Contreras, S.J., García-Contreras, R., 2015. Drug repurposing as an alternative for the treatment of recalcitrant bacterial infections. Frontiers in Microbiology 6 (282), 1–8. https://doi.org/10.3389/fmicb.2015.00282.

Reśliński, A., Dabrowiecki, S., Głowacka, K., 2015. The impact of diclofenac and ibuprofen on biofilm formation on the surface of polypropylene mesh. Hernia 19 (2), 179–185. https://doi.org/10.1007/s10029-013-1200-x.

Reznikov, L.R., Abou Alaiwa, M.H., Dohrn, C.L., Gansemer, N.D., Diekema, D.J., Stoltz, D.A., Welsh, M.J., 2014. Antibacterial properties of the CFTR potentiator ivacaftor. Journal of Cystic Fibrosis 13 (5), 515–519. https://doi.org/10.1016/j.jcf.2014.02.004.

Richter, K., Van den Driessche, F., Coenye, T., 2017. Innovative approaches to treat *Staphylococcus aureus* biofilm-related infections. Essays in Biochemistry 61, 61–70. https://doi.org/10.1042/EBC20160056.

Romling, U., Kjelleberg, S., Normark, S., Nyman, L., Uhlin, B.E., Åkerlund, B., 2014. Microbial biofilm formation: a need to act. Journal of Internal Medicine 276 (2), 98–110. https://doi.org/10.1111/joim.12242.

Sanyal, A.K., Roy, D., Chowdhury, B., Banerjee, A.B., 1993. Ibuprofen, a unique anti-inflammatory compound with antifungal activity against dermatophytes. Letters in Applied Microbiology 17 (3), 109–111. https://doi.org/10.1111/j.1472-765X.1993.tb01436.x.

Scott, E.M., Tariq, V.N., McCrory, R.M., 1995. Demonstration of synergy with fluconazole and either ibuprofen, sodium salicylate, or propylparaben against *Candida albicans in vitro*. Antimicrobial Agents and Chemotherapy 39 (12), 2610–2614. https://doi.org/10.1128/AAC.39.12.2610.

Silva, R.A.C., Silva, C.R., Neto, J.B.A., Silva, A.R., Campos, R.S., Sampaio, L.S., Nascimento, F.B.S.A., Gaspar, B.S., Fonseca, S.G.C., Josino, M.A.A., Grangeiro, T.B., Gaspar, D.M., Lucena, D.F., Moraes, M.O., Cavalcanti, B.C., Nobre Júnior, H.V., 2017. In vitro anti-*Candida* activity of selective serotonin reuptake inhibitors against fluconazole-resistant strains and their activity against biofilm-forming isolates. Microbial Pathogenesis 107, 341–348. https://doi.org/10.1016/j.micpath.2017.04.008.

Skerry, C., Pinn, M.L., Bruiners, N., Pine, R., Gennaro, M.L., Karakousis, P.C., 2014. Simvastatin increases the *in vivo* activity of the first-line tuberculosis regimen. Journal of Antimicrobial Chemotherapy 69, 2453–2457. https://doi.org/10.1093/jac/dku166.

Soo, V.W.C., Kwan, B.W., Quezada, H., Castillo-Juárez, I., Pérez-Eretza, B., García-Contreras, S.J., Martínez-Vázquez, M., Wood, T.K., García-Contreras, R., 2017. Repurposing of anticancer drugs for the treatment of bacterial infections. Current Topics in Medicinal Chemistry 17 (10), 1157–1176. https://doi.org/10.2174/1568026616666160930131737.

Sun, W., Park, Y.-D., Sugui, J.A., Fothergill, A., Southall, N., Shinn, P., McKew, J.C., Kwon-Chung, K.J., Zheng, W., Williamson, P.R., 2013. Rapid identification of antifungal compounds against *Exserohilum rostratum* using high throughput drug repurposing screens. PLoS One 8 (8), e70506. https://doi.org/10.1371/journal.pone.0070506.

Thangamani, S., Mohammad, H., Abushahba, M.F.N., Sobreira, T.J.P., Hedrick, V.E., Paul, L.N., Seleem, M.N., 2016a. Antibacterial activity and mechanism of action of auranofin against multi-drug resistant bacterial pathogens. Scientific Reports 6 (22571), 1–13. https://doi.org/10.1038/srep22571.

Thangamani, S., Mohammad, H., Abushahba, M.F.N., Sobreira, T.J.P., Seleem, M.N., 2016b. Repurposing auranofin for the treatment of cutaneous staphylococcal infections. International Journal of Antimicrobial Agents 47 (3), 195–201. https://doi.org/10.1016/j.ijantimicag.2015.12.016.

Thangamani, S., Younis, W., Seleem, M., 2015a. Repurposing celecoxib as a topical antimicrobial agent. Frontiers in Microbiology 6 (750), 1–12. https://doi.org/10.3389/fmicb.2015.00750.

Thangamani, S., Younis, W., Seleem, M.N., 2015b. Repurposing ebselen for treatment of multidrug-resistant staphylococcal infections. Scientific Reports 5, 13. https://doi.org/10.1038/srep11596.

Thompson, M.G., Truong-Le, V., Alemneh, Y.A., Black, C.C., Anderl, J., Honnold, C.L., Pavlicek, R.L., Abu-Taleb, R., Wise, M.C., Hall, E.R., Wagar, E.J., Patzer, E., Zurawski, D.V., 2015. Evaluation of gallium citrate formulations against a multidrug-resistant strain of *Klebsiella pneumoniae* in a murine wound model of infection. Antimicrobial Agents and Chemotherapy 59 (10), 6484–6493. https://doi.org/10.1128/AAC.00882-15.

Ueda, A., Attila, C., Whiteley, M., Wood, T.K., 2009. Uracil influences quorum sensing and biofilm formation in *Pseudomonas aeruginosa* and fluorouracil is an antagonist. Microbial Biotechnology 2 (1), 62–74. https://doi.org/10.1111/j.1751-7915.2008.00060.x.

Ueda, Y., Saito, A., Fukuoka, Y., Yamashiro, Y., Ikeda, Y., Taki, H., Yasuda, T., Saikawa, I., 1983. Interactions of beta-lactam antibiotics and antineoplastic agents. Antimicrobial Agents and Chemotherapy 23 (3), 374–378.

Vilaplana, C., Marzo, E., Tapia, G., Diaz, J., Garcia, V., Cardona, P.-J., 2013. Ibuprofen therapy resulted in significantly decreased tissue bacillary loads and increased survival in a new murine experimental model of active tuberculosis. The Journal of Infectious Diseases 208, 199–202. https://doi.org/10.1093/infdis/jit152.

Walz, J.M., Avelar, R.L., Longtine, K.J., Carter, K.L., Mermel, L.A., Heard, S.O., 2010. Anti-infective external coating of central venous catheters: a randomized, noninferiority trial comparing 5-fluorouracil with chlorhexidine/silver sulfadiazine in preventing catheter colonization. Critical Care Medicine 38 (11), 2095–2102. https://doi.org/10.1097/CCM.0b013e3181f265ba.

Westermeyer, C., Macreadie, I.G., 2007. Simvastatin reduces ergosterol levels, inhibits growth and causes loss of mtDNA in *Candida glabrata*. FEMS Yeast Research 7, 436–441. https://doi.org/10.1111/j.1567-1364.2006.00194.x.

Wolfart, K., Spengler, G., Kawase, M., Motohashi, N., Molnár, J., Viveiros, M., Amaral, L., 2006. Synergistic interaction between proton pump inhibitors and resistance modifiers: promoting effects of antibiotics and plasmid curing. In Vivo 20 (3), 367–372.

Wu, H., Moser, C., Wang, H.-Z., Høiby, N., Song, Z.-J., 2014. Strategies for combating bacterial biofilm infections. International Journal of Oral Science 7, 1–7. https://doi.org/10.1038/ijos.2014.65.

Younis, W., Thangamani, S., Seleem, M.N., 2015. Repurposing non-antimicrobial drugs and clinical molecules to treat bacterial infections. Current Pharmaceutical Design 21 (28), 4106–4111. https://doi.org/10.2174/1381612821666150506154434.

Yssel, A.E.J., Vanderleyden, J., Steenackers, H.P., 2017. Repurposing of nucleoside- and nucleobase-derivative drugs as antibiotics and biofilm inhibitors. Journal of Antimicrobial Chemotherapy 72 (8), 2156–2170. https://doi.org/10.1093/jac/dkx151.

Zhou, Y., Wang, G., Li, Y., Liu, Y., Song, Y., Zheng, W., Zhang, N., Hu, X., Yan, S., Jia, J., 2012. *In vitro* interactions between aspirin and amphotericin B against planktonic cells and biofilm cells of *Candida albicans* and *C. parapsilosis*. Antimicrobial Agents and Chemotherapy 56 (6), 3250–3260. https://doi.org/10.1128/AAC.06082-11.

Zhou, Y., Yang, H., Zhou, X., Luo, H., Tang, F., Yang, J., Alterovitz, G., Cheng, L., Ren, B., 2018. Lovastatin synergizes with itraconazole against planktonic cells and biofilms of *Candida albicans* through the regulation on ergosterol biosynthesis pathway. Applied Microbiology and Biotechnology 102 (12), 5255–5264. https://doi.org/10.1007/s00253-018-8959-8.

Zimmermann, P., Curtis, N., 2017. Antimicrobial effects of antipyretics. Antimicrobial Agents and Chemotherapy 61 (4), 1–12. https://doi.org/10.1128/AAC.02268-16.

Chapter 15

In silico development of quorum sensing inhibitors

Rita P. Magalhães[1], Tatiana F. Vieira[1], André Melo[2], Sérgio F. Sousa[1]
[1]*UCIBIO/REQUIMTE, BioSIM – Departamento de Biomedicina, Faculdade de Medicina da Universidade do Porto, Porto, Portugal;* [2]*LAQV/REQUIMTE, Departamento de Química e Bioquímica, Faculdade de Ciências da Universidade do Porto Porto, Portugal*

15.1 Biofilms in health

Microbial biofilms are a community of bacteria attached and enclosed in a self-produced extracellular matrix that typically adheres to surfaces (Worthington et al., 2012). These highly organized systems differ from their planktonic (single cells primarily in suspension) (Hall-Stoodley et al., 2004) counterparts, since they present different rates of growth and suffer changes in gene transcription and expression (Donlan and Costerton, 2002). Biofilms typically concentrate at an interface (most commonly solid–liquid), and their association with a surface is mainly irreversible. This means that these structures are not easily removed by mechanical force or rinsing (Donlan, 2002). The matrix that encases the cells is made of an extracellular polymeric substance (EPS), which influences both the formation and the development of the biofilm (Hobley et al., 2015).

Bacteria in biofilms are more resistant to antibiotics and host immune response than when in the planktonic state (Quave et al., 2012; Rasmussen and Givskov, 2006). Biofilm formation, development, and pathogenicity depend on several complex factors and mechanisms. The complexity of the mechanisms combined with their increased drug tolerance elevates the challenge of developing drug therapies against these structures (Koo et al., 2017). Innovative multifactorial treatments against biofilms are urgent, due to the threat they represent to individual, medical, and industrial sets of the society (Worthington et al., 2012).

The large impact of biofilms on public and human health is well reported (Costerton et al., 1987; Musk and Hergenrother, 2006). The US National Institutes of Health announced in 2003 that over 80% of microbial infections in the human body were related with biofilms (Davies, 2003). Most common tissue infections include chronic wounds, kidney stones, endocarditis,

Recent Trends in Biofilm Science and Technology. https://doi.org/10.1016/B978-0-12-819497-3.00015-5

vaginosis, osteomyelitis, urinary and biliary tract infections, lung infections, chronic otitis, sinusitis, tonsillitis, laryngitis, and dental plaque (Davey and O'Toole, 2000; Hoiby et al., 2015; Kamaruzzaman et al., 2018; Lebeaux et al., 2013). Besides, biofilm infections of medical apparatus are virtually impossible to eliminate (Subhadra et al., 2018). Urinary and vascular catheters, contact lenses, endotracheal tubes, orthopedic implants and prosthetic joints, cardiac valves, tissue fillers, pacemakers, ventricular derivations, breast implants, and vascular grafts are some on the more relevant medical devices affected (Kamaruzzaman et al., 2018; Lebeaux et al., 2013). Most tissue- and device-related biofilm infections acquired in a hospital setting are caused by a relatively short list of bacteria: *Staphylococcus aureus* (Chung and Toh, 2014), *Pseudomonas aeruginosa* (Sheikh et al., 2017), *Escherichia coli* (Sharma et al., 2016), *Klebsiella pneumoniae* (Mehta and Das, 2018), *Bacillus subtilis* (Abraham, 2006), *Acinetobacter baumannii, Clostridium difficile*, and *Enterococci* (Kamaruzzaman et al., 2018; Subhadra et al., 2018).

Besides their impact in clinical settings, biofilms also affect industrial settings such as the food industry (Galié et al., 2018) and aquatic systems (Mottola et al., 2015). Altogether with the lack of effective antibiofilm strategies, the impact of biofilms in social, economic, and health-related issues justifies the urge in developing new and operative therapeutic solutions.

15.2 Mechanisms of biofilm formation

The formation of biofilms is complex and characteristic of each system. Some of the main mechanisms involved are motility, quorum sensing (QS), and the formation of an extracellular matrix (Koo et al., 2017). Besides, additional resistance to host and external defenses may be resulting from bacterial dormancy and persistence (Defraine et al., 2018). QS, a process of intercellular communication in bacteria, is commonly responsible for the induction of biofilm development. The alterations in bacterial behavior and genetic code resulting from QS induce biofilm formation (Worthington et al., 2012).

Once the formation is set in motion, there are four major stages identifiable in most biofilm structures (Fig. 15.1), independent of the bacterial species involved in them. These are as follows (Kamaruzzaman et al., 2018; Koo et al., 2017):

Initial attachment—The planktonic cells start to attach to host tissues or medical device surfaces. This attachment is promoted by bacterial eDNA, organelles, and proteins (flagella, pili, fimbriae and other membrane proteins). Bacterial motility is an essential mechanism at this stage.

Early biofilm formation—As cells become attached to a surface, they begin adhering to each other by producing EPS and forming a matrix. EPS components enhance cell-to-cell adhesion and begin to enclose the attached cells;

Biofilm maturation—The EPS matrix develops and accumulates, providing a multifunctional protective scaffold. This allows the bacterial cells to

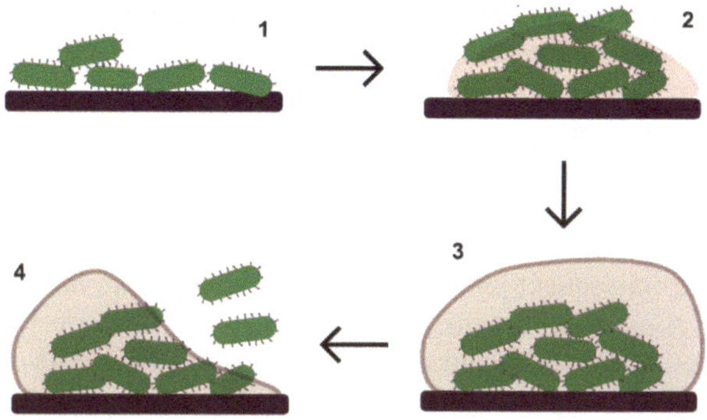

FIGURE 15.1 Biofilm formation, development, and disrupt.

replicate, mature, and interact through QS. From this interaction, the cells optimize nutrient acquisition, conjugation, production of metabolites, pathogenesis, and motility. Maturation results in an increase of the virulence potential.

Dispersion—The final stage implies the cells leaving the biofilm and reentering the planktonic state, detaching from each other and the surface they had previously adhered to. Once the original biofilm is no longer advantageous to their survival and pathogenicity, the bacteria disrupt the structure and leave. The dispersed bacteria can reform the biofilm elsewhere.

Structurally, biofilms are quite simple—they are made up mostly of microbial cells and EPS. Since EPS can represent from 50% to 90% of all organic carbon in the biofilm, it is considered the main component (Flemming et al., 2000). EPS is an essential factor of virulence in biofilms (Sutherland, 2001). The EPS matrix is composed of polysaccharides, nucleic acids (extracellular DNA), lipids, and proteins (Flemming and Wingender, 2010). This structure works as a digestive system by keeping the enzymes closed to the cells, allowing them to dissolve the metabolites available. Besides, it confers mechanical stability and aids in surface adhesion (Xavier and Foster, 2007).

Bacterial motility is involved in several stages of biofilm formation, development, and virulence (Tielen et al., 2005). Bacterial motility is essentially the ability of bacteria to move, either on a surface or free-flowing in a medium (Pollitt and Diggle, 2017). In 1972, Henrichsen (1972) distinguished six categories of motility (swimming, gliding, swarming, sliding, darting, and twitching) after studying over 40 different bacterial species. Some of these forms are flagella dependent (swarming (Harshey, 1994) and swimming (Turner et al., 2000)), others are mediated by type IV pili (twitching and gliding) (Mattick, 2002), and some are passive (sliding and spreading) (Martínez et al., 1999; Matsuyama et al., 1995).

Swarming, gliding, and sliding are typically related with surface colonization. In biofilm forming species and conditions, this is commonly the first step, leading to cell adhesion and attachment. Twitching through type IV pili has been directly correlated with biofilm formation and is present in some of the most pathogenically biofilm-forming species (Harshey, 2003). Flagella are essential in the initial swimming along the surface of attachment, until the desired location for initial cell—surface and cell—cell contact is found. Once attached, in the consequent phases of biofilm development, bacterial cells tend to adopt IV pili-mediated motility (Shimkets, 1999). It is apparent that initial surface contact induces EPS synthesis, allowing for biofilm development, and inhibits flagellar synthesis (Huber et al., 2002). Some motility behaviors have been correlated with QS, suggesting a relationship between the two biofilm-inducing mechanisms (Daniels et al., 2004).

15.3 Quorum sensing

QS is the intercellular chemical communication in bacteria. This mechanism implies the synthesis, detection, and response to signaling molecules called autoinducers (AIs) (Fuqua and Greenberg, 2002). It can be both intergenus and intragenus and allows the bacterial communities to collaboratively alter gene expression in a population density—dependent manner (Mangwani et al., 2017). QS pathways and mechanisms induce and regulate biofilm formation and development in several species (Uroz et al., 2009). Besides, this way of communication is also essential in maintaining typical cell physiology, growth, and nutrient uptake. It also plays a role in other non—biofilm-related virulence and pathogenicity factors (Irie and Parsek, 2008).

QS signaling mechanisms play an important role in biofilm formation (Gamage et al., 2011; Huber et al., 2001). Besides, a high fraction of bacterial genome (4%—10%) and portion of proteome (over 20%) are controlled by QS processes (Schuster et al., 2003). Consequently, several virulence and pathogenicity factors associated with gene expression are also QS dependent (Worthington et al., 2012). Inhibiting and intervening with specific QS signaling pathways is a prospective therapeutic strategy against biofilm formation and development (Sankar Ganesh and Ravishankar Rai, 2018; Shih and Huang, 2002).

QS is present in both gram-negative and gram-positive bacteria (Rutherford and Bassler, 2012). The typical mechanism, schematically represented in Fig. 15.2, requires two proteins and one diffusible molecule. One protein synthesizes (synthase) the diffusible signaling model, referred to as the AI, and the other protein binds to it (receptor). The AI is released into the extracellular medium by the bacteria that produces it. Once the signal reaches a certain threshold, it is captured by other bacteria and binds to the receptor (Mattmann and Blackwell, 2010).

In gram-negative bacteria, the signaling molecule is usually an acyl homoserine lactone (AHL). Once the extracellular signals reach a given

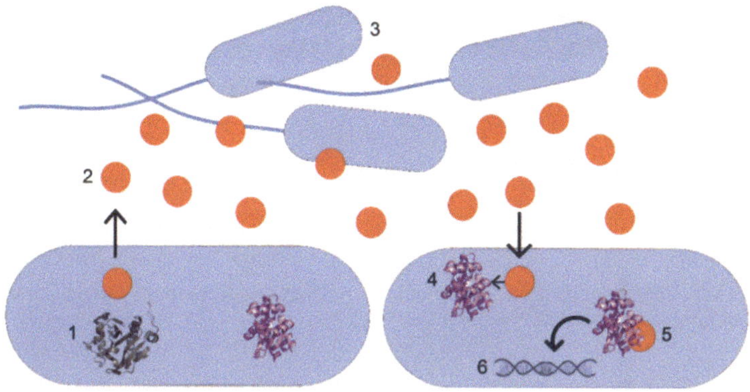

FIGURE 15.2 Typical quorum sensing mechanism. 1, synthase; 2, autoinducer; 3, other planktonic bacteria; 4, receptor; 5, receptor–ligand bound complex; 6, target genes, now ready to bind the complex.

threshold, they enter the cells, binding to the receptors. Receptors are transcriptional regulatory proteins that alter gene expression, inducing virulence factors such as biofilm formation (Abraham, 2006; Skariyachan et al., 2018). In gram-positive bacteria, most molecules are autoinducing peptides (AIPs). The cleaving of an oligonucleotide molecule leads to a functional signaling molecule, which is transported to the outside of the cell. After reaching a given concentration, it binds to receptor proteins on the cell surfaces. The process of AI–protein binding phosphorylates the receptor, which consequently binds a specific site in DNA and actives QS-dependent genes (Bhatt, 2018).

AHLs are typical of gram-negative bacteria and present in over 70 species (Mattmann and Blackwell, 2010) that communicate through QS. The native AHLs found in bacteria are L-isomers and vary in length and oxidation state of the acyl side chain (Finch et al., 1998). AHLs are neutral lipidlike molecules. They are formed by a conserved L-homoserine lactone head group and an acyl side chain, varying both in length and in substituents (Fig. 15.3) (Fuqua et al., 2001). They are synthesized by LuxI-type proteins and bind intracellular LuxR-type receptors. Even though AHLs for different bacterial species are

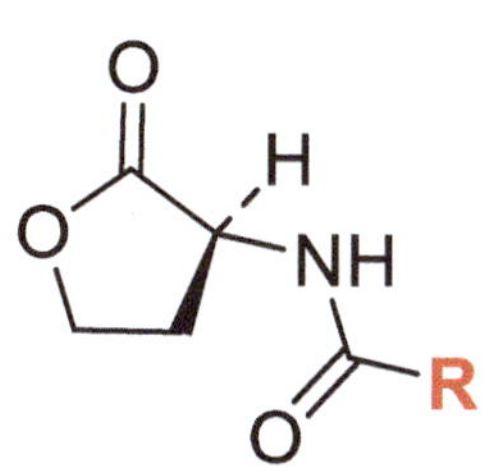

FIGURE 15.3 Conserved L-homoserine lactone head group of acyl homoserine lactones. R represents the substituted acyl side chain.

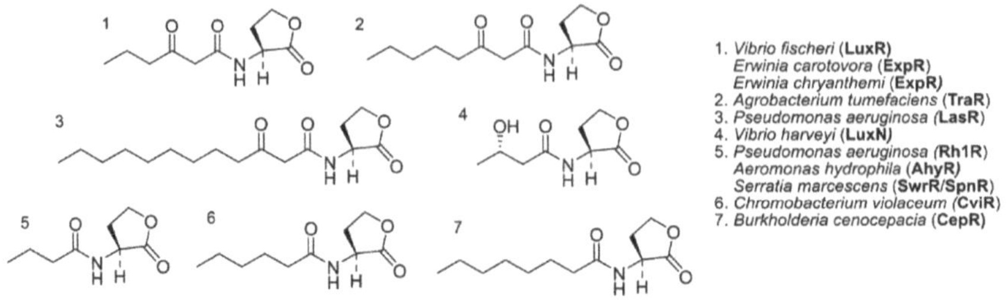

FIGURE 15.4 Different AHLs, corresponding organisms and receptor proteins. *AHLs*, acyl homoserine lactones.

similar (Fig. 15.4), the receptors are usually highly specific. For this reason, they bind the natural AI with a much higher affinity than other AHLs (Janssens et al., 2008).

AIPs are the most common AI for QS in gram-positive bacteria. However, they are not the only AI present in these organisms—other peptides and small thiolactones have also been identified (Mangwani et al., 2017). AIPs are posttranscriptionally processed small peptides with hydrophobic domains crucial for activating the receptor. AIP-1, present in *S. aureus* (Fig. 15.5), is one of the most studied AIPs, due to the pathogenic of this bacteria (Bhattacharya et al., 2015).

Autoinducer-2 (AI-2) is a universal AI—it is involved in QS mechanisms in both gram-positive and gram-negative bacteria and allows for interspecies communication (Ng and Bassler, 2009). This AI (Fig. 15.6) is derived from a precursor molecule, (S)-4,5-dihydroxy-2,3-pentadione (DPD). The synthase enzyme responsible to produce DPD is conserved in over 55 bacterial species, which shows how essential and general AI-2 is. Once the concentration of extracellular AI-2 hits a given threshold, it is imported into the cell. There, it is

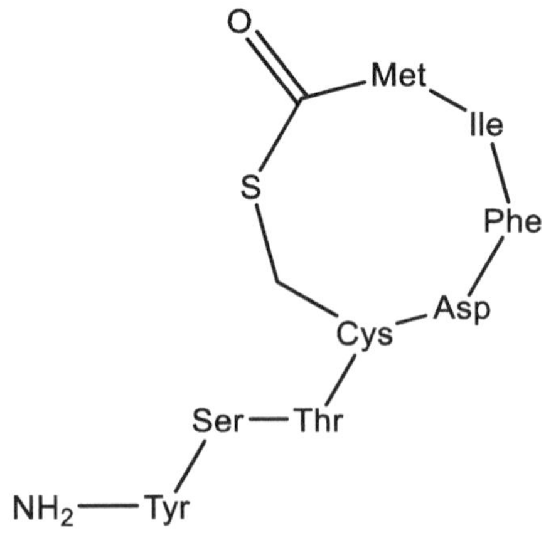

FIGURE 15.5 AIP-I, present in *Staphylococcus aureus*. *AIP*, autoinducing peptide.

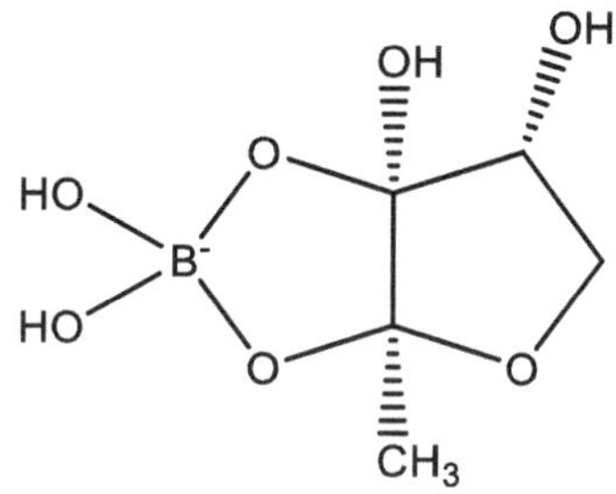

FIGURE 15.6 Chemical structure of AI-2. *AI*, autoinducer.

phosphorylated by kinases and starts binding relevant QS receptors, inducing alterations in genetic expression (Abraham, 2006).

Completely eradicating or killing a biofilm is nearly impossible (Kamaruzzaman et al., 2018). Thus, interfering with cell-to-cell communication mechanisms leading to biofilm formation is a probable therapy to reduce their occurrence (Donlan, 2002). Analyzing the QS mechanism, there are three main strategies to disrupt cell-to-cell communication, leading to biofilm formation: inhibiting the synthetase, so that the production of the AI is not possible (Ahmed et al., 2009; Stickler et al., 1998; Tomlin et al., 2005; Udine et al., 2013; Zhao et al., 2010); degrading the signaling molecule (quorum quenching) (Köse-Mutlu et al., 2019), or inhibiting the receptor, avoiding the binding of the AI (Chen et al., 2011; Kimyon et al., 2016; Koch et al., 2005; Manefield et al., 2002). These therapeutic strategies can be accomplished through drug design and development. Thus, the identification of QS inhibitors through *in silico* methods is a developing field worth investing in to eradicate biofilm formation and development.

15.4 *In silico* methods

15.4.1 Computer-aided drug design: history and methods

Due to the use of computers as an alternative to in vitro or in vivo testing, methods based on the application of computer simulations have been frequently called *in silico*, alluding to the use of computer silicon chips. One of such examples is computer-aided drug design (CADD).

According to the IUPAC, CADD "involves all computer-assisted techniques used to discover, design and optimize biologically active compounds with a putative use as drugs" (McNaught and Wilkinson, 1997). On October 5, 1991, *Fortune Magazine* published a cover article with the title "Next Industrial Revolution: Designer Drugs by Computer at Merck" (Bylinsky, 1981). Several researchers have credited this article with being the genesis of the interest in the potential of CADD (Van Drie, 2007). The main uses of CADD are to optimize experimentally found lead compounds (through rational increasing of binding affinity or improving druglike properties); to screen large databases in search of new active compounds; and to design new drug molecules for a certain receptor or clinical issue (Sliwoski et al., 2014).

CADD popularity is partly due to it being faster and more rational than experimental high-throughput screening (HTS) (Cerqueira et al., 2015). HTS is the quick testing of chemical libraries for biological activity and other experimentally obtained variables (Inglese and Auld, 2008). Even though HTS and experimental techniques are powerful and essential for drug discovery, they are typically very costly and time demanding techniques. Additionally, there is no guarantee of finding new lead compounds in the end of the process (Sousa et al., 2010).

As computers become more powerful and easily accessible, so do the techniques explored in CADD (Ekins et al., 2007). This area of knowledge has been developing fast, mainly due to the advances in both software and hardware. One of the main contributions to improve the research in this field is the large increase of 3D resolved structures of proteins and other macromolecules of clinical interest, especially potential drug targets (Cerqueira et al., 2015). The Protein Data Bank (PDB) currently has ca. 153,601 deposited structures of biological molecules, which created many opportunities for rational drug design (Berman et al., 2000).

There are two general categories of CADD—structure-based (SB) and ligand-based (LB). SB methods are also known as receptor-based, since they depend on the knowledge of the target's structure. Preferably, these would be high-resolution 3D structures (Jorgensen, 2004). SB methods imply the study of the interaction between the ligand and the target. Protein—ligand docking, de novo design of drugs, molecular dynamics, and pharmacophore modeling are some of the most common SB methods. LB methods are mostly used when there are no good structures of the target available or to validate conclusions from SB studies. By performing chemically similarity studies of known active and inactive molecules, one can determine a few key parameters for preferred ligands and use the information to design and optimize novel ligands. The most used LB methods include ligand-based virtual screening (VS), quantitative structure—activity relationships (QSARs), and pharmacophore modeling (Kalyaanamoorthy and Chen, 2011; Sliwoski et al., 2014; Veselovsky and Ivanov, 2003).

Often, protocols for drug discovery with computational strategies combine SB and LB methods. Different workflows and combinations of the existing techniques and software are employed after target identification, with the final goal of lead optimization toward obtaining a drug candidate. This work reports on a typical workflow for the identification of hit drugs to be tested experimentally, as represented in Fig. 15.7. This protocol combines different CADD techniques (such as molecular docking, VS, and QSARs). It has been proven effective in the identification of molecular compounds against biofilm formation and development.

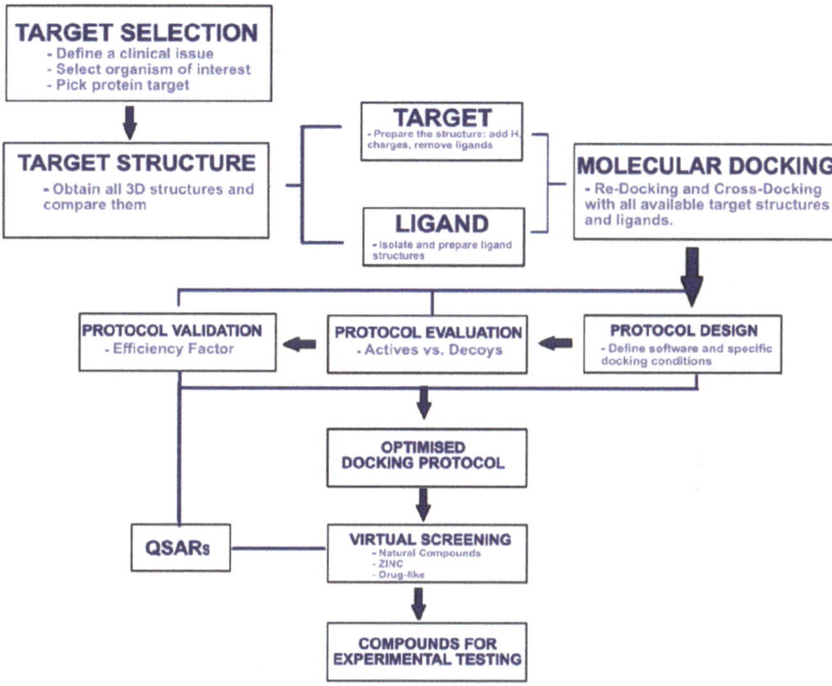

FIGURE 15.7 Typical workflow of a CADD *in silico* experiment. *CADD*, computer-aided drug design.

15.4.2 Molecular docking

Molecular docking is a computational method frequently used in structure-based drug design for its accuracy in predicting the preferred binding pose between two molecules. Depending on the molecules involved, it can be classified as protein–protein or protein–ligand docking. In protein–ligand molecular docking, the two molecules are a smaller one (the ligand) and a larger macromolecule, the receptor—a protein (Brás et al., 2014; Sousa et al., 2013).

This technique allows for the evaluation and ranking of several receptor–ligand poses (the pose can be described as orientation, conformation, and position) (Ferreira et al., 2015). The preferred stable complex formed between ligand and receptor should be distinguished from the other calculated poses by the docking software. The obtained ranking is based on metrics of the binding affinity of these complexes and represents the strength of association between the two molecules. The ranking should correspond to the binding affinities determined experimentally (Vajda and Guarnieri, 2006).

There are currently several molecular docking programs, such as DOCK (Ewing et al., 2001), Glide (Friesner et al., 2004), AutoDock (Morris et al., 1998), AutoDock Vina (Trott and Olson, 2009), CDocker (Wu et al., 2003),

FlexX (Rarey et al., 1996), FlexE (Claussen et al., 2001), GEMDOCK (Yang and Chen, 2004), LigandFit (Venkatachalam et al., 2003), Gold (Jones et al., 1997), MolDock (Thomsen and Christensen, 2006), SODOCK (Chen et al., 2007), and Surflex (Jain, 2003).

To rank the docked ligands in an accurate way, the docking should be able to identify the most favorable binding mode in the active site or binding pocket of the protein (Lengauer and Rarey, 1996). To accomplish the goal of determining and ranking preferred binding poses between a ligand and a protein, molecular docking requires three essential aspects: the definition of a binding center, a search algorithm (SA), and a scoring function (SF) (Tripathi and Bankaitis, 2017).

15.4.2.1 Search algorithm

The SA explores the defined conformational space of the binding center and generates several poses for the bound complex. The final pose is obtained by exploring not only the ligand and protein conformations but also their mutual orientations in the binding pocket. An effective SA should be able to appropriately explore the conformational space while being fast enough to be applied to large databases of compounds (Mohan et al., 2005; Taylor et al., 2002).

This conformational space is a combination of the protein and ligand degrees of freedom and the components of the spatial arrangement of the interaction between the two units. The result is a space too large to be searched accurately in an acceptable time span. Several existing SAs assume certain approximations and simplifications to reduce computational time without compromising on accuracy (Moitessier et al., 2008).

15.4.2.2 Scoring function

SFs are used to rank the poses generated by the searching algorithms employed by the software and are an essential part of molecular docking. It is not enough to obtain good conformations; one also needs to be able to rank them. A good SF should allow for the distinction between valid and invalid binding predictions for the ligand–target complex and even between active and random or inactive compounds (Kitchen et al., 2004). The accuracy of an SF can be evaluated based on their ability to estimate the interaction between the ligand and the target and to rank the generated poses accurately (Ferreira et al., 2015). Besides, the SF should be able to distinguish between ligands that bind and those that do not. This should be reflected on the ranking. Lastly, an SF should be fast enough to use in an acceptable timeline without compromising accuracy (Jain, 2006).

There are currently several different software alternatives available for molecular docking (Pagadala et al., 2017). The programs vary in their algorithms and specifications. However, it is possible to design a typical workflow for a molecular docking experiment. After target selection and preparation, the active center or binding pocket should be defined as the search space to be explored. After ligand preparation, the molecules can be docked onto the receptor, and a ranking of the generated poses according to their likelihood of binding is generated. This ranking, yielded by the SF, allows for the comparison of the binding modes for the protein–ligand complex (Cerqueira et al., 2015; Sousa et al., 2006).

15.4.2.3 Consensus scoring

Accurate predictions of all the binding poses of a ligand into a protein target are still far from perfect. Even though there are a variety of docking programs, there is still no SF capable of considering all the physicochemical phenomena present in protein–ligand binding. Consensus scoring methodology then emerged to minimize these limitations in the SFs (Ren et al., 2018).

Nowadays, consensus scoring is widely used in drug discovery mainly as a postprocessing step, after an initial docking run (Feher, 2006). There are several commercial and publicly available programs that can be used such as CScore (Ouyang et al., 2011), X-Score (Wang et al., 2002), ConsDock (Paul and Rognan, 2002), AutoxX (Wolf et al., 2007), and VoteDock (Plewczynski et al., 2011). Consensus scoring can also be combined with machine learning techniques.

In consensus scoring methodology, different scoring algorithms are combined to improve accuracy. Docking is performed as a first step, to place ligands onto the active site as it would normally be done in a docking protocol. Then, the best docked solutions of each compound are reevaluated with multiple SFs (Wang and Wang, 2001). Enrichment factors are improved by compensating errors in the individual SF, therefore improving the probability of finding the correct solution (Teramoto and Fukunishi, 2008; Wang and Wang, 2001).

The selection of the appropriate SF, for obtaining a new consensus SF, is a very challenging task. An incorrect selection of such functions can lead to an error amplification instead of a desirable error compensation (Wang and Wang, 2001). This effect can be originated from the significant correlated terms associated with different SF. Work done by Charifson et al. (1999) and later confirmed by Stahl et al. (Paul and Rognan, 2002) states that to obtain a successful CS, SFs that perform well individually must be combined (Cerqueira et al., 2015). It is also advantageous to combine different types of SF, for example, combining empirical methods with knowledge-based ones can, in theory, compensate for each of the methods weaknesses (Houston and Walkinshaw, 2013).

The other question is how many SFs to combine. Consensus scoring is based on multiple sampling, and that is the reason why the accuracy increases, as repeated measurements statistically reduce the noise in data collection (Wang et al., 2003). So, it would be expected that the higher the combination of SF, the higher the enrichment. However, some studies have showed that combining only two SF provides the same enrichment, than combining three or more (Cheng et al., 2009; Wang et al., 2003).

Even though it is powerful, this methodology still presents a few limitations. Even though this methodology has been successfully applied in CADD studies, it still presents a few limitations. It has been tested only for relatively small benchmark data sets, some methods are complicated to implement and cannot be easily applied to new targets. Besides, the performance of the methods is strongly dependent on the quality of the docking results (Ren et al., 2018). Consensus scoring is frequently used in combination with VS or large databases of unknown compounds as a way of bettering the confidence of the applied algorithm (Jain, 2006).

15.4.2.4 Examples of application

The role of molecular docking in the discovery of new drugs and therapies against biofilm formation, development, and virulence is well established. Several publications report on new inhibitors and molecules with activity against biofilm-forming pathogenic bacteria. Since an *in silico* approach such as docking requires experimental validation, many of the reported findings include in vitro or even in vivo testing of the identified molecules.

Even though several molecules with QS inhibition potential have been identified, none of them has made it to the clinical use stage. This is usually related with bioavailability and toxicity issues (Imperi et al., 2013). For that reason, Borges et al. (2017) tested Furvina, a broad-spectrum antibiotic commonly used for clinical purposes, for inhibition activity and antimicrobial potential against *P. aeruginosa* QS mechanisms. *In silico* testing was performed through molecular docking with Glide, and in vitro inhibition assays confirmed that Furvina had inhibitory activity against *P. aeruginosa.* The conclusions drawn are useful in the possible design of new drugs against biofilm formation is these bacteria, with prospective clinical approval.

Inspired by previous findings of certain natural products and metabolites having an effect against biofilm formation (Šiman et al., 2016), Rajkumari et al. (2018) performed several in vitro (inhibition and motility assays), *in silico* (molecular docking, molecular dynamics, and postsimulation binding energy calculations), and in vivo experiments to determine the QS inhibition potential of betulin and betulinic acid against *P. aeruginosa.* Molecular docking was performed with AutoDock and AutoDock Vina. The targets were two different QS receptors of *P. aeruginosa*: LasR and Rh1R. LigPlot+ was used to calculate the interactions between the docked ligands and the target proteins. Results revealed potential of the tested compounds in reducing biofilm formation and development and diminishing the expression of virulence factors.

Aliyu et al. (2016) tested sesquiterpene lactones obtained from *Vernonia blumeoides* against *Chromobacterium violaceum*, a bacterium that forms biofilms through QS. Through molecular docking and inhibition assays, it was determined that some of the tested molecules (namely, blumeoidolide A and B) showed inhibition potential against the formation of biofilms by these bacteria. Molecular docking was performed with the Discovery Studio software, using a Flexible Algorithm to generate new poses and a CDOCKER SF to evaluate and rank them. Prior to the docking, the target receptors were prepared and minimized using CHARMm force field. The findings are a good starting point for the development of therapies against this pathogenic species.

15.4.3 Virtual screening

VS is commonly characterized as the computational version of high-throughput compound screening (Kroemer, 2007). VS consists in the screening of a library of compounds through successive and multiple molecular docking experiments, and the evaluation of the ranked results may lead to the discovery of potential drug hits (Walters et al., 1998). One of the main advantages when compared with experimental screening endeavors is that the compounds studied do not have to exist or be acquired/synthesized, and the experiments do not cost valuable physical materials. Other experimental disadvantages, such as optimization of conditions, aggregate formation, or solubility issues, also do not apply in computational studies (Klebe, 2006).

Before applying a VS protocol in a large database, it is necessary to be sure that the protocol is to be trusted and the results are valuable. One could use different protocols and compare the results, or choose a specific protocol and software according to the specific biological molecular targets. However, a more accurate way to evaluate the accuracy of a VS protocol is to test its ability to distinguish active from inactive molecules upon binding (Stumpfe et al., 2012). This is usually done by the creation of a database combining both active and inactive ligands, and screening it several times, possibly applying different VS protocols for testing (Sliwoski et al., 2014).

Active compounds can be obtained from databases such as ChEMBL (Gaulton et al., 2017) or BindingDB (Gilson et al., 2015), which compile biological information such as IC50, percentage of activity or inhibition and organize it according to the desired target. Inactive molecules are less commonly reported, and so it is typical to use randomly generated molecules and assume they will not bind to target. These molecules can be obtained through a specific server available through the Directory of Useful Decoys (DUD) (Huang et al., 2006). This server allows for the submission of a list of the active compound libraries. It generates random ligands that resemble the active ones in terms of physicochemical properties but not topologically and are therefore presumed to be "inactive" against the targets. Currently, DUD generates 50 decoys per submitted active ligand, which allows for the building of a solid testing library for VS—for 50 active compounds, the database would generate 2500 decoys, and the full VS testing library would have 2550 compounds (Cerqueira et al., 2015; Sliwoski et al., 2014).

The VS protocol should be able to distinguish between active molecules and decoys—the active compounds should rank higher than the ones with no experimentally proven activity (Kitchen et al., 2004). To evaluate the performance of the protocol, a correlation of the predicted activities with experimentally proven activities is calculated. This is done through a receiver operating characteristic curve (ROC) that plots the number of true positive predictions in the y-axis, versus the number of false-positive predictions in the x-axis. A random prediction would yield a plot line with a slope of 1. On the other hand, curves with high initial slopes represent high recovery rates of active compounds in the highest scoring poses (Cleves and Jain, 2006; Hristozov et al., 2007).

Given the goal of a VS endeavor, obtaining a good ROC result is not everything. The idea is to generate lists of compounds to test experimentally. Since these protocols are applied to databases with thousands or even millions of compounds, it is only feasible to experimentally test a few hundred molecules out of the large initial set. Thus, if there are no promising ligands in the first few hundreds (1%–5% of the entire database), it is indifferent whether the ROC result is high. In this case, the active ligands will be lost in the middle of the database and never get experimentally tested. It is usually more important to notice what happens with the early stage rankings. Since this can be hard to notice in an ROC graphic, it is very common to calculate the logarithmic curve of this graph, to emphasize the performance in the beginning (Jorgensen, 2004).

Even though the interpretation of a ROC curve is quite simple and immediate, it is useful to have a numerical value to represent the conclusions drawn from the graphical metric. This is AUC (or AU-ROC) the area under the receiver operating characteristic curve. The value for AUC varies between 0 and 1 and represents the probability or ranking actives better than decoys. Just like before, if the selection is completely random, then TPR = FPR, and the AUC would be 0.5. If AUC = 1, then there is a total discrimination between actives and nonactives (Cheng et al., 2012).

Another commonly used metric to assess VS success is the enrichment factor (EF). This value is obtained by dividing the fraction of active compounds found or recovered by the fraction of the library that was screened. EF can be calculated for several steps of the screening (for example, EF1% is calculated for the fraction found in 1% of the screened library) and can also be represented graphically (Ekins et al., 2007).

$$\mathrm{EF} = \frac{\mathrm{actives}_{\mathrm{recovered}}}{\mathrm{actives}_{\mathrm{total}}} \times \frac{N_{\mathrm{Total}}}{N_{\mathrm{screened}}}$$

As previously stated, experimental studies hardly go beyond 1% or 5% of the scanned database, especially when these contain millions of compounds. For that reason, one should aim to optimize a VS protocol so that EF = 1% and the early recognition metrics are the best possible. This is also the reason for the large disproportion between actives and decoys (1:50) (Ekins et al., 2007; Sliwoski et al., 2014).

The optimization of a VS protocol is the attempt to enrich the top of the list with actives, avoiding false negatives as much as possible. Once the protocol is optimized, it can then be applied to large databases of compounds with unknown activities against the specific aim or target one wishes to inhibit. There are several online databases for downloadable (and sometimes commercially available) compounds that can be used in VS. Some of them are summarized in Table 15.1.

Once the VS of the selected library is completed, the top-ranked compounds should be analyzed and experimentally validated to be considered for clinical or therapeutic use. VS, although an extremely useful and powerful technique that ensures a complete exploration of the chemical space at a low cost, is only a starting point in drug discovery and never the final stage (Shoichet, 2004).

15.4.3.1 Examples of application

As previously stated, the list of publications employing the VS of large chemical databases, usually with posterior experimental validation, is long. To represent the significance this technique has acquired in the world of biofilm-inhibition drug discovery, several example studies will be mentioned.

Some studies, like the one published by Kalia et al. (2017), lack experimental validation of the results obtained through *in silico* testing. They report on the VS (with AutoDock and AutoDock Vina) of 2603 compounds from ZINC, of which 410 were filtered through *ADME* druglike filters. SwissADME was used to calculate the ADME properties of the ligands. The obtained descriptors were used to select the most promising molecules out of the 410 analyzed. Finally, the selected 6 compounds were submitted to rescoring and additional molecular docking experiments with AutoDock. However valuable the presented results are, further in vitro testing is required before considering the found compounds for clinical purposes.

A lot of experimentally validated studies focus on *LasR*, a QS receptor from the highly pathogenic bacteria *P. aeruginosa* (Ahumedo Monterrosa et al., 2019). Yang et al. (2009) reported on the VS of 149 compounds, obtained from the SuperNatural and SuperDrug databases. VS was performed with MVD docking program. Of the screened compounds, 6 were tested in vitro for inhibition against *LasR,* and 3 ligands showed high rates of biofilm formation inhibition. Interestingly, one of the compounds had already been discovered as a potent inhibitor (Bandara et al., 2006), which proves the

TABLE 15.1 Compound databases for virtual screening and respective websites.

Database	Website
PubChem	pubchem.ncbi.nlm.nih.gov
ChEMBL	ebi.ac.uk/chembldb/index.php
ChemSpider	chemspider.com
ZINC	zinc.docking.org
ChemBridge	chembridge.com
TimTec	timtec.net/home
Specs	specs.net
PhytoHUB	phytohub.eu/entries
SWMD	swmd.co.in/download.php
NCI	dtp.nci.nih.gov
ACD	acdlabs.com/resources
PDBeChem	ebi.ac.uk/pdbe-srv/pdbechem/
NLDB	nldb.hgc.jp/nldb/top
DrugBank	www.drugbank.ca
ChemDB	cdb.ics.uci.edu
WOMBAT	agbu.une.edu.au/
MDDR	akosgmbh.de/accelrys/databases/
SuperNatural	bioinf-applied.charite.de/supernatural_new
SuperDrug	cheminfo.charite.de/superdrug2/
ChemBridge	chembridge.com
Chemoteca	unicz.chemoteca.it
InterBioScreen	ibscreen.com/
Ambinter	omictools.com/ambinter-tool
Specs	specs.net/

applied protocol was robust and accurate in selecting molecules with a high ability of binding. Skovstrup et al. (2013) created a Pharmacophore Model (in MOE) based on the natural ligand of *LasR* to screen ChemBridge, yielding 2208 compounds that were further submitted to VS. Docking was performed with Glide. From the obtained and ranked poses, 122 hits were selected for in vitro testing, with highly positive results—3 compounds activated the receptor, whereas two showed high inhibition of binding. Furthermore, the group synthesized new inhibitors based on the ones discovered, which also showed inhibition against biofilm formation. Xu et al. (2017) screened through 167,740 compounds from SPECS with combined pharmacophore and molecular docking (with GOLD) techniques, of which three were positively tested for inhibition through in vitro assays.

Soulere et al. (2010) virtually screened 2289 compounds from Chembank against three different yet correlated QS receptors: *LuxR, LasR,* and *TraR* with GADock. Besides analyzing the ranking of the screened compounds, the authors focused on the binding modes and poses of the top-ranking ligands in the binding pockets. Of these, 120 were put through in vitro testing, revealing six possible potent inhibitors. Interestingly, a few of the identified molecules belong to calcium and hormone channel modulators, suggesting a cross-talk between bacteria and host that could be important in the development of new therapeutic drugs. This study resulted not only in new promising antibiofilm compounds and moieties but also in the study of the different binding pockets and modes for the targeted QS receptors.

Vibrio harveyi is a biofilm-forming bacterium that uses several QS mechanisms. One of them employs two different receptors, *LuxP* and *LuxQ*, which sometimes bind each other, yielding *LuxPQ* (Ng and Bassler, 2009). Several VS experiments in search of inhibitors for these receptors have been reported. Zhu et al. (2012) performed VS with DOCK on ca. 0.5 million compounds from SPECS, of which the top 1000 were submitted to consensus scoring and manual curation studies. 42 compounds were studied in vitro, and 7 revealed inhibition properties against *LuxPQ*. Similarly, Li et al. (2008) screened 1.7 million compounds from several databases (including ChemBridge and SPECS) against *LuxP.* VS was performed with DOCK. 42 compounds were submitted to consensus scoring studies, of which only 27 were commercially available and were tested in vitro. Finally, they report on 2 highly promising compounds for inhibition of this receptor.

These are only a few of the available studies on *in silico* and in vitro combined experiments on QS inhibition as a strategy to reduce biofilm formation and virulence. The power of employed combined theoretical and experimental techniques in drug discovery is unquestionable, as shown by the positive results mentioned.

15.4.4 Quantitative structure–activity relationships

QSAR models belong to the LB CADD category of strategies and are based on the principle that chemicals with similar properties and structures must behave the same biologically. QSAR analysis links properties to effects—traditionally, one or more molecular descriptors are linked to activity through statistical and mathematical models. Once a model is developed and validated, one can use it to predict biological properties of compounds that have not yet been tested experimentally (Benfenati, 2012; Neves et al., 2018; Sliwoski et al., 2014; Vilar et al., 2008).

QSARs are thought to have surged in the 1960s in two separate research groups. Hansch and Fujita developed what is now known as the Classic QSAR, the correlation of various electronic, hydrophobic, and steric features with biological activity. Their models provided estimates for the bioactivity and molecules. Around the same time, Free and Wilson created a mathematical approach that linked certain chemical substituents in given molecules to their biologic activity, being that each substituent was responsible for a given contribution. Eventually the two methods were combined. And they are the most erudite basis to what is done currently (Sliwoski et al., 2014).

In its most simplistic form, QSAR models link properties to effects and assume the resulting biological effects to be a function of the compound's characteristics, as described in Eq. (15.1).

$$\text{Effect} = f(\text{properties}) \tag{15.1}$$

An equation of this form is typically the end goal of a QSAR model (Benfenati, 2012).

QSAR models can be roughly classified in two main groups: regression and classification. Regression techniques relate the chemical properties or molecular descriptors with continuous biological properties, such as IC50, EC50, and Ki, among others. On the other hand, Classification methods are used when the biological variables are categorical or binary, instead of numeric (for example, toxic or active) (Neves et al., 2018).

It is important to note the complexity of the current paradigm of QSAR applications—the use of the techniques goes well beyond activity predictions, and each step of the model development is composed of several options and decisions. There are many different chemical descriptors one can generate and consider, various algorithms useable for model development, and different properties (other than activity or toxicity) to consider. It would be naïve and reductive to consider and use QSAR as one single method, since thousands of different models have been reported and successfully used (Benfenati, 2012). Given the variety of QSAR-related methods, the definition of a workflow is subjective to the specific goal of each study, and the techniques available and selected. Even so, a general protocol for the application of this technique would be as described in Fig. 15.8.

QSAR is an extremely powerful technique, and some steps must be taken to ensure the methods are applied correctly and successfully (Benfenati, 2012; Neves et al., 2018; Sliwoski et al., 2014). The quality of the final QSAR model depends on several factors. For example, ensuring that the initial training set is well curated and that the experimental values are validated is essential. Additionally, it is also important to ensure that the set is representative enough of all the search space, so the model will not be limited. It is equally important

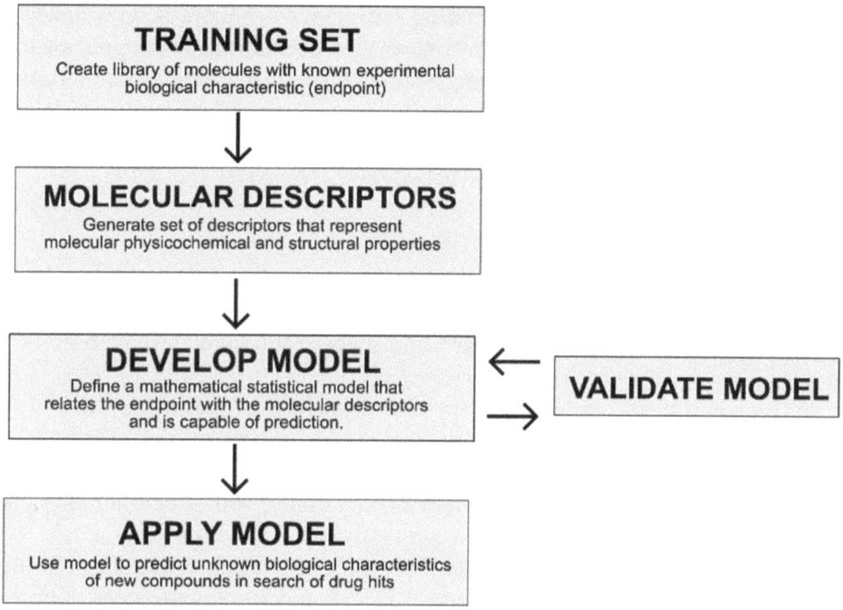

FIGURE 15.8 Typical workflow for a QSAR model developing protocol. *QSAR*, quantitative structure–activity relationship.

to find an appropriate and valid set of descriptors to represent appropriately the mathematical relationship between descriptors and the properties of the training set (Sliwoski et al., 2014).

The Organization for Economic Cooperation and Development (OECD) created a list of advice to guarantee quality and validity of the final QSAR models. The five principles are as follows (Benfenati, 2012): (1) established endpoint, which is the known biological and experimental property one wishes to relate through a mathematical model to the descriptors; (2) well-defined, validated, and unambiguous algorithm; (3) specify the domain of applicability for the model, according to the desired molecular target and the initial training set; (4) accurate validation of the final methodology—measuring the ability of prediction, the robustness, and the goodness-of-fit; and (5) mechanistic interpretation, when feasible.

According to the literature, some extra rules should be considered besides the ones recommended by the OECD, especially with data curation (not only for the training and test sets but also for each step, especially the developed molecular descriptors library) (Benfenati, 2012). QSAR model building algorithms are called pattern recognition models, since their main goal is to construct predictive algorithms capable of distinguishing patterns in the given data set. There are two classes of the methods: supervised (when response data is used—meaning, the endpoint is defined) and unsupervised (the algorithms focus only on the descriptor and training sets data, since there is no response). QSAR algorithms are almost always supervised, since the usual goal of the technique is to relate a biological response with the characteristics of an initial set of compounds. The most common techniques would be multiple linear regression, discriminant analysis, partial least squares, classification and regression trees, or artificial neural networks (Sliwoski et al., 2014).

15.4.4.1 Examples of application

QSARs are a multidisciplinary and powerful technique with multiple possible applications. Their application toward QS and biofilm formation inhibition is different according to the goal, nature, and specific software employed in each study.

A typical QSAR experiment for the discovery of new inhibitors was developed by Haney et al. (2018). In their work, a synthetic immunomodulatory cationic peptide (VRLIVAVRIWRR-NH_2) with experimentally proven antibiofilm activity was used to generate 96 variants with single amino acid residue substitutions. The generated alternatives were synthesized and tested in vitro for their antibiofilm activity. The obtained results were used as the response variable in the development of QSAR models for the prediction of inhibition based on calculated molecular descriptors. The final model considered 7 descriptors essential for the prediction of the inhibition ability of the peptides. To test the model, a library of 100,000 peptides was screened with MOE. 108 peptides with different predicted activities (both from the top, bottom, and middle of the ranking table) were synthesized and tested in vitro. The model was validated at 85% prediction accuracy. The 7 peptides that revealed a higher activity were synthesized, purified, and characterized and can now be used in the development of new therapeutic molecules against biofilm formation.

Shanmugam et al. (2016) performed a purely *in silico* QSAR experiment for the development of a model capable of predicting inhibition activity against SdiA, a QS receptor in *Salmonella* Typhimurium. In this work, a data set of 47 experimentally proven inhibitors was divided in training (35 compounds) and test (12 compounds) set and used in the development of 2D and 3D QSAR models. Three different statistical techniques were employed in the development of 2D QSAR models: multiple linear regression, principal component regression, and partial least squares regression. These models predicted inhibition as a function of several molecular descriptors and revealed that topological and geometric characteristic influenced inhibition greatly. Comparative molecular field analysis 3D QSAR models were also developed, which allowed the authors to understand the role that van der Waals and electrostatic interactions played in inhibition. The combined results allow for the development and design of new potent and selective SdiA inhibitors.

The development of QSAR models for the identification of descriptors and characteristics important in inhibition is one of their most popular uses. Sivakumar et al. (2010) used combined in vitro and *in silico* techniques to study the antibacterial activity of chalcones against biofilms. After testing 47 compounds experimentally, the results were used to develop multiple linear regression models that correlated the response (inhibition) with descriptors calculated on Cerius II. The results showed that the activities of the chalcones were dependent on specific special, electronic, and structural features. The obtained knowledge can now be used in the development of new inhibitors against the studied QS systems.

15.5 Conclusions

Microbial biofilms are an ongoing problem for human and public health. These structures are responsible for an elevated number of infections in clinical and hospital settings, both in human tissues and in medical apparatus. They also affect the food industry, among others, with socioeconomic consequences.

Bacteria in biofilms are highly resistant to host defenses and external therapies. Once a biofilm has formed, it is extremely difficult to eliminate it completely. It is also complicated to avoid relocation and dispersion, or persister cells. There is currently a high need for the development of new effective drugs against biofilm formation and development. Although the development of these structures is complex, a few key mechanisms have been identified, of which QS is the better categorized mechanism of this type. QS can be inhibited in three ways through the development of new druglike molecules, therefore avoiding biofilm formation.

Purely experimental techniques for the identification of new drugs are highly time-consuming and economically expensive and frequently do not yield promising results. For that reason, the use of computational techniques to speed up the process while increasing the efficiency is a growing strategy with already proven results. In the future, with the increase in sophistication of the computational methods and resources available, the use of CADD to combat biofilm formation trough inhibition of QS will continue to yield more and better lead compounds. The optimization of these compounds will allow for a swifter transition into clinical trials and hopefully also the market, aiding in the urgent issue that are biofilm infections in hospital and industrial settings.

Acknowledgment

This work has been supported by the Fundação para a Ciência e a Tecnologia (FCT) UID/Multi/04378/2019 and IF/00052/2014.

References

Abraham, W.R., 2006. Controlling biofilms of gram-positive pathogenic bacteria. Current Medicinal Chemistry 13, 1509–1524.

Ahmed, N.A., Petersen, F.C., Scheie, A.A., 2009. AI-2/LuxS is involved in increased biofilm formation by *Streptococcus* intermedius in the presence of antibiotics. Antimicrobial Agents and Chemotherapy 53, 4258–4263.

Ahumedo Monterrosa, M., Galindo, J.F., Vergara Lorduy, J., Alí-Torres, J., Vivas-Reyes, R., 2019. The role of LasR active site amino acids in the interaction with the acyl homoserine lactones (AHLs) analogues: a computational study. Journal of Molecular Graphics and Modelling 86, 113–124.

Aliyu, A.B., Koorbanally, N.A., Moodley, B., Singh, P., Chenia, H.Y., 2016. Quorum sensing inhibitory potential and molecular docking studies of sesquiterpene lactones from *Vernonia blumeoides*. Phytochemistry 126, 23–33.

Bandara, M.B.K., Zhu, H., Sankaridurg, P.R., Willcox, M.D.P., 2006. Salicylic acid reduces the production of several potential virulence factors of *Pseudomonas aeruginosa* associated with microbial keratitis. Investigative Ophthalmology & Visual Science 47, 4453–4460.

Benfenati, E., 2012. Theory, Guidance and Applications on QSAR and REACH, first ed. Benfenati, Emilio (Istituto di Ricerche Farmacologiche "Mario Negri", Milan. Italy.

Berman, H.M., Westbrook, J., Gilliland, G., Bhat, T.N., Weissig, H., Bourne, P.E., 2000. The protein Data Bank. Nucleic Acids Research 28, 235–242.

Bhatt, V.S., 2018. Quorum sensing mechanisms in gram positive bacteria. In: Implication of Quorum Sensing System in Biofilm Formation and Virulence. Springer, Singapore, pp. 297–311.

Bhattacharya, M., Wozniak, D.J., Stoodley, P., Hall-Stoodley, L., 2015. Prevention and treatment of *Staphylococcus aureus* biofilms. Expert Review of Anti-infective Therapy 13, 1499–1516.

Borges, A., Sousa, P., Gaspar, A., Vilar, S., Borges, F., Simões, M., 2017. Furvina inhibits the 3-oxo-C12-HSL-based quorum sensing system of *Pseudomonas aeruginosa* and QS-dependent phenotypes. Biofouling 33, 156–168.

Brás, N.F., Cerqueira, N.M.F.S.A., Sousa, S.F., Fernandes, P.A., Ramos, M.J., 2014. Protein ligand docking in drug discovery. Protein Modelling 249–286.

Bylinsky, G., 1981. A new industrial revolution os on the way. Fortune 104, 106–114.

Cerqueira, N.M.F.S.A., Gesto, D., Oliveira, E.F., Santos-Martins, D., Brás, N.F., Sousa, S.F., Fernandes, P.A., Ramos, M.J., 2015. Receptor-based virtual screening protocol for drug discovery. Archives of Biochemistry and Biophysics 582, 56–67.

Charifson, P.S., Corkery, J.J., Murcko, M.A., Walters, W.P., 1999. Consensus scoring: a method for obtaining improved hit rates from docking databases of three-dimensional structures into proteins. Journal of Medicinal Chemistry 42, 5100–5109.

Chen, H.M., Liu, B.-F., Huang, H.-L., Hwang, S.-F., Ho, S.Y., 2007. SODOCK: swarm optimization for highly flexible protein-ligand docking. Journal of Computational Chemistry 28, 612–623.

Chen, G., Swem, L.R., Swem, D.L., Stauff, D.L., O'Loughlin, C.T., Jeffrey, P.D., Bonnie, L., Bassler, B.L., Hughson, F.M., 2011. A strategy for antagonizing quorum sensing. Molecular Cell 42 (2), 199–209.

Cheng, T., Li, X., Li, Y., Liu, Z., Wang, R., 2009. Comparative assessment of scoring functions on a diverse test set. Journal of Chemical Information and Modeling 49, 1079–1093.

Cheng, T., Li, Q., Zhou, Z., Wang, Y., Bryant, S.H., 2012. Structure-based virtual screening for drug discovery: a problem-centric review. The AAPS Journal 14, 133–141.

Chung, P.Y., Toh, Y.S., 2014. Anti-biofilm agents: recent breakthrough against multi-drug resistant *Staphylococcus aureus*. Pathogens and Disease 70, 231–239.

Claussen, H., Buning, C., Rarey, M., Lengauer, T., 2001. FlexE: efficient molecular docking considering protein structure variations. Journal of Molecular Biology 308, 377–395.

Cleves, A.E., Jain, A.N., 2006. Robust ligand-based modeling of the biological targets of known drugs. Journal of Medicinal Chemistry 49, 2921–2938.

Costerton, J.W., Cheng, K.J., Geesey, G.G., Ladd, T.I., Nickel, J.C., Dasgupta, M., Marrie, T.J., 1987. Bacterial biofilms in nature and disease. Annual Review of Microbiology 41, 435–464.

Daniels, R., Vanderleyden, J., Michiels, J., 2004. Quorum sensing and swarming migration in bacteria. FEMS Microbiology Reviews 28, 261–289.

Davey, M.E., O'Toole, G.A., 2000. Microbial biofilms: from ecology to molecular genetics. Microbiology and Molecular Biology Reviews 64, 847–867.

Davies, D., 2003. Understanding biofilm resistance to antibacterial agents. Nature Reviews Drug Discovery 2, 114–122.

Defraine, V., Fauvart, M., Michiels, J., 2018. Fighting bacterial persistence: current and emerging anti-persister strategies and therapeutics. Drug Resistance Updates 38, 12–26.

Donlan, R.M., 2002. Biofilms: microbial life on surfaces. Emerging Infectious Diseases 8, 881–890.

Donlan, R.M., Costerton, J.W., 2002. Biofilms: survival mechanisms of clinically relevant microorganisms. Clinical Microbiology Reviews 15, 167–193.

Ekins, S., Mestres, J., Testa, B., 2007. In silico pharmacology for drug discovery: methods for virtual ligand screening and profiling. British Journal of Pharmacology 152, 9–20.

Ewing, T.J.A., Makino, S., Skillman, A.G., Kuntz, I.D., 2001. DOCK 4.0: search strategies for automated molecular docking of flexible molecule databases. Journal of Computer-Aided Molecular Design 15, 411–428.

Feher, M., 2006. Consensus scoring for protein-ligand interactions. Drug Discovery Today 11, 421–428.

Ferreira, L.G., Santos, R.N., Oliva, G., Andricopulo, A.D., 2015. Molecular docking and structure-based drug design strategies. Molecules 20, 13384–13421.

Finch, R.G., Pritchard, D.I., Bycroft, B.W., Williams, P., Stewart, G.S., 1998. Quorum sensing: a novel target for anti-infective therapy. Journal of Antimicrobial Chemotherapy 42, 569–571.

Flemming, H.C., Wingender, J., 2010. The biofilm matrix. Nature Reviews Microbiology 8, 623–633.

Flemming, H.-C., Wingender, J., Griebe, T., Mayer, C., 2000. In: Evans, L.V. (Ed.), Physico-chemical Properties of Biofilms - Biofilms: Recent Advances in Their Study and Control. Harwood Academic Publishers., Amsterdam, pp. 19–34.

Friesner, R.A., Banks, J.L., Murphy, R.B., Halgren, T.A., Klicic, J.J., Mainz, D.T., Repasky, M.P., Knoll, E.H., Shelley, M., Perry, J.K., Shaw, D.E., Francis, P., Shenkin, P.S., 2004. Glide: a new approach for rapid, accurate docking and scoring. 1. Method and assessment of docking accuracy. Journal of Medicinal Chemistry 47, 1739–1749.

Fuqua, C., Greenberg, E.P., 2002. Listening in on bacteria: acyl-homoserine lactone signalling. Nature Reviews Molecular Cell Biology 3, 685–695.

Fuqua, C., Parsek, M.R., Greenberg, E.P., 2001. Regulation of gene expression by cell-to-cell communication: acyl-homoserine lactone quorum sensing. Annual Review of Genetics 35, 439–468.

Galié, S., García-Gutiérrez, C., Miguélez, E.M., Villar, C.J., Lombó, F., 2018. Biofilms in the food industry: health aspects and control methods. Frontiers in Microbiology 9, 898.

Gamage, A.M., Shui, G., Wenk, M.R., Chua, K.L., 2011. N-Octanoylhomoserine lactone signalling mediated by the BpsI–BpsR quorum sensing system plays a major role in biofilm formation of *Burkholderia pseudomallei*. Microbiology 157, 1176–1186.

Gaulton, A., Hersey, A., Nowotka, M., Bento, A.P., Chambers, J., Mendez, D., Mutowo, P., Atkinson, F., Bellis, L.J., Cibrián-Uhalte, E., Davies, M., Dedman, N., Karlsson, A., Magariños, M.P., Overington, J.P., Papadatos, G., Smit, I., Leach, A.R., 2017. The ChEMBL database in 2017. Nucleic Acids Research 45, D945–D954.

Gilson, M.K., Liu, T., Baitaluk, M., Nicola, G., Hwang, L., Chong, J., 2015. BindingDB in 2015: a public database for medicinal chemistry, computational chemistry and systems pharmacology. Nucleic Acids Research 44, D1045–D1053.

Hall-Stoodley, L., Costerton, J.W., Stoodley, P., 2004. Bacterial biofilms: from the natural environment to infectious diseases. Nature Reviews Microbiology 2, 95–108.

Haney, E.F., Brito-Sánchez, Y., Trimble, M.J., Mansour, S.C., Cherkasov, A., Hancock, R.E.W., 2018. Computer-aided discovery of peptides that specifically attack bacterial biofilms. Scientific Reports 8, 1871.

Harshey, R.M., 1994. Bees aren't the only ones: swarming in gram-negative bacteria. Molecular Microbiology 13, 389–394.

Harshey, R.M., 2003. Bacterial motility on a surface: many ways to a common goal. Annual Review of Microbiology 57, 249–273.

Henrichsen, J., 1972. Bacterial surface translocation: a survey and a classification. Bacteriological Reviews 36, 478–503.

Hobley, L., Harkins, C., MacPhee, C.E., Stanley-Wall, N.R., 2015. Giving structure to the biofilm matrix: an overview of individual strategies and emerging common themes. FEMS Microbiology Reviews 39, 649–669.

Hoiby, N., Bjarnsholt, T., Moser, C., Bassi, G.L., Coenye, T., Donelli, G., Hall-Stoodley, L., Hola, V., Imbert, C., Kirketerp-Moller, K., Lebeaux, D., Oliver, A., Ullmann, A.J., Williams, C., 2015. ESCMID guideline for the diagnosis and treatment of biofilm infections 2014. Clinical Microbiology and Infections 21, S1–S25.

Houston, D.R., Walkinshaw, M.D., 2013. Consensus docking: improving the reliability of docking in a virtual screening context. Journal of Chemical Information and Modeling 53, 384–390.

Hristozov, D.P., Oprea, T.I., Gasteiger, J., 2007. Virtual screening applications: a study of ligand-based methods and different structure representations in four different scenarios. Journal of Computer-Aided Molecular Design 21, 617–640.

Huang, N., Shoichet, B.K., Irwin, J.J., 2006. Benchmarking sets for molecular docking. Journal of Medicinal Chemistry 49, 6789–6801.

Huber, B., Riedel, K., Hentzer, M., Heydorn, A., Gotschlich, A., Givskov, M., Molin, S., Eberl, L., 2001. The cep quorum-sensing system of *Burkholderia cepacia* H111 controls biofilm formation and swarming motility. Microbiology 147, 2517–2528.

Huber, B., Riedel, K., Kothe, M., Givskov, M., Molin, S., Eberl, L., 2002. Genetic analysis of functions involved in the late stages of biofilm development in *Burkholderia cepacia* H111. Molecular Microbiology 46, 411–426.

Imperi, F., Massai, F., Ramachandran Pillai, C., Longo, F., Zennaro, E., Rampioni, G., Visca, P., Leoni, L., 2013. New life for an old drug: the anthelmintic drug niclosamide inhibits *Pseudomonas aeruginosa* quorum sensing. Antimicrobial Agents and Chemotherapy 57, 996–1005.

Inglese, J., Auld, D.S., 2008. High throughput screening (HTS) techniques: applications in chemical biology. In: Begley, T.P. (Ed.), Wiley Encyclopedia of Chemical Biology.

Irie, Y., Parsek, M., 2008. Quorum sensing and microbial biofilms. Bacterial Biofilms 322, 67–84.

Jain, A.N., 2003. Surflex: fully automatic flexible molecular docking using a molecular similarity-based search engine. Journal of Medicinal Chemistry 46, 499–511.

Jain, A.N., 2006. Scoring functions for protein-ligand docking. Current Protein and Peptide Science 7, 407–420.

Janssens, J.C.A., Steenackers, H., Robijns, S., Gellens, E., Levin, J., Zhao, H., Hermans, K., De Coster, D., Verhoeven, T.L., Marchal, K., Vanderleyden, J., Vos, D.E., De Keersmaecker, S.C.J., 2008. Brominated furanones inhibit biofilm formation by *Salmonella enterica* serovar Typhimurium. Applied and Environmental Microbiology 74, 6639–6648.

Jones, G., Willett, P., Glen, R.C., Leach, A.R., Taylor, R., 1997. Development and validation of a genetic algorithm for flexible docking. Journal of Molecular Biology 267, 727–748.

Jorgensen, W.L., 2004. The many roles of computation in drug discovery. Science 303, 1813–1818.

Kalia, M., Singh, P.K., Yadav, V.K., Yadav, B.S., Sharma, D., Narvi, S.S., Mani, A., Agarwal, V., 2017. Structure based virtual screening for identification of potential quorum sensing inhibitors against LasR master regulator in *Pseudomonas aeruginosa*. Microbial Pathogenesis 107, 136–143.

Kalyaanamoorthy, S., Chen, Y.P.P., 2011. Structure-based drug design to augment hit discovery. Drug Discovery Today 16, 831–839.

Kamaruzzaman, N.F., Tan, L.P., Yazid, K.A.M., Saeed, S.I., Hamdan, R.H., Choong, S.S., Wong, W.K., Chivu, A., Gibson, A.J., 2018. Targeting the Bacterial protective armour; challenges and novel strategies in the treatment of microbial biofilm. Materials 11, 1–27.

Kimyon, Ö., Ulutürk, Z.I., Nizalapur, S., Lee, M., Kutty, S.K., Beckmann, S., Kumar, N., Manefield, M., 2016. N-acetylglucosamine inhibits LuxR, LasR and CviR based quorum sensing regulated gene expression levels. Frontiers in Microbiology 7, 1–14.

Kitchen, D.B., Decornez, H., Furr, J.R., Bajorath, J., 2004. Docking and scoring in virtual screening for drug discovery: methods and applications. Nature Reviews Drug Discovery 3, 935–949.

Klebe, G., 2006. Virtual ligand screening: strategies , perspectives and limitations. Drug Discovery Today 11, 580–594.

Koch, B., Liljefors, T., Persson, T., Nielsen, J., Kjelleberg, S., Givskov, M., 2005. The LuxR receptor: the sites of interaction with quorum-sensing signals and inhibitors. Microbiology 151, 3589–3602.

Koo, H., Allan, R.N., Howlin, R.P., Stoodley, P., Hall-stoodley, L., 2017. Targeting microbial biofilms: current and prospective therapeutic strategies. Nature Reviews Microbiology 15, 740–755.

Köse-Mutlu, B., Ergön-Can, T., Koyuncu, I., Lee, C.H., 2019. Quorum quenching for effective control of biofouling in membrane bioreactor: a comprehensive review of approaches, applications, and challenges. Environmental Engineering Research 24, 543–558.

Kroemer, R.T., 2007. Structure-based drug design: docking and scoring. Current Protein and Peptide Science 8, 312–328.

Lebeaux, D., Chauhan, A., Rendueles, O., Beloin, C., 2013. From in vitro to in vivo models of bacterial biofilm-related infections. Pathogens 2, 288–356.

Lengauer, T., Rarey, M., 1996. Computational methods for biomolecular docking. Current Opinion in Structural Biology 6, 402–406.

Li, M., Ni, N., Chou, H.-T., Lu, C.-D., Tai, P.C., Wang, B., 2008. Structure-based discovery and experimental verification of novel AI-2 quorum sensing inhibitors against *Vibrio harveyi*. ChemMedChem 3, 1242–1249.

Manefield, M., Rasmussen, T.B., Henzter, M., Andersen, J.B., Steinberg, P., Kjelleberg, S., Givskov, M., 2002. Halogenated furanones inhibit quorum sensing through accelerated LuxR turnover. Microbiology 148, 1119–1127.

Mangwani, N., Kumari, S., Das, S., 2017. Bacterial biofilms and quorum sensing: fidelity in bioremediation technology. Biotechnology and Genetic Engineering Reviews 32, 43–73.

Martínez, A., Torello, S., Kolter, R., 1999. Sliding motility in mycobacteria. Journal of Bacteriology 181, 7331–7338.

Matsuyama, T., Bhasin, A., Harshey, R.M., 1995. Mutational analysis of flagellum-independent surface spreading of *Serratia marcescens* 274 on a low-agar medium. Journal of Bacteriology 177, 987–991.

Mattick, J.S., 2002. Type IV pili and twitching motility. Annual Review of Microbiology 56, 289–314.

Mattmann, M.E., Blackwell, H.E., 2010. Small molecules that modulate quorum sensing and control virulence in *Pseudomonas aeruginosa*. Journal of Organic Chemistry 75, 6737–6746.

McNaught, A.D., Wilkinson, A., 1997. IUPAC – Compendium of Chemical Terminology (The Gold Book), second ed. Blackwell Scientific Publications, Oxford.

Mehta, D.K., Das, R., 2018. Microbial biofilm and quorum sensing inhibition: endowment of medicinal plants to combat multidrug- resistant bacteria. Current Drug Targets 19, 1916–1932.

Mohan, V., Gibbs, A.C., Cummings, M.D., Jaeger, E.P., DesJarlais, R.L., 2005. Docking: successes and challenges. Current Pharmaceutical Design 11, 323–333.

Moitessier, N., Englebienne, P., Lee, D., Lawandi, J., Corbeil, C.R., 2008. Towards the development of universal, fast and highly accurate docking/scoring methods: a long way to go. British Journal of Pharmacology 153, 7–26.

Morris, G.M., Goodsell, D.S., Halliday, R.S., Huey, R., Hart, W.E., Belew, R.K., Olson, A.J., 1998. Automated docking using a Lamarckian genetic algorithm and an empirical binding free energy function. Journal of Computational Chemistry 19, 1639–1662.

Mottola, C., Seixas, R., Tavares, L., Oliveira, M., 2015. Bacterial biofilms: impact on aquatic ecosystems. In: Perspectives in Animal Ecology and Reproduction, tenth ed. Daya Publishing House.

Musk, D.J.J., Hergenrother, P.J., 2006. Chemical countermeasures for the control of bacterial biofilms: effective compounds and promising targets. Current Medicinal Chemistry 13, 2163–2177.

Neves, B.J., Braga, R.C., Melo-Filho, C.C., Moreira-Filho, J.T., Muratov, E.N., Andrade, C.H., 2018. QSAR-based virtual screening: advances and applications in drug discovery. Frontiers in Pharmacology 9, 1–7.

Ng, W.L., Bassler, B.L., 2009. Bacterial quorum-sensing network architectures. Annual Review of Genetics 43, 197–222.

Ouyang, X., Handoko, S.D., Kwoh, C.K., 2011. CScore: a simple yet effective scoring function for protein-ligand binding affinity prediction using modified CMAC learning architecture. Journal of Bioinformatics and Computational Biology 9, 1–14.

Pagadala, N.S., Syed, K., Tuszynski, J., 2017. Software for molecular docking: a review. Biophysical Reviews 9, 91–102.

Paul, N., Rognan, D., 2002. ConsDock: a new program for the consensus analysis of protein–ligand interactions. Proteins: Structure, Function and Bioinformatics 47, 521–533.

Plewczynski, D., Łaźniewski, M., von Grotthuss, M., Rychlewski, L., Ginalski, K., 2011. VoteDock: consensus docking method for prediction of protein-ligand interactions. Journal of Computational Chemistry 32, 568–581.

Pollitt, E.J.G., Diggle, S.P., 2017. Defining motility in the *staphylococci*. Cellular and Molecular Life Sciences 74, 2943–2958.

Quave, C.L., Estevez-Carmona, M., Compadre, C.M., Hobby, G., Hendrickson, H., Beenken, K.E., Smeltzer, M.S., 2012. Ellagic acid derivatives from Rubus ulmifolius inhibit *Staphylococcus aureus* biofilm formation and improve response to antibiotics. PLoS One 7, e28737.

Rajkumari, J., Borkotoky, S., Murali, A., Suchiang, K., Mohanty, S.K., Busi, S., 2018. Attenuation of quorum sensing controlled virulence factors and biofilm formation in *Pseudomonas aeruginosa* by pentacyclic triterpenes, betulin and betulinic acid. Microbial Pathogenesis 118, 48–60.

Rarey, M., Kramer, B., Lengauer, T., Klebe, G., 1996. A fast flexible docking method using an incremental construction algorithm. Journal of Molecular Biology 261, 470–489.

Rasmussen, T.B., Givskov, M., 2006. Quorum-sensing inhibitors as anti-pathogenic drugs. International Journal of Medicinal Microbiology 296, 149–161.

Ren, X., Shi, Y.-S., Zhang, Y., Liu, B., Zhang, L.-H., Peng, Y.-B., Zeng, R., 2018. Novel consensus docking strategy to improve ligand pose prediction. Journal of Chemical Information and Modeling 58, 1662–1668.

Rutherford, S.T., Bassler, B.L., 2012. Bacterial quorum sensing: its role in virulence and possibilities for its control. Cold Spring Harbor Perspectives in Medicine 2.

Sankar Ganesh, P., Ravishankar Rai, V., 2018. Attenuation of quorum-sensing-dependent virulence factors and biofilm formation by medicinal plants against antibiotic resistant *Pseudomonas aeruginosa*. Journal of Traditional and Complementary Medicine 8, 170–177.

Schuster, M., Lostroh, C.P., Ogi, T., Greenberg, E.P., 2003. Identification, timing, and signal specificity of *Pseudomonas aeruginosa* quorum-controlled genes: a transcriptome analysis. Journal of Bacteriology 185, 2066–2079.

Shanmugam, G., Mohamed, S., Natarajan, J., 2016. 2D- and 3D-QSAR study of acyl homoserine lactone derivatives as potent inhibitors of quorum sensor, SdiA in *Salmonella typhimurium*. International Journal of Bioautomation 20, 441–456.

Sharma, G., Sharma, S., Sharma, P., Chandola, D., Dang, S., Gupta, S., Gabrani, R., 2016. *Escherichia coli* biofilm: development and therapeutic strategies. Journal of Applied Microbiology 121, 309–319.

Sheikh, A.H., Raghuram, B., Eschen-lippold, L., Scheel, D., 2017. *Pseudomonas aeruginosa* biofilms – host response and clinical implications in lung infections. American Journal of Respiratory Cell and Molecular Biology 58, 1–36.

Shih, P.C., Huang, C.T., 2002. Effects of quorum-sensing deficiency on *Pseudomonas aeruginosa* biofilm formation and antibiotic resistance. Journal of Antimicrobial Chemotherapy 49, 309–314.

Shimkets, L.J., 1999. Intercellular signaling during fruiting-body development of *Myxococcus xanthus*. Annual Review of Microbiology 53, 525–549.

Shoichet, B.K., 2004. Virtual screening of chemical libraries. Nature 432, 862–865.

Šiman, P., Filipová, A., Tichá, A., Niang, M., Bezrouk, A., Havelek, R., 2016. Effective method of purification of betulin from birch bark: the importance of its purity for scientific and medicinal use. PLoS One 11, e0154933.

Sivakumar, P.M., Prabhawathi, V., Doble, M., 2010. Antibacterial activity and QSAR of chalcones against biofilm-producing bacteria isolated from marine waters. SAR and QSAR in Environmental Research 21, 247–263.

Skariyachan, S., Sridhar, V.S., Packirisamy, S., Kumargowda, S.T., Challapilli, S.B., 2018. Recent perspectives on the molecular basis of biofilm formation by *Pseudomonas aeruginosa* and approaches for treatment and biofilm dispersal. Folia Microbiologica 63, 413–432.

Skovstrup, S., Le Quement, S.T., Hansen, T., Jakobsen, T.H., Harmsen, M., Tolker-Nielsen, T., Nielsen, T.E., Givskov, M., Taboureau, O., 2013. Identification of LasR ligands through a virtual screening approach. ChemMedChem 8, 157–163.

Sliwoski, G., Kothiwale, S., Meiler, J., Lowe, E.W., 2014. Computational methods in drug discovery. Pharmacological Reviews 66, 334–395.

Soulere, L., Sabbah, M., Fontaine, F., Queneau, Y., Doutheau, A., 2010. LuxR-dependent quorum sensing: computer aided discovery of new inhibitors structurally unrelated to N-acylhomoserine lactones. Bioorganic and Medicinal Chemistry Letters 20, 4355–4358.

Sousa, F., Fernandes, P.A., Ramos, J.M., 2006. Protein – ligand docking: current status and future challenges. Proteins 26, 15–26.

Sousa, S.F., Cerqueira, N.M.F.S.A., Fernandes, A.P., Ramos, M.J., 2010. Virtual screening in drug design and development. Combinatorial Chemistry and High Throughput Screening 13, 442–453.

Sousa, S.F., Ribeiro, A.J.M., Coimbra, J.T.S., Neves, R.P.P., Martins, S.A., Moorthy, N.S.H.N., Fernandes, P.A., Ramos, M.J., 2013. Protein-ligand docking in the new millennium – a retrospective of 10 Years in the field. Current Medicinal Chemistry 20, 2296–2314.

Stickler, D.J., Morris, N.S., McLean, R.J., Fuqua, C., 1998. Biofilms on indwelling urethral catheters produce quorum-sensing signal molecules in situ and in vitro. Applied and Environmental Microbiology 64, 3486–3490.

Stumpfe, D., Ripphausen, P., Bajorath, J., 2012. Virtual compound screening in drug discovery. Future Medicinal Chemistry 4, 593–602.

Subhadra, B., Kim, D.H., Woo, K., Surendran, S., Choi, C.H., 2018. Control of biofilm formation in healthcare: recent advances exploiting quorum-sensing interference strategies and multidrug efflux pump inhibitors. Materials 11, 1676.

Sutherland, I., 2001. Biofilm exopolysaccharides: a strong and sticky framework. Microbiology 147, 3–9.

Taylor, R.D., Jewsbury, P.J., Essex, J.W., 2002. A review of protein-small molecule docking methods. Journal of Computer-Aided Molecular Design 16, 151–166.

Teramoto, R., Fukunishi, H., 2008. Consensus scoring with feature selection for structure-based virtual screening. Journal of Chemical Information and Modeling 48, 288–295.

Thomsen, R., Christensen, M.H., 2006. MolDock: A new technique for high-accuracy molecular docking. Journal of Medicinal Chemistry 49, 3315–3321.

Tielen, P., Strathmann, M., Jaeger, K.-E., Flemming, H.-C., Wingender, J., 2005. Alginate acetylation influences initial surface colonization by mucoid *Pseudomonas aeruginosa*. Microbiological Research 160, 165–176.

Tomlin, K.L., Malott, R.J., Ramage, G., Storey, D.G., Sokol, P.A., Ceri, H., 2005. Quorum-sensing mutations affect attachment and stability of *Burkholderia cenocepacia* biofilms. Applied and Environmental Microbiology 71, 5208–5218.

Tripathi, A., Bankaitis, V.A., 2017. Molecular docking: from lock and key to combination lock. Journal of Molecular Medicine and Clinical Applications 2, 1–19.

Trott, O., Olson, A.J., 2009. Software news and update AutoDock Vina: improving the speed and accuracy of docking with a new scoring function , efficient optimization, and multithreading. Journal of Computational Chemistry 31, 455–461.

Turner, L., Ryu, W.S., Berg, H.C., 2000. Real-time imaging of fluorescent flagellar filaments. Journal of Bacteriology 182, 2793–2801.

Udine, C., Brackman, G., Bazzini, S., Buroni, S., Van Acker, H., Pasca, M.R., Riccardi, G., Coenye, T., 2013. Phenotypic and genotypic characterisation of *Burkholderia cenocepacia* J2315 mutants affected in homoserine lactone and diffusible signal factor-based quorum sensing systems suggests interplay between both types of systems. PLoS One 8, e55112.

Uroz, S., Dessaux, Y., Oger, P., 2009. Quorum sensing and quorum quenching: the yin and yang of bacterial communication. ChemBioChem 10, 205–216.

Vajda, S., Guarnieri, F., 2006. Characterization of protein-ligand interaction sites using experimental and computational methods. Current Opinion in Drug Discovery and Development 9, 354–362.

Van Drie, J.H., 2007. Computer-aided drug design: the next 20 years. Journal of Computer-Aided Molecular Design 21, 591–601.

Venkatachalam, C.M., Jiang, X., Oldfield, T., Waldman, M., 2003. LigandFit: a novel method for the shape-directed rapid docking of ligands to protein active sites. Journal of Molecular Graphics and Modelling 21, 289–307.

Veselovsky, A.V., Ivanov, A.S., 2003. Strategy of computer-aided drug design. Current Drug Targets – Infectious Disorders 3, 33–40.

Vilar, S., Cozza, G., Moro, S., 2008. Medicinal chemistry and the molecular operating environment (MOE): application of QSAR and molecular docking to drug discovery. Current Topics in Medicinal Chemistry 8, 1555–1572.

Walters, W.P., Stahl, M.T., Murcko, M.A., 1998. Virtual screening—an overview. Drug Discovery Today 3, 160–178.

Wang, R., Wang, S., 2001. How does consensus scoring work for virtual library screening? An idealized computer experiment. Journal of Chemical Information and Computer Sciences 41, 1422–1426.

Wang, R., Lai, L., Wang, S., 2002. Further development and validation of empirical scoring functions for structure-based binding affinity prediction. Journal of Computer-Aided Molecular Design 16, 11–26.

Wang, R., Lu, Y., Wang, S., 2003. Comparative evaluation of 11 scoring functions for molecular docking. Journal of Medicinal Chemistry 46, 2287–2303.

Wolf, A., Zimmermann, M., Hofmann-Apitius, M., 2007. Alternative to consensus scoring a new approach toward the qualitative combination of docking algorithms. Journal of Chemical Information and Modeling 47, 1036–1044.

Worthington, R.J., Richards, J.J., Melander, C., 2012. Small molecule control of bacterial biofilms. Organic and Biomolecular Chemistry 10, 7457–7474.

Wu, G., Robertson, D.H., Brooks, C.L., Vieth, M., 2003. Detailed analysis of grid-based molecular docking: a case study of CDOCKER-A CHARMm-based MD docking algorithm. Journal of Computational Chemistry 24, 1549–1562.

Xavier, J.B., Foster, K.R., 2007. Cooperation and conflict in microbial biofilms. Proceedings of the National Academy of Sciences 104, 876–881.

Xu, Y., Tong, X., Sun, P., Bi, L., Lin, K., 2017. Virtual screening and biological evaluation of biofilm inhibitors on dual targets in quorum sensing system. Future Medicinal Chemistry 9, 1983–1994.

Yang, J.-M., Chen, C.C., 2004. GEMDOCK: a generic evolutionary method for molecular docking. Proteins 55, 288–304.

Yang, L., Rybtke, M.T., Jakobsen, T.H., Hentzer, M., Bjarnsholt, T., Givskov, M., Tolker-Nielsen, T., 2009. Computer-aided identification of recognized drugs as *Pseudomonas aeruginosa* quorum-sensing inhibitors. Antimicrobial Agents and Chemotherapy 53, 2432–2443.

Zhao, L., Xue, T., Shang, F., Sun, H., Sun, B., 2010. *Staphylococcus aureus* AI-2 quorum sensing associates with the KdpDE two-component system to regulate capsular polysaccharide synthesis and virulence. Infection and Immunity 78, 3506–3515.

Zhu, P., Peng, H., Ni, N., Wang, B., Li, M., 2012. Novel AI-2 quorum sensing inhibitors in *Vibrio harveyi* identified through structure-based virtual screening. Bioorganic & Medicinal Chemistry Letters 22, 6413–6417.

Chapter 16

Challenges and perspectives in reactor scale modeling of biofilm processes

Hermann J. Eberl[1], Matthew J. Wade[2]
[1]*Department of Mathematics and Statistics, University of Guelph, Guelph, ON, Canada;* [2]*School of Engineering, Newcastle University, Newcastle-upon-Tyne, United Kingdom*

16.1 Introduction

Although the microscopist Antoine van Leeuwenhoek first described microbial life growing in dense clusters on the surface of a tooth in the late 17th century, it was not until the mid-20th that microbiologists began to distinguish free-floating microbes suspended in the liquid phase from those that formed colonies attached to surfaces. Furthermore, these communities were formed not as pure cultures but of multiple species, and the properties of the 'biofilm" differed markedly from that of planktonic microbes. However, elucidation of the importance of mixed-species biofilms, in their structure, function, and interactions across a range of environments, is relatively recent (Davey & O'Toole, 2000; Elias and Banin, 2012; Nadell et al., 2009).

Biofilms are prevalent in a diverse range of environments, but their ubiquity remains a source of contention (Bar-On and Milo, 2019; Flemming and Wuertz, 2019). Their prevalence, however, has garnered interest among researchers and thinkers in an equally diverse array of specialisms, from microbiology and ecology to medicine (Percival et al., 2015), marine science (de Carvalho, 2018), and cultural heritage (Pinna, 2014). Biofilm formation is generally conceived as being a negative phenomenon, e.g., medical device fouling, dental plaque, surface corrosion, and physicochemical degradation of materials. However, some of these densely packed, multifunctional communities have found favor as a tool employed by engineers to deliver positive and transformative impacts through varied biochemical and ecological strategies that can be harnessed and controlled in similar ways to other biologically driven processes. As bacteria in aqueous environments always form sessile biofilms when in contact with an immersed surface or as suspended granular

Recent Trends in Biofilm Science and Technology. https://doi.org/10.1016/B978-0-12-819497-3.00016-7

communities that self-aggregate around a dense microbial nucleus, engineering applications of biofilms are specific to processes managing the transformation of substrate-rich liquids, typically municipal or industrial wastewater streams. For drinking water systems, biofilm growth on pipes is undesirable due to the inherent public health risks; subsequently, engineers focus on methods for its removal rather than as a beneficial ecological tool.

Engineered biofilm systems for application in secondary and tertiary wastewater treatment began to emerge in the 1980s (Capdeville and Rols, 1992), although the concept of bacteria attached to support media was not new (Zobell, 1943). However, technological advances in the type and structure of attachment media and a deeper understanding of the biotic functions within the biofilms (Costerton et al., 1978) meant that the design and operation of biofilters was appropriate for large-scale applications, providing effective removal of soluble organic matter or nitrogenous compounds, for example.

Environments in which biofilms are developed, manipulated, controlled, or even measured can be considered biofilm reactors (Lewandowski and Boltz, 2011). As with all bioreactor technologies, biological and physicochemical principles that govern these systems provide the basis for the scope and theoretical limits of their engineering and operation. Biofilm reactors provide the capacity for selective biomass retention and separation of the system into distinct or quasi-distinct phases, i.e., inert attachment surface, biofilm matrix (microorganisms and extracellular polymeric substances), and liquid and gases. An inert fraction of dead biomass is also considered as an abiotic component of the biofilm.

A number of reactor configurations and operating modes have been employed to promote or suppress microbial activity in biofilms, selected to meet the process objectives and associated microbial community characteristics and structural requirements (e.g., fixed-film or granular). These include continuous systems such as simple trickling filters, moving or fixed beds, membrane bioreactors, and sequencing batch reactors (SBRs) in which biofilm—liquid contact time is empirically designed to maximize performance efficiency per fill—reaction—draw cycle, e.g., aerobic granular systems. However, there are considerable trade-offs with the flexibility and smaller footprint of SBRs including operating complexity (e.g., a tighter coupling of process with microbial activity), maintenance requirements, and potential washout of key organisms during reactor emptying. Correctly designed and operated sequencing batch biofilm reactors (SBBRs) combine the benefits attributed to SBRs, the bioconversion potential of biofilm communities, and the ability to theoretically retain and regulate the key biomass communities to a high degree.

While biofilm reactors operating continuously are more widely applied in practice, recent research has focused on addressing some of the significant challenges of SBR design and operation. For example, selective biomass washout using hybrid SBBRs under mainstream wastewater temperatures

(Laureni et al., 2019), protection from shock loading using buffer media, and microbial species differentiation to degrade complex waste streams by promoting functional diversity (Dutta and Sarkar, 2015) are applications facilitated by recent developments in biofilm reactor research, combining a deeper understanding of the microbial ecology and good engineering practice.

Despite the continued and focused research on biofilms among the engineering and microbiological communities, there still remain some unifying challenges and inefficiencies that are particular to such complex and interdisciplinary fields. Mathematical modeling has long been the obvious solution to translate difficult problems into tangible frameworks that are open to analysis and interpretation. Modeling of biofilms, specifically in the context of reactor design and operation, has proved to be a fruitful endeavor for scientists and mathematicians alike. A well-posed model allows for insights into processes that are hard to observe in situ or are impractical to test empirically.

Mathematical models play several roles in science and engineering. Firstly, and most importantly, mathematical models are, and have been since the days of Galileo and even Roger Bacon, the language by which scientific theories are formulated: The process understanding, be it known or conjectured, forms the model assumptions. These are translated into a mathematical language, which allows them to be subjected to thought experiments (which, depending on the mathematical complexity, might need to be computer-aided and involve simulation—this is generally the case in biofilm modeling). In formulating models, as in all of science, Occam's razor should be applied: They should be as complex as necessary and as simple as possible. If the results of the thought experiments do not correctly reflect experimental observations, this suggests that some of the model assumptions were incorrect or that further assumptions are needed that are crucial for the understanding of the question at hand. Thus, a mathematical model can never be used to verify a theory or model assumptions, only for falsification, and to show the limitations of the current process understanding. In this sense, mathematical models are Popperian tools providing a structural basis for theoretical experimentation.

Secondly, in the context of this chapter, mathematical models and their numerical simulation can be used for quantitative prediction, control, and optimization of biofilm reactors. This requires that the model is sufficiently well calibrated. If the underlying process is very complex, as is the case for many biological systems, including biofilms and biofilm reactors, it might be advantageous to use simpler black-box or data-driven models that are designed to relate input to output models but are neither based on nor do they attempt to infer dynamical relationship, i.e., they cannot be used to explain observations by deduction. It can be expected that models of this type will gain importance in biofilm process engineering due to the emergence and broad availability of increasingly more powerful machine learning algorithms employed to elucidate the structure and properties of biofilms. Examples include deep learning for biofilm performance prediction (Shi and Xu, 2018), artificial intelligence to model the physics and chemistry of materials that is highly relevant to biofilms (Butler et al., 2018), and Bayesian emulation of individual-based biofilm models that could provide a means to simulate full-scale biofilm reactors (Oyebamiji et al., 2019).

Thirdly, mathematical models can be tools to aid in determining physical quantities that are not directly accessible to measurements. In the context of biofilms, an important example are diffusion rates of dissolved substrates, which are derived from measurements of concentrations by solving an inverse problem of the basic diffusion equation (e.g, Bryers and Drummond, 1998). Another basic and very simple example from microbiology is the estimation of bacterial growth rates from growth curves. Such inverse problems are usually ill-posed (see Kabanikhin, 2008 for an overview). The outcome will depend critically on the quality and density of the data set that is available. This is the question of observability of a specific model, i.e., the question of how well the internal state of a model can be inferred from measurements. Even if, from the point of view of theoretical observability, good enough data sets are available, computational requirements can be inhibitive, as solving the inverse problem typically requires iterative solutions and successive adjustment of the model.

When developing a biofilm reactor model, consideration must be made as to which processes to include, which to approximate or simplify, and which to ignore. This will dictate the degree of complexity associated with the model. The level of model complexity is always a function of the order of the system described (i.e., degree of nonlinearity), its dimensionality (i.e., number of coupled equations and the nature of their interdependence), and the model structure itself (e.g., discrete, continuous). Selecting the most suitable structure is borne of necessity, but there is typically a trade-off between complexity and accuracy, which typically manifests as a decision between shorter simulation run time and greater precision of solutions.

Transformation and *transport* processes are fundamental to biofilm model development. The former relate to the biochemical reactions determining the overall production and consumption of biotic and abiotic components of the system. Transport processes are intrinsically linked to these through the creation of chemical gradients within the biofilm that in turn produce gradients of microbial growth, resulting in the heterogeneous profile observed throughout the structure. This is modeled by coupling of the reaction and advection describing growth and transport of biomass within the biofilm volume. An adequate description of the transport of nutrients, specifically those substrates required for growth, is also necessary and is typically defined by a diffusion–reaction equation and solved as a two-point boundary problem. Here, rates of nutrient transport in the biofilm are determined by their diffusion coefficients and nutrient consumption rate explicitly via the reaction term. *Transfer* processes describing mass exchange between distinct reactor compartment such as the bulk liquid and biofilm phases or between biofilm compartments are standard to all dimensional biofilm models (i.e., those assuming spatial heterogeneity in the biofilm. Zero-dimensional biofilm models assume spatial homogeneity or a decoupling of the spatial from the temporal dynamics). As may be imagined, the implementation of these three processes in an appropriate and coherent manner is not an inconsiderable task. While much of the structural development relies on a comprehensive biochemical understanding of the system at the scale of implementation, simulation of the model is further complicated by the requirement to numerically solve the nonlinear two-point boundary value problem describing diffusion through the biofilm. This is particularly challenging when considering multisubstrate, multispecies biofilms and is still an open problem for many nonstandard biological systems (Baeten et al., 2017; Dzianach et al., 2019).

Comprehensive reviews of biofilm modeling for the engineering community may be found in the literature and are recommended to readers hoping to gain some introductory insights and initial guidance on model development (Lewandowski and Boltz, 2011; Horn and Lackner, 2014; Tartakovsky and Guiot, 2004). Our intention here is to discuss biofilm modeling from the perspective of mathematical biology, focusing on existing challenges and limitations, some of which are widely accepted and others that may be less well understood. We aim to demonstrate the wide-ranging benefits of having a good model of a biofilm reactor to facilitate design, simulation, control, and analysis but will stress that modeling of biofilms, particularly those found outside the laboratory, is nontrivial. These systems present challenges mathematically, numerically, and conceptually, and this chapter will attempt to provide some clarity and perspective for future research in the field.

16.2 Mathematical modeling of biofilm reactors

Here, we provide an overview of the mathematical modeling components typically used for constructing a representation of biofilm reactors. In Section 16.2.1, we briefly summarize and review the basics of a model describing the standard chemostat as an example of a simple bioreactor for processes involving suspended microbial populations. This model is obtained from mass balances at the reactor or macroscale. In Section 16.2.2, we present the one-dimensional Wanner–Gujer model for biofilms that is well established in engineering applications and illustrate its development using the example of a nitrifying biofilm. This model is formulated at the spatial scale of the biofilm or mesoscale. Then in Section 16.2.3, we combine both into a multiscale model of a biofilm reactor. We do this for a moving bed biofilm reactor, following closely the approach given in Masic and Eberl (2014). The primary purpose of this section is to provide a glimpse into the mathematical complexity of biofilm reactor models and the challenges that arise from it.

16.2.1 Chemostat modeling

In a nutshell, a chemostat is a bioreactor to which growth medium is continuously added. The content of the vessel is completely mixed. Usually the volume inside the reactor is kept constant, i.e., the culture liquid, which contains substrates, metabolites, and microbes, is continuously removed. Although the technical details vary greatly between applications and objectives, this general description suffices normally as the basis of mathematical abstraction. A mathematical model for the chemostat is formulated as a dynamic mass balance. The amount of a substrate or metabolite in the reactor is a balance of substrate that is added with the inflow, removed with outflow, and consumed (or produced) by the microbes in the growth chamber. Similarly, the bacterial biomass concentration in the system is obtained by balancing growth and decay with outflow. For a simple single-species, single-substrate chemostat, one obtains the following ordinary differential equation in terms of substrate concentration in the reactor S and biomass concentration X:

$$\begin{cases} \dot{S} = D(S_0 - S) - \dfrac{\mu}{Y} G(S)X, \\ \dot{X} = \mu G(S)X - DX. \end{cases} \tag{16.1}$$

Here, D is the reactor dilution rate, S_0 is the inflow substrate concentration, _ is the maximum specific growth rate, growth function G(S), and Y is the yield coefficient that infers how much mass of biomass is produced per *G*(*S*) unit mass of substrate. Conventionally, substrate and biomass are expressed in terms of molarity (moles/L) allowing for fairly straightforward stoichiometric balancing of the model. However, in many applications, in particular in environmental engineering, the components may be represented as chemical oxygen demand (COD) equivalent. The function G(S) encodes the dependence of bacterial growth on the substrate concentration. A standard *G*(*S*) model, but not by any means the only possible one, is Monod kinetics,

$$G(S) = \frac{S}{k + S'} \tag{16.2}$$

where parameter k is the half saturation concentration, i.e., the concentration value at which the substrate dependent growth rate is half of the theoretical maximum. This kinetics describes growth with saturation effects, that is, first order (i.e., growth rate is proportional to available substrate) for small concentration ($S << k$) and zeroth order (i.e., growth rate is constant) if substrate is available in abundance ($S >> k$). The above is the simplest and most classical version of a single-species chemostat. It does not account for lysis and other cell loss processes, nor does it account for substrate utilization for cell maintenance. To include such effects is easily possible and does not change the basic structure of the model.

Both in experimental and in theoretical studies, the important question is what happens when the system equilibriates. Mathematically, this is the question of long-term behavior, as time $t \to \infty$. In classical mathematical biology, this question is investigated using the methods of dynamical system theory. Equilibrium states are the steady states of the system, i.e., constant solutions. As these are not exactly attained in finite time, the question is whether asymptotically the state of the system converges to such a steady state or whether it is repelled from a steady state. For our chemostat model discussed earlier, we find always the trivial steady state $(S^*, X^*) = (S_0, 0)$ that is characterized by absence of biomass and a substrate concentration at inflow levels. This steady state is stable and will be attained in the long term if the dilution rate exceeds the maximum growth rate in the reactor, i.e., if

$$D > \mu G(S_0). \tag{16.3}$$

Under these conditions, washout of bacteria dominates growth, i.e., the population is not able to establish itself. If the inequality is reversed, then this trivial equilibrium is unstable and a population will be able to develop. In this case, a second steady state can be found $(S^{**}, X^{**}) = \left(\widehat{S}, Y\left(S_0 - \widehat{S}\right)\right)$.

Here, $\widehat{S}$ is the breakeven concentration, i.e., the substrate concentration level $\widehat{S}$ at which growth is completely balanced by washout or, in other words, the minimum concentration at which growth is large enough to be not dominated by washout. It is found as the value at which

$$\mu G\left(\widehat{S}\right) = D. \tag{16.4}$$

For the Monod kinetics given earlier, one finds

$$\widehat{S} = \frac{Dk}{\mu - D}. \tag{16.5}$$

From a bioreactor perspective, the previous model with a single growth-limiting substrate S and a single bacterial species X is too simple. Many processes, in particular in environmental engineering, involve more complex communities with several species. The simplest interaction between species is competition for a shared substrate. To extend the previous model, we consider now a system with two species:

$$\begin{cases} \dot{S} = D(S_0 - S) - \frac{\mu_1}{Y_1} G(S)X_1 - \frac{\mu_2}{Y} G_2(S)X_2, \\ \dot{X}_1 = \mu_1 G_1(S)X_1 - DX_1, \\ \dot{X}_2 = \mu_2 G_2(S)X_2 - DX_2 \end{cases} \tag{16.6}$$

where $X_{1,2}$ are now the densities of the two species with growth rates $\mu_{1,2}G_{1,2}(S)$ and corresponding yield coefficients $Y_{1,2}$. Note that substrate inflow concentration S_0 and dilution rate D are reactor properties and the same for both species. The substrate concentration S in the reactor is now affected by both species. The Monod functions $G_{1,2}(S)$ are defined in analogy to the above as

$$G_{1,2} = \frac{S}{\kappa_{1,2} + S}. \tag{16.7}$$

The steady-state analysis for this model of competition in a chemostat of two species for a shared resource shows again that the washout steady state (no biomass in the system) is asymptotically stable if $D > \mu_1 G_1(S_0)$ and $D > \mu_2 G_2(S_0)$, i.e., if dilution dominates growth of both species. In the case where this inequality holds for one species, but the reverse holds for the other, say $D > \mu_1 G_1(S_0)$ and $D > \mu_2 G_2(S_0)$, then, as one might expect, X1 will be washed out, whereas X2 will be able to establish itself. More interesting is the case where for both species the maximum growth rate is larger than the dilution rate, i.e., if $D < \mu_1 G_1(S_0)$ and $D < \mu_2 G_2(S_0)$. In this case, one can prove with formal methods that the *Principle of Competitive Exclusion* holds a famous law in ecology (aka Gause's law). In general, Gause's law states that in a scenario where two species compete for one resource, only the one with the lower resource requirements will be able to survive. In our settings, this is the species X_i with the lower break-even concentration $\widehat{S}_i$. If both species, as is commonly assumed, grow according to Monod kinetics, this is the species with the smaller value:

$$\widehat{S}_i = \frac{D_{\kappa_i}}{\mu_i - D}. \tag{16.8}$$

It is important to emphasize that this long-term behavior is independent of the initial state of the system. If at the start of the process both species are present, regardless of their ratio, the one with the lower break-even concentration will eventually outcompete the other one. The initial makeup of the population only affects the duration until extinction of the weaker species.

Only if $\widehat{S}_1 = \widehat{S}_2$, both species can coexist in the long term. In this case from an equilibrium analysis perspective, they cannot be distinguished. This theoretical result, of course, is in complete agreement with our intuitive expectation: At equilibrium, the concentration in the reactor must be at break-even concentration. If both species have different break-even concentrations, this leads to a contradiction. This result can be generalized to more complex populations in which several species compete for several resources, as long as the number of resources is smaller than the number of competitors.

Anecdotal interpretation of our single-species model is that in a more diverse population that consists of several, functionally similar species or strains (e.g., glucose degraders), X can represent the total biomass density, and the model parameters $\mu.Y, \kappa$ are averages of the values of the individual species in some sense. The parameters determine an average behavior or response of the populations and are analogous to the conventional practice in microbiology of making gross calculations of microbial growth rate at the population scale (de Jong et al., 2018). An immediate consequence of our mathematically rigorous and ecologically well-understood result for the chemostat model with two competitors is that this interpretation is wrong. The theory predicts that the population will not converge to some average value but that eventually will only consist of the species with the lowest break-even concentration. The conclusion is that in such a mixed population system, other processes that are not captured by our simple assumption of competition as the only interaction must be at play. In other words, we have here an example where the analysis of mathematical model tells us to revise our assumptions and that our understanding of the process based on which we formulated the model was not complete.

In systems where other effects play a significant role, for example, where one species can gain an additional growth advantage over the other, the break-even concentration of that species might not be given by the above formula anymore, so that the species with higher $\widehat{S}_i$ might survive. In this case, however, the fundamental assumption of our chemostat model that the only interaction between both species is competition for a shared substrate is violated. An example of such a system, which is both microbially and mathematically more complex but can be modeled in our framework, is if one species gains a growth advantage over its competitor in an iron limited system by iron chelation, i.e., by producing siderophores that bind iron in the environment and make it unaccessible to its competitor. An extension of our chemostat model for this scenario can be found in Fgaier and Eberl (2010), for the example of *Pseudomonas fluorescens* that engages in such behavior. Similarly, if in a competitive system one species invests, as resources become limited, in the production of a substance that inhibits the growth of its competitor, it can gain a growth advantage even if it has the higher $\widehat{S}_i$. An example for such an allelopathic system that has been modeled by extension of our chemostat model is the suppression of the food pathogen *Listeria monocytogenes* by a nonpathogenic listeria species, such as *Listeria innocua* or *Listeria welshimeri* (Fgaier et al., 2014). An example of an amensalitic system, in which one species alters the environment to the detriment of both species, but is itself more robust to this change than its competitor, was modeled by extension of the chemostat framework for *L. monocytogenes* and *Lactococcus lactis* in Khassehkhan and Eberl (2016). These few examples illustrate the versatility of the chemostat modeling concept and how it can be adapted to more complex situation. Whereas the basic chemostat model of competition for a shared nutrient was fully accessible to formal mathematical analysis, this is not necessarily true if additional processes are included. In general, the exploration of such a model will then rely on a combination of analytical tools and computer simulations, which in this case involves the numerical integration of ordinary differential equation. For this task, many sophisticated software packages exist.

The fundamental assumption behind the chemostat modeling framework is that the reactor is completely mixed, with no substrate or biomass gradients: The conditions are the same everywhere in the reactor. While they might change dynamically over time, spatially the population is entirely homogeneous. Biofilms are characterized by substrate gradients that possibly induce heterogeneous layering of species and the development of microniches, which even might allow for the coexistence of species that cannot be found simultaneously in completely mixed environments. Another characteristic of bacterial biofilms are diffusive mass transfer limitations, which are not accounted for in a completely mixed system. Therefore, a chemostat model, such as the one described here, cannot be expected to be a good representation of a biofilm reactor, even if the additional assumption is made that bacteria are contained in the reactor and not washed out by dilution but are sessile in the reactor. To account for the specifics of biofilm processes in such a bioreactor, a major extension of the modeling framework is required. Foremost, this requires a mathematical model of the biofilm itself.

16.2.2 One-dimensional biofilm models

Here, as an example, we briefly sketch a standard biofilm model, namely the one-dimensional Wanner–Gujer model (Wanner and Gujer, 1986) applied to describe a nitrifying biofilm. This model framework is suitable for compact, well-developed, and relatively homogeneous biofilms, which substantially stratify parallel to the substratum. These assumptions are valid in many engineered processes that are designed to utilize biofilms, e.g., wastewater engineering and bioremediation. The key processes in this model are bacterial growth due to substrate consumption, diffusion of dissolved substrates and metabolites as well as their degradation and production in the biofilm, and biomass loss to the aqueous phase by detachment. An integral assumption of the model is that the biomass density (ρ) is constant throughout the biofilm. This implies that production of biomass is translated 1:1 into an expansion of the biofilm, and similarly, cell loss leads to a contraction. As is common in biofilm modeling, biomass density here subsumes both cells and extracellular polymeric substances.

The biofilm in our example consists of ammonium-oxidizing bacteria, nitrite-oxidizing bacteria, and inert biomass. The first species converts ammonium to nitrite that the second species converts to nitrate. Oxygen is consumed in both processes. The biofilm is described in terms of the volume fractions f_i of the three particular biomass fractions. These are given by the coupled-system first-order partial differential equations:

$$\frac{\partial f_i}{\partial_t} = f_i \mu_i\left(\overline{C}\right) - \frac{\partial}{\partial z}(f_i v), 0 < z < \lambda, i \, \varepsilon \, I_b \tag{16.9}$$

which are defined within the biofilm, with a thickness $\lambda = \lambda(t)$ that increases as the biofilm grows or decreases as it contracts. Here, $I_b = \{A, N, I\}$ is the index set for biomass, where A denotes ammonium oxidizers, N denotes nitrite oxidizers, and I denotes the inerts. The function $\mu_i(C)$ describes the reaction terms of the biomass groups in dependence of the substrate concentrations in the biofilm, as defined in Table 16.1. These substrate concentrations are variable throughout the biofilm and subject to Fickian diffusion. This will be described in more detail in the following.

In Eq. (16.9), $v = v(t,z)$ is the velocity of biomass in the biofilm due to expansion or contraction. It depends on the reaction terms in the biofilm and is determined by

$$v(t,z) = \int_0^z \sum_{i \in I_b} f_i \mu_i(\overline{C}) dz'. \tag{16.10}$$

The rate of change for the biofilm thickness $\lambda(t)$ depends on the biomass velocity at the interface with the aqueous phase, i.e., at $z = \lambda$, namely, $v(t,\lambda)$. It also accounts for the loss of biomass at the biofilm/liquid interface due to detachment.

$$\dot{\lambda} = \int_0^\lambda (f_A \mu_A(\overline{C}) + f_N \mu_N(\overline{C}) + f_I \mu_I(\overline{C}))dz + d(\lambda)\lambda(f_A(\lambda) + f_N(\lambda) + f_I(\lambda)) \tag{16.11}$$

Here, $d(\lambda)$ is the detachment rate, on which we will comment more in the following. In Eq. (16.10), the integral is taken over the entire biofilm thickness. Hence, the substrate concentrations in the biofilm must be known.

It is common practice in biofilm modeling to take advantage of the disparity of time scales of biomass growth and substrate consumption and diffusion in the biofilms (Wanner et al., 2006), and to consider, therefore, the processes governing the substrate concentrations as being at a quasi−steady state. For the substrates C_k, one has then

$$D_k C_k''(z) = r_k(f_A \rho, f_N \rho, \overline{C}), \quad 0 < z < \lambda, \quad k \in I_S \tag{16.12}$$

with boundary conditions

$$C_k'(0) = 0, C_k(\lambda) = S_k \tag{16.13}$$

where $C_{O2}(z)\left[g/m^3\right]$ denotes oxygen concentration at biofilm thickness z m from the substratum, $C_{NH4}(z)$ ammonium, $C_{NO2}(z)$ nitrite, $C_{NO3}(z)$ nitrate, and $\overline{C} = [C_{O2}, C_{NH4}, C_{NO2}, C_{NO3}]$, and $I_S = \{C_{O2}, C_{NH4}, C_{NO2}, C_{NO3}\}$ is the index set for substrates. In the boundary conditions, the values S_k are the concentration values at the biofilm/liquid interface. In the above formulation, we neglect mass transfer resistance in a biofilm/liquid boundary layer. To include this, the Dirichlet boundary condition at $z = \lambda$ should be replaced by a Robin condition. The no-flux, or Neumann, boundary condition (1.13) at $z = 0$ signifies that diffusive substrates do not leave or enter the biofilm through the substratum.

In our model formulation, we assumed that biomass only detaches from the outer layers of the biofilm, at the interface with the aqueous phase. An alternative assumption would be to permit biomass to be removed from across

TABLE 16.1 Petersen matrix for the growth and conversion processes in the models of Sections 16.2.2 and 16.2.3. A stands for ammonium oxidizers, *N* for nitrite oxidizers, and *I* for inerts. The reactions for biomass components X_i are used in the biofilm $(f_i\rho)$ and for suspended bacteria (u_i). The net reaction rates are obtained as $r_k(t,z) = \sum_{j=1}^{6} v_{k\,j} p_j \left[\frac{g}{m^3 \cdot d}\right]$ and growth rates $\mu_i(t,z) = \sum_{j=1}^{6} \frac{v_{ij} p_j}{f_i \rho}\left[{}^1/_d\right]$. All parameters are positive, typical values can be found, for example in Masic and Eberl (2014).

Process j j = 1,6	Concentrations Sk, $k\varepsilon\{O2, NH4, NO2, NO3\}$				Densities Xi $i\varepsilon\{A, N, I\}$			Process rate Pj j = 1, ….,6
	S_{O2}	S_{NH4}	S_{NO2}	S_{NO3}	X_A	X_N	X_I	
1. Growth of A	$-\frac{(3.43 - Y_A)}{Y_A}$	$-\frac{1}{Y_A} - i_A$	$\frac{1}{Y_A}$	—	1	—	—	$\mu_{A,max} \cdot \frac{S_{O2}}{K_{A,O2} + S_{O2}} \cdot \frac{S_{NH4}}{K_{NH4} + S_{NH4}} \cdot X_A$
2. Endogenous respiration of A	$-(1 - fX1)$	$i_A - i_I fXI$	—	—	−1	—	fX	$b_A \cdot \frac{S_{O2}}{K_{A,O2} + S_{O2}} \cdot \frac{S_{O2}}{K_{O2} + S_{O2}} \cdot X_A$
3. Inactivation of A		-	—	—	−1	—	I 1	$b_A \cdot \eta \cdot \frac{S_{O2}}{K_{A,O2} + S_{O2}} \cdot \frac{S_{O2}}{K_{O2} + S_{O2}} \cdot X_A$
4. Growth of N	$-\frac{(1.14 - Y_N)}{Y_N}$	$-i_A$	$-\frac{1}{Y_N}$	$-\frac{1}{Y_N}$	—	1	—	$\mu_{N,max} \cdot \frac{S_{O2}}{K_{N,O2} + S_{O2}} \cdot \frac{S_{NO2}}{K_{NO2} + S_{NO2}} \cdot X_N$
5. Endogenous respiration of N	$-(1 - fX1)$	$i_A - i_I fXI$	—	—	—	−1	fX	$b_N \cdot \frac{S_{O2}}{K_{N,O2} + S_{O2}} \cdot X_N$
6. Inactivation of N		—	—	—	—	−1	I 1	$b_N \cdot \eta \cdot \frac{S_{O2}}{K_{N,O2} + S_{O2}} \cdot X_N$

the entire biofilm, in which case the detachment terms would appear inside the integral. Both formulations are used, but it is not always clear for which application and biofilm system which option is more suitable. We will discuss this in some more detail in the following.

The biofilm model that we briefly sketched here is mathematically very complex. It is a free boundary value problem of a system of quasilinear hyperbolic balance equations with nonlocal effects (via the definition of v), coupled to a two-point boundary value problem for a system of diffusion–reaction equations. Such biofilm models have been studied by computer simulation for 3 decades, but rigorous mathematical results have only begun to appear in recent years for certain applications (D'Acunto et al., 2015; Klapper and Szomolay, 2011; Szomolay and Cogan, 2015).

Over the years, many biofilm processes have been cast in this model framework, which is rather flexible with regard to the types of ecological interaction of species in the biofilm that can be permitted and to other biofilm processes. One aspect that is not yet included in the framework that we sketched, but which will be important if such a biofilm model is to be embedded in a biofilm reactor model, is (re)attachment of cells from the aqueous phase and invasion into the biofilm. These processes are little understood, and their modeling is only in the early stages, with initial attempts undertaken described in (D'Acunto et al., 2018; D'Acunto et al., 2019; Gaebler and Eberl, 2018; Masic and Eberl, 2012).

16.2.3 A biofilm reactor model

To formulate a biofilm reactor model, we combine ideas laid out above for chemostat like bioreactors with the biofilm model just sketched.

In this multispecies multisubstrate model, bacteria are assumed to detach and become suspended, and suspended bacteria can (re)attach to the biofilm. Biomass growth occurs through consumption of substrate, which is continuously supplied to the reactor, and diffuses into the biofilm as described earlier. Unlike the chemostat setting, where all bacteria experience the same growth conditions, due the complete mixing hypothesis, the diffusion–reaction induced substrate gradients lead now to a model in which bacteria may experience locally different growth conditions. Bacterial cells may become inert matter through inactivation and endogenous respiration. The flow through the reactor washes out suspended biomass as well as substrates and products. Biofilms are assumed to homogeneously cover the colonizable surfaces on which they grow.

Starting point for the development of our reactor model are mass balances for the dissolved substrates in a reactor, akin to the chemostat model, with a completely mixed bulk phase.

$$\widehat{S}_k = D\left(S_k^0 - S_k\right) - \frac{1}{V}\left(r_k\left(u_A, u_N, \overline{S}\right) + AD_{k\,j\,k}\left(\lambda, \overline{S}\right)\right), \quad k \in I_S \tag{16.14}$$

with initial conditions

$$S_k(t_0) = S_k^0. \tag{16.15}$$

As earlier, $I_S = \{O2, NH4, NO2, NO3\}$ is the index set for dissolved substrates, $S_{O2}(t)\left[g/_{m^3}\right]$ denotes the oxygen concentration in the reactor, $S_{NH4}(t)$ ammonium, $S_{NO2}(t)$ nitrite, $S_{NO3}(t)$ nitrate and where $\overline{S} = [S_{O2}, S_{NH4}, S_{NO2}, S_{NO3}]$. These concentrations are assumed to be completely mixed in the aqueous phase. The reactor parameters $D\left[d^{-1}\right]$, $V\left[m^3\right]$, and $A\left[m^2\right]$ denote the dilution rate, and the volume and the colonizable surface area, respectively. The dilution rate is related to the reactor flow rate through $D = Q/V$ where $Q\left[m^3/d\right]$ is the flow rate. $D_k\left[m^2/d\right], k \in I_S$, denotes the diffusion coefficients for the dissolved compounds. Table 16.1 gives the stoichiometric matrix from which the reaction rates r_k can be expressed. In essence, these reaction rates describe uptake or production of substrates by microbial activity. Using the same notation as in Masic and Eberl (2012), the functions $j_k(\lambda, \overline{S}), k \in I_S$, are defined as

$$j_k(\lambda, \overline{S}) = \begin{cases} \dfrac{1}{D}\displaystyle\int_0^{\lambda} r_k(f_{A\rho}, f_{N\rho}, \overline{C})dz & \lambda > 0 \\ 0 & \lambda = 0 \end{cases} \tag{16.16}$$

where (as in Section 16.2.2) $\lambda(t)$ is the biofilm thickness. The suspended biomass types u_A; u_N and the biofilm fractions f_A, f_N will be discussed shortly. Diffusion–reaction equations describe the dissolved compounds within the biofilm. The concentrations $C_k(z)$in the biofilm are given by Eqs. (16.12) and (16.13). Note that the boundary conditions change with time through dependence on $S_k(t)$ and $\lambda(t)$, wherefore solutions to Eq. (16.12) are essentially functions of time.

The three types of biomass that we considered in the biofilm earlier also are present in the reactor in suspended form: ammonium oxidizers (u_A [g]), nitrite oxidizers (u_N [g]), and inerts (u_I [g]). We assume that heterotrophic bacteria are not present in the reactor and consider, therefore, a pure nitrifying system. Suspended biomass in the reactor is assumed to be completely mixed and modeled by chemostat-like reactor mass balances:

$$\dot{u}_A = u_A\left(\mu_A(\overline{S}) - D\right) - \alpha u_A + A\rho d(\lambda)\lambda f_A(\lambda) \tag{16.17}$$

$$\dot{u}_N = u_N\left(\mu_N(\overline{S}) - D\right) - \alpha u_N + A\rho d(\lambda)\lambda f_N(\lambda) \tag{16.18}$$

$$\dot{u}_I = (fXI + \eta)\left(u_A b_A \cdot \frac{S_{O2}}{k_{A,O2} + S_{O2}} + u_N b_N \cdot \frac{S_{O2}}{k_{N.O2} + S_{O2}}\right)$$

$$-u_I D - \alpha u_I + A\rho d(\lambda)\lambda f_I(\lambda) \tag{16.19}$$

with initial conditions

$$u_A(t_0) = u_A^0, u_N(t_0) = u_N^0, u_I(t_0) = u_I^0 \tag{16.20}$$

The details of the growth rates $\mu_A(t)\left[d^{-1}\right]$ and $\mu_N(t)\left[d^{-1}\right]$ are provided in Table 16.1. They account for production of biomass due to substrate utilization, and loss of active biomass due to endogenous respiration and inactivation.$K_{A,O2}, K_{N,O2}\left[g/\mathrm{m}^3\right]$ are the half-saturation Monod coefficients for oxygen, $b_A, b_N\left[\mathrm{d}^{-1}\right]$ are the decay rates, $fXI[-]$ is the fraction of inert biomass produced by endogenous respiration, and $\eta[-]$ is the anoxic reduction factor.

The biomass exchange process between biofilm and suspension occurs in both directions at different rates. As in Masic and Eberl (2012), attachment is described by a linear function of the suspended biomass density with a constant attachment rate $\alpha\left[\mathrm{d}^{-1}\right]$. A standard choice for the detachment function $d(\lambda)\left[\mathrm{d}^{-1}\right]$ in Wanner–Gujer type biofilm models is

$$d(\lambda) = E\lambda. \tag{16.21}$$

with a constant detachment rate $E\left[\mathrm{m}^{-1}\mathrm{d}^{-}\right]$. The biofilm itself, i.e., biofilm thickness λ and volume fractions $f_{A,\ N,\ I}$, is modeled as in Section 16.2.2, i.e., by Eqs. (16.9)–(16.13).

16.3 Modeling challenges and perspectives

Biofilm reactor models, such as the example given here, are mathematically very complex due to spatial gradients that lead to the necessity of solving the nontrivial two-point boundary value problem. If the aim is for a greater level of realism, then an adequate description of physical phenomena such as shear forcing contributing to detachment and mass transport may be needed. However, even in the case of 1D models, where these physical processes are ignored or negated with assumptions of homogeneity, there remain open challenges and shortcomings that hinder their uptake and implementation more widely:

The conventional and routine use of 1D biofilm reactor models has not resulted in a wider standardization in their development, partly due to the reliance on implicit assumptions (e.g., the physical mechanics of detachment/attachment processes), rather than universal principles, narrowing their applicability.

Many potentially important biofilm processes are experimentally not well understood enough to be modeled mechanistically. Existing descriptions are place holders for future development. Where several competing model descriptions are established, it is important to understand and describe why one should be chosen over the other. Model limitations should be stated explicitly in every modeling publication. Many of the issues faced in modeling one are also faced in experimental studies and vice versa. We expand on these points here.

16.3.1 Mathematical and computational challenges

At the reactor scale, chemostat type models as presented in Section 16.2.1 are relatively well understood. These are ordinary differential equations with a particular structure that often can be exploited for the application of methods from dynamical systems theory. This is a rigorous qualitative phenomenological approach that does not aim at actually solving the model but to shed insight into the interplay of processes and long-term behavior. This analysis can be carried out completely for the competition model, but even simple modifications as mentioned earlier, e.g., accounting also for amensalism or allelopathy, can hamper a full analytical treatment. In principle, each such model needs to be investigated on its own merits. When simulations are required, the models themselves are relatively easy to implement in several open-access, platform-independent programming environments that provide prebuilt state-of-the-art solvers that can be reliably applied to simulate the model, such as Julia, python/scipy, and R/deSolve (Bezanson et al., 2017; Jones et al., 2001; Soetaert et al., 2010).

The 1D Wanner—Gujer biofilm model, described in Section 16.2.2, is a free boundary value problem for a quasilinear system of nonlocal balance laws, with a semilinear system of two-point boundary value problems embedded in the reaction terms. Mathematically, these models are very specialized and have only been studied analytically recently (D'Acunto et al., 2015; D'Acunto et al., 2019). Solving the 1D biofilm model requires either a specialized software package such as Aquasim (Wanner and Morgenroth, 2004) or involves the development of complex in-house code that requires rigorous and time-consuming verification. The predominant approach to solve the biofilm model is based on the classical method of characteristics, coupled with a two-point boundary value problem solver, such as collocation methods. In recent decades, there have been substantial developments toward efficient numerical methods for quasilinear first-order partial differential equations. To which extent and at what cost these can be applied to nonlocal hybrid problems, such as the one described here, seems not to have been investigated yet.

Combining chemostat-like models and 1D biofilm models into a biofilm reactor model naturally adds an additional level of complexity and can lead to computationally costly numerical simulations. Our own practical experience shows that a dominating factor for compute time is the solution of the boundary value problem for dissolved components. To what extent accuracy in the solution of this subproblem can be traded off against speed-up, i.e., the question of error propagation, is insufficiently understood. While in principle many numerical methods exist for the solution of this type of problem, nonlinearity of the problem can lead to convergence issues in existing methods, depending on model parameters. For example, our own experience shows that in the case of very small Monod half-saturation concentrations, their magnitude often switches between zeroth and first order on very short length scales, which poses difficulties for the accurate spatial resolution and introduces substantial errors or prevents some methods from converging.

Reactor scale biofilm models such as that presented in Section 16.2.3 are highly complex and can be expensive to solve in terms of compute time. Although generally reliable methods exist for each building block, very little is known about propagation of numerical errors and pitfalls that might come with choosing one algorithm for a subproblem over another. As such, and to ensure fidelity, it will be good practice to carry out and explicitly document thorough validation of the algorithms.

16.3.2 Implicit model assumptions and level of detail in process descriptions

Any mathematical model is based on assumptions, and the interpretation of its results, be they analytical or by simulation, cannot be extended beyond those. Many of these model assumptions are explicitly made by the modeler. For example, whether certain processes should be included or ignored, certain simplifications that are introduced, etc. Other assumptions are implicit and often enter a model when module or subprocess descriptions are adopted from other modeling studies without scrutinizing whether the assumptions of this earlier study are also valid for the task at hand. We briefly give three examples:

(1) An important example is the routine adoption of the Wanner–Gujer 1D biofilm model of Section 16.2.2. This is undoubtedly the most frequently used biofilm model and, among all biofilm models, the one used in most applications. Implicit in this model is the assumption that biofilms are dense and that horizontal stratification, parallel to the substratum, takes place both in bacterial constituents and in dissolved substrates. This is a valid assumption for many biofilm systems, in particular those arising in wastewater engineering where this model originated. But it might not be the best description for other biofilms, such as biofilm communities that are formed by sparse arrays of colonies each of which consists of a single species, e.g., certain biofilm communities in soils (Zaccaria et al., 2017). In the latter case, it might be more suitable to model the multispecies biofilm by several single-species biofilm models, which are connected indirectly via the substrate concentrations in the aqueous phase but do not include local competition for space (cf. Gaebler). In another study (Klapper and Szomolay, 2011), it was argued that, in the Wanner–Gujer model, the initial data, i.e., the initially assumed distribution of biomass on the substratum, can affect the distribution of biomass in the biofilm predicted by the model and that, in some cases, other mechanism might need to be included, which allow propagation of certain species against the growth direction of the biofilm. Also, the Wanner–Gujer model in its original form, due to the strong assumption that biomass fractions always add up to unity, does not allow the formation of cavities in a biofilm, which, for example, are observed in dispersal events. This can be remedied by introducing additional bacterial populations that are freely moving in the biofilm, also along the lines suggested in Klapper and Szomolay (2011), e.g., to describe dispersal and attachment phenomena

(D'Acunto et al., 2015; D'Acunto et al., 2018; D'Acunto et al., 2019). In other situations, it might be more appropriate to assume *a priori* a particular biofilm structure, e.g., by imposing horizontal or vertical layering for the component fractions f_i instead of the dynamic spatio-temporal mass balances (Eq. 16.9) (cf. the pseudoanalytical 1D model in Wanner et al., 2006). A possible point for confusion here is that, in some cases, it might not be understood which one-dimensional description would best describe a particular biofilm. It is, therefore, often difficult to estimate how much uncertainty is introduced by choosing a particular type of biofilm model.

(2) In 1D biofilm models, one commonly distinguishes between surface detachment and volume detachment; the former assumes biomass is removed at the biofilm/water interface only, while the latter considers that bacteria are removed across the entire depth of the biofilm. The question of which detachment behavior is more realistic depends on the nature of the biofilm. For a homogeneous dense biofilm layer, removal of biomass from the surface appears more natural, whereas for a biofilm consisting of arrays of individual colonies, detachment of single colonies might appear a more reasonable assumption. This indicates that biofilm detachment might be a too complicated phenomenon for a detailed, physics-based one-dimensional description. This is exacerbated if the biofilm model is to be embedded in a reactor scale model. There is broad consensus that detachment from biofilms depends heavily on local physical conditions, both structural properties of the biofilm itself and local hydrodynamic properties, such as flow velocity and shear forces exerted onto the biofilm, which in turn depend on biofilm morphology and the shape of the colonies (Horn and Lackner, 2014). These cannot be incorporated in a straightforward manner in the highly simplified, coarse description of reactor scale models. There have been several modeling efforts to understand biofilm mechanics, in particular deformation and detachment, on the biofilm scale using a spatially refined 2D or 3D description (Picioreanu et al., 2018; Sudarsan et al., 2016; Xavier et al., 2005), often at very short time scales. Not much is known how this information can be included in reactor scale models in a manner that does not increase computational requirements within reasonable computational resource and data requirement limits. Therefore, whichever choice of detachment description one includes in a reactor scale model will depend to a large extent on personal preferences and may include a trade-off with complexity. Obviously, the choice of detachment function used will affect the outcome of the model simulation, in particular the predicted biofilm thickness and the total amount of biomass in the system. However, studies also have shown that the choice of detachment model and parameters might have only minor effect on biofilm performance, in particular if the biofilm is thicker than the active layer (Masic and Eberl,

2012, 2014). This suggests that, in such a complex, hybrid model, not all output parameters need to have the same level of fidelity and that it will be up to the modeler to ensure that the model components chosen are those that are most appropriate for the purpose of the study.

(3) In reactor scale models of biofilm processes, such as the one presented in Section 16.2.3, substrate fluxes between the biofilm and the aqueous phase are important as they connect the biofilm scale with the reactor scale. Computationally, they are very expensive, which is why sometimes attempts are made to replace the detailed solution of the diffusion–reaction system (1.12) by simplified models or empirical relationships based on experimental data (Laureni et al., 2019; Rauch et al., 1999). However, as convenient as they may sometime seem, such experimental results only hold for the system for which they were obtained and under the conditions in which the experiment was conducted. Therefore, adopting such empirical formulae imposes the requirement of an implicit assumption that the same conditions are found in the modeled system as in those experiments. In complex reactor scale models, it is often difficult to estimate how large the errors are that are introduced by these choices.

What these three examples have in common is that in formulating a model for a complex multiscale, multiphysics process such as biofilm reactors, one has to make choices as to which processes to include and which to ignore, and at which level of details they are accounted for. This is a trade-off decision that is influenced by the purpose of the modeling study and the level of complexity one is prepared to handle. Adopting existing building blocks might lead to introducing model assumptions that might strictly not be valid for the system at hand. All interpretation of simulation results must account for this. A good practice in modeling studies is to explicitly state and argue the assumptions that go into model formulation, also for those parts of the model that are adopted from earlier studies.

16.3.3 Some parallels and connections between modeling and experimental studies

Here, we present some concepts relevant to biofilm reactor modeling that have synergies with experimental practice and the conceptual and practical challenges faced therein.

(1) Biofilm and biofilm reactor models are notoriously difficult to parameterize. This is an inverse problem that exacerbates the highly involved mathematical structure of the model. Such inverse problems are ill-posed even for much simpler models and are a significant numerical challenge. In part, these difficulties are also caused by experimental limitations to provide both spatially and temporally detailed enough data set, which would be required to identify the typically large number of parameters with high fidelity. Therefore, in simulation studies, some parameters are adapted from other studies in the published literature, and others need to be assumed. These degrees of freedom, which are sometimes perceived as arbitrary, mirror the degrees of freedom an experimenter has in setting up their experiments, including choice of model organisms and substrates, operating conditions for reactors, etc. To the same extent, as the results of one series of experiments only give quantitative information for exactly that system under the set conditions, the quantitative information obtained from simulations is only valid for the simulation scenario considered. In simulation studies, in contrast to experimental work for which this would be prohibitively expensive, one often can assess the propagation of uncertainties in model parameters in a statistical sense, which, however, is rarely done practically.

(2) In experimental research, the potential challenges and limitations to transfer quantitative and qualitative insight from *in vitro* studies to *in situ* or *in vivo* situations, or *vice versa*, are well understood and accepted. In modeling and simulation studies of biofilm processes, this is mirrored by the question of micro versus meso versus macro scale, or cell versus colony versus reactor models, i.e., the multiscale nature of biofilm processes. As such, the question "Which type of model is most appropriate?" relates to the problem at hand, the insight one wishes to gain, and the practical constraints, for example, in terms of computing resources and ability to parameterize. In vitro experiments might not allow one to control and monitor all aspects of the biofilm system and processes. Similarly, in reactor scale simulation studies, certain subprocess cannot be controlled in detail, for example, instead of detailed models of biofilm detachment one has to introduce a coarse description of the process.

(3) Replication of experiments is a cornerstone of empirical science, providing a platform for rigor and reliability. However, in recent years, there has been growing attention to a so-called "reproducibility crisis," suggested as emerging in tandem with the growing quantity and complexity of scientific output that is not being matched in its reporting (Baer and Gilmore, 2019). However, this is by no means a generally held opinion, with others suggesting that modern science is entering a new epoch of opportunity and challenge, where science will develop new ways to ensure that methods and results are reproducible and reliable (Fanelli, 2018). The 3 Rs that define the gold standard for experimental research are defined as in Plesser (2018): (1) Repeatability: the same workgroup, using the same experimental setup, is able to obtain the same results; (2) Replicability: a different team taking the same approach is able to confirm the results; (3) Reproducibility: the results can be confirmed using a different approach and a different team. The same discussion on reproducibility of modeling studies has recently arisen in mathematical biology (cf. Laubenbacher and Hastings, 2017) and the articles it references. With respect to biofilm reactor modeling, and mathematical biology in general, reporting in the scientific literature often falls short of ensuring adherence to the 3 Rs. This may be rooted in the fact that such work often finds its home in publishing literature that requires mathematical rigor or scientific transparency, but sometimes not both. One factor here may be that, in many instances, reviewers and editors recommend to omit details of a model, its computational implementation, and validation of the computational tools, because they are assumed to take away emphasis from the scientific message. In other journals, it has become practice to relegate model formulation into supplementary material, even for studies in which the models are a major aspect. Often, modeling work that is derived from earlier research fails to acknowledge the differences in computational tools, versioning or details of numerical solvers used. Even when the mathematics is described precisely, this can lead to a significant disparity between the published work and attempts to reproduce the results. Also, to ensure reproducibility, as biofilm reactor modeling continues to develop and be used more broadly, it is imperative that the multitude of modeling approaches not only are able to exist in parallel, but it is required that several approaches are tried to see whether they lead to comparable results. Biofilm modeling itself offers a very successful example of this: In the early 2000s, several mathematical model concepts were pursued to understand the formation of heterogeneous biofilms structures at the mesoscale, ranging from deterministic partial differential equation models based on either a continuum mechanistic description of biofilms or on a description of biofilms as spatially structured microbial populations, to stochastic individual-based and cellular automation–like models. Although the mathematical concepts utilized were very different, these models were found on very similar assumptions and led to very similar conclusions (cf. Wanner et al., 2006), i.e., they achieved reproducibility. This gives credence to all these models, and supports that the structural heterogeneities predicted by one model are inherent in the system and not an artifact introduced by the choice of model.

16.4 Conclusion

Mathematical modeling has a long tradition in microbiological and microbial bioprocess research. This can take, for example, the form of simple growth curve and inactivation models, e.g., in food microbiology, chemostat-like models for continuous culture processes, or spatiotemporal models for biofilms, or complex regulatory network models. In particular in biofilm research, mesoscopic modeling, i.e., on the actual biofilm length scale, has a tradition that spans over four decades. Biofilm reactor models that focus on macroscale parameters, such as mesoscale processes, have typically been simplified. However, with better computational tools (hardware and software), more detailed multiscale approaches are within reach, and existing macroscopic and mesoscopic descriptions can be combined. How modeling errors that are introduced by certain assumptions and simplification propagate across these length scales is not yet sufficiently understood, i.e., the limitations of such approaches still need to be explored. Importantly, it is necessary to understand that one cannot expect that a single model exists that serves all purposes best, but that each model formulation needs to be tailored to the purpose of a particular study. Adopting existing building blocks requires a reexamination of the underlying assumptions. All models, like all experimental techniques, have their limitations. Knowing these limitations is paramount to benefiting from models. In science, mathematical models are formulations of theoretical concepts. Hence, understanding model limitations helps in identifying where knowledge gaps still exist. Or, returning to Galileo's metaphor of mathematics being the scientific language: If processes and their interplay cannot be formulated as models they are not yet sufficiently understood.

Acknowledgments

M.J.W. acknowledges the support from the European Union's Horizon 2020 research and innovation program under the Marie Skłodowska-Curie Grant Agreement No. 702408 (DRAMATIC). H.J.E. acknowledges financial support received from the Natural Sciences and Engineering Research Council of Canada under the Discovery Grant program.
The discussions and failed attempts at replicability that eventually lead to this essay began during the Thematic Semester on Emerging Challenges in Mathematical Biology at the Fields Institute for Research in the Mathematical Sciences in Toronto, ON, that we both attended as long-term visitors.

References

Baer, D.R., Gilmore, I.S., 2019. Responding to the growing issue of research reproducibility. Journal of Vacuum Science and Technology A 36 (6), 068502.

Baeten, J.E., van Loosdrecht, M.C.M., Volcke, E.I.P., 2017. Improving the accuracy of granular sludge and biofilm reactor simulations in Aquasim through artificial diffusion. Biotechnology and Bioengineering 114 (9), 2131–2136.

Bar-On, Y.M., Milo, R., 2019. Towards a quantitative view of the global ubiquity of biofilms. Nature Reviews Microbiology 17, 199–200.

Bezanson, J., Edelman, A., Karpinski, S., Shah, V.B., 2017. Julia: a fresh approach to numerical computing. SIAM Review 59 (1), 65–98.

Bryers, J.D., Drummond, F., 1998. Local macromolecule diffusion coefficients in structurally non-uniform bacterial biofilms using fluorescence recovery after photobleaching (FRAP). Biotechnology and Bioengineering 60 (4), 462–473.

Butler, K.T., Davies, D.W., Cartwright, H., Isayev, O., Walsh, A., 2018. Machine learning for molecular and materials science. Nature 559, 547–555.

Capdeville, B., Rols, J.L., 1992. Introduction to biofilms in water and wastewater treatment. In: Melo, L.F., Bott, T.R., Fletcher, M., Capdeville, B. (Eds.), Biofilms — Science and Technology, vol. 223. Springer, Dordrecht, pp. 13–20.

Costerton, J.W., Geesey, G.G., Cheng, K.-J., 1978. How bacteria stick. Scientific American 238, 86–95.

D'Acunto, B., Frunzo, L., Klapper, I., Mattei, M.R., 2015. Modeling multispecies biofilms including new bacterial species invasion. Mathematical Biosciences 259, 20–26.

D'Acunto, B., Frunzo, L., Luongo, V., Mattei, M.R., 2018. Invasion moving boundary problem for a biofilm reactor model. European Journal of Applied Mathematics 29 (6), 1079–1109.

D'Acunto, B., Frunzo, L., Luongo, V., Mattei, M.R., 2019. Free boundary approach for the attachment in the initial phase of multispecies biofilm growth. Zeitschrift für Angewandte Mathematik und Physik 70 (3), 1–16.

Davey, M.E., O'Toole, G.A., 2000. Microbial biofilms: from ecology to molecular genetics. Microbiology and Molecular Biology Reviews 64 (4), 847–867.

de Carvalho, C.C.C.R., 2018. Marine biofilms: a successful microbial strategy with economic implications. Frontiers in Marine Science 5, 126.

de Jong, R., Casagranda, S., Giordano, N., Cinquemani, E., Ropers, D., Geiselmann, J., Gouzé, J.-L., 2018. Editorial for the topical collection reproducibility in computational biology research. Bulletin of Mathematical Biology 80 (12), 3069–3070.

Dutta, A., Sarkar, S., 2015. Sequencing batch reactor for wastewater treatment: recent advances. Current Pollution Reports 1 (3), 177–190.

Dzianach, P.A., Dykes, G.A., Strachan, N.J.C., Forbes, K.J., Pérez-Reche, F.J., 2019. Challenges of biofilm control and utilization: lessons from mathematical modelling. Journal of The Royal Society Interface 16 (155), 20190042.

Elias, S., Banin, E., 2012. Multi-species biofilms: living with friendly neighbors. FEMS Microbiology Reviews 36 (5), 990–1004.

Fanelli, D., 2018. Opinion: is science really facing a reproducibility crisis, and do we need it to? Proceedings of the National Academy of Sciences of the United States of America 115 (11), 2628–2631.

Fgaier, H., Eberl, H.J., 2010. A competition model between pseudomonas fluorescens and pathogens via iron chelation. Journal of Theoretical Biology 263 (4), 566–578.

Fgaier, H., Kalmokoff, M., Ellis, T., Eberl, H.J., 2014. An allelopathy based model for the listeria overgrowth phenomenon. Mathematical Biosciences 247 (13–26).

Flemming, H.-C., Wuertz, S., 2019. Bacteria and archaea on earth and their abundance in biofilms. Nature Reviews Microbiology 14, 563–575.

Gaebler, H.J., Eberl, H.J., 2018. A simple model of biofilm growth in a porous medium that accounts for detachment and attachment of suspended biomass and their contribution to substrate degradation. European Journal of Applied Mathematics 29 (6), 1110–1140.

Gaebler, H.J., in. prep. Modeling Uranium Bioreduction by Multi Species Biofilms (Tentative). (Ph.D. thesis), University of Guelph.

Horn, H., Lackner, S., 2014. Modeling of biofilm systems: a review. Advances in Biochemical Engineering 146, 53–76.

Jones, E., Oliphant, T., Peterson, P., et al., 2001. SciPy: Open Source Scientific Tools for Python.

Kabanikhin, S.I., 2008. Definitions and examples of inverse and ill-posed problems. Journal of Inverse and Ill-Posed Problems 16, 317–357.

Khassehkhan, H., Eberl, H.J., 2016. A computational study of amensalistic control of listeria monocytogenes by lactococcus lactis under nutrient rich conditions. Foods 5 (3).

Klapper, I., Szomolay, B., 2011. An exclusion principle and the importance of mobility for a class of biofilm models. Bulletin of Mathematical Biology 73 (9), 2213–2230.

Laubenbacher, R., Hastings, A., 2017. Mathematical modelling of microbes: metabolism, gene expression and growth. Journal of The Royal Society Interface 14 (136), 20170502.

Laureni, M., Weissbrodt, D.G., Villez, K., Robin, O., de Jonge, N., Rosenthal, A., Wells, G., Lund Nielsen, J., Morgenroth, E., Joss, A., 2019. Biomass segregation between biofilm and flocs improves the control of nitrite-oxidizing bacteria in mainstream partial nitritation and anammox processes. Water Research 154, 104–116.

Lewandowski, Z., Boltz, J., 2011. Biofilms in water and wastewater treatment. In: Wilderer, P. (Ed.), Treatise on Water Science, vol. 4. Academic Press, Oxford, pp. 529–570.

Masic, A., Eberl, H.J., 2012. Persistence in a single species CSTR model with suspended flocs and wall attached biofilms. Bulletin of Mathematical Biology 74 (5), 1001–1026.

Masic, A., Eberl, H.J., 2014. A modeling and simulation study of the role of suspended microbial populations in nitrification in a biofilm reactor. Bulletin of Mathematical Biology 76 (1), 27–58.

Nadell, C.D., Xavier, J.B., Foster, K.R., 2009. The sociobiology of biofilms. FEMS Microbiology Reviews 33, 206–224.

Oyebamiji, O.K., Wilkinson, D.J., Li, B., Jayathilake, P.G., Zuliani, P., Curtis, T.P., 2019. Bayesian emulation and calibration of an individual-based model of microbial communities. Journal of Computer Science 30, 194–208.

Percival, S.L., Suleman, L., Vuotto, C., Donelli, G., 2015. Healthcare-associated infections, medical devices and biofilms: risk, tolerance and control. Journal of Medical Microbiology 64, 323–334.

Picioreanu, C., Blauert, F., Horn, H., Wagner, M., 2018. Determination of mechanical properties of biofilms by modelling the deformation measured using optical coherence tomography. Water Research 145, 588–598.

Pinna, D., 2014. Biofilms and lichens on stone monuments: do they damage or protect? Frontiers in Microbiology 5, 133.

Plesser, H.E., 2018. Reproducibility vs. replicability: a brief history of a confused terminology. Frontiers in Neuroinformatics 11, 76.

Rauch, W., Vanhooren, H., Vanrolleghem, P.A., 1999. A simplified mixed-culture biofilm model. Water Research 33 (9), 2148–2162.

Shi, S., Xu, G., 2018. Novel performance prediction model of a biofilm system treating domestic wastewater based on stacked denoising auto-encoders deep learning network. Chemical Engineering Journal 347, 280–290.

Soetaert, K., Petzoldt, T., Setzer, R.W., 2010. Solving differential equations in R: package deSolve. Journal of Statistical Software 33 (9), 1–25.

Sudarsan, R., Ghosh, S., Stockie, J.M., Eberl, H.J., 2016. Simulating biofilm deformation and detachment with the immersed boundary method. Communications in Computational Physics 19 (3), 682–732.

Szomolay, B., Cogan, N.G., 2015. Modelling mechanical and chemical treatment of biofilms with two phenotypic resistance mechanisms. Environmental Microbiology 17 (6), 1870–1883.

Tartakovsky, B., Guiot, S.R., 2004. Biofilm modelling. In: Nedovic, V., Willaert, R. (Eds.), Fundamentals of Cell Immobilisation Biotechnology, Focus on Biotechnology, vol. 8. Springer, Dordrecht, pp. 531–545.

Wanner, O., Gujer, W., 1986. A multispecies biofilm model. Biotechnology and Bioengineering 28 (3), 314–328.

Wanner, O., Morgenroth, E., 2004. Biofilm modeling with AQUASIM. Water Science and Technology 49 (11–12), 137–144.

Wanner, O., Eberl, H., Morgenroth, E., Noguera, D., Picioreanu, C., Rittmann, B., van Loosdrecht, M., 2006. Mathematical Modeling of Biofilms. IWA Publishing.

Xavier, J.D., Picioreanu, C., van Loosdrecht, M.C.M., 2005. A general description of detachment for multidimensional modelling of biofilms. Biotechnology and Bioengineering 91 (6), 651–669.

Zaccaria, M., Dedrick, S., Momeni, B., 2017. Modeling microbial communities: a call for collaboration between experimentalists and theorists. Processes 5 (4), 53.

Zobell, C.E., 1943. The effect of solid surfaces on bacterial activity. Journal of Bacteriology 46 (39–56).

Index

'*Note*: Page numbers followed by "t" indicate tables and "f" indicate figures.'

A

Abbott-Firestone curve, 243–244
Abrasion, 9, 111
Acinetobacter baumannii, 309
Acyl homoserine lactones (AHLs), 6–7, 280–281, 332–334
Adhesins, 3–6
AHLs. *See* Acyl homoserine lactones (AHLs)
AIPs. *See* Autoinducing peptides (AIPs)
Alkaloids, 276
Antibiofilm enzymes
  amylase and protease, 259
  biofilm disruption
    biofilm removal, 258
    enzymatic formulations, 257
    extracellular polymeric substances (EPS) break down, 258–259
    extracellular polymeric substances (EPS) structural components, 257–258
  cell metabolism, 256
  extracellular polymeric substances (EPS) matrix, 257
  food industry, 262
  hydrolytic enzymes, 256, 261
  industrial applications, 256–257
  isomerization, 256
  lactonase, 260
  oxidoreductases, 256, 260–261
  savinase, 259–260
  temperature, 256
  and ultrasonic waves, 261–262
Antibiotic resistance genes (ARGs), 10–12
Antimicrobial activity
  antineoplastic drugs
    floxuridine, 309
    5-fluorouracil, 309
    fluoxetine with fluconazole, 310
    gallium compounds, 309
  antipsychotic and antidepressant drugs
    fluoxetine with fluconazole, 310
    Galleria mellonella infection model, 310
    phenothiazines, 310–311
    phenothiazines and tricyclic antidepressants, 310
    sertraline, 310
  antipyretic and antiinflammatory drugs
    celecoxib, 311–312
    diclofenac, 312
    ibuprofen, 312–313
  auranofin, 313
  drug development, 307–308
  ebselen, 314
  investigational drugs, 307–308
  ivacaftor, 313
  minocycline, 313–314
  multidrug-resistant (MDR) pathogens, 307
  statins, 311
  systematic approaches, 308
  terfenadine, 313
Antimicrobial photodynamic therapy (APDT), 255–256, 298–299, 299f
  antimicrobial resistance, 301–302
  chitosan films (CFs), 299–300
  chitosan nanoparticles (CSNPs), 301
  Escherichia coli sessile cells eradication, 301
  food industry, 298
  gallic acid (GA) and UV-A light treatment, 301–302
  light-activated molecules, 302
  Listeria innocua biofilms, 299–300
  Listeria monocytogenes biofilms, 299
  mesoporous silica nanoparticles, 300–301
  Na-chlorophyllin, 302
  nano-sized vehicles, 300
  Staphylococcus aureus biofilm, 301, 302f
  Staphylococcus aureus biofilms, 300
  synthetic porphyrin, 300
APDT. *See* Antimicrobial photodynamic therapy (APDT)
Aspirin, 318–319
Atorvastatin, 311
Auranofin, 313

Autoflocculation, 134
Autoinducers (AIs), 6–7
Autoinducing peptides (AIPs), 6–7, 332–334, 334f
Autotrophic organism, 127

B

Bacterial drinking water biofilm formation, 110
Bacterial-fungal interactions (BFIs), 118–119, 119f
Bacteriocins, 219–222
Bacteriophages, 254–255
Batch reactors, 71–72
Biocide dosing strategy, 79–80
Biofilm Annular Reactor, 73
Biofilm control
  antibiofilm approaches, 254
  antibiofilm enzymes. *See* Antibiofilm enzymes
  antimicrobial photodynamic therapy (APDT), 255–256
  bacteriophages, 254–255
  biosurfactants, 255
  chemical products, 254
  nanoparticles, 255
  problems associated, 249–251
  quorum sensing (QS), 254
Biofilm, definition, 1
Biofilm formation and resistance, 249, 331f
  active adhesion, 3–6
  antimicrobials
    biofilm cell heterogeneity, 13–14
    ESKAPE pathogens, 9–10
    horizontal gene transfer, 10–12
    innate resistance, 12–13
    multidrug resistance (MDR), 9–10
    Mycobacterium tuberculosis resistance, 12–13
    persister cells, 13–14
    toxin-antitoxin (TA) system, 13–14
  bacterial motility, 331–332
  biofilm maturation, 330–331
  diseases associated, 250
  dispersion, 331
  early biofilm formation, 330
  extracellular polymeric substance (EPS) matrix, 331
  flagellar mechanosensing, 3–6
  genes related, 4t–5t, 11t–12t
  growth modes
    adhesins, 3–6
    Bacillus subtilis flagellum, 3–6
    bacterial membrane vesicles (MVs), 3–6
    biofilm structure, 8–9
    environmental conditions, 1–2
    environmental stress/nutrient deficit, 9
    enzymes, 2
    extracellular polymeric substances (EPS), 1–2
    flagellar gene expression, 3–6
    irreversible attachment, 6
    Listeria monocytogenes serotype 1/2a, 8f
    multicellular, 1
    nonenzymatic proteins, 2
    passive adhesion, 6
    quorum sensing (QS), 6–7
    self-produced matrix, 1–2
    signaling molecules, 6–7
    unicellular, 1
  initial attachment, 330
Biofilm formation and structure
  against antimicrobial therapy, 26f
  biofilm cells, 27
  cell-to-cell signaling, 27
  cooperative interactions, 27
  environmental parameters, 37–38
  extracellular polymeric substances (EPS)
    biofilm architecture, 28
    characterization, 29
    constituents and function, 30t–31t
    contact surface and environmental properties, 28
    3D patterns, 29
    heterogeneous structure, 29
    matrix functions, 28
    microcolonies, 29
    sample extraction methods, 28–29
    in silico docking analysis, 29
  hostile environmental conditions, 27
  imaging techniques
    confocal laser scanning microscopy (CLSM), 38–39
    confocal Raman microscopy, 38–39
    electron microscopy, 38
    fluorescence in situ hybridization and microautoradiography, 38
    fluorescent microscopy, 38
  irreversible attachment, 25–27
  living/nonliving substance, 25–27
  magnetic resonance imaging (MRI), 32, 41–42

persister cells, 27
shear stress, 37–38
three-dimensional (3D) communities, 25–27
Biofilm reactor
aerobic research, 71–72
anaerobic systems, 71–72
batch reactors, 71–72
Biofilm Annular Reactor, 73
biomass retention and separation, 360
configurations and operating modes, 360
Constant Depth Film Fermenter, 73
continuously stirred tank reactors (CSTRs), 72
continuous systems, 360
design process, 74, 74f
attributes, 76
fluid dynamics, 76
gap analysis, 74–75
growth surface, 76
injection molding, 77
material selection, 76–77, 77t
part fabrication, 77
plastics and metal, 77t–78t
stakeholder, 75
time course studies, 76
experimental studies
challenges and limitations, 378
reproducibility crisis, 378–379
simulation studies, 377–378
hard nonporous surfaces, 71
industrial surfaces biofilm reactor (ISBR). *See* industrial surfaces biofilm reactor (ISBR)
mathematical modeling
assumptions, 375–377
biomass exchange process, 373
biomass types, 372
challenges, 373–375
chemostat modeling, 364–368
complex reactor scale models, 377
dissolved substrate's diffusion rates, 362
1D Wanner-Gujer biofilm model, 374–376
and experimental studies, 377–379
homogeneous dense biofilm layer, 376–377
machine learning algorithms, 361–362
model complexity, 362
model formulation, 361
multispecies multisubstrate model, 371
one-dimensional biofilm models, 368–371
reaction rates, 372
surface and volume detachment, 376–377
suspended biomass, 372
open systems, 72
plug flow reactor (PFR), 72
porous surfaces, 71
residence time, 72
standardized methods, 73
transformation and transport process, 362–363
Biofilms, 1, 109
bacteria, 329
bacterial resistance, 251–253
biochemical reactions and reactivity, 24
definition, 249
drinking water distribution systems (DWDS), 251
drug tolerance, 329
engineered systems, 360
extracellular polymeric substances (EPS) matrix, 24
biofilm-associated antimicrobial resistance, 252
biosynthesis, 252
persister cells, 253
protective barrier, 251–252
reaction-diffusion limitation capacity, 251–252
in health, 329–330
holistic analytical approach, 25
human microbial infections, 23
industrial settings, 330
mesoscopic modeling, 379–380
multifunctional communities, 359–360
phenotypic traits, 23–24
prevalence, 359–360
sessile biofilms, 359–360
spatial and metabolic interactions, 250
tissue infections, 329–330
Bioflocculation, 134–135
Biofouling, 160
BioSurface Technologies Corporation (BST), 74, 74f
Biosurfactants, 224, 255
Bis-(3′-5′)-cyclic dimeric guanosine monophosphate (c-di-GMP), 3–6

C

CADD. *See* Computer-aided drug design (CADD)
Carnosic acid, 277–278
Carotenoids, 276
Carvacrol, 277
Catechins, 277
Celecoxib, 311–312
Cell-to-cell communication mechanism, 280–281
Centrifugation, 136
Chemical flocculation, 134
Chemostat modeling
  allelopathic system, 367
  bacterial biomass concentration, 364
  break-even concentration, 366
  chemostat, 364
  classical mathematical biology, 364–365
  diffusive mass transfer limitations, 368
  environmental engineering, 365–366
  long-term behavior, 366
  minimum concentration, 365
  Monod functions, 365–366
  Monod kinetics, 364–365
  Pseudomonas fluorescens, 367
  single-species model, 366–367
  steady state, 364–365
Chitosan nanoparticles (CSNPs), 301
Chloramines and chlorine dioxide, 116
Chlorination, 115
Clomipramine, 310
Colistin, 309
Computational fluid dynamics (CFD), 234–236, 236f
Computer-aided drug design (CADD)
  and high-throughput screening (HTS), 336
  ligand-based (LB) methods, 336
  Protein Data Bank (PDB), 336
  structure-based (SB) methods, 336
  uses of, 335–336
  workflow, 336
Confocal laser scanning microscopy (CLSM), 38–39
Consensus scoring, 339–340
Constant Depth Film Fermenter, 73
Continuously stirred tank reactors (CSTRs), 72
Correlation spectroscopy (COSY), 34–35
Cross-polarization- magic angle spinning (CP-MAS), 36
Crude bacteriocin fermentate (CBF), 220
Cryoprobes, 34
CSTRs. *See* Continuously stirred tank reactors (CSTRs)

D

Dental caries, 92–93
Dental plaque
  coaggregation phenomenon, 90–92
  colonization, 90–92, 91f
  pellicle formation, 90–92
Derjaguin-Landau-Verwey-Overbeek (DLVO) theory, 242–243
Desalination, 157
Desorption, 9
Detachment, 9
2,2-Dibromo-3-nitrilopropionamide (DBNPA), 164
Diclofenac, 312
Diffusion-ordered (DOSY) NMR spectroscopy, 45–47
Diffusion-reaction equation, 362–363
4,5-Dihydroxy-2,3-pentadione (DPD), 334–335
Direct electron transfer (DET), 180
Dispersin B (DspB), 9
Double-layer capacitance, 182–183
Drinking water (DW) concerns
  chemical contamination, 102
  microbial contamination, 101–102
  regulations, 101
  safe drinking water, 101
Drinking water distribution systems (DWDS), 251
  bacterial-fungal interactions (BFIs), 118–119, 119f
  bench-top laboratory devices, 103–104
  biofilms
    adhesion phase, 110
    bacterial cultures, 109–110
    bacterial DW biofilm formation, 110
    cell attachment, 109
    detachment, 111
    maturation phase, 111
    microbial biofilms, 110–111
    pipe surface preconditioning, 110
    surfaces, 109
  disinfection
    chloramines and chlorine dioxide, 116
    chlorine, 115–117
    disinfectants, 115
    ozone, 116–117
    UV radiation, 117
  environments, 102

factors affecting, 103
filamentous fungi (ff)
bioactive actions, 105
biofilm formation, 111–114
fungal contamination, 104
growth problems, 104
organoleptic changes, 104
patulin, 105
recovery, 104
tastes and odors changes, 105
maintenance, 114–118
microbial heterogeneity, 103f
pipe walls, 103
total biomass, 103
Drinking water systems, 359–360
Drug repurposing
anthelmintic drug, 319
antibiofilm effects, 315t–317t
antidepressant and antipsychotic drugs, 318
antineoplastic drugs
finasteride, 318
5-fluorouracil, 314
gallium nitrate, 314
uracil analogs, 314
antipyretic and antiinflammatory drugs, 318–319
aspirin, 318–319
auranofin, 319
ebselen, 319
ibuprofen and diclofenac, 319
terfenadine, 319
gama aminobutyric acid (GABA) receptor agonists, 320
statins, 318
valproic acid, 320
1D Wanner-Gujer biofilm model, 374
DWDS. *See* Drinking water distribution systems (DWDS)

E

Ebselen, 314, 319
Electrocoagulation-flocculation (ECF), 135
Electromotive force (EMF), 182
Enterococcus casseliflavus, 208–209
Equipment design
air-liquid-material (ALM) interfaces
Bacillus cereus biofilms, 239, 240f
bacterial contamination, 238–239
bacterial properties, 238–239
environmental parameters, 238–239
ringlike structure, 238–239
storage tanks and piping systems, 239
food contact surfaces, 233
immersed surfaces
adhesion process, 236–237
cleaning in place (CIP) installations, 237–238
hydrodynamic conditions, 236–237
particle image velocimetry (PIV) measurements, 237
poor design features, 237
semiclosed equipment, 237
physicochemistry, 242–243
simple equipment geometries
boundary layer detachment, 234–236
cleanability experiments, 234–236
computational fluid dynamics (CFD), 234–236
electrochemical measurements, 234–236
flow disturbance, 234–236
hydrodynamics, 233–234
mean wall shear stress, 234–236
misalignment, 234–236
particles size and density, 234
soiling experiments, 234–236
turbulence generation and transport, 234
splash areas, 239–242
substratum topography, 243–244
surface disinfections, 233
Equivalent circuit, 192f
Erosion, 9, 111
Exogenous mediators, 180
Extracellular DNA (eDNA), 2
Extracellular enzymes, 30t–31t
Extracellular polymeric substances (EPS), 1–2, 24
insoluble and complex nature
cross-polarization- magic angle spinning (CP-MAS), 36–37
magic angle spinning (MAS-NMR) spectroscopy, 36–37
solid-state NMR (SS-NMR), 36–37
lactic acid bacteria (LAB), antibiofilm activity, 222–223
microalgae and cyanobacteria, 139
oral biofilms, 89
solubilized
concentration step, 33
conventional 2D techniques, 34–35
cryoprobes, 34

Extracellular polymeric substances (EPS) (*Continued*)
extraction yield, 33
mass spectrometry (MS), 33–34
microcoil probes, 34
nuclear magnetic resonance (NMR) spectroscopy, 33–34
Extracellular polymers, 109
Extracellular polysaccharides, 30t–31t

F

Faradaic charge transfer process
Butler-Volmer-type equation, 181
oxidation reactions, 181–182
Feed channel pressure (FCP) drop, 160, 161f, 163f
Fick's law, 193
Filamentous fungi (ff), drinking water
bioactive actions, 105
biofilm formation
active attachment, 113
dispersal/planktonic phase, 113–114
dispersive forms, 111–112
fungal aerial morphogenesis, 111–112
hyphal bundles, 113
maturation phase, 113
microcolony formation, 113
pellet formation, 114
phenotypic changes, 112
preliminary model, 112, 113f
propagule adsorption, 112
structural features, 112
surface-associated growth, 111–112
fungal contamination, 104
growth problems, 104
organoleptic changes, 104
patulin, 105
recovery, 104
tastes and odors changes, 105
Finasteride, 318
Flagellar mechanosensing, 3–6
Flat configuration microbial fuel cell, 186, 188f
Flavonoids, 275–276, 278
Flotation, 135–136
Floxuridine, 309
5-Fluorouracil, 309, 314
Foodborne pathogenic bacteria, 205
Free-induction decay (FID) signal, 42
Fusobacterium nucleatum, 90–92

G

Galangin, 277
Galleria mellonella infection model, 310
Gallium nitrate, 314
Gama aminobutyric acid (GABA) receptor agonists, 320
Glatiramer acetate, 319
Glucose oxidase, 258
Gravitational sedimentation, 135
Grazing, 9

H

Heteronuclear single-quantum correlation (HSQC), 34–35
High-throughput screening (HTS), 336
Human microbial infections, 23
Hydraulic pressure-driven membrane filtration process, 158f
Hydrolases, 256
Hydrophobic interactions, 242–243
Hydrophobins, 111–112
Hydroxybenzoic acids, 276
Hydroxycinnamic acids, 276

I

Ibuprofen, 312–313
Immobilization techniques, 131, 132t–133t
Industrial surfaces biofilm reactor (ISBR)
commercialization, 84–85
cooling towers
aerosols, 79–80
biocide dosing strategy, 79–80
challenge, 79–80
design, 79–80, 79f
design attributes, 80–81
packing material, 79–80, 80f
final design testing
coupon-to-coupon variability, 84, 85t
drinking water microbes, 84
laboratory protocol, 84
log densities, 86f
repeatability standard deviation, 84, 85t
run-to-run variability, 84, 85t
laboratory validation and design optimization, 83
prototypes, 81–83, 81f–82f
Intradiffusion coefficients, 41–42
ISBR. *See* Industrial surfaces biofilm reactor (ISBR)
Isomerization, 256

Isothiocyanate, 276
Ivacaftor, 313

J

Jablonski scheme, 296f

L

Lactic acid bacteria (LAB), antibiofilm activity, 207
 biosurfactants, 224
 cell-free culture supernatants
  antibacterial and antiadherence activities, 213–214
  antibiofilm activities, 215–216, 218–219
  antibiofilm efficiency, 217–218
  antilisterial strains, 214
  artisanal milk cheese isolates, 215
  bacteriocin-containing crude extract, 215
  biofilm removal potential, 216–217
  fecal samples, 216
  fermented milk product strains, 215
  Lactobacillus monocytogenes strains, 212–213
  Lactobacillus sakei strain, 218
  milk tanks and milking machine surfaces, 213
  traditional Tunisian fermented food isolates, 214
 exopolysaccharides (EPS), 222–223
 genera, 205–206
 mechanisms, 205, 206f
 metabolites, 207
 protective role, 205–206
 purified lactic acid bacteriocins, 219–222
 quorum sensing (QS), 205–206, 224–225
 in situ lactic acid bacteria presence
  bacteriocinogenic strain, 208–209
  competitive-exclusion isolates, 212
  Enterococcus species, 208–209
  Lactobacillus casei, 209–210
  Lactobacillus delbrueckii, 207
  Lactobacillus monocytogenes, 210–212
  Lactobacillus paraplantarum cheese isolate, 211–212
  Lactobacillus plantarum, 209–210
  minimum biofilm eradication concentration (MBEC) biofilm, 209
  nonbacteriocinogenic strain, 208–209
  protected designation of origin (PDO) French cheese making, 208
  Salmonella Enteritidis strain, 210
Lactobacillus delbrueckii, 207
Lactonase, 260
Laser Doppler anemometry, 41
Layer-by-layer assembly approach, 167–168
Light amplification by stimulated emission of radiation (LASERs), 297–298
Light-emitting diodes (LEDs), 297–298
Lignan, 276
Lipids, 30t–31t
Low-field benchtop 1H nuclear magnetic resonance (1H LF-NMR) relaxometry, 43

M

Matrix-assisted laser desorption ionization imaging high-resolution mass spectrometry (MALDI-imaging-HRMS), 40
Mediated electron transfer (MET), 180
Membrane filtration
 bacterial growth potential bioassays, 161–162
 biofouling, 160, 170
 biofouling control strategies, 170
  acid/alkali treatments, 169
  activated carbon, 162–163
  air/water flushing, 169
  biodegradable nutrients, 163, 163f
  chemical dosage risk, 164–166
  chlorination and dechlorination, 163–164
  curative chemical cleaning, 166, 166f
  feed spacer modification, 168
  formulated cleaning products, 169
  hydraulic cleaning, 166–167
  membrane coating, 167–168
  membrane modification, 167–168
  microfiltration (MF) pretreatment, 162–163
  nonoxidizing biocide, 164
  pneumatic cleaning, 169
  prevention strategies, 162–164
  reverse osmosis (RO) membrane, 162
  ultrafiltration (UF) pretreatment, 162–163
  ultraviolet (UV) irradiation, 164
  urea cleaning, 169
 conventional chemical cleanings, 170–171

Membrane filtration (*Continued*)
feed pressure, 159
feed spacers, 159
feedwater's biofilm growth potential, 160–162
feedwater stream, 157–158
fouling, 159–160
industrially applied membrane process, 157–158
inoculum selection, 161–162
microfiltration (MF), 158–159
nanofiltration (NF), 158–159
pressure-driven membrane filtration process, 157–158
pure cultures, 161–162
reverse osmosis (RO), 158–159
spiral-wound membrane modules, 158–159
ultrafiltration (UF), 158–159
water pretreatment process, 170
Membrane modification, 167–168
MFC. *See* Microbial fuel cell (MFC)
Microalgae and cyanobacteria
applications, 127, 128t–129t
cultivation
biomass removal, 130–131
immobilized system, 130–131, 132t–133t
open pond systems, 130
photobioreactors (PBRs), 130
factors affecting
CO_2 concentration, 137
extracellular polymeric substances (EPS), 139
hydrodynamic conditions, 139
light, 137
nutrients, 138
pH, 137–138
surface properties, 138
temperature, 138
harvesting techniques
autoflocculation, 134
bioflocculation, 134–135
centrifugation, 136
chemical flocculation, 134
electrocoagulation-flocculation (ECF), 135
filtration, 136
flotation, 135–136
gravitational sedimentation, 135
production costs, 131
marine and freshwater environments, 127
microorganism interactions, 140
species selection, 139–140
wastewater treatment process, 140–141, 142t–144t
Microbial flocculation, 134–135
Microbial fuel cell (MFC)
achievable maximum power, 196, 197f
bacteria-catalyzed reactions, 179–180
charge transfer, 188–189
diffusion parameter, 197f
dimensional electrodes
boundary conditions, 200
charge transfer, 200–201
discrete sequence, 200
porous electrode, 199f
problem formulation, 198
spherical substrate diffusion, 199
substrate diffusion, 198–200
total diffusion, 200
double-layer capacitance, 182–183
electron transference
biological oxidation, 180–181, 181f
direct electron transfer (DET), 180
electrochemical oxidation, 180–181
mediated electron transfer (MET), 180
Faradaic and double-layer current, 182
faradaic charge transfer process, 181–182
flat configuration microbial fuel cell, 186, 188f
internal resistance, 197f
mass transfer
boundary condition, 184–185
flux and faradaic current, 185
spatial discretization, 183–184, 184f
substrate diffusional transport, 183
vector-matrix formulation, 185
mass transport process, 188–189
materials used, 186–187
maximum steady-state power delivery
Fick's law, 193
parameter vector, 194
resistance, 194
stable systems, 193
Thévenin electric circuit, 193, 193f
negative electrode, 179
open-circuit voltage (OCV), 182
open-circuit voltage and constant, 197f
parameter identification, 189–190
positive electrode, 179
potential relationship, 182
proton exchange system, 186–187
schematic representation, 189f

sensitivity analysis, 194–195
simulation results, 201–202
single-chamber microbial fuel cell, 185, 186f
small-signal operation
complete linear model equations, 192
equivalent circuit, 192f
first-order approximation, 192
output voltage, 190–191
pseudolinear function, 191–192
stacked microbial fuel cell, 186
two-chamber microbial fuel cell, 185, 187f
up-flow mode microbial fuel cell, 186
Microcoil probes, 34
Microfiltration (MF), 158–159
Minimum biofilm eradication concentration (MBEC) biofilm, 209
Mobile genetic elements (MGEs), 10–12
Modified Robbins Device, 73
Molecular docking
consensus scoring, 339–340
Discovery Studio software, 341
natural products and metabolites, 340
programs, 337–338
protein-ligand molecular docking, 337
receptor-ligand complex, 337
scoring function (SF), 338–339
search algorithm (SA), 338
sesquiterpene lactones, 341
Multidrug-resistant (MDR) pathogens, 23, 307
Multispecies multisubstrate model, 371
Multivariate data analysis (MVDA) techniques, 48, 48f, 53–54
Mycobacterium tuberculosis, 309
Mycotoxins, 105

N

Nafion, 186–187
Nanofiltration (NF), 158–159
Nanowires, 180
Naringenin, 277
Nosocomial pneumonia, 95
Nuclear magnetic resonance (NMR)
advantages, 32
biofilm formation and structure
combined CLSM and NMR spectroscopy, 40
combined SEM and MALDI-imaging-HRMS, 40
electron and fluorescence microscopy, 40
NMR relaxation time, 39
spin mapping, 39
diffusion and mass transport
CSLM velocimetry, 41
diffusion-ordered (DOSY) NMR spectroscopy, 45–47
fluorescence recovery, 41
metabolic activity, 40–41
microelectrodes, 41
porous systems, 41
position-dependent measurements, 41
pulsed-field gradient nuclear magnetic resonance, 43–45
transport measurement methods, 41
disadvantage, 32
metabolomics, 55t–57t
acquisition, 51–52
applications, 32
cell quenching, 49–50
chemical shift variations, 51
chromatography, 47–48
community proteomics and transcriptomics, 47
extraction solvent, 50
intra- and extracellular metabolites, 47
metabolite extraction technique, 50
metabolome profiling, 49
multivariate data analysis (MVDA) techniques, 48, 48f, 53–54
pH variation, 50–51
sample collection and preparation, 49–51
spectral analysis, 52–53
relaxation data analysis, 42–43
relaxation measurement methods, 32
solid-state NMR (SS-NMR), 32
Nuclear overhauser effect spectroscopy (NOESY), 35

O

Oleuropein, 274–275
One-dimensional biofilm models
boundary conditions, 369
coupled-system first-order partial differential equations, 368–369
detachment rate, 369
integral assumption, 368
substrate concentrations, 368–369
Wanner-Gujer model, 368
Open-circuit voltage (OCV), 182
Oral biofilms
disease-associated

Oral biofilms (*Continued*)
caries and periodontitis, 92
dental caries, 92–93
microbial infections, 92
nonoral infections, 92
oral bacteria, 93t
periimplantitis, 94
periodontitis, 93–94
extracellular polymeric substances (EPS), 89
nonoral infections
carcinogenesis, 95
dental plaque, 95
nosocomial pneumonia, 95
periodontal and cardiovascular diseases, 95
systemic diseases, 94–95
oral cavity, 90
resident microbial community protection, 89
Organosulfur compounds, 276
Orthogonal projection to latent structures (OPLS) models, 54
Oxidoreductases, 256
Ozone, 116–117

P

Paramecium caudatum, 295
Partial least squares (PLS), 54
Pathogen-associated molecular patterns (PAMPs), 93–94
Pathogen recognition receptors (PRRs), 93–94
Patulin, 105
PDT. *See* Photodynamic therapy (PDT)
Penicillin-binding protein (PBP), 310–311
Penicillium expansum, 105, 114
Periimplantitis, 94
Periodontitis, 93–94
Peroxidases, 258
Persister cells, 13–14
Phenolic acids, 276
Phenothiazines, 310–311
2-Phenyl-1,4-benzopyrone, 274–275
Photobioreactors (PBRs), 130
Photodynamic therapy (PDT)
antimicrobial photodynamic therapy (aPDT). *See* Antimicrobial photodynamic therapy (aPDT)
biofilm formation, 298
photosensitization
conventional gas discharge and incandescent lamps, 297–298
Jablonski scheme, 296f
light amplification by stimulated emission of radiation (LASERs), 297–298
light-emitting diodes (LEDs), 297–298
photosensitizer (PS), 295–298, 297t
principle, 295–296
Photosensitizer (PS), 295–298, 297t
Phytochemicals
antimicrobial properties
alkaloids, 276
bacterial multiresistance, 273–274
carotenoids, 276
cytoplasmic membrane function destabilization, 277–278
DNA synthesis interference, 277
drug tolerance, 273
flavonoids, 275–276
metabolic pathway destabilization, 278
modes of action, 276–279
phenolic acids, 276
plant-derived natural compounds, 277–278
plant extract, 274
polyphenols, 275–276
solvents choice, 274
techniques, 274
biofilm-control, 282t–286t
antibiofilm strategies, 279
bacterial adhesion, 279–280
bacterial motility, 280
biofilm matrix, 281
cell-cell interactions, 280–281
chemical classification, 274–275, 275f
medicinal properties, 278–279
Plug flow reactor (PFR), 72
Pneumatic cleaning, 169
Polybeads microspheres, 240–241
Polyelectrolyte coating, 167–168
Polyphenols, 275–276
Porous electrode, 199f
Porphyromonas gingivalis, 90–92
Pressure-driven membrane filtration process, 157–158, 158f
Principal component analysis (PCA), 53–54
Proteases, 258
Pseudomonas aeruginosa, 309
Pulsed-field gradient spin echo nuclear magnetic resonance (PGSE NMR), 34–35

double stimulated echo (DSTE), 44
linear magnetic field, 43
longitudinal eddy current delay (LED), 44
Pseudomonas aeruginosa biofilm, 44–45
spin echo (SE), 44
stimulated echo (STE), 44

Q

QS. *See* Quorum sensing (QS)
QSAR analysis. *See* Quantitative structure-activity (QSAR) analysis
Quantitative structure-activity (QSAR) analysis
algorithms, 347–348
classification techniques, 346
compound's characteristics, 346
2D QSAR models, 348
3D QSAR models, 348
immunomodulatory cationic peptide, 348
mathematical approach, 346
pattern recognition models, 347–348
quality and validity, 347
regression techniques, 346
workflow, 346, 347f
Quorum sensing (QS), 6–7, 280–281
acyl homoserine lactone (AHL), 332–334
autoinducer-2 (AI-2), 334–335
autoinducers (AIs), 332, 335f
autoinducing peptides (AIPs), 332–334
cell-to-cell communication, 335
lactic acid bacteria (LAB), antibiofilm activity, 205–206, 224–225
mechanism, 333f
virulence and pathogenicity factors, 332

R

Residence time, 72
Reverse osmosis (RO), 158–159
Rhamnolipid, 255

S

Saliva, 90
Saponins, 277–278
Savinase, 259–260
Scoring function (SF), 338–339
Seawater desalination, 157
Self-diffusion coefficients, 41–42
Sequencing batch biofilm reactors (SBBRs), 360
Sequencing batch reactors (SBRs), 360
Sertraline, 310
Sesquiterpene lactones, 341
Signal-to-noise ratio (SNR), 34
SIMPLEX numerical optimization method, 189–190
Simvastatin, 311
Single-chamber microbial fuel cell, 185, 186f
Sloughing, 111
Sophoraflavanone G, 277
Spherical substrate diffusion, 199
Spiral-wound membrane modules, 158–159, 159f
Spiral-wound reverse osmosis (RO) membrane module, 158–159, 159f
Stacked microbial fuel cell, 186
Standardized Biofilm Methods Laboratory (SBML), 74
Statins, 311, 318
Statistical total correlation spectroscopy (STOCSY), 53
Streptococcus mutans, 90
Streptococcus oligofermentans, 90
Structural proteins, 30t–31t

T

Terfenadine, 313
Thévenin electric circuit, 193, 193f
Thioridazine and triflupromazine, 310
Thymol, 277
Total correlation spectroscopy (TOCSY), 34–35
Two-chamber microbial fuel cell, 185, 187f

U

Ultrafiltration (UF), 158–159
Up-flow mode microbial fuel cell, 186

V

Valproic acid, 320
Virtual screening (VS)
accuracy, 341
active compounds, 341–342
advantages, 341
compound databases, 343, 344t
disadvantages, 341
enrichment factor (EF), 342
inactive compounds, 341–342
Pharmacophore Model, 343–345

Virtual screening (VS) (*Continued*)
receiver operating characteristic curve (ROC), 342
SwissADME, 343
Vibrio harveyi, 345
VS. *See* Virtual screening (VS)

W

Wanner-Gujer 1D biofilm model, 374–376
Wanner-Gujer model, 368

Z

Zero-dimensional biofilm models, 362–363

FSC
www.fsc.org
MIX
Papier | Fördert
gute Waldnutzung
FSC® C083411

Kolibri 360 GmbH
Ferdinand-Jühlke-Straße 7
99095 Erfurt, Deutschland
produktsicherheit@kolibri360.de